BLUEBERRIES, 2ND EDITION

CROP PRODUCTION SCIENCE IN HORTICULTURE SERIES

This series examines economically important horticultural crops selected from the major production systems in temperate, subtropical and tropical climatic areas. Systems represented range from open field and plantation sites to protected plastic and glass houses, growing rooms and laboratories. Emphasis is placed on the scientific principles underlying crop production practices rather than on providing empirical recipes for uncritical acceptance. Scientific understanding provides the key to both reasoned choice of practice and the solution of future problems.

Students and staff at universities and colleges throughout the world involved in courses in horticulture, as well as in agriculture, plant science, food science and applied biology at degree, diploma or certificate level, will welcome this series as a succinct and readable source of information. The books will also be invaluable to progressive growers, advisers and end-product users requiring an authoritative, but brief, scientific introduction to particular crops or systems. Keen gardeners wishing to understand the scientific basis of recommended practices will also find the series very useful.

The authors are all internationally renowned experts with extensive experience of their subjects. Each volume follows a common format covering all aspects of production, from background physiology and breeding, to propagation and planting, through husbandry and crop protection, to harvesting, handling and storage. Selective references are included to direct the reader to further information on specific topics.

Titles Available:
 1. **Ornamental Bulbs, Corms and Tubers**   A.R. Rees
 2. **Citrus**   F.S. Davies and L.G. Albrigo
 3. **Onions and Other Vegetable Alliums**   J.L. Brewster
 4. **Ornamental Bedding Plants**   A.M. Armitage
 5. **Bananas and Plantains**   J.C. Robinson
 6. **Cucurbits**   R.W. Robinson and D.S. Decker-Walters
 7. **Tropical Fruits**   H.Y. Nakasone and R.E. Paull
 8. **Coffee, Cocoa and Tea**   K.C. Willson
 9. **Lettuce, Endive and Chicory**   E.J. Ryder
10. **Carrots and Related Vegetable *Umbelliferae***   V.E. Rubatzky, C.F. Quiros and P.W. Simon
11. **Strawberries**   J.F. Hancock
12. **Peppers: Vegetable and Spice Capsicums**   P.W. Bosland and E.J. Votava
13. **Tomatoes**   E. Heuvelink
14. **Vegetable Brassicas and Related Crucifers**   G. Dixon
15. **Onions and Other Vegetable Alliums, 2nd Edition**   J.L. Brewster
16. **Grapes**   G.L. Creasy and L.L. Creasy
17. **Tropical Root and Tuber Crops: Cassava, Sweet Potato, Yams and Aroids**   V. Lebot
18. **Olives**   I. Therios
19. **Bananas and Plantains, 2nd Edition**   J.C. Robinson and V. Galán Saúco
20. **Tropical Fruits, 2nd Edition Volume 1**   R.E. Paull and O. Duarte
21. **Blueberries**   J. Retamales and J.F. Hancock
22. **Peppers: Vegetable and Spice Capsicums, 2nd Edition**   P.W. Bosland and E.J. Votava
23. **Raspberries**   R.C. Funt
24. **Tropical Fruits, 2nd Edition Volume 2**   R.E. Paull and O. Duarte
25. **Peas and Beans**   A.J. Biddle
26. **Blackberries and Their Hybrids**   H.K. Hall & R.C. Funt
27. **Tomatoes, 2nd Edition**   E. Heuvelink
28. **Grapes, 2nd Edition**   G.L. Creasy and L.L. Creasy
29. **Blueberries, 2nd Edition**   J. Retamales and J.F. Hancock

BLUEBERRIES
2ND EDITION

Jorge B. Retamales

International Consultant, Chile

and

James F. Hancock

Michigan State University, USA

CABI is a trading name of CAB International

CABI
Nosworthy Way
Wallingford
Oxfordshire OX10 8DE
UK

Tel: +44 (0)1491 832111
Fax: +44 (0)1491 833508
E-mail: info@cabi.org
Website: www.cabi.org

CABI
745 Atlantic Avenue
8th Floor
Boston, MA 02111
USA

T: +1 (617)682-9015
E-mail: cabi-nao@cabi.org

A catalogue record for this book is available from the British Library, London, UK.

Library of Congress Cataloging-in-Publication Data

Names: Retamales, J. B. (Jorge B.), author. | Hancock, James F., author.
Title: Blueberries / Jorge B. Retamales and James F. Hancock.
Description: 2nd edition. | Boston, MA : CABI, [2018] | Series: Crop production
    science in horticulture series ; 28 | Includes bibliographical references and index.
Identifiers: LCCN 2018002404 (print) | LCCN 2018004005 (ebook) |
    ISBN 9781780647272 (ePDF) | ISBN 9781780647289 (ePub) |
    ISBN 9781780647265 (pbk: alk. paper)
Subjects: LCSH: Blueberries. | Blueberry industry.
Classification: LCC SB386.B7 (ebook) | LCC SB386.B7 R48 2018 (print) | DDC
381/.414737--dc23

ISBN-13: 978 1 78064 726 5 (pbk)

Commissioning editor: Rachael Russell
Editorial assistant: Emma McCann
Production editor: Shankari Wilford

Typeset by AMA DataSet Ltd, Preston, UK
Printed and bound in the UK by CPI Group (UK) Ltd, Croydon, CR0 4YY

CONTENTS

INTRODUCTION TO SECOND EDITION

What we reported in the first edition of this book in 2012 still holds very true. Blueberry production and commerce continues to grow by leaps and bounds. From a crop that was mainly consumed and cultivated in North America, we have come to an era of worldwide blueberry cultivation. This expansion has encompassed plantings in a greater diversity of environments and has required the development of innovative cultural practices and conditions. It was only 40 years ago that highbush blueberry cultivation was restricted to temperate climates with more than 1000 chilling hours. Now cultivars are available that can be grown in subtropical and tropical environments with no chilling hours. This situation has challenged researchers to increase both the scope and depth of their activities. The number of articles and meetings on blueberries has grown at a steady pace, and even in the 5 years since the first edition of this book, the literature has expanded dramatically. The information available for people involved in blueberry research, culture or marketing is widely dispersed and often difficult to access or interpret. This book was developed to provide readers with the current status of knowledge on blueberry science and management.

The second edition of this book is still structured in nine chapters like the first. In Chapter 1, the industry is described with information on the history of cultivation, the most important locations, the species and the cultural practices employed in the different production regions. The dramatic changes in the worldwide industry in the last few years are incorporated. Chapter 2 deals with the taxonomy of blueberry species, the history of improvement and current breeding efforts, tools and goals, and describes the most important blueberry cultivars grown worldwide. The emerging genomic information on the blueberry is highlighted. In Chapter 3, the anatomy and morphology of the highbush and rabbiteye blueberry are discussed, along with vegetative and reproductive growth and development. Chapter 4 deals with the generation and distribution of carbohydrates, and the factors involved in dry-matter production and partitioning among the various plant organs. Chapter 5 concentrates on the mineral nutrition of blueberries, the factors that affect the availability of nutrients, and the methods to establish and supply the nutrients to satisfy crop demands. Chapter 6 covers various management practices that are important in blueberry cultivation, including mulching, irrigation, soilless culture, cultivation under high tunnels, pruning, pollination and harvest. There has been much new information published on pruning, grafting and irrigation. Chapter 7 examines plant growth regulators with regard to their application and the factors that affect their performance. Current and potential uses of these substances are presented. The most relevant pests, diseases and weeds that attack blueberries are covered in Chapter 8. The information on diseases (viruses, bacteria and fungi) is presented based on the organs affected, and the symptoms associated with diseases and pests are described. Chapter 9 discusses the pre- and postharvest management of fruit quality. The attributes and factors affecting fruit quality are defined, as well as the factors that influence the postharvest life of the fruit and the approaches used to extend fruit quality. Much new information has been gathered on fruit quality and consumer acceptance.

This edition, like the first, is meant to be an overview of the various aspects of blueberry science and culture. It is targeted towards blueberry researchers and students in horticulture, but it should also be useful for growers and people in the industry who want to update their knowledge on this crop. Our approach has been to explain in an understandable manner the basic science behind the growth and development of blueberries, their botanical characteristics, as well as the implications and effects of various management practices and environmental conditions.

The authors are grateful for the encouragement and assistance of many people who made the work possible. The University of Talca financed sabbatical leaves for the senior author to start the writing of the first edition and work on the second. Our wives, Beatriz and Ann, along with JBR's children (Beatriz, Jorge and Gabriela), were tremendously supportive throughout the preparation of this book. Several people provided help in the first edition, and we built on their input in the second. These contributors were:

- Carlos Araya, Universidad de Talca: performed literature searches.
- Randy Beaudry, Michigan State University: reviewed Chapter 9.
- Pilar Bañados, Pontificia Universidad Católica de Chile: reviewed sections of Chapters 3 and 6.
- Ridley Bell, Mountain View Orchards: reviewed sections of Chapter 1 and provided information for Chapter 2.
- Peter Caligari, Universidad de Talca: reviewed the book proofs.
- Reinaldo Campos, Universidad Andrés Bello: reviewed Chapter 9.
- Bill Cline, University of Georgia: reviewed sections of Chapter 8.
- Nicolás Cobo, Universidad de Talca: prepared drawings.

- Sandra Carrizo, Universidad de Talca, reference search.
- Rebecca Darnell, University of Florida: reviewed Chapter 6.
- Bruno Defilippi, Instituto de Investigaciones Agropecuarias: reviewed Chapter 9.
- Chad Finn, US Department of Agriculture–Agricultural Research Service: reviewed Chapter 2.
- Mark Greef, Nelson Mandela Metropolitan University: reviewed sections of Chapter 1.
- Eric Hanson, Michigan State University: reviewed sections of Chapter 8.
- Jane Hoyle, Technical editing.
- Rufus Isaacs, Michigan State University: reviewed sections of Chapter 8.
- Raúl S. Lavado, Universidad de Buenos Aires: provided information on Argentinean soils.
- Oscar Liburd, University of Florida: reviewed sections of Chapter 8.
- Cheng Liu, Liaoning Institute of Pomology: reviewed sections of Chapter 1.
- Gustavo Lobos, Universidad de Talca: prepared drawings.
- José Manuel López-Aranda, Junta de Andalucía (Spain): provided soil data.
- Scott NeSmith, University of Georgia: reviewed sections of Chapter 2.
- Samuel Ortega, Universidad de Talca: reviewed Chapter 6.
- Maria José Palma, Universidad de Talca: did reference searches and prepared drawings.
- Narandra Patel, Gourmet Group of Companies: reviewed sections of Chapter 1.
- Alejandro del Pozo, Universidad de Talca: reviewed Chapter 4.
- Julio Retamales, private consultant: reviewed Chapter 7.
- Sebastián Romero, Universidad de Talca: prepared drawings and tables.
- Takato Tamada, Japan Blueberry Association: reviewed sections of Chapter 1.
- Claudio Valdes, private consultant: reviewed Chapter 5.
- Ed Wheeler, MBG Marketing: reviewed sections of Chapter 2.
- Garry Wright, BerryExchange: reviewed sections of Chapter 1.
- Wei Yang, Oregon State University: provided soil data for Chapter 6.

1

THE BLUEBERRY INDUSTRY

INTRODUCTION

The predominant cultivated blueberry species are *Vaccinium corymbosum* L. (highbush blueberry), *Vaccinium virgatum* Ait. (rabbiteye blueberry; syn. *Vaccinium ashei* Reade) and native stands of *Vaccinium angustifolium* Ait. (lowbush blueberry). Highbush cultivars are further separated into northern, southern and intermediate types depending on their chilling requirements and winter hardiness. Half-high types are also grown that are hybrids of highbush and lowbush species.

Where the different types of blueberry are grown is to a large extent determined by their chilling requirement and winter cold hardiness. All blueberries require a well-drained, acid soil with ample moisture. The lowbush types require at least 1000 chilling hours (<7°C) for normal floral development and can tolerate temperatures as low as −30°C. Rabbiteye cultivars require about 600h of chilling, and their floral buds cannot tolerate temperatures much below freezing. Northern highbush cultivars (NHBs) are adapted to quite cold mid-winter temperatures below −20°C and grow well anywhere there are 800 to 1000h of chilling. Southern highbush cultivars (SHBs) do not tolerate winter temperatures much below freezing and require chilling hours under about 550h. Intermediate highbush cultivars have a wide range of chilling requirements from 400 to 800h. They generally fail in the colder climates because they bloom too early and are too slow to harden in the autumn, resulting in freeze damage to the flower buds.

Most of the commercial production of blueberries now comes from highbush and lowbush types, although rabbiteyes are important in south-east North America and hybrids of highbush × lowbush (half-highs) have made a minor impact in Upper Mid-west USA. Some rabbiteye cultivars are grown in the US Pacific Northwest and Chile for their very late-ripening fruit and wider soil adaptability. NHBs are grown primarily in Australia, France, Germany, Italy, New Zealand, the USA (Pacific Northwest, Michigan, New Jersey), Poland

and Chile. SHBs are grown predominantly in Australia, Argentina, the USA (California, Florida, Georgia), Chile, Peru, Colombia and southern Spain. The intermediate highbush types are grown mostly in Chile and the USA (Arkansas, New Jersey, North Carolina and Pacific Northwest).

Highbush blueberries have become a major international crop, with over 65,000 ha planted in North America, 23,000 ha in South America, 27,000 ha in Asia and the Pacific, 16,000 ha in Europe and about 1000 ha in the Mediterranean and North Africa and sub-Saharan Africa (Brazelton and Young, 2017). Overall, world production of highbush is over 650 million t annually, while lowbush blueberry production exceeds 250 million t.

EARLY HISTORY OF HIGHBUSH BLUEBERRY CULTIVATION

Many of the wild, edible *Vaccinium* spp. have been harvested for thousands of years by indigenous peoples (Moerman, 1998). Native Americans in western and eastern North America intentionally burned native stands of blueberries and huckleberries to renew their vigour and eliminate competition. Highbush and rabbiteye blueberries were domesticated at the end of the 19th century. Plants were initially dug from the wild and transplanted into New England and Florida fields.

The NHB *V. corymbosum* was first domesticated in 1908 by Frederick Coville of the US Department of Agriculture (USDA). He was the first to establish the fundamental requirements of these plants, determining that blueberries need acid, well-drained soils, have no root hairs and require a low-temperature rest period (Coville, 1916). He also learned how to propagate blueberries by stem cuttings and established that bumble bees were the best pollinators. In addition, Coville learned that some genotypes are self-unfruitful, and that the highbush blueberry is a tetraploid (Coville, 1927).

Coville began breeding highbush blueberries in 1908, with the help of a private grower in New Jersey, Elizabeth White (Ehlenfeldt, 2009). White grew out Coville's hybrid populations and was particularly helpful in identifying elite wild clones by offering a reward to individuals sending her samples of blueberries that were unusually large. The best clones identified by Coville and White were named after their discoverers and included 'Adams', 'Brooks', 'Dunfee', 'Grover', 'Harding', 'Rubel', 'Russell', Sam' and 'Sooy'. 'Rubel' is still grown today, being favoured as a processed berry.

In an address to the Philadelphia Society for the Promotion of Agriculture in 1934, White stated:

> The best of the hundred bushes located was that found by Rube Leek. In my notes on the variety, I at first used his full name, but Dr. Coville said Leek savored of onions and used in his notes the name Rube. Rube! What a name for such an

aristocratic a plant! Finally, on Dr. Coville's suggestion, a happy solution was found in the name 'Rubel'; the finder's first name plus the initial of his surname.

(White, 1934)

Coville bred blueberries from 1908 to 1937, and left 30,000 seedlings for his replacement, George Darrow, to sort through (Mainland, 1998). His first releases came in 1920, 'Pioneer' and 'Katherine' ('Brooks' × 'Sooy'). Coville remains the most successful breeder, as over half of the current blueberry hectarage is still composed of his selections, including 'Rubel' (1911), 'Jersey' (1928), 'Weymouth' (1936), 'Bluecrop' (1952), 'Croatan' (1954) and 'Blueray' (1955). However, none of these cultivars is widely planted anymore.

GROWTH OF THE BLUEBERRY INDUSTRY IN NORTH AMERICA

Through the 1970s and early 1980s, the NHB area continued to expand in Michigan, and large new plantings began to appear in British Columbia and Oregon (Table 1.1). The area covered by NHB blueberries increased from about 370 to 1,540 ha (+316%) in the Pacific Northwest and from 3,100 to 4,900 ha (+58%) in Michigan. Rabbiteye hectarage also grew by about 40% in the 1970s and early 1980s, primarily in Florida and Georgia. Only limited planting was done in North Carolina and New Jersey.

In the 1980s, steady growth continued in the northern production regions of Oregon (+168%), Washington (+53%), British Columbia (+100%) and Michigan (+41%), and major new plantings of NHBs also began to appear in New York and Indiana. Significantly less planting was done in New Jersey (+11%) and North Carolina (0%) (Table 1.1). In the south, the surface area of blueberries increased by 112% in Florida and 39% in Georgia, and Mississippi and Texas emerged as significant blueberry-producing states. The first plantings of SHBs were made during this period in Florida and Georgia, and by 1992 southern blueberry hectarage was about 7% SHBs and 93% rabbiteyes.

Growth in the blueberry industry generally slowed during the 1990s in the primary northern growing states of Michigan, New Jersey, Oregon and Washington; however, much more substantial gains were observed in British Columbia (148%) and the newer producing states of New York (+105%) and Indiana (+132%). Growth across the southern states was relatively stagnant, except in the newer Gulf States of Mississippi (+82%) and Texas (+41%), where rabbiteye hectarage began to expand.

From 2003 to 2014, blueberry hectarage in the USA and Canada continued to dramatically expand. California emerged for the first time as a major blueberry-growing state, along with the country of Mexico. There was tremendous growth in the Pacific Northwest, the south-eastern USA and the

Table 1.1. Growth patterns of major centres of highbush and rabbiteye blueberry production in North America. (Adapted from Moore, 1994; Strik and Yarborough, 2005; Brazelton, 2009, 2015.)

Region	State	Area (ha)				Change (%)		
		1982	1992	2003	2014	1982–1992	1992–2003	2003–2014
Atlantic	North Carolina[a]	1,600	1,580	1,600	2,955	−1	+1	+85
Mid-west	Michigan[b]	4,900	6,890	6,980	9,109	+41	+1	+31
	Indiana[b]	105	125	290	401	+19	+132	+38
	All	5,362	7,419	7,870	10,152	+38	+6	+29
North-east	New Jersey[b]	3,000	3,320	3,600	3,320	+11	+8	−8
	New York[b]	125	200	410	526	+60	+105	+28
	All	4,600	4,670	4,375	4,528	+2	−6	+3
North-west	British Columbia[b]	970	1,820	4,510	10,931	+88	+148	+142
	Oregon[b]	250	670	680	4,296	+168	+1	+532
	Washington[b]	320	490	495	5,303	+53	+1	+971
	All	1,540	2,980	7,330	20,530	+94	+146	+180

South-east	Arkansas[b]	143	210	145	231	+47	-31	+59
	Florida[a]	400	850	860	2,037	+113	+1	+137
	Georgia[d]	1,200	1,670	1,650	9,514	+39	-1	+477
	Mississippi[d]	40	450	820	1,700	+1025	+82	+107
	Texas[d]	30	220	310	555	+633	+41	+79
	All	4,030	5,845	6,765	17,368	+45	+16	+157
South-west	California[b]	–	–	50	2,951	–	–	+5802
	Mexico[b]	0	0	325	2,579	–	–	+693
	All	0	0	375	5,539	–	–	+1377
North America	All	17,132	22,4944	26,390	61,072	+35	+26	+120

[a]Mostly highbush.
[b]All highbush.
[c]Highbush and rabbiteye.
[d]Mostly rabbiteye.

south-west. Much more modest gains occurred in the Mid-western and Atlantic states. Growth in California (+5802%) Washington (+971%), Mexico (+693%), Oregon (+532%), Georgia (+477%) and British Columbia (+145%) led the way. By the early 2000s, in southern USA, SHBs became more widely planted than rabbiteyes (Table 1.1).

Today, the total hectarage in North America is over 65,000 (Brazelton and Young, 2017), and the surface area covered by blueberries increased by 120% between 2003 and 2014. The western region accounted for the majority of the area in 2014 at 40%, while the southern region represented 35%, the mid-west 17% and the north-east 6% (Table 1.2). This distribution pattern is quite a contrast from 25 years ago, when the western region represented about 14% of the hectarage in North America, while the mid-west was 35%. British Columbia and Georgia now have more blueberry hectarage than Michigan, the industry leader for the previous 50 years.

GROWTH OF THE BLUEBERRY INDUSTRY OUTSIDE NORTH AMERICA

The first planting of highbush blueberries outside of North America was made in 1923 by a Mr Borgesius in Assen, the Netherlands. Dr Piort Hoser, the founder of the Faculty of Horticulture of Warsaw Agricultural University, also imported some blueberries from the USA in 1924, but they were killed by winter cold in 1929. The NHB was introduced into Germany in the 1930s by Dr Walter Heermann, who also started breeding blueberries and introduced commercial production techniques. His fields encompassed 50 ha by 1951. Other early plantings were made by David Trehane in the UK in 1959 and Wilhelm Dierking in Germany in 1962. The first commercial plantings in Poland, the Netherlands and Italy were made in the 1970s. Planting of blueberries began in France in the 1980s and in Spain in the 1990s.

Table 1.2. Proportion of total highbush and rabbiteye hectarage found in different regions of North America; see Table 1.1 for the states included in each region. (Adapted from Moore, 1994; Strik and Yarborough, 2005; Brazelton, 2009, 2015.)

Region	Percentage			
	1982	1992	2003	2014
North-east	30	22	17	7
Mid-west	35	35	30	17
South	26	28	26	35
West	10	14	28	40

Blueberry hectarage remained small across Europe until 1990, with most planting being done in Italy, France, Germany and the Netherlands. In the 1990s, hectarage rose from about 1,000 to 4,000 ha across Europe, with the greatest growth occurring in Poland (1,520 ha) and Germany (1,370 ha) (Table 1.3). Between 2003 and 2014, the European hectarage continued to expand dramatically to over 11,400 ha, an increase of 187%. The greatest growth in that decade was in Spain/Portugal (+954%), Italy (+199%), Poland (+149%) and the Netherlands (+136%). The blueberry hectarage in Germany,

Table 1.3. Growth patterns of major centres of blueberry production across the world. (Adapted from Eck and Childers, 1966; Brazelton, 2009, 2015.)

Region	Country	First plantings	Major growth period	Area (ha) 2003	Area (ha) 2014	Change (%)
Africa	South Africa	1970s	1990s	355	520	+46
	Morocco	1990s	2010s	0	766	–
	Total			355	1,286	+46
Asia	China	1980s	2010s	51	14,858	+29,033
	Japan	1950s	2000s–now	355	1,303	+267
	Total			406	16,161	+29,300
Europe	Poland	1970s	Mid-1990s–now	1,520	3,785	+149
	Germany	1960s	Late 1990s–early 2000s	1,370	2,344	+71
	France	1980s	Late 1980s	415	421	0
	Netherlands	1970s	Late 2000s	300	708	+136
	Spain/Portugal	1990s	2010s	215	2,267	+954
	Italy	1970s	Early 1980s	160	478	+199
	UK	1950s	2000s	0	385	–
	Total			3,980	10,388	+187
Pacific Rim	Australia	1960s	1990s, 2010s	520	1,085	+109
	New Zealand	1970s	Late 1980s	405	737	+82
	Total			925	1,822	+97
South America	Chile	1980s	2000s	2,135	15,559	+629
	Argentina	1990s	2000s	710	3,004	+323
	Brazil	2000s	2010s	0	400	–
	Peru	2000s	2010s	0	1,073	–
	Uruguay	2010s	2000s	0	368	–
	Total			2,845	20,404	+610
World				8,511	50,061	+520

Poland and the Netherlands is spread across them fairly well, while it is concentrated in the south-eastern corners of France and Spain. NHBs are grown at all locations in Europe, except in Spain/Portugal where SHBs predominate.

The first blueberries were planted in Australia and New Zealand in the 1960s and 1970s, primarily as a crop for export markets (Table 1.3). The industry in New Zealand had its greatest growth period during the late 1980s to 1990s until its export markets became oversupplied. The industry in Australia has undergone relatively steady growth since the 1980s. The industries in New Zealand and Australia are dominated by a few larger grower-marketers, some of which have close ties with northern hemisphere producers and marketers, as they pursue a year-round supply model for the Asia Pacific and European regions. From 2003 to 2014, hectarage in Australia rose from 520 to 1,085 ha (+109%), and in New Zealand from 405 to 737 ha (+82%).

In Australia, blueberries were historically grown in New South Wales, but in recent years their culture has expanded dramatically into Tasmania, Far North Queensland and Western Australia (Bell, 2006). The biggest blueberry region is around Coffs Harbour. In New Zealand, most of the hectarage is in the northern Waikato region, with some new plantings being made on the east coast of the North Island (Furniss, 2006). NHBs predominate in northern New Zealand and southern Australia. SHBs predominate in northern New South Wales and southern New Zealand.

The first blueberries were planted in Asia in the 1950s (Table 1.3). A significant industry emerged in Japan in late 1980s (180 ha), but only a few blueberries were planted in China until recently. From 2003 to 2014, the surface area of blueberries increased from 51 to 14,858 ha (+29,033%) in China and from 355 to 1,303 ha (+267%) in Japan. Production in China has been highest in Jilin Province, but significant hectarage has now also been planted in Shandong, Liaoning, Yunnan and Zhejiang (Li *et al.*, 2006; Yu *et al.*, 2012). Currently, hectarage in Yunnan Province is growing rapidly. NHBs and half-highs are grown in the colder regions of China, while SHBs and rabbiteyes are found in Shangdong, Yunnan and Zhejiang. Japanese production is scattered in small hectarages across the country (Tamada, 2006). NHBs and half-highs are grown in the north of Japan, NHBs, SHBs and rabbiteyes in central Japan, and SHBs and rabbiteyes in southern Japan.

The blueberry was first brought to South Africa in the 1970s, but was not widespread until the 1990s (Table 1.3). In 2016, a total 1,300 ha were reported to be planted in that country. Of this total, nearly 50% were planted between 2015 and 2016 (Sikuka, 2017). South African blueberry production is scattered across the country, with concentrations along the Cape, Eastern Free State and Lydenburg/Nelspruit (Greeff and Greeff, 2006). SHBs predominate in the southern coastal areas, while NHBs are grown in the more inland areas with higher chilling hours. In the last few years, a major blueberry industry has emerged in Morocco. With over 766 ha now planted, it is the ninth largest blueberry-producing country outside the USA (Brazelton and Young, 2017).

The first South American blueberry plantings were made in Chile in the 1980s and Argentina in the 1990s (Table 1.3). In 2014, there were 15,559 ha of blueberries in Chile and 3,004 ha in Argentina. Blueberry hectarage grew in Chile by 629% between 2003 and 2014 and by 323% in Argentina. Significant hectarages of blueberries have also emerged recently in Peru (3,800 ha), Brazil (400 ha) Uruguay (368 ha) and Colombia (220 ha). Growth of the blueberry industry in Peru is currently growing by leaps and bounds.

In Chile, blueberries are grown from region IV to region XIV, with concentrations in regions VII and VIII (over 49% of the hectarage). Over the last 10 years, by far the greatest growth in Chile has occurred in the south-central regions from 34°50′ to 38°15′. In the north-central regions IV and V, only SHBs are grown, while in the south-central regions IX and X, NHBs predominate. In the middle regions VII and VIII, there is a transitional zone where both SHB and NHB types are cultivated. There are three major production regions in Argentina that all grow predominantly SHBs: Tucumán, Entre Ríos, and Buenos Aires and San Luis (Taquini, 2006).

In 2016, the top ten highbush blueberry-producing countries in the world were (in million t): the USA (270), Chile (125), Canada (73), Spain (30), China (28), Argentina (18), Poland (16), Peru (16), Mexico (17) and Morocco (12) (Brazelton and Young, 2017).

CLIMATES OF MAJOR PRODUCTION REGIONS

Highbush blueberries are grown across a broad range of climatic zones including: (i) climatic type I: mild, moist summers and very cold winters; (ii) climatic type II: mild, moist summers and moderate winters; (iii) climatic type III: hot, wet summers and mild winters; and (iv) climatic type IV: hot, dry summers and mild winters.

Jilin Province in China, northern Italy, Germany, Michigan and New Jersey, the Netherlands and Poland fall into the first climatic class with mild, wet summers and very cold winters (Tables 1.4 and 1.5). At these locations, winter temperatures commonly fall below 0°C and summer temperatures are generally below 30°C. Chilling hours exceed 1000 h and the number of frost-free days ranges from 130 to 180. Most of the soils in this climatic type are rich organic sands or loams that do not require acidification. NHB cultivars are commonly grown, with 'Bluecrop', 'Duke', 'Elliott' and 'Jersey' predominating, and 'Aurora', 'Draper' and 'Liberty' being widely planted. In the coldest zones of China, half-highs are also grown.

France, northern Japan, northern New Zealand, south-central Chile, the Pacific Northwest and southern Australia have climates that fall into climatic type II with mild, moist summers and moderate winters (Tables 1.4 and 1.5). At these locations, chilling hours generally exceed 600 h, average winter low temperatures are above freezing and the number of frost-free days ranges from

Table 1.4. Climates of major North American highbush and rabbiteye blueberry production regions. (Adapted from Lyrene, 2008; National Centers for Environmental Information, https://www.ncdc.noaa.gov; accessed 15 January 2015.)

Region	State	City	Rainfall (mm)		Temperature (°C)				Frost-free days	Chilling hours (<7°C)
			Annual	Summer	Mid-summer high	Mid-summer low	Mid-winter high	Mid-winter low		
Atlantic	North Carolina	Wilmington	1378	478	33.0	21.0	14.0	1.5	246	500–800
North-east	New Jersey	Hammonton	1097	284	30.0	19.0	5.0	-4.5	182	1000+
Mid-west	Michigan	Holland	1021	294	27.0	14.0	-2.0	-10.0	156	1000+
South-east	Florida, north	Gainesville	1234	495	33.5	22.5	22.0	10.0	285	150–350
	Florida, central	Orlando	1228	528	33.0	21.5	19.0	6.0	315	400–500
	Georgia	Alma	1248	432	33.5	22.0	17.0	5.0	250	450–600
	Mississippi	Poplarville	1606	414	33.5	22.0	15.5	3.5	256	450–600
North-west	British Columbia	Vancouver	1201	134	21.5	13.0	6.0	0.5	170	1000+
	Oregon	Corvallis	1168	92	28.0	13.5	9.0	2.0	190	1000+
	Washington	Vancouver	1267	71	25.0	12.0	7.5	0.0	177	1000+
South-west	California	Bakersfield	163	5	36.0	21.0	13.5	4.0	277	450–550
	Mexico	Guadalajara	927	676	32.4	16.8	26.5	10.2	365	0

Table 1.5. Climates of major global highbush and rabbiteye blueberry production regions. (Adapted from Novoa et al., 1989; Lyrene, 2008; World Meteorological Organization, https://www.wmo.int/pages/index_en.html; accessed 15 January 2018.)

Region	State	City	Rainfall (mm)		Temperature (°C)				Chilling hours (<7°C)
			Annual	Summer	Mid-summer high	Mid-summer low	Mid-winter high	Mid-winter low	
Africa	South Africa	Cape Town	515	52	26.1	15.7	17.5	7.0	400–600
	Morocco	Casablanca	300	1	26.7	18.6	17.1	10.5	0–200
Asia	China	Dalian	632	405	26.1	20.7	-0.9	-7.7	1000+
		Kunming	560	380	25.0	16.7	18.9	3.9	0–100
	Japan	Tokyo	1465	481	30.8	24.2	9.8	2.1	1000+
Europe	Poland	Warsaw	520	203	23.6	12.9	0.4	-4.8	1000+
	Germany	Hamburg	773	224	22.1	12.7	3.5	-1.4	1000+
	France	Bordeaux	986	179	22.6	15.2	10.0	2.8	1000+
	Spain	Huelva	490	16	29.6	21.4	16.1	7.0	200–400
	Netherlands	Amsterdam	778	194	21.8	12.5	5.4	0.2	1000+
	Italy	Venice	810	154	27.5	17.8	5.8	-0.9	1000+
Pacific Rim	Australia	Coffs Harbour	1704	570	27.0	19.0	19.1	7.0	400–500
		Melbourne	665	154	24.8	14.5	13.4	5.9	800+
	New Zealand	Auckland	1135	246	25.0	14.0	16.0	7.0	800+
South America	Chile	Santiago	311	3	29.7	13.0	14.9	3.9	800+
	Chile	Osorno	1383	160	23.8	8.6	11.3	3.8	800+
	Peru	Trujillo	15	2	26.2	18.2	22.0	15.6	0
	Argentina	Buenos Aires	1215	348	30.4	20.4	14.9	7.4	300–400

170 to 200 (or more). The soils vary from sands and loams with high organic content and low pH to mineral soils requiring acidification. NHBs predominate with 'Aurora', 'Bluecrop', 'Brigitta', 'Draper', 'Duke', 'Elliott', 'Legacy', 'Liberty' and 'Reka' being the most common.

Argentina, Mexico, northern New South Wales, southern China, north-central Chile, south-eastern USA and Uruguay represent climatic type III with hot, wet summers and mild winters (Tables 1.4 and 1.5). At these locations, available chilling hours (<7°C) are low, ranging from 0 (Mexico) to 500–800 h in North Carolina. Low winter temperatures generally remain above freezing, summer temperatures average above 28–30°C and the number of frost-free growing days exceeds 250. Mexico has the most unique climate of the group, being the only one with no chilling hours and mostly dry winters. These regions generally have mineral soils with a high clay content that require acidification. SHBs predominate in these zones, with 'Emerald', 'Jewel' and 'Star' being the most widely planted. The intermediate highbush cultivar 'Legacy' has made recent inroads in North Carolina. The SHB 'Biloxi' is the most widespread cultivar planted in Mexico but is being replaced by newer public and proprietary cultivars.

North-central Chile, Peru, Morocco, Colombia, southern South Africa and Spain represent climatic type IV with hot, dry summers and mild winters (Table 1.5). Chilling hours are generally between 250 and 450 h, mean winter low temperatures rarely fall below freezing, summer temperatures average above 30°C, and growing seasons exceed 250 days. Most of these regions have mineral soils with a high clay content that require acidification, although fields in northern Peru (near Trujillo) are planted on almost pure sand. Only SHBs are grown in this climate type, with University of Florida cultivars often dominating, although several proprietary cultivars now compete for hectarage. In Peru, 'Biloxi' has been the most widely planted cultivar, but its importance is in decline.

CULTURAL CONDITIONS OF MAJOR PRODUCTION REGIONS

There is considerable variation across the blueberry-growing regions in production systems. In eastern and mid-western USA, south-central Chile, Japan and the cold climates of Europe, plants are typically planted in the spring at in-row spacings of 1.0–1.2 m with 3 m between rows ('pick-your-own' in Japan is at 2 m spacing). Most plantings are on natural acidic soils with high levels of organic matter (>3%). Pine chips or sawdust is sometimes used for mulching, most commonly in Chile. Overhead irrigation is more common than trickle irrigation in these areas where frost protection is critical. Dormant pruning is done annually or biannually by removing the least productive canes; only limited fine pruning in the canopy is employed, except in Chile. Fertilizer is generally broadcast on the soil in the eastern USA and Europe, while fertigation is most common in Chile.

In the Pacific Northwest, southern USA, north-central Chile, Argentina and Spain, higher-density plantings are generally used (0.7–0.9 m within rows and 3 m between rows), on raised beds, with considerable attention often paid to pH management. Plants are commonly set in the autumn or early winter. Plastic mulches are common, along with sawdust and pine bark. Trickle irrigation is more prevalent than overhead irrigation, and fertigation is commonly employed.

Growth regulators (e.g. Dormex®) are commonly used in the southern USA to enhance leaf development in the spring and advance ripening in SHB blueberries. Gibberellic acid has been used to increase fruit set and to 'rescue' frost-damaged rabbiteyes in Georgia. Rabbiteye and highbush blueberries are commonly hedged to control plant size, encourage branching and enhance fruit set.

While most of the highbush hectarage across the world is grown in open fields, protected culture is becoming increasingly important in the earliest production regions such as Spain, Mexico, Chile and Florida to speed up harvesting and protect against rain damage (Santos and Salame-Donoso, 2012). Hail-proof netting is used in Argentina and Mexico for season extension and hail protection. Netting is also used in Australia to protect against hail and birds. Shading nets were tested in Chile and the USA to delay the harvest and prevent sunburn of leaves and fruit (Lobos *et al.*, 2012), but have not been widely used.

WORLDWIDE PRODUCTION PATTERNS

Blueberry fruit is now available all year around across the world (Tables 1.6 and 1.7). In North America, the season starts in Florida in March, followed soon by California, then Georgia and the Gulf Coast in mid-April. North Carolina begins harvesting in mid-May, followed by New Jersey, Oregon and Washington in mid-June and finally Michigan in July. The last fruit comes out of Michigan in late September and Washington, and from British Columbia in mid-October. The fruit is exported to North America from South America in the northern hemisphere's autumn and winter, starting with Peru and then Argentina in September, and finishing with southern Chile in mid-March. The fruit is harvested in Mexico from October to February.

Most of the fruit coming out of Florida, Mexico, California, New Jersey and North Carolina is sold fresh (Table 1.6). Close to half of the fruit harvested in Georgia, Michigan and Oregon goes to the fresh market, while 29–39% of the fruit produced in British Columbia and Washington is sold fresh. Most of the fruit going to the fresh market is harvested by hand, although the use of machines is now common. Almost all of the processed fruit is harvested by machine.

Table 1.6. Production and marketing patterns in major North American highbush blueberry production regions. (Adapted from Brazelton, 2015.)

Region	State	Season	Utilization (%)	
			Fresh	Processed
Atlantic	North Carolina	Mid-May to August	75	25
Mid-west	Michigan	July to late September	48	52
North-east	New Jersey	Mid-June to early August	91	9
North-west	British Columbia	Mid-July to mid-October	39	61
	Oregon	Mid-June to mid-September	45	55
	Washington	Mid-June to mid-October	29	61
South-east	Florida	March to mid-June	100	0
	Georgia	Mid-April to mid-August	58	42
South-west	California	March to mid-July	83	17
	Mexico	October–February	100	0

In Europe, the season starts in March in Spain and Portugal, followed by Italy beginning in June, France and the Netherlands in July, Germany in mid-July and Poland in August. The last fruit in Europe is picked in late September. Most of the European fruit is hand harvested and goes to the fresh market (Table 1.7).

In South America, Argentina produces fruit from mid-September to January. Chile starts a little later, from October to mid-March. Peru can produce fruit in two seasons – March–April and September–December. Most of the Argentinean, Chilean and Peruvian fruit is sold fresh and hand harvested (Table 1.7).

Australia begins shipping fruit to Europe and Asia in August and continues until February. New Zealand begins harvesting in November and its season ends in March. The season begins in China in April, but most fruit is produced in the autumn. Japan produces blueberry fruit from May to mid-August. The fruit from South Africa is produced from August to January and that from Morocco from December to June. The majority of the African, Asian and Pacific Rim fruit is hand harvested and goes to the fresh market (Table 1.7).

CONCLUSIONS

While highbush and rabbiteye blueberries are native to the USA, they have now become an international crop, being widely planted in North America, South America, Europe, China and the Pacific Rim. Most of this growth has come in the last 15–25 years. Traditionally, the greatest amount of hectarage

Table 1.7. Production and marketing patterns in major global highbush blueberry regions. (Adapted from Brazelton, 2015.)

Region	State	Season	Utilization (%)	
			Fresh	Processed
Africa	South Africa	August–January	92	7
	Morocco	December–June	97	3
Asia	China	April to mid-October	80	20
	Japan	May to mid-August	82	18
Europe	Poland	August–September	90	10
	Germany	Mid-July to September	93	7
	France	July–August	98	2
	Netherlands	July–September	91	9
	Spain/Portugal	March–June	100	0
	Italy	June–September	96	4
Pacific Rim	Australia	August–February	90	10
	New Zealand	November–March	73	27
South America	Chile	October to mid-March	83	17
	Argentina	Mid-September to January	81	19
	Peru	March–April, September–December	100	0

has been in Michigan, but the amount of land planted to blueberries is now larger in Chile, Georgia and British Columbia. The blueberry industries China, Peru, Mexico, Spain and Morocco are growing at a fast rate. Highbush blueberries are grown across a broad range of environmental conditions ranging from hot, dry climates with limited chilling hours (<7°C) to cold, wet climates with considerable chilling hours. SHB types are grown where chilling hours are less than about 550 h, while NHBs are grown in regions with more chilling hours. Highbush blueberries are typically grown at in-row spacings of 1.0–1.2 m with 3 m between rows. Overhead irrigation is more common than trickle irrigation, and annual dormant pruning is typical. SHBs are generally grown at closer spacings, with trickle irrigation being most common, and the bushes are typically hedged after harvesting for size control. Most of the highbush hectarage across the world is in open fields, although protected systems are common in many early production regions. Blueberry fruit is now available all year around. In the northern hemisphere, harvesting begins in Florida, Spain and North Africa in March and ends in the Pacific Northwest and Poland in late September. In the southern hemisphere, harvesting starts in August in Australia and ends in Chile in March.

REFERENCES

Bell, R. (2006) Australian blueberry industry. In: Childers, N.F. and Lyrene, P.M. (eds) *Blueberries for Growers, Gardeners and Promoters.* Dr Norman F. Childers Publications, Gainesville, Florida, pp. 250–254.

Brazelton, C. (2009) *World Blueberry Acreage and Production Report.* US Highbush Blueberry Council, Folsom, California.

Brazelton, C. (2015) *World Blueberry Acreage and Production Report.* US Highbush Blueberry Council. Folsom, California.

Brazelton, C. and Young, K. (2017) World blueberry statistics and global market analysis. International Blueberry Organization, El Dorado Hills, California. Available at: http://www.internationalblueberry.org/ (accessed 18 November, 2017).

Coville, F.V. (1916) The wild blueberry tamed. *National Geographic* 29, 535–546.

Coville, F.V. (1927) Blueberry chromosomes. *Science* 66, 565–566.

Eck, P. and Childers, N.F. (1966) *Blueberry Culture.* Rutgers University Press, New Brunswick, New Jersey.

Ehlenfeldt, M.K. (2009) Domestication of the highbush blueberry at Whitesbog, New Jersey, 1911–1916. *Acta Horticulturae* 810, 147–152.

Furniss, G. (2006) New Zealand blueberry industry. In: Childers, N.F. and Lyrene, P.M. (eds) *Blueberries for Growers, Gardeners and Promoters.* Dr Norman F. Childers Publications, Gainesville, Florida, pp. 248–249.

Greeff, M.P. and Greeff, P.F. (2006) Blueberries in South Africa. In: Childers, N.F. and Lyrene, P.M. (eds) *Blueberries for Growers, Gardeners and Promoters.* Dr Norman F. Childers Publications, Gainesville, Florida, pp. 255–256.

Li, Y., Yang, W.Q., Lu, W. and Zhang, Z. (2006) Blueberries in China. In: Childers, N.F. and Lyrene, P.M. (eds) *Blueberries for Growers, Gardeners and Promoters.* Dr Norman F. Childers Publications, Gainesville, Florida, pp. 243–247.

Lobos, G.A., Retamales, J.B., Hancock, J.F., Flore, J.A., Cobo N. and del Pozo, A. (2012) Spectral irradiance, photosynthetic characteristics and leaf traits of *Vaccinium corymbosum* L. 'Elliott' grown under photo-selective nets. *Environmental and Experimental Botany* 75, 142–149.

Lyrene, P.M. (2008) Breeding southern highbush blueberries. *Plant Breeding Reviews* 30, 354–414.

Mainland, C.M. (1998) Frederick Coville's pioneering contributions to blueberry culture and breeding. In: Cline, W.O. and Ballington, J.R. (eds) *Proceedings of the 8th North American Blueberry Research and Extension Workers Conference.* North Carolina State University, Raleigh, North Carolina, pp. 74–79.

Moerman, D.E. (1998) *Native American Ethnobotany.* Timber Press, Portland, Oregon.

Moore, J.N. (1994) The blueberry industry in North America. *HortTechnology* 4, 96–102.

Novoa, R., Villaseca, S., del Canto, P., Rouanet, J.L., Sierra, C. and del Pozo, A. (1989) *Mapa Agroclimático de Chile [Agroclimatic Map of Chile].* Instituto de Investigaciones Agropecuarias (INIA), Santiago, Chile.

Santos, B.M. and Salame-Donoso, T.P. (2012) Performance of southern highbush blueberry cultivars under high tunnels in Florida. *HortTechnology* 22, 700–704.

Sikuka, W. (2017) The budding blueberry industry in South Africa. Global Agricultural Information Network, Washington, DC. Available at: https://gain.fas.usda.gov/

Recent%20GAIN%20Publications/The%20Budding%20Blueberry%20 Industry%20in%20South%20Africa_Pretoria_South%20Africa%20-%20 Republic%20of_10-5-2017.pdf (accessed 1 November 2017).

Strik, B. and Yarborough, D. (2005) Blueberry production trends in North America, 1992 to 2003, and predictions for growth. *HortTechnology* 15, 391–398.

Tamada, T. (2006) Blueberries in Japan. In: Childers, N.F. and Lyrene, P.M. (eds*) Blueberries for Growers, Gardeners and Promoters.* Dr Norman F. Childers Publications, Gainesville, Florida, pp. 257–260.

Taquini, I.L. (2006) Argentine blueberries. In: Childers, N.F. and Lyrene, P.M. (eds) *Blueberries for Growers, Gardeners and Promoters.* Dr Norman F. Childers Publications, Gainesville, Florida, pp. 257–260.

White, E.C. (1934) The Development of Blueberry Culture. Address to the Philadelphia Society for the Promotion of Agriculture. Whitesbog Preservation Trust archives, Whitesbog, New Jersey.

Yu, H., Gu, Y., Jiang, V. and He, S. (2012) An update on blueberry growing in China. *International Journal of Fruit Science* 12, 100–105.

2

BLUEBERRY TAXONOMY AND BREEDING

TAXONOMY OF BLUEBERRIES

The genus of blueberries, *Vaccinium*, is widespread, with species being found in the Himalayas, New Guinea and the Andean region of South America. The origin of the genus is thought to be South American, and estimates of species numbers range from 150 to 450. Crop species are found in the sections *Cyanococcus* (blueberries), *Oxycoccus* (cranberries), *Vitis-idaea* (lingonberry) and *Myrtillus* (bilberry, whortleberry). Many of the species in the genus are polyploid and carry multiple sets of chromosomes. Chromosome numbers range from diploid $(2n = 2x = 14)$ to tetraploid $(2n = 4x = 28)$ to hexaploid $(2n = 6x = 42)$.

Most blueberry production comes from cultivars derived from tetraploid *V. corymbosum* L. (highbush blueberry), hexaploid *V. virgatum* Ait. (rabbiteye blueberry; syn. *V. ashei* Reade) and native stands of tetraploid *V. angustifolium* Ait. (lowbush blueberry). The cultivated hexaploid blueberry has long been referred to by horticulturists as *V. ashei* but is more correctly called *V. virgatum*, as a type specimen of this taxon was originally described as *V. virgatum* in *Hortus Kewensis* in 1789 (US National Germplasm System).

The identification of species in the blueberry subgenus *Cyanococcus* has been problematic due to polyploidy, overlapping morphologies, extensive hybridization and a general lack of chromosome differentiation. In the first detailed taxonomy of the group, Camp (1945) described nine diploid, 12 tetraploid and three hexaploid species, but Vander Kloet (1980, 1988) reduced this list to six diploid, five tetraploid and one hexaploid taxa. He included all the crown-forming species into *V. corymbosum* with three chromosome levels $(2x, 4x$ and $6x)$.

Most blueberry workers feel that the variation patterns in *V. corymbosum* are distinct enough to retain Camp's diploid *Vaccinium elliottii* Chapm., *Vaccinium caesariense* Mack. and *Vaccinium fuscatum* Ait., tetraploid *Vaccinium simulatum* Small and hexaploid *V. ashei* Reade and *Vaccinium constablaei* A. Gray (Luby *et al.*, 1991; Galletta and Ballington, 1996; Table 2.1). This is the classification system used by the USDA National Resources Conservation Service. Others have chosen to use the designation 'forma' to represent these taxa; i.e. *V. corymbosum* forma *caesariense* (Ehlenfeldt and Ballington, 2012).

There is also an ongoing debate on the correct species name of the cultivated tetraploids. Camp considered the progenitor of the cultivars to be what he recognized as *Vaccinium australe*. Uttal (1986) argued that Camp's *V. australe* was more appropriately named *Vaccinium formosum* and considered *V. formosum* to be the cultivated species epithet. Utall's conclusion was based on Camp's apparent mistaken identity of a type specimen. Vander Kloet (1988) did not recognize either *V. australe* or *V. formosum*, and thus in his taxonomy the cultivars would be represented by *V. corymbosum*. *V. corymbosum* is now most commonly listed as the progenitor species of the cultivars, but *V. formosum* is still used in some taxonomic treatments (Uttal, 1987; Weakley, 2007).

Further confusion about the correct species name for the cultivars has been added in the last few decades as more and more different species have been introgressed into the cultivar background by breeders (see section on Use of Native Germplasm Resources in Blueberry Breeding). For this reason, Butkus and Plizka (1993) suggested that the cultivars be given a separate species name, *Vaccinium* × *covilleanum*. Ballington *et al.* (1997) argued that native *V. corymobosum* was actively hybridizing with other species itself, so remained an appropriate epitaph for the cultivated forms.

The tetraploid highbush blueberry *V. corymbosum* is genetically an autopolyploid, with two sets of similar chromosomes (Soltis *et al.*, 2007). The primary mode of speciation in *Vaccinium* has been through unreduced gametes (Qu and Hancock, 1995; Vorsa and Rowland, 1997; Qu and Vorsa, 1999).

HISTORY OF IMPROVEMENT

Blueberry breeding is a recent development (Lyrene, 1998; Hancock, 2006a,b). Highbush breeding began in the early 1900s in New Jersey, with the first hybrid being released in 1908 by Frederick Coville of the USDA. He conducted the fundamental life history studies of the blueberry that served as the basis of cultivation, such as soil pH requirements, cold and day-length control of development, pruning strategies and modes of propagation. Working with Elizabeth White and others, he collected several outstanding wild clones of *V. corymbosum* and *V. angustifolium*, which he subsequently used for breeding improved types. A high percentage of the current blueberry hectarage is still composed of his hybrids, most notably 'Berkeley', 'Bluecrop', 'Blueray', 'Croatan', 'Jersey', 'Rubel' and 'Weymouth' (Mainland, 1998, 2011).

George Darrow assumed the USDA programme after Coville died in 1937 and made important contributions on the interfertility and phylogeny of the native *Vaccinium* spp. in cooperation with the taxonomist W.H. Camp

Table 2.1. Important native species of the genus *Vaccinium*. (Adapted from Hancock *et al.*, 2008.)

Section	Species	Ploidy	Location
Batodendron	*V. arboreum* Marsh	2x	South-east North America
Cyanococcus	*V. angustifolium* Ait.	4x	North-east North America
	V. virgatum Ait. (syn. *V. ashei* Reade)	6x	South-east North America
	V. boreale Hall & Aald.	2x	North-east North America
	V. constablaei Gray	6x	Mountains of south-east North America
	V. corymbosum L.	2x	South-east North America
	V. corymbosum L.	4x	East and mid-western North America
	V. darrowii Camp	2x	South-east North America
	V. fuscatum Ait.	2x	Florida
	V. myrtilloides Michx.	2x	Central North America
	V. pallidum Ait.	2x, 4x	Mid-Atlantic North America
	V. tenellum Ait.	2x	South-east North America
	V. elliottii Chapm.	2x	South-east North America
	V. hirsutum Buckley	4x	South-east North America
	V. myrsinites Lam	4x	South-east North America
	V. simulatum Small	4x	South-east North America
Oxycoccus	*V. macrocarpon* Ait.	2x	North America
	V. oxycoccos L.	2x, 4x, 6x	Circumboreal
Vitis-Idaea	*V. vitis-idaea* L.	2x	Circumboreal
Myrtillus	*V. cespitosum* Michx.	2x	North America
	V. chamissonis Bong.	2x	Circumboreal
	V. deliciosum Piper	4x	North-west North America
	V. membranaceum Dougl. ex Hook	4x	West North America
	V. myrtillus L.	2x	Circumboreal
	V. ovalifolium Sm.	4x	North-west North America
	V. parvifolium Sm.	2x	North-west North America
	V. scoparium Leiberg ex Coville	2x	North-west North America
Polycodium	*V. stamineum* L.	2x	Central and E. North America
Pyxothamnus	*V. consanguineum* Klotzch	2x	South Mexico and Central America
	V. ovatum Pursh	2x	North-west North America
	V. bracteatum Thunb.	2x	East Asia, China and Japan
Vaccinium	*V. uliginosum* L.	2x, 4x, 6x	Circumboreal

(Hancock, 2006a). He formed a large collaborative testing network that encompassed private growers and Agricultural Experiment Station scientists in Connecticut, Florida, Georgia, Maine, Massachusetts, Michigan, New Jersey and North Carolina. From 1945 to 1961, he sent out almost 200,000 seedlings to his cooperators for evaluation.

Arlen Draper followed Darrow and focused on incorporating the genes of most wild *Vaccinium* spp. into the cultivated highbush background (Draper, 1995; Hancock, 2006b). He maintained and strengthened Darrow's collaborative network and released a prodigious number of SHB and NHB cultivars with improved fruit colour and firmness, smaller pedicel scars and higher productivity (Hancock and Galletta, 1995). His NHBs 'Duke' and 'Elliott' have been huge successes, along with his last release, 'Legacy'. The blueberry industries in Mexico and Peru were built on his SHB 'Biloxi'. Mark Ehlenfeldt assumed the USDA–Agricultural Research Service (ARS) programme in 1998.

Ralph Sharp began working in the 1950s in Florida on the development of SHB types in collaboration with Darrow (Sharp and Darrow, 1959; Lyrene, 1998). He was the first collector of *V. darrowii* for breeding, and, until recently, all SHB cultivars contained genes from his wild clones. Sharp and his colleague, Wayne Sherman, developed 'Sharpblue', which was the first commercially successful low-chill cultivar. Paul Lyrene took over the breeding programme in 1977 and released a group of landmark, low-chill cultivars. These proved critical to the worldwide expansion of the blueberry industry into subtropical and Mediterranean climates.

Jim Ballington in North Carolina was a leader in incorporating wild southern species into the highbush background and released several important cultivars including 'O'Neal' and 'Reveille'. Jim Moore at the University of Arkansas and his successor John Clark were also intent on generating highbush types with complex species backgrounds, and released the once widely planted SHB 'Ozarkblue'.

Stanley Johnston at Michigan State University spent a considerable amount of time in the 1950s and 1960s improving the cold tolerance of highbush by crossing it with *V. angustifolium*. Out of this work came the half-high cultivar 'Northland' and the mostly pure highbush type 'Bluejay', which was released by his successor Jim Moulton. The programme was abandoned in 1978 but renewed in 1990 by Jim Hancock. He released 'Aurora', 'Draper' and 'Liberty', which have become dominant wherever northern highbush are grown. 'Draper' has also proven to be an exceptional parent.

In the Pacific Northwest, Joseph Eberhart, in Olympia, Washington, released three cultivars – 'Olympia', 'Pacific' and 'Washington' – in the 1920s and 1930s. 'Olympia' is still grown today.

Outside the USA, blueberry breeding work was conducted in Australia, Germany and New Zealand. Johnston sent open-pollinated seed to David Jones and Ridley Bell in Australia in the 1960s that generated the important cultivar 'Brigitta Blue' (commonly known as 'Brigitta') along with several others.

Narandra Patel at HortResearch Inc. in New Zealand released the cultivars 'Nui', 'Puru' and 'Reka' from breeding material initially provided by the University of Arkansas and the USDA at Beltsville in the 1960s and 1970s. Walter Heermann in Germany working with seed provided by Frederick Coville released several cultivars in the 1940s and 1950s including 'Ama', 'Blauweiss-Goldtraube', 'Blauweiss-Zukertraube', 'Heerma', 'Gretha' and 'Rekord'.

Rabbiteye breeding was initiated in 1939 by George Darrow in collaboration with Otis J. Woodard at the Georgia Coastal Plain Experiment Station (Tifton, Georgia) and Emmett B. Morrow at the North Carolina Experiment Station, although a collection of wild selections from Florida and Georgia had been planted at Tifton in the 1920s (Austin, 1994). This work was continued by Max Austin and then Scott NeSmith in Georgia, Gene Galletta followed by Jim Ballington in North Carolina, and Ralph Sharp, Wayne Sherman and then Paul Lyrene in Florida. These breeding programmes have resulted in significant improvements in fruit colour, size, texture and appearance over the original wild selections. The most important cultivars have been 'Tifblue' (1955) and 'Brightwell' (1971) from Georgia, 'Bluegem' (1970) and 'Bonita' (1985) from Florida, and 'Powderblue' and 'Premier' (1978) from North Carolina. Rabbiteye cultivars were also bred in the New Zealand HortResearch Inc. programme of Narandra Patel. Several releases came from this programme in the 1990s including 'Maru' and 'Rahi'.

Lowbush blueberries have been hybridized with *V. corymbosum* to produce half-high cultivars. The major releases of this type were 'Northland' developed by Stanley Johnston in Michigan and 'Chippewa', 'Northblue', 'Northcountry', 'Northsky', 'St. Cloud' and 'Polaris' released by Jim Luby in Minnesota. The half-highs have much higher yields and larger fruit than lowbush but have low enough stature to be protected by snow in areas with extreme winter cold.

CURRENT BREEDING EFFORTS

The current goals of SHB breeders are to obtain early-ripening cultivars with high plant vigour, improved disease resistance and a later flowering (particularly in the south-east USA, where late freezes are a problem). Higher yields, better flavour and characteristics favourable for mechanical harvesting are also being sought. There is a growing interest in developing cultivars that fruit on 1-year-old wood without chilling (primocane fruiting). These can produce spring and autumn crops, or can be manipulated through pruning to only produce an autumn crop. Cultivars and advanced lines are being used to breed SHBs, along with hybrids derived from native, low-chill highbush selections from Florida and Georgia (*V. darrowii*, *V. virgatum*, *V. elliottii* and *V. atrococcum*). Because of their low chilling requirement and the influence of genes from the evergreen species *V. darrowii*, many SHB cultivars can be grown as evergreens that avoid dormancy in areas with mild winters, with a harvest season that extends for several months through the winter and early spring (Darnell and Williamson, 1997; Lyrene, 2008). Rabbiteye breeders hope to improve berry size and fruit quality, expand harvest dates, reduce susceptibility to rain cracking and extend storage life.

SHB cultivars are being developed at several public institutions including North Carolina State University, the University of Florida, the University of Georgia, and the USDA-ARS in Poplarville, Mississippi. At the University of Florida, Jim Olmstead replaced Paul Lyrene in 2009 and was the first blueberry breeder to incorporate molecular approaches into his genetic improvement efforts. His focus was on very early, low- to no-chill types that can be grown in an evergreen management system. He recently left the university to manage the global blueberry berry improvement programme of Driscoll's and has been replaced by Patricio Muñoz. Scott NeSmith at the University of Georgia has generated several new early, mid-chill cultivars including 'Camellia', 'Palmetto' and 'Rebel'. He also has an active rabbiteye breeding programme and his 'Ochlockonee' and 'Titan' have generated considerable interest. Steve Stringer at the USDA-ARS in Mississippi has developed a number of promising new cultivars, including the SHBs 'Gupton' and 'Pearl' and rabbiteye 'Prince'. Jim Ballington in North Carolina has successfully incorporated a vast array of southern species germplasm into the highbush background and has generated a number of cultivars, including 'New Hanover', 'O'Neal' and 'Reveille'. He has retired and has been replaced by Hamid Ashrafi, who hopes to fully integrate traditional and genomic approaches to blueberry improvement.

A number of large, private breeding programmes are also actively breeding SHB types including Atlantic Blue in Spain (Juan Luis Navarro and Ulf Hayler), Berry Blue LLC in Michigan, Chile and Peru (Ed Wheeler and Jozer Mangandi), Driscoll's in California (Jim Olmstead), Costa in Australia (Gary Wright), Fall Creek Genetics (Paul Sandefur and Dave Brazelton) and Mountain Blue Orchard in Australia (Ridley Bell). The genetics of all these SHB breeding programmes came from germplasm provided contractually by the University of Florida.

The Berry Blue programme is actively mixing southern- and northern-bred germplasm to develop a new generation of cultivars for Florida, Georgia, California, Mexico, China, Chile and Peru with chilling requirements from 0 to 750 h. The programmes of Driscoll's, Costa and Mountain Blue are more focused on very low-chill, evergreen types, but they still have their eyes on global expansion. Costa licenses its cultivars to Driscoll's, who are planting them in California, Mexico, Chile and Peru. Driscoll's and Costa have established joint farming ventures in Australia, Morocco and China, near Shiping in the Yunnan Province. Costa and Driscoll's have licensed Mountain Blue Orchard cultivars for their expansion into China. Fall Creek Genetics has released only one SHB ('Ventura'), but their nursery is doing most of the propagation for the joint efforts of Driscoll's and Costa.

NHB breeders are concentrating on flavour, longer-storing fruit, expanded harvest dates, disease and pest resistance, and machine harvestability. Established breeding lines are being used in these efforts, along with complex hybrids made up of *V. darrowii*, *V. angustifolium*, *V. constablaei* and most of the other wild species. Even though it has limited winter hardiness, *V. darrowii* has proved to be an interesting parent in colder climates because it passes on a powder blue colour, firmness, high flavour, heat tolerance and potential upland adaptation (Hancock, 1998).

NHBs are currently being bred in public breeding programmes in New Jersey, Michigan, Oregon and Chile. Jim Hancock at Michigan State University was replaced in 2015 by Pat Edger who will continue to focus on late, long-storing cultivars. He plans to systematically incorporate genomics into his programme. Mark Ehlenfeldt of the USDA programme in New Jersey is focusing on identifying genotypes with high disease resistance and tolerance to winter cold, and has released several cultivars including 'Chanticleer' and 'Hannah's Choice'. Nicholi Vorsa at the Cranberry and Blueberry Research Station of Rutgers University is working to develop locally adapted NHB cultivars with machine harvestability and high fruit quality. Chad Finn of the USDA-ARS has recently released two novel cultivars: 'Baby Blue', the first small-fruited type released primarily for the processed market, and 'Perpetua', which has early summer and autumn harvest seasons. Jessica Scalzo is coordinating a joint breeding effort in New Zealand (HortResearch Inc.) and Germany (Dierking Blueberries), which is focusing on nutraceuticals and fruit quality.

Private programmes breeding NHBs include Berry Blue LLC, Fall Creek Genetics and Driscoll's Fall Creek Genetics has released a stable of cultivars that are well adapted to almost all global environments where NHBs thrive, except the coldest regions. The NHB efforts of Driscoll's have produced cultivars best adapted to the milder areas of the Pacific Northwest and the UK. Berry Blue has released several NHB cultivars that are well adapted to both moderate and very cold winters.

BLUEBERRY BREEDING GOALS

Fruit and flowering characteristics

Among the most important characteristics being sought after by blueberry breeders are flavour, large size, light blue colour (a heavy coating of wax), a small scar where the pedicel detaches, easy fruit detachment for hand or machine harvesting, firmness and a long storage life (Hancock *et al.*, 2008). Most people prefer a sweet, crunchy fruit with a trace of acidity; however, high-acid fruits tend to store longer than low-acid ones. The best compromise is to develop cultivars with high sugar and moderate amounts of acid. Other important fruit characteristics are uniform shape, size and colour, high aroma and the ability to retain texture in storage. Much genetic improvement has been made in all of these traits through conventional breeding. *V. darrowii* has been a particularly important source of powder blue colour, intense flavour and fruit that remain in good condition in hot weather.

Interest remains high in the antioxidant capacity of blueberry fruit, although little breeding has been undertaken for this trait (Hancock *et al.*, 2008). Blueberries are among the most antioxidant-rich fruit crops. Genetic improvement could be rapid, as considerable amounts of variability have been observed in this trait.

Plant architecture

The most sought-after bush habit is one that is upright, open and vase shaped, with a bush height of 1.5–2.0 m and a modest number of renewal canes. This is the ideal bush shape for both hand and mechanical harvesting. Many cultivars have been developed that meet this ideotype. In general, plant height appears to be inherited quantitatively, although the short stature of *V. angustifolium* and *V. darrowii* can be dominant to highbush in many interspecific crosses. High percentages of dwarf plants are found in many SHB breeding populations.

Another common breeding goal is to identify genotypes that can easily be picked, with open fruiting clusters and fruits that are well separated. Long pedicels and peduncles are the major components of this feature. There is considerable genetic variability for this characteristic in the current highbush breeding populations, but Paul Lyrene in Florida has also found the wild species *V. arboreum* to be a valuable donor of this characteristic.

Due to limited labour availability and its cost, many highbush breeding programmes are developing cultivars that can be harvested mechanically for the fresh market. A number of traits must be incorporated to achieve this goal, including an upright, open bush habit, loose fruit clusters, easy detachment of mature berries compared with immature berries, no stem retention, a small stem scar, a persistent wax layer and firm fruit (Olmstead and Finn, 2014).

Physiological adaptations

Expanding the range of adaptation of NHBs by reducing their chilling requirement has been an important breeding goal for over 50 years (Hancock *et al.*, 2008). This has been successfully accomplished by incorporating genes from the southern diploid species *V. darrowii* into *V. corymbosum* via unreduced gametes, although hybridizations with native southern *V. corymbosum* and *V. virgatum* have also played a role. Cultivars are now available with an almost continuous range of chilling requirements, from 0 to 1000 h. The genetics of

the chilling requirement has not been formally determined; however, segregation patterns suggest that it is inherited largely quantitatively with the low
chilling requirement showing some dominance.

Most blueberry breeding programmes are working on expanding the
harvest season. Earliness is at a particular premium in the southern parts of
the USA, Spain, Argentina, Morocco, Colombia, Peru and north-central Chile,
while lateness is extremely important in Michigan, the Pacific Northwest and
southern Chile. Increases in earliness have been successfully achieved by
selecting for earlier bloom dates and shorter ripening periods, while lateness
has been increased primarily by selecting individuals with very slow rates of
fruit development. Bloom date is strongly correlated with ripening date, but
early-ripening cultivars have been developed that have later-than-average
flowering dates, such as 'Duke'.

There is now major interest in developing cultivars that are specifically
adapted to regions with few to no chilling hours, such as southern Florida,
Mexico, Morocco and Peru. Two kinds of cultivars are being developed for
these climatic regions: (i) those that remain evergreen throughout the winter
and have a very early production season; and (ii) those that flower and fruit on
first-year wood without dormancy (primocane fruiting) and can produce both
spring and autumn crops. The latter can also be manipulated through pruning
to only produce an autumn crop.

A major focus of many blueberry breeders has also been adaptation to
heat. Most blueberry species are negatively impacted by high temperature and
drought; however, SHBs are generally superior to NHBs. Breeders have had
some success in producing more heat-tolerant cultivars, although the hottest
temperatures of summer still have a major impact on the storage life of
harvested fruit in all areas of blueberry production.

Spring frosts commonly damage flower buds in most production regions.
The stage of floral development when a frost occurs appears to be much more
important than relative bud hardiness. Those cultivars with late bloom dates
tend to suffer less frost damage than those flowering earlier because frosts are
less common later in the season. As previously mentioned, breeders have produced a number of early-ripening cultivars with later bloom dates that can
avoid frosts.

Winter cold often causes severe damage to blueberry flower buds and
young shoots in the colder production regions. In general, NHB types survive
much colder mid-winter temperatures than SHB cultivars, although considerable variability within groups exists that has been exploited by breeders.

Among the other abiotic factors limiting blueberries, high pH and tolerance to mineral soils are very important. The *Vaccinium* are 'acid-loving' plants
and as such generally require soils below pH 5.8 for high vigour. Most blueberry breeders have not focused on this characteristic, even though a number
of interspecific hybrids have been generated by Arlen Draper and Jim Ballington
that have considerable upland adaptations.

Pest resistance

The most important problems in highbush and rabbiteye blueberries are mummy berry (*Monilinia vaccinii-corymbosi* (Reade)), blueberry red ringspot virus, blueberry stunt phytoplasma, blueberry scorch virus, blueberry shoe-string virus, blueberry shock virus, tomato ringspot virus, bacterial leaf scorch (*Xylella fastidiosa*), stem blight (*Botryosphaeria dothidea* (Moug.: Fr.) Ces and de Not.), stem or cane canker (*Botryosphaeria corticis* Demaree and Wilcox), *Phytophthora* root rot (*Phytophthora cinnamomi* Rands), *Phomopsis* canker (*Phomopsis vaccinii* Shear), *Botrytis* fruit rot (*Botrytis cinerea* Pers.: Fr.) and anthracnose fruit rot [*Colletotrichum gloeosporioides* (Penz.) Penz. and Sacc.]. Resistant or tolerant cultivars have been produced for most of the fungal diseases in high-bush blueberries; however, only limited sources of resistance have been found to most virus diseases. See Chapter 8 (this volume) for more detailed descriptions of cultivar resistances.

The most important insect and arthropod pests of highbush and rabbiteye blueberries include the blueberry maggot (*Rhagoletis pomonella* Walsh), blueberry gall midge (*Dasineura oxycoccana* Johnson), blueberry bud mite (*Acalitus vaccinii* Keifer), flower thrips (*Franklinellia* ssp.), Japanese beetle (*Popillia japonica* Newman), sharp-nosed leaf hopper (stunt vector) (*Staphytopius magdalensis* Prov.), blueberry aphid (vector of the blueberry shoestring and scorch viruses) (*Illinoia pepperi* Mac. G.), cranberry fruit worm (*Acrobasis vaccinii* Riley), cherry fruit worm (*Grapholita packardi* Zell), plum curculio (*Conotrachelus nenuphar* Herbst) and spotted wing *Drosophila* (*Drosophila suzukii* Matsumura). Little variation in resistance has been reported to most of these pests in *Vaccinium* spp., except for the sharp-nosed leaf hopper, blueberry aphid, bud mite and gall midge. See Chapter 8 (this volume) for more detailed descriptions of cultivar resistances to these pests.

USE OF NATIVE GERMPLASM RESOURCES IN BLUEBERRY BREEDING

Interspecific hybridization within the *Vaccinium* section *Cyanococcus* has played a major role in the development of highbush blueberries. Most species with similar chromosome numbers hybridize freely, and crosses between species with different chromosome numbers are frequently successful, through unreduced gametes. Even pentaploid hybrids of diploid × hexaploid crosses have been shown to cross relatively easy to tetraploids.

Numerous interspecies crosses have been made by breeders within the section *Cyanococcus*, including: (i) tetraploid *V. corymbosum* × tetraploid *V. angustifolium*; (ii) tetraploid *V. myrsinites* Lam. × tetraploid *V. angustifolium* and *V. corymbosum*; (iii) colchicine-doubled diploid hybrids of *V. myrtilloides* Michx. × tetraploid *V. corymbosum*; (iv) diploid *V. darrowii* × hexaploid

V. virgatum, (v) hexaploid *V. constablaei* × tetraploid *V. corymbosum* and hexaploid *V. virgatum*; and (vi) diploid *V. elliottii* × tetraploid highbush cultivars (Lyrene and Ballington, 1986; Hancock *et al.*, 2008). Probably the most widely employed interspecific hybrid has been US 75, a tetraploid derived from the cross of diploid *V. darrowii* selection 'Fla 4B' × tetraploid highbush cultivar 'Bluecrop' (Fig. 2.1). In spite of its being a hybrid of an evergreen, diploid species crossed with a deciduous, tetraploid highbush, US 75 is completely fertile and is the source of the low chilling requirement of many SHB cultivars.

A number of important characteristics have been associated with the various native species (Ballington, 1990, 2001; Luby *et al.*, 1991; Galletta and Ballington, 1996; Lyrene, 2008). *V. angustifolium* is known for its winter hardiness, early ripening, blossom frost tolerance, adaptation to high pH, stem blight and *Phytophthora* root rot resistance, light blue fruit colour, small scar, high soluble solids and low acidity. *V. virgatum* possesses drought tolerance, a low chilling requirement, an upright plant habit, late ripening, a long flowering-to-ripening period, fruit firmness, a small scar, loose fruit clusters, resistance to cane canker, stem blight and *Phytophthora* root rot, and resistance to sharpnosed leaf hopper. *V. constablaei* has strong winter hardiness, a high chilling requirement and a light blue fruit colour. *V. darrowii* has a low chilling requirement, heat tolerance, resistance to mummy berry, adaptation to high pH,

Fig. 2.1. Morphological differences in hybrids of *V. darrowii* and *V. corymbosum*. Left to right: *V. corymbosum*; backcrossed hybrid; F₁ hybrid (US 75); *V. darrowii*.

tolerance to mineral soils, late flowering, late ripening, a long flowering-to-ripening period, fruit firmness, excellent complex flavour, a small scar, a light blue fruit colour, fruit that hold well in heat, high soluble solids and low acidity, and a loose fruit that clusters. *V. elliottii* has drought tolerance, adaptation to high pH, tolerance to mineral soils, a low chilling requirement, an upright plant habit, late flowering, early ripening, small fruit scar, excellent flavour, resistance to cane canker, stem blight, *Phytophthora* root rot and sharp-nosed leaf hopper. *V. myrsinites* has a low chilling requirement, a small scar, low acidity and firm fruit. *V. myrtilloides* has strong winter hardiness, early ripening, blossom frost tolerance, resistance to mummy berry, a small scar, high soluble solids and low acidity.

Intersectional crosses have generally proved difficult, although partially fertile hybrids have been derived from *V. tenellum* and *V. darrowii × V. stamineum*, *V. darrowii* and *V. tenellum × V. vitis-idaea*, *V. darrowii × V. ovatum*, *V. arboreum × V. stamineum*, *V. uliginosum ×* highbush cultivars, (*V. darrowii × V. arboreum*)× highbush cultivars, colchicine-doubled *V. arboreum ×* highbush cultivars, and colchicine-treated *V. staminium ×* highbush cultivars (Hancock *et al.*, 2008; Lyrene, 2011, 2016). Ehlenfeldt and Ballington (2012) produced seedlings of *V. darrowii* (2x)× *V. cylindraceum*, *V. corymbosum* (4x)× *V. cylindraceum*, *V. corymbosum* forma *ashei* (6x)× *V. smallii*, and (*V. darrowii × V. cylindraceum*) (4x)× *V. corymbosum*. Tsuda *et al.* (2013) have generated hybrids between colchicine-induced tetraploid *V. bracteatum* and the NHB cultivar 'Spartan'.

Of all the wide species crosses, those containing *V. arboreum* are probably the most interesting as this species is known for its drought tolerance, adaptation to basic mineral soils, open flower clusters, upright bush habit, stem blight resistance and resistance to sharp-nosed leaf hopper. Lyrene (2011, 2013) has produced numerous fertile F_1 hybrids of *V. arboreum × V. corymbosum*.

Many of the highbush cultivars that have been released in the last 25 years are complex hybrids (Table 2.2). Among the SHB types, almost all of the native southern species are represented. *V. darrowii* is in most pedigrees, but some also have high proportions of *V. elliottii* ('Carteret', 'Flicker' and 'Snowchaser'), *V. arboreum* ('Meadowlark') and *V. fuscatum* ('Biloxi' and 'Millennia'). 'Carteret' stands out as an SHB in that it has no genes of *V. darrowii*. The NHB types are not as diverse as the SHB ones, although many now have a significant proportion of *V. darrowii* genes in their background, and minor contributions from *V. tenellum* and *V. virgatum*. Most SHBs have some genes of the northern species *V. angustifolium*.

Many cultivars have very complex backgrounds. Some of the most dramatic examples are 'O'Neal', which contains genes from four species (*V. corymbosum*, *V. darrowii*, *V. virgatum* and *V. angustifolium*) and 'Biloxi', 'Cara's Choice' and 'Sierra', which are composed of the genes of five species (*V. corymbosum*, *V. darrowii*, *V. virgatum V. constablaei* and *V. angustifolium*). 'Biloxi' actually has fewer *V. corymbosum*-derived genes than non-*V. corymbosum* genes in its genome.

Table 2.2. Species background of representative NHB and SHB cultivars.

Type and cultivar	Programme[a]	Year	Northern species (%)[b]			Southern species (%)[b]					
			COR	ANG	CON	DAR	ELL	FUS	TEN	VIR	ARB
NHB											
'Rubel'	USDA	1911	100								
'Jersey'	USDA	1928	100								
'Bluecrop'	USDA	1952	87	13							
'Croatan'	USDA	1954	62	38							
'Elliott'	USDA	1973	100								
'Duke'	USDA	1986	96	4							
'Sierra'	USDA	1988	48	2	15	20					15
'Reveille'	NCSU	1990	90	4		3				<1	2
'Chandler'	USDA	1994	97	3							
'Cara's Choice'	USDA	2005	48	2	15	20					15
'Draper'	MSU	2003	90	6		2				<1	1
'Liberty'	MSU	2003	100								
'Huron'	MSU	2012	75			25					
'Osorno'	MSU	2014	95	3		1				<1	<1
'Calypso'	MSU	2014	87			13					
'Top Shelf'	FCG	2014	83	6		6			1		4
'Last Call'	FCG	2014	88	2		6					4
SHB											
'Avonblue'	UF	1976	84	1		5					8
'Sharpblue'	UF	1976	53	2		29					15
'Star'	UF	1981	78	8		7			1		6
'Millennia'	UF	1986	80	5		1		13			2
'O'Neal'	USDA/ NCSU	1987	83	10		2			5		
'Legacy'	USDA	1988	75			25					
'Misty'	UF	1989	85	1		6			1		6
'Emerald'	UF	1991	82	2		14				<1	2
'Ozarkblue'	UA	1996	76	4		12					8
'Biloxi'	USDA	1998	47	2		33		7			11
'Sampson'	NCSU	1998	76	11		13					
'Lenoir'	NCSU	2003	84	3		13					
'Carteret'	NCSU	2005	71	4			25				
'New Hanover'	NCSU	2005	78	2		14					6
'Camelia'	UG	2005	74	2		20					4

Table 2.2. *continued.*

Type and cultivar	Programme[a]	Year	Northern species (%)[b]			Southern species (%)[b]					
			COR	ANG	CON	DAR	ELL	FUS	TEN	VIR	ARB
'Rebel'	UG	2006	78	5		15				2	
'Snowchaser'	UF	2007	65	5		8	19		1	2	
'Flicker'	UF	2010	66	<1		12	19		1	1	
'Meadowlark'	UF	2010	75			13					12

[a]FCG, Fall Creek Genetics; MSU, Michigan State University; NCSU, North Carolina State University; UA, University of Arkansas; UF, University of Florida; UG, University of Georgia; USDA, US Department of Agriculture.
[b]ANG, *V. angustifolium*; ARB, *V. arboreum*; CON, *V. constablaei*; COR, *V. corymbosum*; DAR, *V. darrowii*; ELL, *V. elliottii*; FUS, *V. fuscatum*; TEN, *V. tenellum*; VIR, *V. virgatum*.

BREEDING TECHNIQUES

Blueberries are propagated asexually through cuttings and tissue culture, so elite genotypes can be utilized directly without the need to develop pure lines. Self-pollinations are rarely used in *Vaccinium* breeding due to reduced seed set and germination, and because seedlings from selfing tend to be weak. Most breeding programmes have relied primarily on pedigree breeding where elite parents are selected in each generation for intercrossing. The Florida SHB and rabbiteye breeding programmes have also employed recurrent selection.

Historically, in most NHB breeding programmes, evaluation begins 2 years after planting and selections are made over the next 2 years. Traditionally, the selected seedling plants were dug and moved to further spacing distances and evaluated for another year or two, before the most elite types were propagated and tested in rows of 25–50 plants for several years. The most promising selections from this row trial were then again propagated and tested in small numbers (five to ten plants) in replicate designs across multiple sites. The whole process took 15–20 years for the release of a new cultivar from the original cross.

More accelerated programmes are now being conducted by some NHB breeders, where the selected plants in the original planting are propagated and tested directly in replicated plantings at multiple sites. It is expected that about 1% of the progeny plants will go into this trial. After 3–5 years, the elite types will be released as cultivars. This approach can speed the release time to 8–10 years, even though it will result in the final testing of a larger number of ultimately rejected genotypes.

Most breeding programmes set the plants at spacings of about 60 cm apart in the row, although many use a higher density, with the extreme being the

Florida breeding programme, which does its primary selections in a 'fruiting nursery' at much closer spacings of 10,000–15,000 seedlings in a 0.2 ha field nursery. In the close-spaced SHB programme, the first selections are made within 12 months of planting (stage I). Ninety per cent of the seedlings are removed and the remaining plants are left in place for a further 3 years (stage II). Each year, they are evaluated for possible advancement to stage III, with about 300 selections being advanced each year into 15-plant plots. These plantings are observed for 10 years, with about 15 clones being selected each year and propagated for planting at multiple locations in larger plots (stage IV). Cultivars are ultimately selected from these blocks at a rate of about one geno-type from each stage IV test (Lyrene, 2008). The fastest-moving cultivars can go through this system in 10–12 years, although many are evaluated for much longer.

One problematic issue in the selection process of highbush cultivars for large commercial plantings of one genotype is the self-fruitfulness of a selec-tion. Inherently, breeders' trials are heterogeneous in the composition of geno-types, selections and standards being tested, which usually facilitates the opportunity for cross-pollination, resulting in a quite different environment from that of commercial plantings. Efforts must be made to test the self-fruitfulness of cultivars before release. This can be done by comparing the performance of selfed versus outcrossed hand pollinations.

BIOTECHNOLOGICAL APPROACHES TO BLUEBERRY GENETIC IMPROVEMENT

Micropropagation

Blueberries are now routinely micropropagated for commercial sale using tissue culture techniques (Hancock *et al.*, 2008). Axillary meristems are used as explants. The basic micropropagation steps include surface sterilization, proliferation and rooting. Lloyd and McCown's (1980) woody plant medium (WPM) or modified WPM is the most important basal medium used for all *Vaccinium* spp. The cytokinins, 6-(γ,γ-dimethylallylamino)-purine (2-isopentenyladenine (2iP)) and zeatin, are generally used for shoot pro-liferation. Most *Vaccinium* cultures are maintained at 20–25°C under a 16 h photoperiod of 10–75 µmol/m^2/s. Rooting is generally done under mist in the greenhouse or in covered flats.

Genetic linkage maps

The first genetic maps of blueberries have emerged and will set the ground-work for marker-assisted breeding. Rowland's group at the USDA-ARS in

Beltsville, Maryland, developed the first blueberry map using a diploid population segregating for chilling requirement (Rowland and Levi, 1994). Their population was a cross between an F_1 interspecific hybrid (*V. darrowii* × *V. elliottii*) and another clone of *V. darrowii*. The map had 72 randomly amplified polymorphic DNA (RAPD) markers mapped to 12 linkage groups, which is in agreement with the basic chromosome number of blueberry.

Later, her group constructed genetic maps of an F_1 hybrid of a low chilling, freezing-sensitive *V. darrowii* selection and a hybrid of two high chilling, freezing-tolerant diploid *V. corymbosum* selections. The goal was to develop populations that segregate for chilling requirement and cold tolerance. First, RAPD markers and more recently simple sequence repeat (SSR) (Rowland *et al.*, 2003a; Boches *et al.*, 2005, 2006) and expressed sequence tag polymerase chain reaction (EST-PCR) markers were added to this map. A quantitative trait locus (QTL) was identified that explained about 20% of the genotypic variance associated with cold hardiness (Rowland *et al.*, 2003b).

Rowland has continued to add markers to this map and, at the last report, it spanned 1448.7 cM and included 280 RAPD, SSR, EST-PCR and single nucleotide polymorphism (SNP) markers. The estimated map coverage is 85.7%, and the average distance between markers is 5.6 cM (Rowland *et al.*, 2014).

This updated mapping population was used to identify QTLs for mid-winter bud cold hardiness and chilling requirement under controlled conditions. The authors discovered high broad-sense heritability for both cold hardiness (0.88) and chilling requirement (0.86). They identified one QTL for cold hardiness and one for chilling requirement that were consistent over 2 years, and a second weaker QTL for chilling requirement that was detected in only 1 year.

Hancock's group at Michigan State University constructed an RAPD-based genetic map of a tetraploid population resulting from the cross of US 75 × tetraploid *V. corymbosum*, 'Bluecrop' (Qu and Hancock, 1997). A total of 140 markers were mapped to 29 linkage groups. The map was essentially that of *V. darrowii*, as US 75 was produced from an unreduced gamete of *V. darrowii* and only unique markers for 'Fla 4B' were used. 'Fla 4B' was included in the parentage of Rowland's map (Brevis *et al.*, 2007).

More recently, McCallum *et al.* (2016) developed a linkage map of the tetraploid cross 'Jewel' (SHB) × 'Draper' (NHB), as part of a grant funded through the Specialty Crop Research Initiative (SCRI) of the USDA (Rowland *et al.*, 2011). In total, 1794 SNPs and 233 SSRs were identified that exhibited segregation patterns consistent with an autotetraploid. Of these, 700 SNPs and 85 SSRs were used to construct a genetic map of 'Draper' and 450 SNPs and 86 SSRs for a 'Jewel' map. The 'Draper' map comprises 12 linkage groups and totals 1621 cM, while the 'Jewel' map comprises 20 linkage groups and totals 1610 cM.

The 'Jewel' × 'Draper' map is being used to identify QTL for a broad range of developmental and fruit characteristics in collaboration between the public sector principle investigators of the original SCRI grant, and researchers at General Mills and Berry Blue. The 'Jewel' × 'Draper' family was planted and phenotyped at Corvallis, Oregon (Chad Finn), Gainesville, Florida, and Manor, Georgia (Jim Olmstead and Rachel Itle), Grand Junction, Michigan (Ed Wheeler) and Inverness, UK (Susan McCallum and Julie Graham). Over a 2-year period, data were collected at all the sites for plant size, yield, rate of flower, bud and fruit development, leafing ability, fruit quality (firmness, scar, flavour, size, soluble solids and acidity) and chilling requirement.

Genomic resources

Several research groups have been actively developing genomic resources for blueberry crop improvement. The first few thousand ericaceous ESTs were generated and made publicly available for the family *Ericaceae* about a decade ago (Rowland *et al.*, 2008, 2011). The ESTs from blueberry (5000) were from non-acclimatized and cold-acclimatized flower bud libraries (Dhanaraj *et al.*, 2004, 2007). Around another 16,000 ESTs were generated from blueberry fruit by the New Zealand Institute for Plant & Food Research Ltd (formerly HortResearch Inc.) but were not made publicly available.

Rowland's group has utilized transcriptome sequencing to find the genes regulating dormancy induction and release (Die and Rowland, 2013). Her group has annotated 454 sequence assemblies from two blueberry cDNA libraries that represent flower buds in the first and second stages of cold acclimation and have identified transcripts related to carbohydrate metabolism and lipid metabolism that are associated with different stages of cold acclimation (Die and Rowland, 2014). They have generated transcriptome sequences from blueberry fruit at different stages of development (Rowland *et al.*, 2012) and have investigated the proteome-level changes that occur in flower buds with increasing exposure to chilling temperatures (Die *et al.*, 2016).

A draft genome of the diploid *V. corymbosum* selection 'W8520' has been generated at North Carolina State University by a *de novo* hybrid approach utilizing Roche 454 and Illumina GAIIx libraries. The draft represents more than 25,000 genes and covers 500 million base pairs. This effort was spearheaded by Allen Brown at the Plants for Human Health Institute in Kannapolis, North Carolina, and is being carried on with a number of collaborators led by Robert Reid in the Bioinformatics Department at the University of North Carolina at Charlotte (Reid *et al.*, 2016). A total of 43,594 SSRs were identified in the draft 'W8520' sequence, with dinucleotide repeats being the most abundant repeat types in all genomic regions except in probable gene-coding sequences (Bian *et al.*, 2014). A sample of these new genomic SSR and previously available EST-SSR markers was used to evaluate genetic diversity and population structure across 150 blueberry accessions. Rabbiteye blueberry proved to be distinct from other species and three subpopulations were detected including NHBs and SHBs and accessions related to the NHB 'Weymouth'.

Pat Edger at Michigan State University is in the process of sequencing the tetraploid cultivar 'Draper'. The Loraine laboratory at the University of North Carolina has worked on identifying candidate genes in the cultivated tetraploids that are involved in synthesis of bioactive compounds and other biosynthetic pathways (Gupta *et al.*, 2015). To characterize gene expression patterns, Illumina and 454/Roche sequencing of cDNA was done on samples of green and ripe fruit, from plants of the SHB 'O'Neal'. As expected, dynamic gene expression changes were found to accompany blueberry growth, maturation and ripening.

Several websites have been developed to provide access to expression profiles and gene sequences of blueberries. Rowland and her colleagues have produced the blueberry genomic database BBGD454, which allows the research community to identify genes involved in flower bud and fruit development, cold acclimation and chilling accumulation (Darwish *et al.*, 2013). This database is hosted by the Bioinformatics server at Towson University in Maryland (http:// bioinformatics.towson.edu/BBGD454/). The Genome Database for Vaccinium (http://www.vaccinium.org/) provides genomic, genetic and breeding data for blueberry, cranberry and other *Vaccinium* spp. This database is supported by the Plants for Human Health Institute, North Carolina State University and Washington State University, and contains the genome sequencing work on the diploid blueberry. The Plant Genome Network website (PGN: http:// identifiers.org/pgn/) provides public access to EST library statistics, details of Unigene builds and EST chromatograms, and permits Floral Genome Project (FGP) taxon-specific BLAST searches of gene sequences related to variations in floral architecture. This website is hosted by the FGP centred at Penn State University, Pennsylvania.

Recombinant DNA techniques

Two groups have reported on the transformation of blueberry using *Agrobacterium* (Graham *et al.*, 1996; Song and Sink, 2004). The screenable reporter gene *gusA* driven by either the cauliflower mosaic virus (CaMV) *35S* or a chimeric superpromoter, *(Aocs)3AmasPmas*, and each terminated by T-*nos*, has been transformed into blueberry cultivars using *nptII* as a selectable marker (Ni *et al.*, 1995; Graham *et al.*, 1996; Song and Sink, 2004). After selection with the herbicide glufosinate (GS), three chimeric *bar* genes with the promoter nopaline synthase (*nos*), CaMV *35S* or CaMV *34S* yielded transgenic plants, whereas the synthetic *(Aocs)3AmasPmas* superpromoter did not lead to successful regeneration of transgenic plants. The herbicide GS (Rely; Bayer Crop-Science) was applied at five levels using a track sprayer (GS in mg/l: 0, 750, 1500, 3000 and 6000) on 3-month-old plants in the laboratory, representing three separate transgenic events each for the *35S* and *nos* promoters. Evaluations of leaf damage 2 weeks after spraying indicated that all transgenic plants exhibited much higher herbicide resistance than non-transgenic plants. After application of eight times the standard level of GS (6000 mg/l) in the field, over 90% of the leaves on transgenic plants with *35S:bar* showed no symptoms of herbicide damage, whereas 95% of the leaves on non-transgenic plants were abscised. Transgenic plants with *35S:bar* showed higher herbicide resistance than those with *nos:bar*, in which 19.5–51.5% of the leaves had no damage (Song *et al.*, 2007, 2008).

A C-repeat binding factor (CBF)/dehydration-responsive element-binding (DREB) transcription factor gene identified from *V. corymbosum* (GenBank accession no. FJ222601) was transformed into the relatively cold-sensitive cultivar 'Legacy' (Walworth *et al.*, 2012). Almost 60 independent transgenic events were produced. Transgenic lines showed an increase in freezing tolerance in leaves and dormant buds. Expression of putative downstream components of the blueberry CBF regulon was increased in non-acclimated transgenic lines, and, in some cases, to a level similar to that of acclimated control plants. Following low temperature exposure, blueberry CBF-over-expressing transgenics and controls expressed these genes at similar levels.

Song *et al.* (2013) found that the blueberry *FLOWERING LOCUS T* (*FT*)-like gene (*VcFT*) cloned from the cDNA of highbush blueberry could reverse normal photoperiodic and chilling requirements and caused early and continuous flowering. Expression of *35S:VcFT* in 'Aurora' resulted in an extremely early flowering phenotype, which flowered not only during *in vitro* culture but also in 6–10-week-old, soil-grown transgenic plants that had not received any chilling hours. In related work, Song's group carried out transcript profiling of genes regulating flowering in *VcFT*-overexpressing blueberry plants (Walworth *et al.*, 2016) and found that overexpression of *VcFT* altered the expression of phytohormone-related genes (Gao *et al.*, 2016). They also found evidence of gene networks associated with the overexpression of a blueberry *DWARF AND DELAYED FLOWERING 1* gene in transgenic blueberry plants (Song and Gao, 2017).

Thomas Colquhoun and his associates at the University of Florida (Bizzio *et al.*, 2016) have begun a programme to identify volatile compounds that have a major impact on blueberry flavour. They are developing PCR primers from candidate sequences from other species, determining if they are present in blueberry and testing each gene's role in flavour volatile biosynthesis *in vivo* using RNAi transgenic blueberry constructs. They have identified several candidate genes of the oxylipin biosynthetic pathway, and at press were awaiting maturation of transgenic blueberry plants to verify their role.

Patenting and licensing

Currently, all new blueberry cultivars are being patented and licensed. What is called the 'plant patent' is being used in the USA and applies to plants that can be asexually reproduced (and cannot be reproduced by seed). In the Townsend-Purnell Plant Patent Act of 1930 it is stated that:

> Whoever invents or discovers and asexually reproduces any distinct and new variety of plant, including cultivated sports, mutants, hybrids, and newly found seedlings, other than a tuber propagated plant or a plant found in an uncultivated state, may obtain a patent therefore, subject to the conditions and requirements of this title.

An amendment was made to the Plant Patent Act in 1998 that adds:

> In the case of a plant patent, the grant shall include the right to exclude others from asexually reproducing the plant, and from using, offering for sale, or selling the plant so reproduced, or any parts thereof, into the US.

It has been suggested that such 'plant parts' includes gametes, and as a result, patented cultivars cannot be used in breeding by anyone except the inventor. However, this issue has not been challenged legally and thus is unresolved. This revision also restricts the importation of plant parts from patented cultivars into the USA.

The international protection offered for blueberries is termed 'plant breeders' rights'. The International Union for the Protection of New Varieties of Plants was established by the UPOV (Union Internationale pour la Protection des Obtentions Végétales) convention in 1961 (with additional conventions following in 1978 and 1991 providing additional provisions), and the UPOV system provides for cultivar protection using plant breeders' rights in 65 member countries. Plant breeders' rights have no restriction on breeding activity using a protected cultivar.

Both general and restricted licences have been awarded to nurseries for the propagation and sale of patented cultivars. General licences are usually made available to a group of companies without territorial limitations, while restricted licences are awarded to only one or a few companies by territory. In a few instances, partnerships have been developed prior to licensing that include trialling of advanced selections before cultivar release. In these trialling arrangements, a number of nurseries spanning the range of probable cultivar adaptation have been awarded testing rights and are required to provide production and quality data.

Both plant royalties and production royalties have been paid for blueberry cultivars. In the plant royalties, a set fee is paid per plant that is generally passed on to the grower by the propagator. This is the standard practice so far with blueberries, and ranges from US$0.20 to US$0.75 per plant. In the production royalties, the grower or fruit production company pays a royalty based on fruit produced or more often on the sales price of the fruit. This has been done in a few instances. Plant rental fees and plant production area fees have been discussed but have not yet been implemented.

CHARACTERISTICS OF THE MOST IMPORTANT BLUEBERRY CULTIVARS GROWN WORLDWIDE

Most popular SHB cultivars

'Chickadee' (2009, University of Florida) is very early with a chilling requirement of around 100 h. The bush is upright, vase shaped and blooms very early, requiring frost protection. The bush can uproot if there are heavy ice loads during frost protection. The berries are large, sweet, low acid and firm. It is increasing in importance in very low-chill areas.

'Emerald' (1991, University of Florida) is early with a chilling requirement of approximately 250 h. The bush is spreading, vigorous and highly productive, although leaf bud break can be poor in spring. 'Emerald' has a very early bloom date and can be subject to frost. The berries are very large, firm and medium blue with a medium scar and excellent flavour. It is resistant to cane canker and stem blight. It is widely planted in regions with low chilling hours.

'Farthing' (2008, University of Florida) is early with a chilling requirement of around 300 h. The bush is slightly spreading, vigorous and high yielding, and tends to overcrop. It leafs well in spring and is late blooming. The fruit are medium to large sized and firm with good colour, a slightly tart flavour and a small scar. There is potential for mechanical harvesting. It is widely planted in regions with low chilling hours.

'Flicker' (2010, University of Florida) has a chilling requirement of around 200 h. It is adaptable to early-season, deciduous or evergreen production. It has had problems with leafing in some years. The fruit are large, light blue, sweet and very firm, and have a small, dry picking scar and a long storage life. The fruit clusters are very loose. It has above-average resistance to root rot (*P. cinnamomi*), average resistance to stem blight (*Botryosphaeria* spp.) and has shown no signs of cane canker (*B. corticis*). It is widely planted in central Florida, with potential for all regions with low chilling hours.

'Jewel' (1998, University of Florida) is early mid-season with a chilling requirement of approximately 200 h. The bush is slightly spreading and very vigorous. 'Jewel' can be slow to leaf in spring and is highly subject to foliar diseases. The berries are large, moderately firm and light blue, with a small scar and slightly tart flavour. It is resistant to cane canker and stem blight, although it flowers heavily and can overbear. It is popular in low-chill areas across the world.

'Kestrel' (2010, University of Florida) is very early and has high evergreen fruiting potential. The fruit are large, firm, aromatic and sweet, even at early stages of ripening. The berry clusters are medium loose and the fruit are easily detached. There is increasing interest in 'Kestrel' in very low chilling regions.

'Legacy' (1988, USDA, Maryland) is late mid-season with a chilling requirement of 400–600 h. The bush is upright and very vigorous. 'Legacy' has medium- to large-sized fruit that are powder blue in colour with a good flavour, firmness and scar. It machine harvests well, and is widely planted across the world in areas with mild and moderate winter climates.

'Misty' (1989, University of Florida) is mid-season with a chilling requirement of 150 h. The bush is slightly spreading and very vigorous. Leaf bud break in spring can be poor, but it responds well to Dormex®. The picking season is long, and the bush needs to be heavily pruned to avoid overbearing. The berries are medium sized and very firm, with a small scar and mild flavour. It is widely planted in California and is important locally in Chile.

'Rocio' (2009, Atlantic Blue, Spain) is very early with a very low chilling requirement. Its fruit are attractive, medium sized and medium blue in colour, extremely firm, and exhibit a pleasant balance of acid and sweetness. The plant is evergreen, self-fertile and has an upright growth habit. It is widely planted in regions with little to no chilling hours.

'Springhigh' (2005, University of Florida) is very early with a chilling requirement of around 200 h. It has strong upright growth and leafs well in spring. The berries are large, dark blue, medium firm with a small scar and good flavour. The berries can get soft in hot weather. It is locally important across the world in areas with very few chilling hours but is diminishing in interest.

'Star' (1981, University of Florida) is early with a chilling requirement of approximately 400 h. The bush is upright and slightly spreading with moderate vigour. It leafs well in spring, and blooms later than 'O'Neal' but is harvested at about the same time. The berries are large and medium blue with good firmness, a small scar and good flavour. It has excellent postharvest fruit quality. The fruit can crack after heavy rains. It is widely planted in low-chill areas across the world, but its popularity is gradually declining.

'Ventura' (2013, Fall Creek Genetics, Oregon) is very early with a chilling requirement of 200 h or less. The bush is upright and vigorous. The berries are large, firm and medium blue with good flavour. It has been grown successfully in evergreen culture in locations with few chilling hours. It is rising in importance in areas with minimal chilling hours.

Older locally important SHB cultivars

'Abundance' (2006, University of Florida) is early mid-season with a chilling requirement of around 300 h. The bush is upright, very vigorous and leafs out well in spring. It has very high yield potential. The berries are large, medium blue, crisp textured and excellent tasting with a small, dry scar. It has not been widely planted.

'Biloxi' (1998, USDA, Mississippi) is early with a chilling requirement below 500 h. It may be adapted to evergreen culture with very little chilling. The plant habit is bushy with high vigour and productivity. The berries are small to medium sized, with only a medium scar, and are light blue, very firm and well flavoured. It is popular in Mexico and Peru.

'Bluecrisp' (1997, University of Florida) is mid-season with a chilling requirement of 500–600 h. The bush is slightly spreading, moderately vigorous and leafs out well in spring. Production is average, with some autumn blooming. It has little winter hardiness. The berries are medium size and extremely firm with a medium blue colour and small scar. The flavour is excellent, with a crisp texture. It is not widely planted.

'Bobolink' (2009, University of Florida) is early with a chilling requirement of around 200 h. The bush is upright and vigorous with excellent yield potential. The berries are large, sweet and of high quality. 'Bobolink' was released as potential replacement for 'Star' but has not been widely adopted.

'Camellia' (2005, University of Georgia) is early mid-season with a chilling requirement of 450–500 h. The bush is upright with moderate to high vigour and only modest winter hardiness. The berries are large, sky blue and firm with an excellent flavour and small picking scar. It is of modest importance in Georgia.

'Corona' (2009, Atlantic Blue, Spain) is mid-season with a very low chilling requirement. The fruit are extremely large, attractive, medium to dark blue and have a medium scar and good flavour. The plant is evergreen, vase shaped, very vigorous and grows well in a wide array of soil types. It requires cross-pollination. It is of modest importance in Chile.

'Daybreak' (2012, Berry Blue, Michigan and Chile) is early with a chilling requirement of approximately 250 h. The bush is vigorous and moderately upright with a medium-sized crown. It has good mechanical harvesting potential. The berries are very large, medium light blue and firm, with excellent flavour and a small picking scar. It is locally important in Chile, and is being planted in a number of low chilling areas across the world.

'Eureka' (2014, Mountain Blue Orchards, Australia) is an early to mid-season, evergreen cultivar with a chilling requirement of 200–250 h. The bush is round, upright and very high yielding. The fruit are jumbo sized, very firm and sweet with a good scar and shipping characteristics. It is not very self-fertile and needs a pollinizer. It is being planted in a number of low chilling areas across the world.

'First Blush' (2014, Mountain Blue Orchards, Australia) is early to mid-season, evergreen, upright and very vigorous with a chilling requirement of about 250 h. Its berries are very large, sweet and crisp with a good flavour and scar. It has very good storage characteristics. It was originally selected as a pollinator of 'Eureka' and is being planted along with 'Eureka' in many low-chill areas across the world.

'Gupton' (2005, USDA, Mississippi) is mid-season with a chilling requirement of around 500 h. The bush is vigorous and upright. The fruit are medium to large with good colour, firmness, flavour and picking scar. It is of modest importance in Mississippi.

'Meadowlark' (2010, University of Florida) is extremely early ripening and upright, with a very low chilling requirement. It produces very open clusters of berries that detach with medium force. It may have potential for mechanical harvesting. The fruit have a mild flavour with a good balance of sugar and acid, and the mature berries maintain quality for a long time when hanging on the plant. It is of modest importance in Georgia and Florida.

'Millennia' (1986, University of Florida) is early with a chilling requirement of around 300 h. The bush is spreading, vigorous and very productive. It is slow to leaf in spring. The berries are large and firm with a tiny scar, medium blue colour and mild flavour. It is locally important in Florida and Georgia.

'New Hanover' (2005, University of North Carolina) is early mid-season with a chilling requirement of 600–800 h. It has good self-fruitfulness. It is upright but can be floppy with a heavy crop. The berries are medium to large and firm with excellent colour. The flavour is a little tart, and the scar is small but can tear. It is of modest importance in North Carolina and Chile.

'Opi' (2014, Mountain Blue Orchards, Australia) is very early ripening, productive, upright, evergreen and leafs out vigorously in early spring. Its chilling requirement is less than 200 h. and it is well adapted to tunnels. The berries are large, medium to dark in colour and firm with a medium scar and sweet flavour. They may have a limited storage life. 'Opi' is being planted in a number of low chilling areas across the world.

'O'Neal' (1987, University of North Carolina) is early with a chilling requirement of approximately 400 h. The bush is erect but slightly spreading. The berries are large sized, firm, sweet and medium blue in colour. It is early blooming and subject to frost, and is resistant to stem canker. It is important locally in North Carolina, California, Georgia, Chile and Argentina, but its popularity is waning.

'Ozarkblue' (1996, University of Arkansas) is a late-season cultivar with a chilling requirement of 600–800 h. The bush is vigorous and upright. The berries are medium to large, light coloured, firm and sweet with small scar. It was popular in areas with mild and moderate climates but is no longer being planted.

'Palmetto' (2003, University of Georgia/USDA) is early with a chilling requirement of 300–450 h. The bush is open and spreading with medium vigour. Blooms very early so it may need frost protection and it has only modest winter hardiness. The berries are medium sized and medium blue with good firmness, good flavour and a medium scar. It is of modest importance in Georgia.

'Primadonna' (2007, University of Florida) is very early with a chilling requirement of around 200 h. The bush is upright and round with medium vigour. It may leaf out poorly in the spring. The berries are large, firm and medium blue with excellent flavour. Fruit size can be irregular. Only modest yields. It is locally most important in Florida.

'Raven' (2009, University of Florida) is early with a chilling requirement of 300 h. The bush has medium vigour and leafs well in spring. The berries are very large and firm, but have a large stem scar. Trialled extensively in Florida but has not been widely adapted.

'Rebel' (2006, University of Georgia) is very early with a chilling requirement of 250–350 h. The bush is spreading and vigorous. It leafs well following mild winters. The berries are very large, medium light blue and very firm with a bland flavour. It is important locally in Georgia.

'Robust' (2012, Berry Blue, Michigan, and Chile) is early with a chilling requirement of around 300 h. The bush is vigorous and moderately upright with a medium-sized crown. It has good mechanical harvesting potential. The berries are very large, medium light blue, firm and slightly tart, with a small picking scar. It is important locally in Georgia and Chile.

'Romero' (2008, Royal Berries SL, Spain) is very early and evergreen, with a very low chilling requirement. The bush is upright with high vigour. The berries are dark blue with a pleasant acid flavour and long postharvest fruit quality. It is self-fertile and an excellent pollen producer. It is important locally in Spain.

'San Joaquin' (2008, University of Florida) is early ripening with a chilling requirement of 400–500 h. The bush is very vigorous and upright. The fruit are large, sweet and firm with a good colour and an excellent picking scar. It is more upright than 'Star' with a higher yield potential. It is important locally in the Central Valley of California.

'Sapphire' (1980, University of Florida) is early with a chilling requirement of approximately 200 h. The bush is semi-spreading with medium vigour. Leaf bud break in spring is good. The berries are medium sized, light blue and very firm with a small scar and excellent flavour. It is no longer widely planted.

'Scintilla' (2008, University of Florida) is very early with a chilling requirement of around 200 h. The bush is upright and vigorous with modest yields and an early blooming date. Plant longevity is an issue. The fruit are large, light blue and firm, with a small scar and good flavour. It is important locally in areas with very low chilling hours but is decreasing in importance in Florida.

'Sharpblue' (1976, University of Florida) is early with a chilling requirement of less than 150 h. It has been grown successfully in evergreen culture. The bush is slightly spreading and extremely vigorous. The berries are very large with a medium scar, colour and firmness, and the flavour is excellent. Fruit quality is sensitive to hot temperatures. It was once popular in Florida but is no longer planted.

'Snowchaser' (2007, University of Florida) is very early with a chilling requirement of 200 h or less. The bush is upright and round with medium vigour. It tends to flower in autumn and blooms very early in the spring when frost is likely. The berries are medium sized, firm and medium blue with good flavour and a tiny scar. It is highly susceptible to stem blight. It is important locally across the world in regions with very low chilling hours.

'Suziblue' (2009, University of Georgia) is early with a chilling requirement of 250–350 h. The bush is vigorous and semi-spreading with a medium crown. The berries are large and medium light blue with good firmness, flavour and a small, dry picking scar. It is of modest importance in Georgia.

'Sweetcrisp' (2007, University of Florida) is early with a chilling requirement of around 200 h. The bush is upright, fast growing and leafs out early in spring. It may be well adapted to machine harvesting. The berries are medium sized with excellent firmness, flavour and scar. The texture is crisp. It is of modest importance in Georgia and Florida.

'Twilight' (2015, Mountain Blue Orchards, Australia) is mid- to late season with a chilling requirement of about 400 h. The bush is upright, evergreen, vigorous and high yielding. The fruit are very large, light flavoured and firm with a slightly acidic but sweet flavour. It has excellent storage and shipping qualities. It is important locally in low and intermediate chilling areas of Australia.

'Windsor' (2000, University of Florida) is early mid-season with a chilling requirement of 300–500 h. The bush is spreading and vigorous. Leaf bud break in spring is very good. The berries are large, firm and good flavoured, but the scar is wet and tears. It is no longer widely planted.

New SHB releases

'Arcadia' (2015, University of Florida) is mid-season with a long harvest season and a chilling requirement of less than 200 h. The bush is vigorous and spreading with excellent survival and leaf disease tolerance. It produces well in the evergreen management system. The berries are large and sweet with small scars and fair firmness.

'Avanti' (2015, University of Florida) is very early with a chilling requirement of around 100 h. It blooms early, so frost protection is necessary. The bush is moderately upright and vigorous. The fruit are medium sized, good flavoured and firm with a small, dry picking scar. It produces well in the evergreen management system. It is susceptible to *Botrytis* fruit rot.

'Bliss' (2012, Berry Blue, Michigan and Chile) is late mid-season with a chilling requirement of approximately 500 h. The bush is very vigorous and moderately upright with a medium-sized crown. The berries are very large, medium blue, firm and crunchy with an excellent flavour and small picking scar. It leafs well in spring and has good mechanical harvesting potential.

'Endura' (2015, University of Florida) is mid- to late maturing with a long harvest period and a chilling requirement of around 150 h. The bush is vigorous and moderately upright. The fruit are large and firm with small scars and excellent colour but are tart. It produces well in the evergreen management system. It is susceptible to leaf rust and *Phytophthora* root rot.

'Georgia Dawn' (2011, University of Georgia) is very early ripening with a chilling requirement of 100–150 h. It has good plant vigour, an upright growth habit and a narrow crown. The fruit are medium to large with a good flavour, scar and firmness. It is recommended as a pollinizer of 'Rebel'.

'Indigocrisp' (2013, University of Florida) is early with a chilling requirement of around 300 h. The bush is upright and vigorous. The fruit are large, very firm, crisp, sweet and very flavourful. It has good mechanical harvesting potential.

'Keecrisp' (2016, University of Florida) is mid- to late season with a chilling requirement of about 300 h. The bush is vigorous and upright with long, somewhat whippy canes. The fruit are large and exceptionally firm with a small, dry picking scar and a mild very sweet flavour. It has high potential as a machine-harvestable cultivar.

'Miss Alice Mae' (2015, University of Georgia) is a mid-season cultivar with a chilling requirement of 450–550 h. The bush is semi-upright, compact and moderately vigorous. It blooms a little later than 'Star' but harvests in about the same season. Yields are very good, and the berry firmness, flavour, and quality are excellent.

'Miss Jackie' (2015, University of Georgia) is a later-season cultivar with a chilling requirement of 450–550 h. The bush is upright, compact and moderately vigorous. The fruit are high quality with good size, firmness and flavour.

'Miss Lilly' (2015, University of Georgia) is an early mid-season cultivar with a chilling requirement of 500–600 h. The bush is narrow, upright and moderately vigorous, allowing it to be planted at high densities. The fruit are large and of high quality. It flowers 12–15 days after 'Star', allowing it to be grown without frost protection.

'Optimus' (2017, University of Florida) is an early cultivar released specifically for machine harvesting. The fruit are medium sized, light blue, very firm and good tasting.

'Patrecia' (2016, Straughn Farms and University of Florida) is very early ripening with a short fruit development period. The bush is vigorous and spreading. The fruit is large and firm, with good quality and a dry stem scar.

'Pearl' (2011, USDA/ARS, Mississippi) is early with a chilling requirement of 400–450 h. The bush is upright and vigorous with a narrow crown. The fruit are large, light blue and firm and have good flavour.

'Prelude' (2012, Berry Blue, Michigan and Chile) is very early with a chilling requirement of around 150 h. The bush is upright, moderately vigorous, with good leafing potential in spring. It has good mechanical harvesting potential. The berries are very large, medium blue, very firm and sweet with a small dry picking scar.

'Presto' (2016, Berry Blue, Michigan, and Chile) is very early with a chilling requirement of less than 200 h. The bush is medium upright and vigorous. It blooms later than other very early SHBs and may have some tolerance to frost. The fruit are very large, medium light blue and firm with a small scar and very good flavour.

'Southern Splendor' (2010, University of Georgia) is early with a chilling requirement of 450–500 h. The bush is semi-upright and vigorous with a narrow crown. It has good mechanical harvesting potential and a very short bloom-to-ripening time. The berries are medium to medium-large in size, with a medium light blue colour, excellent flavour and small, dry picking scar.

'Stellar' (2017, Berry Blue, Michigan and Chile) is early with a chilling requirement of less than 200 h. The bush is upright and vigorous. The fruit are very large, light blue and firm with a small scar and very good flavour.

'Temptation' (2012, Berry Blue, Michigan and Chile) is early with a chilling requirement of about 400 h. The bush is semi-upright and vigorous. The fruit size is large, medium blue in colour, very firm with a small picking scar and good flavour.

Most popular NHB cultivars

'Aurora' (2003, Michigan State University) is very late with a chilling requirement of more than 800 h. It is vigorous and bushy with high productivity and excellent winter hardiness. The berries are large, light blue and firm with a tiny scar and slightly tart flavour. The fruit are susceptible to sunburn under high temperatures from the late green fruit stage through to ripening. It is widely planted in regions with high chilling hours.

'Bluecrop' (1952, USDA, New Jersey) is mid-season with a chilling requirement of more than 800 h. The bush is upright but flops when carrying a heavy crop. It is high yielding with good winter hardiness. The berries are medium sized and firm with a small scar. It is the most widely planted cultivar, although its popularity is diminishing.

'Brigitta Blue', commonly known as 'Brigitta' (1980, Victorian State Department of Agriculture, Australia) is late mid-season with a chilling requirement of more than 800 h. The bush is upright with moderate winter hardiness. The berries are large, firm and sweet with a small scar and long storage life. It machine harvests well. It is important locally across the world in areas with mild winters.

'Draper' (2003, Michigan State University) is early mid-season with a chilling requirement of more than 800 h. It is an upright, slow-growing bush with excellent winter hardiness. The fruit are large, light blue and very firm with excellent flavour, a tiny scar and a superior shelf-life. It has the potential to be mechanically harvested for the fresh market. It is widely planted in regions with high chilling hours.

'Duke' (1986, USDA, New Jersey) is very early with a chilling requirement of more than 800 h. The bush is upright and open with good winter hardiness. It machine harvests well. Its vigour declines over time without expert culture. The fruit are medium sized, firm and medium coloured with a small scar and weak flavour. It is resistant to mummy berry. It is the most widely planted early cultivar in regions with high chilling hours.

'Elliott' (1973, USDA, Michigan) is late with a chilling requirement of more than 800 h. The bush is upright, bushy with good winter hardiness. The fruit are medium sized, firm, medium blue with a tart flavour. It machine harvests well, and is resistant to mummy berry, *Phomopsis* canker and anthracnose fruit rot. It has been very widely planted because of its late season, but interest is diminishing.

'Jersey' (1928, USDA, New Jersey) is late mid-season with a chilling requirement of more than 800 h. The bush is tall, upright and very winter hardy. The berries are medium sized, dark and somewhat soft, with a moderate scar and good flavour. It machine harvests well and has very broad soil adaptations. It is important locally in Michigan, but is seldom planted any more.

'Liberty' (2003, Michigan State University) is very late with a chilling requirement of more than 800 h. It is an upright, vigorous bush with excellent winter hardiness. It needs extensive pruning and trellising in some areas. The fruit are large, light blue and firm, with excellent flavour and a tiny scar. It is susceptible to cane diseases in northern production regions. It is probably machine harvestable, and is widely planted in regions with high chilling hours.

'Rubel' (1911, USDA, New Jersey) is late mid-season with a chilling requirement of more than 800 h. The bush is tall and upright with excellent winter hardiness. It harvests well by machine and is used primarily in the processed market. The fruit are small and firm with a fair flavour, medium scar and high levels of antioxidants. It is important locally in Michigan as a processed berry.

Older locally important NHB cultivars

'Bluegold' (1988, USDA, New Jersey) is late mid-season with a chilling requirement of more than 800 h. The bush is low growing with many branches and good winter hardiness. The berries are medium in size with small dry scars and good flavour and firmness. It is very susceptible to blueberry shock virus. It machine harvests well. There are limited plantings across high-chill regions.

'Bluejay' (1978, Michigan State University) is early mid-season with a chilling requirement of more than 800 h. It is an upright, open, rapidly growing bush that produces moderate yields of medium-sized, firm fruit with a small stem scar and mild, slightly tart fruit. It has field resistance to blueberry shoestring virus. It is important locally as a machine-picked, processed berry.

'Blueray' (1959, USDA, New Jersey) is early mid-season with a chilling requirement of more than 800 h. Upright, spreading habit and is very winter hardy. The berries are large, dark blue, firm and of excellent flavour. May overproduce if not regularly pruned. Once important locally, but not planted commercially any more.

'Bluetta' (1968, USDA, New Jersey) is very early with a chilling requirement of more than 800 h. The bush is small, low growing and spreading. It produces moderate yields of medium-sized, dark fruit. The flavour and firmness are only fair and the stem scar is broad. It is resistant to *Phomopsis* canker but very susceptible to *Botryosphaeria* canker. It was once widespread but is no longer planted commercially.

'Chandler' (1994, USDA, Oregon) is late mid-season with a chilling requirement of more than 800 h. The bush is spreading with good winter hardiness. The berries are exceptionally large and medium blue with excellent flavour and a long ripening season. There are very limited plantings across high-chill regions.

'Chanticleer' (1997, USDA, New Jersey) is very early with a chilling requirement of more than 800 h. It is an upright and moderately tall bush. It has good winter hardiness with modest yields. The fruit are medium sized with good colour, firmness and flavour (mild). There are limited plantings in areas with high chilling hours.

'Coville' (1949, USDA, New Jersey) is late mid-season with a chilling requirement of more than 800 h. The bush is moderately upright with limited winter hardiness. It has very large, firm fruit with a medium scar and good, tart flavour. It is not often planted any more.

'Croatan' (1954, USDA, North Carolina) is early with a chilling requirement of more than 800 h. It is a very productive, erect bush with only medium fruit quality and modest winter hardiness. The fruit are soft with a mild flavour and ripen very quickly in hot weather. It is resistant to stem canker. It was once important in North Carolina but is no longer planted.

'Darrow' (1965, USDA, New Jersey) is mid-season with a chilling requirement of more than 800 h. The bush is low and bushy with only limited winter hardiness. The fruit are very large, light blue, firm and flavourful when fully mature. It is little planted any more.

'Earliblue' (1952, USDA, New Jersey) is very early with a chilling requirement of more than 800 h. The bush is vigorous, upright and moderately winter hardy. The fruit are medium sized with medium colour, firmness and flavour. The scar is medium and tends to hold the fruit pedicel. It is resistant to powdery mildew. There are limited plantings in the Pacific Northwest.

'Hardyblue' or '1613A' (early 1900s, USDA, New Jersey) is mid-season with a chilling requirement of more than 800 h. The bush is upright and vigorous. The berries are medium sized, light blue and very sweet. It is a machine-harvested, processed berry. There are limited plantings in the Pacific Northwest.

'Hannah's Choice' (2005, USDA, New Jersey) is early with a chilling requirement of more than 800 h. It is upright and vigorous with good winter hardiness. Yields are modest. The fruit are medium to large, medium blue in colour and very sweet with a good scar and excellent firmness. There are very limited plantings in regions with high chilling hours.

'Huron' (2009, Michigan State University) is early with a chilling require-ment of more than 800 h. It is upright and vigorous with excellent winter har-diness. Yields are excellent. The fruit are large, light medium coloured and very sweet with an excellent scar and firmness. There are limited plantings in Europe in areas with high chilling hours.

'Nelson' (1988, USDA, Michigan) is mid-season with a chilling require-ment of more than 800 h. The bush is very productive, upright and very winter hardy. The berries are large, firm and good flavoured with a small picking scar. It machine harvests well, and is important locally in Michigan.

'Nui' (1989, Ruakura Research Centre, New Zealand) is very early with a chilling requirement of more than 800 h. The bush is spreading with moderate vigour and medium yields. It has good winter hardiness. The berries are very large and light blue with good firmness and excellent flavour. It is important locally in the Pacific Northwest.

'Patriot' (1976, USDA, Maine) is early mid-season with a chilling require-ment of more than 800 h. The bush is small to medium in height and slightly spreading. It has very good winter hardiness but blooms very early and is sub-ject to frost. The fruit are large and firm with a small scar and excellent flavour. It is not widely planted.

'Reka' (1985, Ruakura Research Centre, New Zealand) is early with a chilling requirement of more than 800 h. The bush is upright and vigorous with modest winter hardiness. It reaches adult productivity very quickly and has broad soil adaptations. The berries are medium to large and dark blue with an excellent flavour. It is important locally in the Pacific Northwest as a machine-harvested, processed berry.

'Reveille' (1990, North Carolina State University) is very early with a chilling requirement of more than 800 h. The bush is upright and suitable for mechanical harvesting. The fruit are small, firm and light blue with an excel-lent flavour. An early bloom is subject to frost, fruit cracking during rain can be a problem and some berries are slow to turn blue at the stem end. It is resistant to stem canker. It was important locally in North Carolina and Georgia but is little planted any more.

'Spartan' (1978, USDA, New Jersey) is early with a chilling requirement of more than 800 h. The bush is upright and open. It blooms unusually late for an early cultivar. The fruit are firm and very large with a medium scar and are highly flavoured. It has a narrow soil adaptive range. There are limited plantings in the Pacific Northwest.

'Toro' (1987, USDA, New Jersey) is mid-season with a chilling requirement of more than 800 h. The bush is upright, open and has good winter hardiness. The fruit are medium in size, powder blue and firm with a good flavour and small scar. It has not been widely planted.

'Weymouth' (1936, New Jersey) is very early with a chilling requirement of more than 800 h. The bush is low growing with good winter hardiness. The fruit are soft, dark blue and weakly flavoured. It is locally grown in New Jersey and Washington but is rarely planted.

New NHB releases

'Baby Blues' (2016, USDA, Oregon) is late mid-season with a chilling requirement of 600–800 h. The bush is vigorous and spreading with dense foliage and only modest winter hardiness. The berries are very small, very light blue and firm with a small scar and excellent flavour. It is machine harvestable and probably ideal for the processed market.

'Barbara Ann' or 'DrisBlueNine' (2014, Driscoll's, California) is late season with a chilling requirement of more than 800 h. The bush is semi-upright with medium to strong vigour. The berries are medium blue, firm and very large with medium sweetness and acidity. It was originally selected and trialled in the Pacific Northwest.

'Blue Ribbon' (2013, Fall Creek Genetics, Oregon) is early mid-season with a chilling requirement of 800–1000 h. The bush is vigorous and spreading, with moderate winter hardiness. The berries are medium sized, very light blue and firm. It is important locally in areas with reasonably mild winters.

'Calypso' (2014, Michigan State University) is late season with a chilling requirement of 800–1000 h. The bush is vigorous and upright with excellent winter hardiness. The berries are large, have small dry picking scars and are medium light blue colour with excellent firmness and flavour. It is growing in importance in Oregon and Michigan.

'Cargo' (2013, Fall Creek Genetics, Oregon) is late with a chilling requirement of 800–1000 h. The bush is very vigorous and upright with a narrow crown. It has only modest winter hardiness. It has good machine-harvesting potential. The berries are medium sized, very light blue and firm with a mild tart flavour. It is important locally in areas with reasonably mild winters.

'Clockwork' (2013, Fall Creek Genetics, Oregon) is mid-season with a chilling requirement of 800–1000 h. It has only modest winter hardiness. The bush is upright with a narrow crown. The berries are medium sized with an excellent, sweet flavour. It has good machine-harvesting potential for the processed market; it does not store well enough for the fresh market.

'Envy' (2016, Berry Blue, Michigan and Chile) is very early with a chilling requirement of 800–1000 h. The bush is medium upright with a small crown. It has excellent winter hardiness. The berries are jumbo sized and firm with excellent uniformity of size and eating quality.

'Granite' (2015, Oregon Blueberry) is mid- to late season with a chilling requirement of 800–1000 h. The bush is upright, compact and vigorous. The berries are light blue, medium sized and very firm.

'Keepsake' (2014, Berry Blue, Michigan and Chile) is late mid-season with a chilling requirement of 800–1000 h. The bush is medium upright and vigorous with a narrow crown and excellent winter hardiness. The berries are light blue, large and very firm with good flavour and have small dry picking scars. It is important locally in Michigan.

'Last Call' (2014, Fall Creek Genetics, Oregon) is very late with a chilling requirement of 800–1000 h. It has only modest winter hardiness. The bush is very vigorous and upright. The berries are medium to large sized and very light blue with an aromatic flavour. The berries can shrivel and crack with heat and late summer rains. It is important locally in areas with reasonably mild winters.

'Jolene' or 'DrisBlueFourteen' (2015, Driscoll's, California) is late mid-season with a chilling requirement of more than 800 h. The bush is semi-upright with medium to strong vigour. The berries are medium blue, large, firm and flavourful with medium sweetness and low acidity. It was originally selected and trialled in the Pacific Northwest.

'MegasBlue' (2015, Oregon Blueberry, Oregon) is mid-season with a chilling requirement of 800–1000 h. The bush is vigorous and spreading with good machine-harvesting potential. The berries are large and light blue with a mild flavour.

'Osorno' (2014, Michigan State University) is mid-season with a chilling requirement of 800–1000 h. The bushes are vigorous and upright, although the canes can be lax when loaded with fruit. It has only modest winter hardiness. The berries are large, have small dry picking scars and are a light blue colour with excellent firmness and a superior flavour. The fruit hold up extremely well in hot weather.

'Razz' (2011, USDA, Maryland) is early mid-season with a chilling requirement of 880–1000 h. The berries are medium to large in size and medium blue coloured. It is a little soft and has only a fair but unique flavour with raspberry overtones. It has been released for the home market.

'Sensation' (2014, Berry Blue, Michigan and Chile) is late season with a chilling requirement of 800–1000 h. The bush is very vigorous and medium upright with a medium-large crown. The berries are light blue, very large and firm with a small picking scar and a good, somewhat tart flavour. It is growing in importance in Michigan.

'Sweetheart' (2011, USDA, Maryland) is very early and requires 800–1000 chilling hours. The berries are small with very good firmness and a superior flavour. It has a tendency to re-fruit in the autumn. The bush can overcrop and produce dark, bicoloured fruit.

'Sweet Jane' or 'DrisBlueTen' (2014, Driscoll's, California) is early season with an intermediate chilling requirement. The bush is semi-upright with medium to strong vigour. The berries are medium blue, large and very firm with medium sweetness and acidity. It was originally selected in California but was trialled in the Pacific Northwest.

'Titanium' (2014, Oregon Blueberry, Oregon) is early mid-season with a chilling requirement of 800–1000 h. The bush is vigorous with good machine-harvesting potential. The berries are large, crisp and light blue with an excellent flavour.

'Top Shelf' (2013, Fall Creek Genetics, Oregon) is a mid-season cultivar with a chilling requirement of 800–1000 h. It has only modest winter hardiness. The bush is vigorous and vase shaped. It has only modest winter cold hardiness. The berries are very large, firm and sweet but develop colour slowly. It is important locally in areas with reasonably mild winters.

'Valor' (2017, Fall Creek Genetics, Oregon) is a mid-season cultivar with a chilling requirement of 800–1000 h and good cold hardiness. The bush is upright with good vigour. The berries are light blue, large, very firm and have a good flavour.

Half-high cultivars

'Chippewa' (1997, University of Minnesota) is mid-season with a chilling requirement of more than 800 h. It is a medium-stature bush with excellent winter hardiness. The berries are medium sized, very light blue and medium firm with a medium scar and good flavour. It is important locally in areas with extreme cold.

'Northblue' (1986, University of Minnesota) is early mid-season with a chilling requirement of more than 800 h. It is a medium-stature bush with superior cold hardiness. The berries are medium to large sized and dark blue with a medium scar, firmness and flavour (a little acid). It is resistant to mummy berry. It is important locally for pick-your-own and farm sales where winters are extremely cold.

'Northcountry' (1986, University of Minnesota) is early mid-season with a chilling requirement of more than 800 h. The bush is medium stature with superior cold hardiness. The berries are small, light blue and soft with a medium scar and good sweet flavour. It is important locally for pick-your-own and farm sales where winters are extremely cold.

'Northsky' (1986, University of Minnesota) is mid-season with a chilling requirement of more than 800 h. The bush is of very low stature with superior cold hardiness. The berries are very small, light blue and soft with a medium scar and good sweet flavour. It is resistant to mummy berry. It is important locally for pick-your-own and farm sales where winters are extremely cold.

'Polaris' (1996, University of Minnesota) is early with a chilling requirement of more than 800 h. It is a low-stature plant with superior cold hardiness. The fruit are medium sized and light blue with a good firmness and flavour. It is important locally for pick-your-own and farm sales where winters are extremely cold.

'St. Cloud' (1991, University of Minnesota) is early with a chilling requirement of more than 800 h. It is a medium-stature bush with excellent winter hardiness. The berries are firm, flavourful and medium sized, with a small scar. It is important locally for pick-your-own and farm sales where winters are extremely cold.

'Superior' (2008, University of Minnesota) is late with a chilling requirement of more than 800 h. It is a medium-stature bush with a spreading habit and mature height of 3.3–3.5 m. It is very productive and has extreme cold tolerance. The fruit are small to medium sized and light to medium blue coloured with a good colour, scar and flavour. It is recommended for trialling in areas with extremely cold winters.

Most popular rabbiteye cultivars

'Alapaha' (2001, University of Georgia) is an early-ripening cultivar with a chilling requirement of 450–500 h. It is a vigorous, productive, upright bush that leafs very well in spring. It flowers after 'Climax' but ripens at the same time. It has medium-sized berries with an excellent colour, firmness and flavour and a small, dry scar. It has a considerable degree of self-fruitfulness, and resists fruit cracking. It is a new cultivar being widely trialled in Georgia.

'Austin' (1996, University of Georgia) is early, productive and upright with a chilling requirement of 450–500 h. It is widely adapted. It flowers and ripens a few days after 'Climax'. The fruit are large and light blue with a good scar, firmness and flavour. It needs a pollinator. Modest hectarage has been planted in Georgia.

'Brightwell' (1981, University of Georgia) is an early mid-season cultivar with a chilling requirement of 400–450 h. The bush is vigorous, productive and upright. It is widely adapted and machine harvestable. It is at least partially self-fertile, and has a medium-sized berry that is medium blue in colour with a good scar, firmness and flavour. It is susceptible to fruit cracking under rainy conditions. It has been the most widely planted rabbiteye blueberry in the last 15 years.

'Briteblue' (1969, University of Georgia) is late harvesting with a chilling requirement of around 600 h. The bush is moderately vigorous and upright. The berries are light blue, large and very firm with a good flavour. It was once popular for pick-your-own operations but is no longer marketed much.

'Centurion' (1978, North Carolina State University) is late with a chilling requirement of 600–700 h. It ripens 1 or 2 weeks after 'Tifblue'. The bush is vigorous, productive and upright. The fruit are dark blue with a good scar and medium firmness but can crack after heavy rains. It is now planted primarily to extend the harvest season as a pick-your-own cultivar.

'Climax' (1974, University of Georgia) is early with a chilling requirement of 400 h. The bush has medium vigour and is slightly spreading. The fruit are medium sized and coloured, with excellent firmness, a good scar and a nice flavour. It has concentrated ripening and may be suitable for machine harvesting for the fresh market. It was once the most popular early cultivar in southeast USA but is now less popular due to its high susceptibility to spring freezes, gall midge and flower thrips.

'Delite' (1969, University of Georgia) is mid-season with a chilling requirement of 500 h. It ripens with 'Tifblue'. The bush is moderately vigorous and upright. The fruit are medium to large in size and medium coloured with a good scar and firmness, a sweet flavour and high levels of aromatics. It is highly susceptible to blueberry rust and is no longer recommended for planting.

'Ira' (1997, North Carolina State University) is late with a chilling requirement of 700–800 h. The bush is upright with medium to high vigour. It has a late bloom that minimizes freeze damage in the spring, and has good self-fertility. The fruit are medium in size and colour with a good scar, firmness and flavour. The fruit are aromatic and store well. It is recommended as a pick-your-own cultivar in the Piedmont and mountain regions of North Carolina.

'Maru' (1992, HortResearch, New Zealand) is very late with a chilling requirement of 600–750 h. The bush is slightly spreading with high vigour and productivity. The fruit are large, firm, medium blue and mild flavoured. It is important locally in New Zealand and is being trialled in the Pacific Northwest.

'Ochlockonee' (2002, University of Georgia) is very late with a chilling requirement of 600–700 h. The bush is moderately upright and vigorous with narrow crowns and very high productivity. It flowers late enough to miss most frosts in south Georgia. The fruit are large, medium blue and medium firm with a small scar and sweet flavour. It requires cross-pollination. It is being widely trialled in south-east USA and the Pacific Northwest, and has shown good resistance to fruit cracking in rainy conditions.

'Powderblue' (1978, North Carolina State University and USDA/ARS, North Carolina) is late mid-season cultivar with a chilling requirement of 550–600 h. The bush is upright, vigorous and productive. It is not very self-fertile. The fruit are medium to large and light blue with a good firmness, scar and flavour. The fruit are resistant to cracking. This has been a popular cultivar in recent years.

'Premier' (1978, North Carolina State University and USDA/ARS, North Carolina) is early with a chilling requirement of 550 h. It is vigorous and productive but has poor self-fertility. The fruit are large with a good scar, medium firmness, good flavour and dark blue colour. It is very susceptible to blueberry gall midge. It is still widely planted but needs frequent harvesting to retain adequate firmness. Late-season flowers are often malformed with partial or no corolla.

'Rahi' (1992, HortResearch, New Zealand) is late with a chilling requirement of 600–750 h. The bush is spreading and vigorous with medium yields. The fruit are medium sized, very firm and light blue with excellent flavour. It is important locally in New Zealand and is being trialled in the Pacific Northwest.

'Tifblue' (1955, University of Georgia) is a late mid-season cultivar with a chilling requirement of 600–700 h. The bush is vigorous, upright and productive. It has poor self-fertility. The fruit are medium sized and light blue with a good scar, firmness and flavour. It is susceptible to rain cracking. It is well adapted to machine harvesting. Until the early 1990s, this was the most planted rabbiteye but is now little planted.

'Woodard' (1960, University of Georgia) is early mid-season with a chilling requirement of 350–400 h. The bush is vigorous, spreading and productive. The fruit are large and dark blue, with soft to medium firmness, an excellent flavour and a medium scar. It is poor for shipping and freezing, and is often damaged by spring frosts. It was once widely grown but is now popular only for pick-your-own and farm markets.

Newly released rabbiteye cultivars

'Centra Blue' (2008, Institute For Plant And Food Research, New Zealand) is a very late cultivar with a chilling requirement of 600–750 h. The bush is semi-erect and has medium vigour. The fruit are large and light blue with a small scar and good firmness and flavour. The fruit have minimal grittiness.

'Columbus' (2003, North Carolina State University) is early mid-season to mid-season with a chilling requirement of more than 600 h. It ripens a little before 'Tifblue'. The bush is semi-upright, with medium vigour and good productivity. The fruit are large and powder blue, with an average scar, high aroma and good flavour. It has good storage life and is resistant to cracking but is not self-fruitful.

'DeSoto' (2007, USDA/ARS, Mississippi) fruits in the late mid-season with a chilling requirement of 600 h or more. The bush is vigorous, semi-dwarf and somewhat spreading. It will not grow taller than 2 m at maturity, removing the need for top pruning. The fruit are medium to large, light blue, firm and flavourful with a small picking scar.

'Ocean Blue' (2010, Institute For Plant And Food Research, New Zealand) is a mid-season cultivar with a chilling requirement of 600–750 h. The bush has medium vigour and is upright. It has medium-sized fruit that are medium blue with little grittiness, a small scar, good firmness and a sweet flavour. It is recommended for trial as a fresh market, mid-season cultivar.

'Onslow' (2001, North Carolina State University) is a late mid-season cultivar with a chilling requirement of 600 h or more. The bush is upright, vigorous and self-fruitful. The fruit are large, medium blue and firm with a good scar and pleasant aromatic flavour.

'Prince' (2010, USDA/ARS, Mississippi) is early ripening with a chilling requirement of approximately 600 h. The bush is very vigorous and upright, and flowers for an extended period. The fruit are medium sized, with excellent firmness, a dry scar and good flavour.

'Robeson' (2007, North Carolina State University) is early ripening with a chilling requirement of 400–600 h. It is vigorous and upright, and is unusually adapted to soils with a higher pH. The fruit are medium sized with a good colour and scar, although they are a little soft. This is a pentaploid.

'Savory' (2004, University of Florida) is early with a chilling requirement of 300 h. The bush habit is between upright and spreading, and tends to overbear. The fruit are large and light blue with a good scar, firmness and flavour. It is very susceptible to flower thrips and gall midge. Early flowering in the spring often results in freeze damage.

'Titan' (2010, University of Georgia) is early with a chilling requirement of 500 to 550 h. The bush is upright, productive and vigorous. It has good mechanical harvest potential. The fruit are extremely large, very firm, and hang well on the plant when ripe; berry colour and flavour are good, and dry scars contribute to good shelf-life. Fruit can crack under wet conditions.

'Vernon' (2004, University of Georgia) is early with a chilling requirement of 450–500 h. The bush is vigorous and open. The fruit are large, light blue and firm with a sweet flavour. It needs a pollinator. It has become popular in recent years due to good fruit quality for harvesting, handling and shipping.

CONCLUSIONS

Most blueberry production comes from cultivars derived from *V. corymbosum* L. (highbush blueberry), *V. ashei* Reade (rabbiteye blueberry; syn. *V. virgatum* Ait.) and native stands of *V. angustifolium* Ait. (lowbush blueberry). Highbush cultivars are further separated into northern (NHB) and southern (SHB) types depending on their chilling requirements and winter hardiness. The identification of wild species in the subgenus *Cyanococcus* has been problematic due to polyploidy, overlapping morphologies, extensive hybridization and a general lack of chromosome differentiation.

Among the most important characteristics being sought by blueberry breeders are flavour, a large fruit size, a light blue colour (a heavy coating of wax), a small scar where the pedicel detaches, easy fruit detachment for hand or machine harvesting, firmness and a long storage life. Expanding the range of adaptation of the highbush blueberry by reducing its chilling requirement has been an important breeding goal, along with season extension and winter cold tolerance. The chilling requirement has been reduced by incorporating genes from the southern diploid species *V. darrowii* into *V. corymbosum* via unreduced gametes, although hybridizations with native southern *V. corymbosum* and *V. virgatum* have also played a role. Cultivars are now available with an almost continuous range of chilling requirements from 0 to 1000 h.

Most breeding programmes have relied primarily on pedigree breeding where elite parents are selected in each generation for intercrossing. The Florida SHB and rabbiteye breeding programmes have also relied on recurrent selection. There are now a number of blueberry breeding programmes found across the world that are releasing a steady stream of new cultivars. All new blueberry cultivars are being patented and licensed.

Blueberries are now routinely micropropagated for commercial sale using tissue culture techniques. Other biotechnological techniques have not been widely utilized with blueberries, although the first genetic maps of blueberries are beginning to emerge that will set the groundwork for marker-assisted breeding.

REFERENCES

Austin, M.E. (1994) *Rabbiteye Blueberries: Development, Production and Marketing.* Agscience, Inc., Auburndale, Florida.

Ballington, J.R. (1990) Germplasm resources available to meet future needs for blueberry cultivar improvement. *Fruit Varieties Journal* 44, 54–62.

Ballington, J.R. (2001) Collection, utilization and preservation of genetic resources in Vaccinium. *HortScience* 36, 213–220.

Ballington, J.R., Rooks, S.D., Cline, W.O., Meyer, J.R. and Milholland, R.D. (1997) The North Carolina State University blueberry breeding program – toward *V. × covilleanum? Acta Horticulturae* 446, 243–250.

Bian, Y., Ballington, J., Raja, A., Brouwer, C., Reid, R., Burke, M., Wang, X., Rowland, L.J., Bassil, N.V. and Brown, A. (2014) Patterns of simple sequence repeats in cultivated blueberries (*Vaccinium* section *Cyanococcus* spp.) and their use in revealing genetic diversity and population structure. *Molecular Breeding* 34, 675–689.

Bizzio, L., Gonzalez, B., Kim, J.W., Cho, K. and Colquhoun, T. (2016) Towards genomic resources for the molecular breeding of enhanced blueberry flavor: the oxylipin volatile biosynthetic pathway as a test case for a new gene-identification methodology. In: *Abstract Book. XI International Vaccinium Symposium.* University of Florida, Gainesville, Florida, p. 101.

Boches, P., Bassil, N. and Rowland, L. (2005) Microsatellite markers for *Vaccinium* from EST and genomic libraries. *Molecular Ecology Notes* 5, 657–660.

Boches, P., Bassil, N.V. and Rowland, L.J. (2006) Genetic diversity in the highbush blueberry evaluated with microsatellite markers. *Journal of the American Society for Horticultural Science* 131, 674–686.

Brevis, P., Hancock, J.F. and Rowland, L.J. (2007) Development of a genetic linkage map for tetraploid highbush blueberry using SSR and EST-PCR markers. *HortScience* 42, 963–970.

Butkus, V. and Plizka, K. (1993) The highbush blueberry – a new cultivated species. *Acta Horticulturae* 346, 81–85.

Camp, W.H. (1945) The North American blueberries with notes on other groups of *Vaccinium*. *Brittonia* 5, 203–275.

Darnell, R.L. and Williamson, J.G. (1997) Feasibility of blueberry production in warm climates. *Acta Horticulturae* 446, 251–256.

Darwish, O., Rowland, L.J. and Alkharouf, N.W. (2013) BBGD454: a database for transcriptome analysis of blueberry using 454 sequences. *Bioinformation* 9, 883–886.

Dhanaraj, A.L., Slovin, J.P. and Rowland, L.J. (2004) Analysis of gene expression associated with cold acclimation in blueberry floral buds using expressed sequence tags. *Plant Science* 166, 863–872.

Dhanaraj, A.L., Alkharouf, N.W., Beard, H.S., Chouikha, I.B., Matthews, B.F., Wei, H., Arora, R. and Rowland, L.J. (2007) Major differences observed in transcript profiles of blueberry during cold acclimation under field and cold room conditions. *Planta* 225, 735–751.

Die, J.V. and Rowland, L.J. (2013) Advent of genomics in blueberry. *Molecular Breeding* 32, 493–504

Die, J.V and Rowland, L.J. (2014) Elucidating cold acclimation pathway in blueberry by transcriptome profiling. *Environmental and Experimental Botany* 106, 87–98.

Die, J.V., Arora, R., and Rowland, L.J. (2016) Global patterns of protein abundance during the development of cold hardiness in blueberry. *Environmental and Experimental Botany* 124, 11–21.

Draper, A.D. (1995) In search of the perfect blueberry variety. *Journal of Small Fruits and Viticulture* 3, 17–20.

Ehlenfeldt, M.K. and Ballington, J.R. (2012) *Vaccinium* species of section *Hemimyrtillus*: their value to cultivated blueberry and approaches to utilization. *Botany* 90, 347–353.

Galletta, G.J. and Ballington, J.R. (1996) Blueberries, cranberries and lingonberries. In: Janick, J. and Moore, J.N. (eds) *Fruit Breeding*. Vol. II. *Vine and Small Fruit Crops*. Wiley, New York, New York, pp. 1–108.

Gao, X., Walworth, A., Mackie, C. and Song, G-Q. (2016) Overexpression of a blueberry *FLOWERING LOCUS T* is associated with changes in the expression of phytohormone-related genes in blueberry plants. *Horticulture Research* 3, 16053.

Graham, J., Greig, K. and McNicol, R.J. (1996) Transformation of blueberry without antibiotic selection. *Annals of Applied Biology* 128, 557–564.

Gupta, V., Estrada, A.D., Blakley, I., Reid, R., Patel, K., Meyer, M.D., Andersen, S.U., Brown, A.F., Lila, M.A. and Loraine, A.E. (2015) RNA-Seq analysis and annotation of a draft blueberry genome assembly identifies candidate genes involved in fruit ripening, biosynthesis of bioactive compounds, and stage-specific alternative splicing. *GigaScience* 4: 1–22.

Hancock, J.F. (1998) Using southern blueberry species in Northern highbush breeding. In: Cline, W.O. and Ballington, J.R. (eds) *Proceedings of the 8th North American Blueberry and Extension Workers Conference*. North Carolina State University, Raleigh, North Carolina, pp. 91–94.

Hancock, J.F. (2006a) Northern highbush blueberry breeding. *Acta Horticulturae* 715, 37–40.

Hancock, J.F. (2006b) Highbush blueberry breeders. *HortScience* 41, 20–21.

Hancock, J.F. and Galletta, G.J. (1995) Dedication: Arlen D. Draper: blueberry wizard. *Plant Breeding Reviews* 13, 1–10.

Hancock, J.F., Lyrene, P., Finn, C.E., Vorsa, N. and Lobos, G.A. (2008) Blueberries and cranberries. In: Hancock, J.F. (ed.) *Temperate Fruit Breeding: Germplasm to Genomics.* Springer Science+Business Media, Dordrecht, The Netherlands, pp. 115–150.

Lloyd, G. and McCown, B. (1980) Commercially feasible micropropagation of mountain laurel, *Kalmia latifolia*, by the use of shoot tip culture. *Proceedings of the Plant Propagation Society* 30, 421–427.

Luby, J.J., Ballington, J.R., Draper, A.D., Pliszka, K. and Austin, M.E. (1991) Blueberries and cranberries (*Vaccinium*). In: Moore, J.N. and Ballington, J.R. (eds) *Genetic Resources of Temperate Fruit and Nut Crops.* International Society for Horticultural Science, Wageningen, The Netherlands, pp. 391–456.

Lyrene, P.M. (1998) Ralph Sharpe and the Florida blueberry breeding program. In: Cline, W.O. and Ballington, J.R. (eds) *Proceedings of the 8th North American Blueberry Research and Extension Workers Conference.* North Carolina State University, Raleigh, North Carolina, pp. 1–7.

Lyrene, P.M. (2008) Breeding southern highbush blueberries. *Plant Breeding Reviews* 30, 354–414.

Lyrene, P.M. (2011) First report of *Vaccinium arboreum* hybrids with cultivated highbush blueberry. *HortScience* 46, 563–566.

Lyrene, P.M. (2013) Fertility and other characteristics of F_1 and backcross$_1$ progeny from an intersectional blueberry cross [(highbush cultivar × *Vaccinium arboreum*) × highbush cultivar]. *HortScience* 48, 146–149.

Lyrene, P.M. (2016) Phenotype and fertility of intersectional hybrids between tetraploid highbush blueberry and colchicine-treated *Vaccinium stamineum*. *HortScience* 15, 15–22.

Lyrene, P.M. and Ballington, J.R. (1986) Wide hybridization in *Vaccinium*. *HortScience* 21, 52–57.

Mainland, C.M. (1998) Frederick Coville's pioneering contributions to blueberry culture and breeding. In: Cline, W.O. and Ballington, J.R. (eds) *Proceedings of the 8th North American Blueberry Research and Extension Workers Conference.* North Carolina State University, Raleigh, North Carolina, pp. 74–79.

Mainland, C.M. (2011) Frederick V. Coville and the history of North American highbush blueberry culture. *International Journal of Fruit Science* 12, 4–13.

McCallum, S., Graham, J., Jorgensen, L., Rowland, L.J., Bassil, N.V., Hancock, J.F., Wheeler, E.J., Vining, K., Poland, J.A., Olmstead, J.W., Buck, E., Wiedow, C., Jackson, E. Brown, A. and Hackett, C.A. (2016). Construction of a SNP and SSR linkage map in autotetraploid blueberry using genotyping by sequencing. *Molecular Breeding* 36, 1–24.

Ni, M., Cui, D., Einstein, J., Narasimhulu, S., Vergara, C.E., and Gelvin, S.B. (1995) Strength and tissue-specificity of chimeric promoters derived from the octopine and mannopine synthase genes. *The Plant Journal* 7, 661–676.

Olmstead, J.W. and Finn, C.E. (2014) Breeding highbush blueberry cultivars adapted to machine harvest for the fresh market. *HortTechnology* 24, 290–294.

Qu, L. and Hancock, J.F. (1995) Nature of 2n gamete formation and mode of inheritance in interspecific hybrids of diploid *Vaccinium darrowii* and tetraploid *V. corymbosum*. *Theoretical and Applied Genetics* 91, 1309–1315.

Qu, L. and Hancock J.F. (1997) RAPD-based genetic linkage map of blueberry derived from an interspecific cross between diploid *Vaccinium darrowii* and tetraploid *V. corymbosum*. *Journal of the American Society for Horticultural Science* 122, 69–73.

Qu, L. and Vorsa, N. (1999) Desynapsis and spindle abnormalities leading to 2n pollen formation in *Vaccinium darrowii*. *Genome* 42, 35–40.

Reid, R.W., Lin, Y.-C., Gharaibeh, R., Brouwer, C., Rowland, J., Maine, D., Walstead, R., Lila, M.A. and Brown, A. (2016) "Blueprints for Blueberry" – the current status of assembly and annotation of the blueberry genome. In: *Plant and Animal Genome Conference XXVI*. San Diego, California, Poster P31131.

Rowland, L.J. and Levi, A. (1994) RAPD-based genetic linkage map of blueberry derived from a cross between diploid species (*Vaccinium darrowii* and *V. elliottii*). *Theoretical and Applied Genetics* 87, 863–868.

Rowland, L.J., Mehra, S., Dhanaraj, A., Ogden, E.L. and Arora, R. (2003a) Identification of molecular markers associated with cold tolerance in blueberry. *Acta Horticulturae* 625, 59–69.

Rowland, L.J., Smriti, M., Dhanaraj, A.L., Ehlenfeldt, M., Ogden, E.L. and Slovin, J.P. (2003b) Development of EST-PCR markers for DNA fingerprinting and mapping in blueberry (*Vaccinium*, Section *Cyanococcus*). *Journal of the American Society for Horticultural Science* 128, 682–690.

Rowland, L.J., Dhanaraj, A.L., Naik, D., Alkharouf, N., Matthews, B. and Arora, R. (2008) Study of cold tolerance in blueberry using EST libraries, cDNA microarrays and subtractive hybridization. *HortScience* 43, 1975–1981.

Rowland, L.J., Alkharouf, N., Bassil, N., Beers, L., Bell, D.J., Buck, E.J., Drummond, F.A., Finn, C.E., Graham, J., Hancock, J.F., McCallum, S.M. and Olmstead, J.W. (2011) Generating genomic tools for blueberry improvement. *International Journal of Fruit Science* 12, 276–287.

Rowland, L.J., Alkharouf, N., Darwish, O., Ogden, E.L., Polashock, J.J., Bassil, N.V., Main, D. (2012) Generation and analysis of blueberry transcriptome sequences from leaves, developing fruit, and flower buds from cold acclimation through deacclimation. *BMC Plant Biology* 12, 46.

Rowland, L.J., Ogden, E.L., Bassil, N.V., Buck, E.J., McCallum, S. (2014) Construction of a genetic linkage map of an interspecific diploid blueberry population and identification of QTL for chilling requirement and cold hardiness. *Molecular Breeding* 34, 2033–2048.

Sharp, R.H. and Darrow, G.M. (1959) Breeding blueberries for the Florida climate. *Proceedings of the Florida State Horticulture Society* 72, 308–311.

Soltis, D.E., Soltis, P.S., Schemske, D.W., Hancock, J.F., Thompson, J.N., Husband, B.C., and Judd, W.S. (2007) Autopolyploidy in angiosperms: have we grossly underestimated the number of species? *Taxon* 56, 13–30.

Song, G.-Q. and Gao, X. (2017) Transcriptomic changes reveal gene networks responding to the overexpression of a blueberry *DWARF AND DELAYED FLOWERING 1* gene in transgenic blueberry plants. *BMC Plant Biology* 17(1), 106. doi: 10.1186/s12870-017-1053-z

Song, G.-Q. and Sink, K.C. (2004) *Agrobacterium tumefaciens*-mediated transformation of blueberry (*Vaccinium corymbosum* L.). *Plant Cell Reporter* 23, 475–484.

Song, G.-Q., Roggers, R.A., Sink, K.C., Particka, M. and Zandstra, B. (2007) Production of herbicide-resistant highbush blueberry 'Legacy' by *Agrobacterium*-mediated transformation of the *bar* gene. *Acta Horticulturae* 738, 397–407.

Song, G.-Q., Sink, K.C., Callow, P.W., Baugham, R. and Hancock, J.F. (2008) Evaluation of different promoters for production of herbicide-resistant blueberry plants. *Journal of the American Society for Horticultural Science* 133, 605–611.

Song, G.-Q., Walworth, A., Zhao, D., Jiang, N. and Hancock, J.F. (2013) The *Vaccinium corymbosum FLOWERING LOCUS T*-like gene (*VcFT*): a flowering activator reverses photoperiodic and chilling requirements in blueberry. *Plant Cell Reports* 32, 1759–1769.

Tsuda H., Kunitake, H., Yamasaki, M. Komatsu, H. and Yoshioka, K. (2013) Production of intersectional hybrids between colchicine-induced tetraploid Shashanbo (*Vaccinium bracteatum*) and highbush blueberry 'Spartan'. *Journal of the American Society of Horticultural Science* 138, 317–324.

Uttal, L.J. (1986) An older name for *Vaccinium australe* Small (Ericaceae). *Castanea* 51, 221–224.

Uttal, L.J. (1987) The genus *Vaccinium* L. (Ericaceae) in Virginia. *Castanea* 52, 231–255.

Vander Kloet, S.P. (1980) The taxonomy of highbush blueberry, *Vaccinium corymbosum*. *Canadian Journal of Botany* 58, 1187–1201.

Vander Kloet, S.P. (1988) *The Genus* Vaccinium *in North America*. Publication No. 1828, Research Branch of Agriculture Canada, Ottawa, Canada.

Vorsa, N. and Rowland, L.J. (1997) Estimation of 2*n* megagametophyte heterozygosity in a diploid blueberry (*Vaccinium darrowii* Camp) clone using RAPDs. *Journal of Heredity* 88, 423–426.

Walworth, A.E., Rowland, L.J., Polashock, J.J., Hancock, J.F., and Song, G.-Q. (2012) Overexpression of a blueberry-derived *CBF* gene enhances cold tolerance in a southern highbush blueberry cultivar. *Molecular Breeding* 30, 1313–1323.

Walworth, A., Chai, B. and Song, G.-Q. (2016) Transcript profile of flowering regulatory genes in *VcFT*-overexpressing blueberry plants. PLOS ONE 11: e0156993.

Weakley, A.S. (2007) Flora of the Carolinas, Virginia, Georgia, and surrounding areas. University of North Carolina, Chapel Hill, North Carolina. Available at: http://www.herbarium.unc.edu/WeakleysFlora.pdf (accessed 22 January 2018).

Growth and Development of Blueberries

INTRODUCTION

In this chapter, the anatomy and morphology of the highbush and rabbiteye blueberry are discussed, followed by a discussion of vegetative and reproductive growth and development. Environmental effects on growth and development are also presented.

ANATOMY AND MORPHOLOGY

Plant habit

All species of *Vaccinium* are woody perennials, and stature is one of the most striking differences among the various cultivated blueberries. Lowbush blueberries range from 0.1 to 0.15 m in height, while highbush plants can reach 1.8–4.0 m and rabbiteyes may grow to 6 m tall.

The blueberry shrub is composed of shoots that emerge from newly formed buds or previously formed dormant buds located in the crown. The shoots emerging from the base of plants are called canes and become woody in the second season of growth.

A dormant 1-year-old blueberry shoot typically has inflorescence buds at the top, with vegetative buds below (Fig. 3.1). The flower buds are large and round, while the vegetative buds are smaller, narrow and pointed. The dormant vegetative bud is about 4 mm long, with a single apex (Gough and Shutak, 1978).

The number of flowers found in the inflorescence buds is negatively correlated with distance from the tip. In 'Bluecrop', the primary buds at the tip of the shoots average nine or ten flowers, while the tertiary ones have eight and the quaternary ones have seven (Gough, 1994). There is usually only one

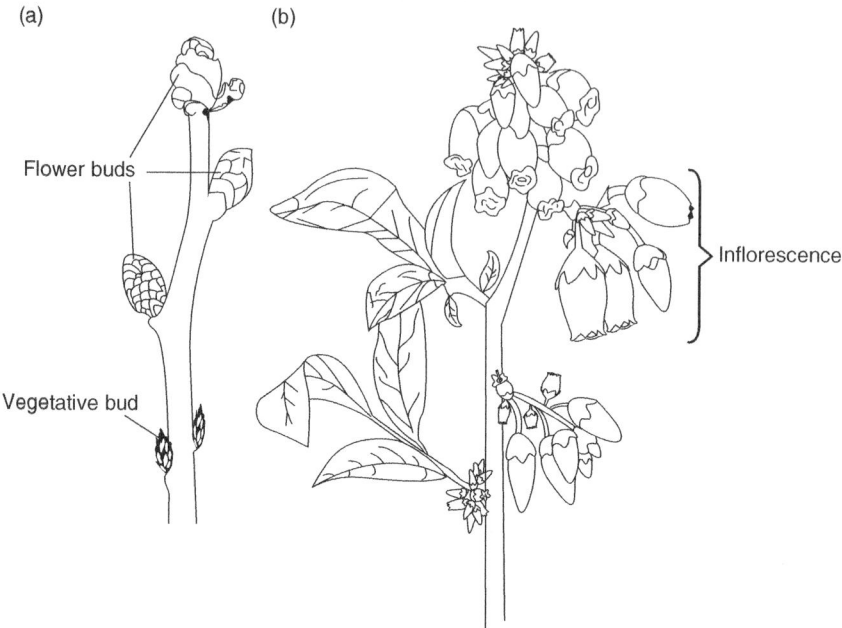

Fig. 3.1. (a) A blueberry shoot with the flower buds at the top of the current season's growth (adapted from Handley, 2016), and (b) the inflorescence of the blueberry. (Adapted from Darnell, 2006.)

flower bud at a node, although some of the upper nodes have a secondary bud, with only one to five flowers. The number of flower buds on a shoot is related to shoot thickness, cultivar and light penetration. There are large differences among cultivars in flowers per bud, buds per cane, laterals per cane and canes per bush (Table 3.1).

Table 3.1. Mean yield components in adult plants of six NHB blueberry cultivars at Grand Junction, Michigan, over 2 years. (Adapted from Hancock, 1989.)

Cultivar	Flowers per bud	Buds per cane	Laterals per cane	Canes per bush
'Elliott'	10.8	4.8	9.6	24.9
'Spartan'	7.7	4.2	7.4	20.6
'Jersey'	7.8	6.1	8.2	26.3
'Bluejay'	7.1	4.7	8.2	23.9
'Bluecrop'	7.8	4.0	7.2	23.9
'Rubel'	8.1	6.7	7.3	21.8
Standard error	1.3	1.2	0.9	4.7

Leaves

Blueberry leaves are simple, entire to serrated and alternately arranged along the stem. Most highbush and rabbiteye species are deciduous, although some of the lower chilling cultivars can be evergreen if temperatures remain above freezing. Leaf shape ranges widely from elliptic, spatulate and oblanceolate to ovate. Highbush and rabbiteye cultivars have varying amounts of pubescence and numbers of glands on the underside of the leaves.

Roots

Highbush and rabbiteye blueberries have two major types of root: thick storage roots (up to 11 mm in diameter) and fine, thread-like roots (as small as 1 mm in diameter). The former anchor the plants and perform a storage function, while the latter are primarily responsible for water and nutrient absorption. Blueberries do not have root hairs and are inhabited by an endotropic mycorrhiza (Coville, 1910; Jacobs *et al.*, 1982). In general, about 50% of the roots are located within 30 cm of the crown and 80–85% are within 60 cm (Paltineanu *et al.*, 2017) (Table 3.2). Over 80% of the root dry mass is found in the top 36 cm. Mulching tends to concentrate roots near the surface. Abbott and Gough (1987) found that highbush plants with mulch had 83% of their roots in the upper 15 cm of soil compared with 40% in plants that were not mulched. They also found that high rates of irrigation tended to increase root depth.

Paltineanu *et al.* (2017) found that in the mature NHBs 'Bluecrop' and 'Blueray' growing in a sandy-loam soil, the highest root density (more than 60 roots per 100 cm^3) was for roots of less than 0.1 mm in diameter. Abbott and

Table 3.2. Amount of roots (percentage of dry weight) at various depths and distances from the crown of 13-year-old 'Coville' NHB plants in Bridgehampton fine sandy loam soil with sawdust mulch. (Adapted from Gough, 1980.)

Depth of soil (cm)	Distance from crown (cm)						
	31	61[a]	94	122	153	183	Total
23	26	15	5	3	T	T	49
36	11	11	5	3	0	0	30
58	11	5	1	1	0	0	18
81	2	3	T	T	0	0	5
Total	50	34	11	7	T	T	0

T, trace amounts.
[a]Position of drip line.

Gough (1987) studied the size distribution of highbush roots and placed roots in seven categories of size from 40 μm (first order) to approximately 1 mm (seventh order). He found that blueberry roots are one-fifth to one-tenth of the diameter of those of other temperate fruit crops. Length decreased steadily as root width decreased, with first- and second-order roots representing nearly 75% of total root length. The root fresh weight followed the opposite trend, with third-, fourth- and fifth-order roots representing approximately 25, 29 and 29% of the total root fresh weight. He also determined that first- and second-order roots were the ones primarily involved in nutrient absorption. Fifth- and higher-order roots were used primarily for conduction and anchorage, while third- and fourth-order roots were transitional.

   Vaccinium spp. absorptive roots can have diameters of less than 50 μm ('hair roots') compared with a typical diameter of more than 200 μm in most other woody species. Valenzuela-Estrada *et al.* (2008) used minirhizotrons to study the root system of mature NHB 'Bluecrop' plants and established that the ephemeral portion of the root system was mainly in the first three root orders. First- and second-order roots were nearly anatomically identical, with similar mycorrhizal colonization and diameter, and also, despite being extremely fine, had median lifespans that were quite long (115–120 days). The more permanent portion of the root system occurred in fourth- and higher-order roots. Roots in these orders had radial growth; the lowest specific root length, nitrogen:carbon (C:N) ratios and levels of mycorrhizal colonization; the highest tissue density and vessel number; and the coarsest root diameter.

Flowers and inflorescences

The inflorescence of the blueberry is a raceme. The corolla of the blueberry is united, with four or five lobes, and is solid white to pink fringed in colour. The corolla is inverted and shaped like a globe or urn (Fig. 3.1). The pistil can be slightly longer or slightly shorter than the corolla. The ovary is inferior and has four to five cells (locules), with several to many ovules in each locule.

   There are eight to ten stamens per flower that insert at the base of the corolla and circle the style. The stamens are composed of an anther and filament; the anther has two awns, which have pores at their end through which the pollen emerges. The blueberry pollen grain is a tetrad, although it rarely produces multiple germ tubes (Brewer and Dobson, 1969).

   The fruit is a true berry with many seeds, and it ripens 2–3 months after pollination, depending on cultivar and environmental conditions. High temperatures tend to advance fruit ripening. Blueberry fruit range in colour from light blue to black and have a waxy cuticle layer that is about 5 μm thick. Pigments are found in the epidermal and hypodermal layers, which are separated from the rest of the cortex by a ring of vascular bundles. The majority of the blueberry flesh is white. At the centre of the fruit is a carpel with five lignified placentas with numerous seeds attached. Stone cells are found sporadically throughout the mesocarp but are most prevalent just below the epidermis.

GROWTH AND DEVELOPMENT

Vegetative and floral growth

Initiation along the cane proceeds basipetally, and florets within the individual racemes are initiated acropetally (Gough *et al.*, 1978; Lyrene, 1984; Huang *et al.*, 1997). Floral initiation in NHBs begins in late July in New Jersey, and by October all floral parts are developed (Gough *et al.*, 1978). In SHBs, floral bud initiation in the southern USA begins in early September, depending on cultivar, and continues into December (Huang *et al.*, 1997; Kovaleski *et al.*, 2015). Inflorescence bud initiation occurs a month earlier in 'Emerald' than in 'Jewel'. In most climates, differentiation of flower buds starts on 1-year-old wood in mid- to late summer (Gough *et al.*, 1978). Pollen development is arrested in NHBs after the formation of microspore mother cells in November, while in SHBs the development of pollen grains and ovules continues throughout the winter (Huang *et al.*, 1997). As photoperiods become short in the autumn and temperatures diminish, blueberries become dormant and subsequently require a certain number of chilling hours to begin normal floral and leaf growth in the spring.

When plants begin to grow in the spring, flower buds start to crack and then, except in more tropical areas, open over 3–4 weeks, depending on cultivar and temperature. Several stages of bloom development are generally recognized including bud swell and crack, tight cluster, early pink, late pink, early bloom, full bloom, petal drop and fruit expanding (for examples, see http:// www.canr.msu.edu/blueberries/growing_blueberries/growth-stages).

Vegetative buds begin to swell in the early spring as the leaves begin to develop within the buds. Vegetative bud break tends to occur more slowly than floral bud break depending on cultivar, chilling duration and temperatures in the spring. As the vegetative buds open, the leaves are closely clustered around the stem, but over time the internodes expand and the leaves become separated. Up to six leaf primordia are present in vegetative buds, and as the shoots grow, additional leaves are initiated by the shoot apex every 5 days (Gough and Shutak, 1978).

Growth of the shoots is sympodial and episodic. Individual shoots initially grow rapidly and then stop due to apical abortion, which is called 'black tip'. Shoots can have one, two or multiple growth flushes depending on cultivar and environmental conditions (Shutak *et al.*, 1980). Growth is renewed when an axillary bud is released from dormancy and the black tip is sloughed off. Generally, only one axillary bud is released from dormancy, leaving the shoot unbranched; however, it is not uncommon to have two or three buds break. Typically, there are two or three growth flushes in NHBs. There is a tendency for earlier-ripening cultivars to have more growth flushes than later ones, but this is not always the case. For example, 'Lateblue' tends to have as many flushes as a mid-season cultivar (Gough, 1994).

When a new shoot breaks from the base of the plant, it generally remains unbranched in the first year and all growth flushes arise from a single vegetative bud. After fruiting in the second year, two or more vegetative buds below the inflorescence break dormancy and begin to grow, resulting in the first branching. In subsequent years, multiple vegetative buds break each year after fruiting, resulting in increased branching and 'twiginess' of the shoot over time. Fruit size and yield per cane diminish as the fruiting canes become more twiggy.

Root growth

Studies done in young containerized highbush plants growing in sawdust showed that root growth has two peaks during the season (Abbott and Gough, 1987). The first and weaker peak occurs in spring, starting near the time of fruit set and extending to the immature green stage of fruit development. The second peak occurs after fruit harvest has started and ends before the plants go into dormancy.

Abbott and Gough (1987) found that the lifespan of blueberry roots ranges from 115 to 120 days for first- and second-order roots, while third-order roots have a lifespan of 136–155 days. Mycorrhizal colonization was highest in youngest roots (first- and second-order roots), decreased steadily in third- and fourth-order roots, and was not detectable in higher-order roots.

Pollination/fruit set

The receptivity of stigmas to pollen varies from 5–8 days in highbush blueberries (Moore, 1964) to 5–6 days in rabbiteyes (Young and Sherman, 1978). However, the percentage fruit set drops dramatically in highbush and lowbush blueberries if pollination is delayed by 3 or 4 days (Merrill, 1936; Wood, 1962; Kirk and Isaacs, 2012). In contrast, Young and Sherman (1978) found high levels of fruit set in rabbiteye plants pollinated 6 days after anthesis, while Brevis *et al.* (2005, 2006) documented that stigmatic receptivity in the rabbiteye cultivars 'Brightwell' and 'Tifblue' actually increased from 0 to 6 days before levelling off. Cultivars can also vary in their periods of stigma receptivity; Moore (1964) showed that, in NHBs, the pistils of 'Blueray' are receptive for longer than the pistils of 'Coville'. Growth of pollen tubes is favoured by warm temperatures (Knight and Scott, 1964).

Seed number has a significant influence on final fruit size in NHBs (White and Clark, 1939; Darrow, 1958; Moore *et al.*, 1972; Krebs and Hancock, 1988), SHBs (Lang and Danka, 1991) and rabbiteye blueberries (Moore *et al.*, 1972; Kushima and Austin, 1979). Cultivars can vary greatly in their response to increased seed numbers (Eaton, 1967). While seed numbers are important in determining final fruit size, more than 50% of the variation in fruit size is accounted for by other factors, including the amount of pollinator activity, air temperature, crop load and water availability (Brewer and Dobson, 1969; Eck, 1988).

Typically, only a fraction of the ovules develop into seeds. Highbush and rabbiteye blueberries have in excess of 110 ovules per fruit (Darrow, 1941; Parrie, 1990), but developed seed numbers rarely exceed half that number. Darrow (1958) found NHB cultivars had 16–74 seeds per fruit, while rabbiteye cultivars had from 38–82 seeds per fruit. Normally developed seeds are plump and brown, while those that have aborted are small and collapsed. Most seeds abort during late stage I and early stage II of fruit development (Edwards *et al.*, 1972), depending on the level of self-fertility. Huang *et al.* (1997) found that most ovule abortion occurred 5–10 days after pollination in the SHB 'Sharpblue'. Vander Kloet (1991) observed the first deterioration in highbush embryos about 20 days after pollination.

Many of the seeds that abort in highbush and rabbiteye blueberries do so because of early-acting inbreeding depression. Seeds abort as deleterious alleles are expressed during seed development (Krebs and Hancock, 1988, 1990, 1991). Several lines of evidence have been developed that support this in highbush blueberry including: (i) a range in self-fertility between different cultivars; (ii) a significant correlation between self- and outcross fertility; and (iii) a significant correlation between the percentage of aborted ovules and the inbreeding coefficient (Krebs and Hancock, 1988, 1990, 1991). In similar work, Vander Kloet and Lyrene (1987) and Vander Kloet (1991) also found an association between the level of relatedness and seed set in diploid, tetraploid and hexaploid races of *V. corymbosum*.

Fewer outcrossed pollen tetrads are needed for stigmatic saturation of blueberries than selfed ones. Parrie and Lang (1992) discovered that cross-pollination resulted in a cessation of stigmatic fluid production at lower tetrad densities than self-pollination. The number of pollen tetrads necessary for stigmatic saturation ranged from 295 (selfed 'Gulfcoast') to 201 (outcrossed 'O'Neal') in SHBs, from 256 (selfed 'Meader') to 195 (outcrossed 'Bluechip') in NHBs, and from 218 (selfed 'Northland') to 186 (outcrossed 'Northland') in half-high blueberries.

Once germinated, selfed pollen tubes grow in the style at the same rate as outcrossed ones. When Krebs and Hancock (1988), using fluorescence microscopy, compared rates of pollen tube growth in selfed NHB 'Spartan' versus 'Spartan' × 'Bluejay', they found that both selfed and crossed pollen reached the base of the style at day 2 after pollination, and at day 6 after pollination both types of pollen were entering the ovules. El-Agamy *et al.* (1981) found that the percentage of pollen that travelled the full length of the style was higher after 48 h in outcrossed versus selfed pollen of SHBs and rabbiteye blueberries, but by 72 h both classes had travelled the full length. Vander Kloet and Lyrene (1987) also found that selfed pollen could eventually fertilize ovules.

Fruit drop occurs about 3–4 weeks after flowering, and is less common in highbush blueberries than in rabbiteyes. The fruit that drop usually do not expand during the initial phase of fruit growth and have an abnormal red coloration. There is considerable variability among highbush cultivars in fruit set, ranging from about 50% to nearly 100%. Lyrene and Goldy (1983) observed a range in fruit set among open-pollinated rabbiteye cultivars from 36% in 'Tifblue' to 75% in 'Southland'. Davies (1986) also found that 'Tifblue' set only 21–27% of its flowers compared with 46–60% in 'Woodard' and 55% in 'Bluegem'. The position of flowering shoots in a bush had no consistent influence on fruit set.

Fruit development

All blueberry fruit exhibit a double-sigmoidal growth curve (Fig. 3.2). Stage I is characterized by rapid cell division and dry weight gain (Birkhold *et al.*, 1992; Cano-Medrano and Darnell, 1997) and lasts from 25 to 35 days depending on cultivar and environmental conditions. Little fruit growth is observed in stage II, but it is an active period of seed development (Edwards *et al.*, 1972). This period lasts from 30 to 40 days depending on cultivar and environment, and also on the number of viable seeds (Darnell, 2006). Highbush cultivars tend to

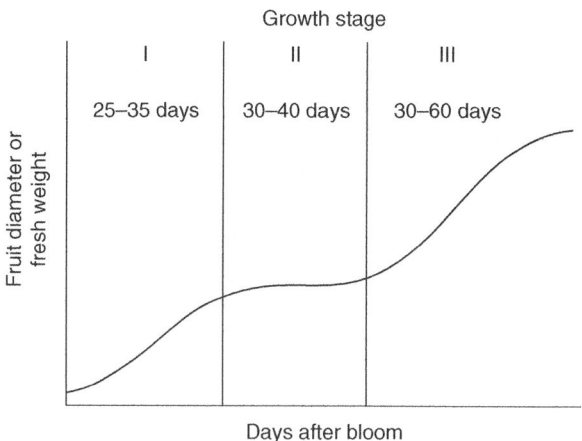

Fig. 3.2. Fruit growth in blueberry exhibits a double sigmoid curve. (Adapted from Darnell, 2006.)

have a shorter stage II than rabbiteyes, but there is considerable overlap (Edwards *et al.*, 1972). Stage III is characterized by very rapid fruit growth through cell enlargement (Eck, 1986; Birkhold *et al.*, 1992; Cano-Medrano and Darnell, 1997). Stage III lasts for 30–60 days, again depending on species, cultivar and environment. During stage III, sugars accumulate and the berry turns from green to blue as anthocyanins accumulate. The total length of the fruit development period ranges from 42 to 90 days in NHBs, from 55 to 60 days in SHBs, and from 60 to 135 days for rabbiteyes (Darnell, 2006).

There is some question as to whether blueberry fruit are climacteric. Bergman (1929), Ismail and Kender (1967, 1969) and Windus *et al.* (1976) measured a rise in carbon dioxide (CO_2) evolution during fruit development in lowbush and highbush blueberries that peaked during stage III. Lipe (1978) also observed an increase in ethylene production at the red berry stage in rabbiteye blueberries. However, Hall and Forsyth (1966) and Frenkel (1972) could not find any surge in respiration or ethylene production during ripening in their studies of lowbush and highbush blueberries. Janes *et al.* (1978) also could not induce respiration surges in highbush blueberries through treatments with acetaldehyde or ethylene.

A number of studies have been conducted to follow changes in organic composition in blueberry fruit as it matures. Perhaps the most complete analysis was done by Woodruff *et al.* (1960) on field-grown NHB 'Jersey' fruit in Michigan (Table 3.3). The intensity of colour increased over the first 6 days after the fruit began to colour and then stabilized. The percentages of lipids and

Table 3.3. Composition of 'Jersey' NHB fruit at different days after red coloration (percentage of dry weight). (Adapted from Woodruff *et al.*, 1960.)

Constituent	Days after red coloration						
	0	3	6	9	12	16	20
Reducing sugars	54.9	58.5	62.2	63.8	64.0	64.0	63.1
Non-reducing sugars	4.6	6.7	4.8	6.9	6.9	7.8	7.2
Total sugars	59.5	65.2	67.0	70.7	70.9	71.8	70.3
Titratable acid (as citric)	9.0	4.4	2.6	2.0	1.6	1.3	1.1
Sugar:acid ratio	6.6	14.8	25.8	35.3	44.3	55.2	63.9
Acid-hydrolysable polysaccharides	4.1	3.5	3.4	3.0	4.3	4.2	4.6
Starch	6.6	6.8	6.2	6.9	6.8	6.8	7.1
Cellulose	4.5	3.8	4.1	3.4	3.7	3.5	3.5
Lignin	6.8	5.4	4.6	4.2	4.3	5.4	4.9
Soluble pectin	1.1	1.0	1.0	0.9	0.9	0.6	0.7

waxes decreased in the early stages of ripening and then remained constant. Starch and other complex carbohydrates were relatively stable throughout maturation. Soluble pectin decreased throughout development as pectin methylesterase activity increased. They found that the percentage of total sugars increased for 9 days after colour change and then levelled off. The level of non-reducing sugars increased during the latter stages of development, but the concentration of reducing sugars decreased, keeping overall sugar levels constant. Titratable acidity decreased continually during berry ripening, resulting in a steady increase in the ratio of sugar:acid during ripening; others have noted similar patterns (Woodruff *et al.*, 1960). Sugar accumulation has been found to stop when berries are detached (Shutak *et al.*, 1957).

Ballinger *et al.* (1963) and Kushman and Ballinger (1963) measured the effect of a number of management practices on the sugar and acid composition of NHBs. They found that increases in crop load decreased sugar levels in the fruit but had no effect on acidity levels or fruit storage quality. Higher nitrogen decreased acidity but had little effect on sugar levels or postharvest quality. Expanding the number of days between harvests increased sugar content and reduced titratable acidity, as well as shelf-life. Fruit from the third harvest had higher sugar levels and lower acidity than those from the first two harvests, and had reduced shelf-life.

As fruit ripen, there are changes in the specific types of sugars and acids. Kushman and Ballinger (1968) and Ballinger (1966) found that as 'Wolcott' fruit ripen, there are increases in glucose and fructose and decreases in citric acid. Malic and quinic acids also decrease slightly during ripening. Markakis *et al.* (1963) found that ripening NHB 'Jersey' and 'Rubel' blueberries also had decreased levels of citric acid over time, and that 'Rubel' had higher overall levels of citric acid than 'Jersey'.

Blueberries become softer during ripening through the enzymatic digestion of the cell wall components, pectin, cellulose and hemicelluloses (Eskin, 1979; Proctor and Miesle, 1991). This process accelerates as the fruit become overripe, with concomitant increases in sugar content and decreases in acidity. Thus, the fruit get sweeter with the advancement of ripening but also softer. Blueberry cultivars vary greatly in their ability to maintain firmness after ripening (Ehlenfeldt and Martin, 2002; Hancock *et al.*, 2008).

FINAL FRUIT COMPOSITION

An average blueberry fruit is composed of approximately 83% water, 0.7% protein, 0.5% fat, 1.5% fibre and 15.3% carbohydrate (Hancock *et al.*, 2003). Blueberries have 3.5% cellulose and 0.7% soluble pectin, while cranberries contain 1.2% pectin. The total sugars in blueberries amount to more than 10% of the fresh weight, and the predominant reducing sugars are glucose and fructose, which represent 2.4%.

The overall acid content of *Vaccinium* spp. fruit is relatively high, with blueberries falling in the range of 0.5–1.5%. The primary organic acid in blueberries is citric acid (1.2%). They also contain significant amounts of ellagic acid, a compound thought to reduce the risk of cancer (Maas *et al.*, 1991). Compared with other fruits and vegetables, blueberries have intermediate to low levels of vitamins, amino acids and minerals (Hancock *et al.*, 2003). Blueberries contain 22.1 mg vitamin C per 100 g fresh weight; blueberries are unusual in that arginine is their most prominent amino acid. The major volatiles contributing to the characteristic aroma of blueberry fruit are *trans*-2-hexanol, *trans*-2-hexanal and linalool (Parliment and Kolor, 1975).

In general, blueberries are one of the richest sources of antioxidant phytonutrients among the fresh fruits, with total antioxidant capacity among cultivars ranging from 13.9 to 45.9 μmol Trolox equivalents/g fresh berry (Ehlenfeldt and Prior, 2001; Connor *et al.* 2002a,b) and 29.9 to 83.4 mmol Trolox equivalents/g fresh weight FRAP (ferric reducing ability of plasma) (Gündüz *et al.*, 2015). Berries from the various *Vaccinium* spp. contain relatively high levels of polyphenolic compounds, with chlorogenic acid predominating. Total anthocyanins in blueberry fruit range from 85 to 270 mg per 100 g, and species in the subgenus *Cyanococcus* carry the same predominant anthocyanins, aglycones and aglycone sugars, although the relative proportions vary (Ballington *et al.*, 1988). The predominant anthocyanins are delphinidin monogalactoside, cyanidin monogalactoside, petunidin monogalactoside, malvidin monogalactoside and malvidin monoarabinoside. The composition of various highbush and rabbiteye cultivars is similar, but the proportions of each compound are cultivar dependent (Lohachoompol *et al.*, 2008). Rabbiteye cultivars have significantly higher total anthocyanin content than highbush cultivars.

ENVIRONMENTAL EFFECTS ON GROWTH AND DEVELOPMENT

Temperature and photoperiod

A number of studies have shown that flower bud initiation in highbush and rabbiteye blueberries is induced by short-day photoperiods. In NHBs, 8 weeks of 8, 10 or 12 h photoperiods at a constant 21°C in the greenhouse resulted in much greater flower bud initiation than 14 or 16 h photoperiods (Table 3.4; Hall *et al.*, 1963). Darnell (1991) found that 6 weeks of 8 h photoperiods initiated more flower buds than day lengths of 11–12 h in rabbiteyes ('Beckyblue' and 'Climax') in Gainesville, Florida. In SHBs ('Misty' and 'Sharpblue') and *V. darrowii*, Spann *et al.* (2003) found that many more flower buds were induced under a constant temperature of 21°C than at 28°C for 8 weeks, although

Table 3.4. Average number of flowers per plant on three cultivars of NHB blueberry during 8 weeks of photoperiod treatment at 18°C night and 21°C day temperatures. (Adapted from Hall et al., 1963.)

Photoperiod (h)	No. of flowers per plant		
	'Coville'	'Earliblue'	'Jersey'
8	14[a,b]	64[b]	103[a]
10	41[a]	96[a]	133[a]
12	13[a,b]	61[b]	104[a]
14	12[a,b]	26[c]	15[b]
16	0[b]	2[c]	0[b]

[a,b,c]Mean values within a column with non-identical superscript letters are significantly different at $P=0.05$.

plant dry weight and cane height were not affected by these temperatures (Spann et al., 2004).

The number of flower buds initiated in highbush blueberries generally increases with the time of exposure to short days (Hall and Ludwig, 1961; Hall et al., 1963; Darnell, 1991). The full induction of flowering requires 5–6 weeks of shortening day lengths; however, Bañados and Strik (2006) found that some flower buds were initiated in the NHBs 'Bluecrop' and 'Duke' after only 2 weeks of 8 h photoperiods.

In some climates with long growing seasons, floral initiation in SHBs occurs in both early and late summer on new growth. In New South Wales, Australia (Wright, 1993), and in central Mexico and northern coastal Peru, floral initiation occurs on 1-year-old shoots after the first harvest (spring shoots) and then again on new growth later in the summer (summer shoots). These two periods of floral induction result in multiple crops, although pruning is often employed to limit production to only one crop in the year (Bañados, 2009).

Temperature also has a dramatic effect on root, shoot and fruit growth in highbush and rabbiteye blueberries. Abbott and Gough (1987) found that peaks in the growth of white unsuberized roots for mature highbush blueberries grown in sawdust mulch occurred when temperatures were between 14 and 18°C. Spiers (1995) reported that root, shoot and total dry weight in containerized SHB and rabbiteye cultivars was negatively correlated with root temperatures from 16 to 38°C.

Bloom date, ripening interval and harvest dates vary greatly in highbush and rabbiteye blueberries (Lyrene, 1985; Hancock et al., 1991; Finn et al., 2003), and there is a strong interaction with temperature (Carlson and Hancock, 1991). High spring temperatures generally accelerate bloom date and hasten petal drop. Bloom date is strongly correlated with ripening date, but early-ripening cultivars have been developed that have later-than-average flowering dates such as the NHBs 'Duke', 'Huron' and 'Spartan', and the SHBs 'Santa Fe' and 'Star'. There is a positive relationship between ripening interval and crop load in half-high cultivars, although genotypes with high yield potential and uniform ripening can be found (Finn and Luby, 1986; Luby and Finn, 1987). Fruit set, size and harvest period in NHB blueberries were greater in a cool greenhouse (8–24°C) than in a warm one (16–27°C) (Knight and Scott, 1964). Fruit set, size and harvest period were lower in rabbiteye blueberries grown under warm night conditions (21°C) compared with cool ones (10°C) (Williamson *et al.*, 1995).

Chilling requirement

Blueberries will flower if maintained under long days for extended periods, even if they receive no chilling hours. In highbush blueberries held under 16 h photoperiods, floral bud break occurred eventually after floral initiation, although it was not as uniform as in plants held under a normal dormancy cycle (Hall *et al.*, 1963). Rabbiteye blueberries have been shown to flower and fruit normally under long days without dormancy, if the plants are not defoliated and are vigorous (Sharpe and Sherman, 1971).

Once blueberries enter dormancy, they require a period of low temperatures for normal growth and development to occur (Table 3.5). Highbush cultivars are now available whose chilling requirements range from 0 to

Table 3.5. Chilling requirement of selected SHB and rabbiteye cultivars.

Chilling requirement (h)	Cultivar(s)
<300	SHB: 'Chickadee', 'Emerald', 'Flicker', 'Jewel', 'Kestrel', 'Misty', 'Primadonna', 'Scintilla', 'Snowchaser', 'Springhigh', 'Ventura'
300–400	SHB: 'Abundance', 'Farthing', 'Rebel'
	Rabbiteye: 'Woodard'
400–500	SHB: 'Biloxi', 'O'Neal', 'Star'
	Rabbiteye: 'Alapaha', 'Brightwell', 'Climax', 'Vernon'
500–600	SHB: 'Legacy'
	Rabbiteye: 'Columbus', 'Powderblue', 'Premier'
600–700	SHB: 'Ozarkblue'
	Rabbiteye: 'Ochlockonee', 'Tifblue'
700–800	SHB: 'New Hanover'
	Rabbiteye: 'Ira'
800–900	SHB: 'Reveille'

1000 h (Norvell and Moore, 1982; Darnell and Davies, 1990), and rabbiteye cultivars are available requiring 300–700 h (Williamson *et al.*, 2002). Too little chilling results in delayed, irregular bud break (Norvell and Moore, 1982; Darnell and Davies, 1990). Spiers (1976) found in the rabbiteye 'Tifblue' that floral bud break was more influenced by insufficient chilling hours than was vegetative bud development.

There is some controversy as to which temperatures are most effective in satisfying the chilling requirement of highbush and rabbiteye blueberries. The optimal chilling temperatures for buds of rabbiteye blueberries and SHBs are thought to be higher than those for NHBs (Mainland, 1985; Darnell, 2006), although comparative data are limited.

Mainland *et al.* (1977) determined that a constant 0.5°C satisfied the chilling requirement of floral and vegetative buds of the NHBs 'Croatan' and 'Wolcott', but 6°C was more effective in the rabbiteyes 'Tifblue' and 'Woodard'. They also found that intermittent temperatures above 10.5°C had a negative effect on the number of accumulated chilling hours. Norvell and Moore (1982) found that temperatures ranging from 1 to 12°C satisfied the requirements of NHB blueberries for vegetative bud break in 'Coville', but that 6°C was most effective. Alternating 1 to 6°C and 6 to 12°C at weekly intervals had little impact on leaf bud break, compared with a constant 6°C.

Gilreath and Buchanan (1981) found that the rate of floral bud break was similar at 0.6, 3.3, 7.0 and 10°C for 'Bluegem', 'Tifblue' and 'Woodard'. At 15°C, the rate of bud break in 'Bluegem' was similar to that at the other temperatures, but the higher temperature slowed the rate of development in the other two cultivars. Diurnal fluctuations of 0/7°C and 7/15°C for 14/10 h had little impact on days to terminal flower bud break compared with constant temperatures. A period of 14 days at 30°C in the middle of the chilling period did not affect the final level of floral and vegetative bud break in plants of 'Woodard', but floral bud break did occur faster in the high-temperature interruption treatment. Spiers (1976) found that an alteration of 10 h at 18°C with 14 h at 7°C delayed floral and vegetative bud break in 'Tifblue' but did not completely nullify the effect of low temperature.

Mainland *et al.* (1977) and Spiers (1976) recommended that the chilling requirement of blueberries be estimated using a modification of the Utah chill unit model for peach (Table 3.6). This was proposed to take into consideration that the chilling requirement of highbush and rabbiteye blueberries is at least partially satisfied by temperatures below 1.4°C and up to 12.4°C. More fine-tuning is likely to be necessary for individual cultivars, as responses to higher temperatures may vary. Gilreath and Buchanan (1981) found that fewer chilling hours were required for lateral floral bud break in the rabbiteye 'Tifblue' at 3.3°C (450–650 h) than at 7.0 or 10°C (650–850 h), while there was little difference in chilling requirement across the same temperatures in 'Bluegem'. Shine and Buchanan (1982) also found that the chilling temperature optimum and effective range of 'Aliceblue' (7.2°C optimum, range −2.5 to 15.9°C) and

Table 3.6. Conversion of selected temperatures to chill units for peach, highbush blueberry and rabbiteye blueberry. (Adapted from Spiers, 1976; Norvell and Moore, 1982.)

Temperature (°C)	Chill units[a]		Temperature (°C)	Chill units[b]
	Peach	Highbush		Rabbiteye
<1.4	0.0	0.5	<2	0.0
1.5–2.4	0.5	0.5	3–5	0.5
2.5–9.1	1.0	1.0	6–15	1.0
9.2–12.4	0.5	0.5	15–18	0.5
12.5–15.9	0.0	0.0	19–21	0.0
16–18	−0.5	−0.5	22–24	−0.5
>18	−1.0	−1.0	>25	−1.0

[a]Data from Norvell and Moore (1982).
[b]Data from Spiers (1976).

'Woodard' (11.0°C, −2.5 to 13.8°C) were higher and broader than those of 'Tifblue' (6.7°C, −1.2 to 12.9°C). SHB cultivars with complex ancestry may be particularly variable in their temperature thresholds, although this has not been documented.

Damage from cold

Winter cold often causes severe damage to blueberry flower buds and young shoots in the colder production regions. The key to survival in very cold temperatures is the ability to limit ice crystal formation to bud scales, floret scales and bud bracts (Flinn and Ashworth, 1994). In general, NHB types survive much colder mid-winter temperatures than rabbiteye and SHB cultivars, although considerable variability exists (Hancock *et al.*, 1987; Ehlenfeldt *et al.*, 2003, 2006; Hanson *et al.*, 2007). In full dormancy, NHB genotypes have been found to range in tolerance from −20 to −30°C, while rabbiteye genotypes range from −14 to −22°C. Few SHBs have been evaluated, although 'Legacy' has been found to tolerate temperatures down to −17°C and 'Ozarkblue' to −26°C. 'Sierra', which is composed of 50% southern germplasm, has tolerated temperatures below −32°C (Hanson *et al.*, 2007). The wood of half-high cultivars such as 'Northblue' can survive to −40°C and their flower buds can tolerate −36°C (C.E. Finn, USDA/ARS, Corvallis, Oregon, personal communication). In a comparison of the mid-winter cold hardiness of 25 rabbiteye cultivars Ehlenfeldt *et al.* (2006) found ranges in 50% lethal temperature (LT_{50}: temperature that kills 50% of the samples) values from −24.9 °C for 'Pearl River' to −13.7 °C for 'Chaucer'.

Spring frosts commonly damage flower buds of all blueberry species. Over-all, SHB flower buds and developing flowers appear to be more cold tolerant than rabbiteye flower buds (Lyrene, 2008), and NHB flower buds tend to be more tolerant than SHB types. Those cultivars with late bloom dates tend to suffer less frost damage than those flowering earlier because frosts are less common and the stage of floral development is correlated with relative bud hardiness (Hancock *et al.*, 1987; Lin and Pliszka, 2003). Terminal flower buds also tend to be less hardy than median or basal buds (Biermann *et al.*, 1979; Cappiello and Dunham, 1994), and styles are more sensitive to cold than corol-las. In controlled experiments, NeSmith *et al.* (1999) showed that ovaries in opened flowers of the rabbiteye 'Brightwell' could withstand −4.4°C, while styles survived −3.4°C and corollas −3.8°C. Rowland *et al.* (2013) found that the average sensitivity of plant parts followed the order: corolla < fila-ment < anther < style < external   ovary < stigma < ovules < internal   ovary < placenta.

Variation exists in the spring frost tolerance of NHB cultivars. After a −6°C evening, Bailey (1949) found a wide range of damage (10–74%) among nine cultivars in Massachusetts when blossoms were distinctly separated but corol-las were still closed. Johnston (1939) observed damage ranging from 10 to 58% in seven cultivars after a similar freeze in Michigan. Rejman (1977) found 'Blueray' and 'Darrow' to have more buds killed (7–10%) than 'Bluecrop', 'Jersey' and 'Lateblue' (0–1%) after an evening of −8.5°C in Poland. Lin and Pliszka (2003) found that 'Lateblue' had significantly less damage (11%) than seven other cultivars that bloomed much earlier (52–84%) after a night of −6°C in Poland. When Hancock *et al.* (1987) assessed flower bud injury in 17 NHB cultivars after two spring frosts in Michigan, they found significant differ-ences in the proportion of brown ovaries among cultivars, ranging from 26 to 94% (Table 3.7). Most of the variation was associated with the stage of bud development.

Spiers (1978) found that the temperature required to damage floral buds in rabbiteye blueberries was also inversely related to their development, similar to highbush blueberries. Swollen buds with individual florets still enclosed withstood temperatures of −6°C, those with individual flowers exposed after the bud scales abscised were killed at −4°C, those with well-separated flowers before corolla expansion survived to −2°C and fully opened flowers were killed at 0°C. A comparison of the percentage of buds killed in rabbiteye blueberries after a night of −9°C in Mississippi found that 'Delite' (98%) and 'Woodard' (85%) showed the greatest damage, while 'Climax' (53%), 'Briteblue' (56%), 'Southland' (63%) and 'Tifblue' (63%) had the least (Spiers, 1981). These plants were in a later stage of floral development where the individual flowers could be discerned but were not distinctly separated. Gupton (1983) found that fully opened flowers of 'Southland' were much hardier at −2°C than those of 'Climax', 'Delite', 'Tifblue' and 'Woodard'.

Table 3.7. Developmental stage and proportion of brown ovaries in terminal flower buds of 17 NHB cultivars in Michigan after spring freezes in 1983 and 1986. (Adapted from Hancock *et al.*, 1987.)

Cultivar	Developmental stage[a]		Proportion of brown ovaries (%)		
	1983	1986	1983	1986	Mean
'Elliott'	2.0	3.7	05	47	26
'Lateblue'	2.0	3.2	19	60	40
'Rubel'	2.0	4.6	36	77	57
'Coville'	2.2	4.2	46	72	59
'Bluejay'	2.7	4.8	54	81	68
'Berkeley'	2.5	4.3	51	88	70
'Jersey'	2.0	4.5	62	78	70
'Spartan'	2.5	4.8	65	89	77
'Bluecrop'	2.2	5.0	68	94	81
'Collins'	2.0	6.0	62	100	81
'Blueray'	2.7	4.8	73	93	83
'Darrow'	2.5	4.5	66	100	83
'Earliblue'	2.7	5.4	80	97	89
'Meader'	2.0	4.0	77	100	89
'Bluetta'	2.5	5.6	80	98	89
'Patriot'	2.5	5.6	80	88	84
'Bluehaven'	3.0	4.9	90	98	94
Standard error	0.5	0.8	28	21	25

[a]Stages: 1, first swell; 2, scales separated; 3, terminal florets exposed; 4, all florets exposed; 5, florets separated; 6, corollas expanding.

The rate of deacclimation probably plays an important role in early spring flower bud tolerance. Ehlenfeldt *et al.* (2003) found that the NHB 'Duke' deacclimated fastest in a mixed group of 12 cultivars, while the SHB 'Magnolia', the NHB × rabbiteye pentaploid hybrid 'Pearl River', the rabbiteye × *V. constablaei* cultivar 'Little Giant' and the half-highs 'Northcountry' and 'Northsky' were the slowest. The NHBs 'Bluecrop' and 'Weymouth', SHBs 'Legacy' and 'Ozarkblue' and rabbiteye 'Tifblue' were intermediate.

In another comparison of the rate of acclimation and deacclimation in blueberry genotypes with differing species backgrounds (Fig. 3.3), Ehlenfeldt *et al.* (2012, 2015) found in New Jersey that floral buds of highbush 'Bluecrop' and 'Legacy', rabbiteye 'Tifblue' and two rabbiteye hybrids (US 1043 and US 1056) all reached maximum cold hardiness by late December ranging from −22 to −27°C, while the half-high 'Northsky' and a *V. constablaei* × *V. virgatum* hybrid ('Little Giant') achieved cold acclimation of −28°C by the end of

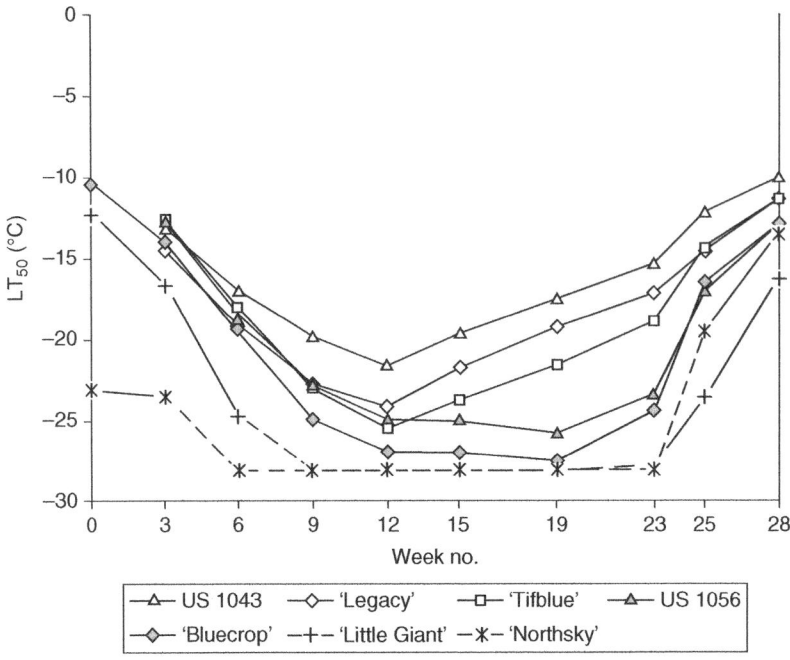

Fig. 3.3. Bud cold-hardiness values as LT_{50} (temperature causing 50% lethality) of blueberry genotypes averaged across the 2006/07 and 2007/08 seasons in Chatsworth, New Jersey. (Adapted from Ehlenfeldt *et al.*, 2012.)

November. 'Legacy', 'Tifblue' and US 1043 began an almost linear deacclimation in late December, while 'Bluecrop', 'Little Giant', 'Northsky' and US 1056 did not begin deacclimating for another 2 months. 'Little Giant' and 'Northsky' had no LT_{50} values higher (warmer) than –25°C until late March (Ehlenfeldt and Rowland, 2006).

Dehydrin concentration in floral buds may play a role in the level of cold tolerance achieved by blueberries. The level of three dehydrin-like proteins of 65, 60 and 14 kDa in dormant buds of the NHB 'Bluecrop' and rabbiteye 'Tifblue' have been shown to increase within 300 h of chilling and decrease to the pre-chilling level with the initiation of bud break (Arora *et al.*, 1997; Rowland *et al.*, 2004; Dhanaraj *et al.*, 2005). The increases in dehydrin levels were associated with increased bud hardiness in both cultivars, and overall levels were higher in the more cold-tolerant 'Bluecrop'.

Flower buds can also be damaged by rapid freezes in the autumn. The flower buds of rabbiteye and SHB cultivars are generally considered to acclimatize more slowly in the autumn than those of NHB cultivars and as a result are more subject to late autumn freezes (Rowland *et al.*, 2005; Hanson *et al.*, 2007). Hanson *et al.* (2007) found that leaf retention in autumn was not a good predictor of rate of deacclimation, as 'Ozarkblue' and US 245 retain their leaves until the very late autumn but are just as hardy as the mid-season standard 'Bluecrop'. Bittenbender and Howell (1975) also found no correlation between flower bud hardiness and autumn leaf retention.

Blueberry fruit set is also very sensitive to cold damage, although data are limited in highbush blueberry. Hall *et al.* (1971) found decreases in fruit set in lowbush blueberries ranging from 42 to 77% by holding plants for 4 h at −2°C or for 2 h at −3°C. The results were similar whether the cold was presented immediately after pollination or 6 days later. NeSmith *et al.* (1999) showed that plants exposed to −1°C for 1 h after flowering had significantly reduced fruit set without any visible damage.

CONCLUSIONS

All species of *Vaccinium* are woody perennials, ranging in height from 0.1–0.15 m in lowbush to 1.8–4.0 m in highbush blueberries and up to 6 m in rabbiteyes. Blueberry shoots emerge from buds located in the crown. One-year-old blueberry shoots typically form flower buds at the top, with the vegetative buds located below. Winter cold often causes severe damage to blueberry flower buds and young shoots in colder production regions. In full dormancy, NHB genotypes range in cold tolerance from −20 to −30°C, while rabbiteye and SHB genotypes range from −14 to −26°C. Spring frosts commonly damage flower buds of all blueberry species. Root and shoot growth occurs in cycles; root growth is greatest in the early spring and autumn, while shoot growth occurs in two or three flushes during the growing season. Individual shoots initially grow rapidly and then stop due to apical abortion, which is called 'black tip'. Over 80% of the root dry mass is found in the top 36 cm of soil. Most floral initiation occurs under short days, and the chilling requirement of cultivars varies greatly. Cultivars are now available with an almost continuous range of chilling requirements: from 0–800 h in SHBs to 800–1200 h in NHBs and 300–600 h in rabbiteye cultivars. There is some controversy on which temperatures are most effective in satisfying the chilling requirement of highbush and rabbiteye blueberries, but most researchers use a modification of the Utah chill unit model for peach. A pollinator is necessary in blueberries and a wide range of self-fertility is found among blueberry cultivars due to early-acting inbreeding depression. Rabbiteye blueberries are less self-fruitful than highbush types. The fruit is a true berry with many seeds and ripens 2–3 months after pollination. Seed number has a significant influence on final fruit size. All blueberry fruit exhibit a double-sigmoidal growth curve with a fruit development period ranging from 42 to 92 days in NHBs, from 55 to 60 days in SHBs and from 60 to 135 days in rabbiteye cultivars. After fruit colour change, the percentage of total sugars increases, while titratable acidity decreases, resulting in a steady increase in the sugar:acid ratio during ripening.

REFERENCES

Abbott, J.D. and Gough, R.E. (1987) Seasonal development of highbush blueberry roots under sawdust mulch. *Journal of the American Society for Horticultural Science* 112, 60–62.

Arora, R., Rowland, L.J. and Panta, G.R. (1997) Chill-responsive dehydrins in blueberry: are they associated with cold hardiness or dormancy transitions? *Physiologia Plantarum*, 101, 8–16.

Bailey, J.S. (1949) Frost injury to blueberries. *Fruit Varieties and Horticultural Digest* 44, 98.

Ballinger, W.E. (1966) *Seasonal Trends in Wolcott Blueberry (Vaccinium corymbosum) Leaf and Berry Composition.* North Carolina Agricultural Experiment Station Technical Bulletin No. 173.

Ballinger, W.E., Kushman, L.J. and Brooks, J.F. (1963) Influence of crop load and N applications on the yield and fruit quality of Wolcott blueberries. *Proceedings of the American Society for Horticultural Science* 88, 264–276.

Ballington, J.R., Kirkman, W.B., Ballinger, W.E. and Maness, E.P. (1988) Anthocyanin, aglycone and aglycone-sugar content in the fruits of temperate North American species of four sections in *Vaccinium. Journal of the American Society for Horticultural Science* 113, 746–749.

Bañados, M.P. (2009) Expanding blueberry production into non-traditional production areas: northern Chile and Argentina, Mexico and Spain. *Acta Horticulturae* 810, 439–445.

Bañados, M.P. and Strik, B. (2006) Manipulation of the annual growth cycle of blueberry using photoperiod. *Acta Horticulturae* 715, 65–71.

Bergman, H.F. (1929) Changes in the rate of respiration of the fruits of the cultivated blueberry during ripening. *Science* 70, 15.

Biermann, J., Stushnoff, C. and Burke, M.J. (1979) Differential thermal analysis and freezing injury in cold hardy blueberry flower buds. *Journal of the American Society for Horticultural Science* 104, 444–449.

Birkhold, K.T., Koch, K.E. and Darnell, R.L. (1992) Carbon and nitrogen economy of developing rabbiteye blueberry fruit. *Journal of the American Society for Horticultural Science* 117, 139–145.

Bittenbender, H.C. and Howell, G.S. (1975) Interactions of temperature and moisture content on spring de-acclimation of flower buds of highbush blueberry. *Canadian Journal of Plant Science* 55, 447–452.

Brevis, P.A., NeSmith, D.S., Seymore, L. and Hausman, D.B. (2005) A novel method to quantify transport of self- and cross-pollen by bees in blueberry plantings. *HortScience* 40, 2002–2004.

Brevis, P.A., NeSmith, D.S., Wetzstein, H.Y. and Hausman, D.B. (2006) Production and viability of pollen and pollen–ovule ratios in four rabbiteye blueberry cultivars. *Journal of the American Society for Horticultural Science* 131, 181–184.

Brewer, J.W. and Dobson, R.C. (1969) Seed count and berry size in relation to pollinator level and harvest date for the highbush blueberry, *Vaccinium corymbosum. Journal of Economic Entomology* 62, 1353–1356.

Cano-Medrano, R. and Darnell, R.L. (1997) Cell number and cell size in parthenocarpic vs. pollinated blueberry fruits. *Annals of Botany* 80, 419–425.

Cappiello, P.E. and Dunham, S.W. (1994) Seasonal variation in low-temperature tolerance of *Vaccinium angustifolium* Ait. *HortScience* 29, 302–304.

Carlson, J.D. and Hancock, J.F. (1991) A methodology for determining suitable heat-unit requirements for harvest of highbush blueberry. *Journal of the American Society for Horticultural Science* 116, 774–779.

Connor, A.M., Luby, J.J. and Tong, C.B.S. (2002a) Variability in antioxidant activity in blueberry and correlations among different antioxidant assays. *Journal of the American Society for Horticultural Science* 127, 238–244.

Connor, A.M., Luby, J.J., Finn, C.E. and Hancock, J.F. (2002b) Genotypic and environmental variation in antioxidant activity, total phenolics and anthocyanin content among blueberry cultivars. *Journal of the American Society for Horticultural Science* 127, 89–97.

Coville, F.V. (1910) Experiments in blueberry culture. *US Department of Agriculture Bulletin No. 193*.

Darnell, R.L. (1991) Photoperiod, carbon partitioning and reproductive development in rabbiteye blueberry. *Journal of the American Society for Horticultural Science* 116, 856–860.

Darnell, R.L. (2006) Blueberry botany/environmental physiology. In: Childers, N.F and Lyrene, P.M. (eds) *Blueberries for Growers, Gardeners and Promoters.* Dr Norman F. Childers Publications, Gainesville, Florida, pp. 5–13.

Darnell, R.L. and Davies, F.E. (1990) Chilling accumulation, budbreak, and fruit set of young rabbiteye blueberry plants. *HortScience* 25, 635–638.

Darrow, G.M. (1941) Seed size in blueberry and related species. *Proceedings of the American Society for Horticultural Science* 38, 438–440.

Darrow, G.M. (1958) Seed number in blueberry fruit. *Proceedings of the American Society for Horticultural Science* 72, 212–214.

Davies, F.S. (1986) Flower position, growth regulators, and fruit set of rabbiteye blueberries. *Journal of the American Society for Horticultural Science* 111, 338–341.

Dhanaraj A.L., Slovin J.P. and Rowland L.J. (2005) Isolation of a cDNA clone and characterization of expression of the highly abundant, cold acclimation-associated 14 kDa dehydrin of blueberry. *Plant Science* 168, 949–957.

Eaton, G.W. (1967) The relationship between seed number and berry weight in open pollinated highbush blueberry. *HortScience* 2, 14–15.

Eck, P. (1986) Blueberry. In: Monselise, S.P. (ed.) *Handbook of Fruit Set and Development.* CRC Press, Boca Raton, Florida, pp. 75–85.

Eck, P. (1988) *Blueberry Science.* Rutgers University Press, New Brunswick, New Jersey.

Edwards, T.W. Jr, Sherman, W.B. and Sharpe, R.E. (1972) Seed development in certain Florida tetraploid and hexaploid blueberries. *HortScience* 7, 127–128.

Ehlenfeldt, M.K. and Martin, R.B. Jr (2002) A survey of fruit firmness in highbush blueberry and species-introgressed blueberry cultivars. *HortScience* 37, 386–389.

Ehlenfeldt, M.K. and Prior, R.L. (2001) Oxygen radical absorbance capacity (ORAC) and phenolic and anthocyanin concentrations in fruit and leaf tissues of highbush blueberry. *Journal of Agricultural and Food Chemistry* 49, 2222–2227.

Ehlenfeldt, M.K. and Rowland, R.L. (2006) Cold-hardiness of *Vaccinium ashei* and *V. constablaei* germplasm and the potential for northern-adapted rabbiteye cultivars. *Acta Horticulturae* 715, 77–80.

Ehlenfeldt, M.K., Rowland, L.J. and Arora, R. (2003) Bud hardiness and deacclimation in blueberry cultivars with varying species ancestry: flowering time may not be a good indicator of deacclimation. *Acta Horticulturae* 626, 39–44.

Ehlenfeldt, M.K., Ogden, E.L., Rowland, L.J., and Vinyard, B. (2006) Evaluation of midwinter cold hardiness among 25 rabbiteye blueberry cultivars. *HortScience* 41, 579–581.

Ehlenfeldt, M.K., Rowland, L.J., Ogden, E.L. and Vinyard, B.T. (2012) Cold-hardiness, acclimation, and deacclimation among diverse blueberry (*Vaccinium* spp.) genotypes. *HortScience* 137, 31–37.

Ehlenfeldt, M.K., Rowland, L.J., Ogden, E.L. and Vinyard, B.T. (2015) LT_{25}, LT_{50}, and LT_{75} floral bud cold hardiness determinations for a diverse selection of *Vaccinium* genotypes. *Canadian Journal of Plant Sciences* 95, 491–494.

El-Agamy, S.Z.A., Sherman, W.B. and Lyrene, P.M. (1981) Fruit set and seed number from self- and cross-pollinated highbush (4X) and rabbiteye (6X) blueberries. *Journal of the American Society for Horticultural Science* 106, 443–445.

Eskin, N.A.M. (1979) The plant cell wall. In: Eskin, N.A.M. (ed.) *Plant Pigments, Flavors and Textures: Textural Components of Food*. Academic Press, New York, New York, pp. 123–138.

Finn, C.E. and Luby, J.J. (1986) Inheritance of fruit development interval and fruit size in blueberry progenies. *Journal of the American Society for Horticultural Science* 111, 784–788.

Finn, C.E., Hancock, J.F., Mackey, T. and Serce, S. (2003) Genotype × environment interactions in highbush blueberry (*Vaccinium* sp. L.) families grown in Michigan and Oregon. *Journal of the American Society for Horticultural Science* 128, 196–200.

Flinn, C.L. and Ashworth, E.N. (1994) Seasonal changes in ice distribution and xylem development in blueberry flower buds. *Journal of the American Society for Horticultural Science* 119, 1176–1184.

Frenkel, C. (1972) Involvement of peroxidase and indole 3-acetic acid oxidase isozymes in pear, tomato and blueberry fruit ripening. *Plant Physiology* 49, 757–763.

Gilreath, P.R. and Buchanan, D.W. (1981) Temperature and cultivar influences on the chilling period of rabbiteye blueberry. *Journal of the American Society for Horticultural Science* 106, 625–628.

Gough, R.E. (1980) Root distribution of 'Coville' and 'Lateblue' highbush blueberry under sawdust mulch. *Journal of the American Society for Horticultural Science* 105, 576–578.

Gough, R.E. (1994) *The Highbush Blueberry and its Management*. Food Products Press, New York, New York.

Gough, R.E. and Shutak, V.G. (1978) Anatomy and morphology of cultivated highbush blueberry. *Rhode Island Agricultural Experiment Station Technical Bulletin No. 423*.

Gough, R.E., Shutak, V.G. and Hauke, R.L. (1978) Growth and development of highbush blueberry. II. Reproductive growth, histological studies. *Journal of the American Society for Horticultural Science* 103, 476–479.

Gündüz, K., Serçe, S. and Hancock, J.F. (2015) Variation among highbush and rabbiteye cultivars of blueberry for fruit quality and phytochemical characteristics. *Journal of Food Composition and Analysis* 38, 69–79.

Gupton, C.L. (1983) Variability among rabbiteye blueberry cultivars for tolerance of flowers to frost. *HortScience* 18, 713–714.

Hall, I.V. and Forsyth, F.R. (1966) Respiration rates of developing fruits of the lowbush blueberry. *Canadian Journal of Plant Science* 47, 157–159.

Hall, I.V. and Ludwig, R.A. (1961) The effects of photoperiod, temperature, and light intensity on the growth of the lowbush blueberry (*Vaccinium angustifolium* Ait.). *Canadian Journal of Botany* 39, 1733–1739.

Hall, I.V., Craig, D.L. and Aalders, L.E. (1963) The effect of photoperiod on the growth and flowering of the highbush blueberry (*Vaccinium corymbosum* L.). *Proceedings of the American Society for Horticultural Science* 82, 260–263.

Hall, I.V., Aalders, L.E. and Newberry, R.J. (1971) Frost injury to flowers and developing fruits of the lowbush blueberry as measured by impairment of fruit set. *Canadian Naturalist* 98, 1053–1057.

Hancock, J.F. (1989) Why is 'Elliott' so productive? A comparison of yield components in 6 highbush blueberry cultivars. *Fruit Varieties Journal* 43, 106–109.

Hancock, J.F., Nelson, J.W., Bittenbender, H.C., Callow, P.W., Cameron, J.S., Krebs, S.L., Pritts, M.P. and Schumann, C.M. (1987) Variation among highbush blueberry cultivars in susceptibility to spring frost. *Journal of the American Society for Horticultural Science* 112, 702–706.

Hancock, J.F., Sakin, M. and Callow, P.W. (1991) Heritability of flowering and harvest dates in *Vaccinium corymbosum*. *Fruit Varieties Journal* 45, 173–176.

Hancock, J.F., Luby, J.J. and Beaudry, R. (2003) Fruits of the *Ericaceae*. In: Trugo, L., Fingas, P. and Caballero, B. (eds) *Encyclopedia of Food Science, Food Technology and Nutrition*. Academic Press, London, pp. 2762–2768.

Hancock, J.F., Callow, P., Serçe, S., Hanson, E. and Beaudry, R. (2008) Effect of cultivar, controlled atmosphere storage, and fruit firmness on the long term storage of highbush blueberries. *HortTechnology* 18, 199–205.

Handley, D.T. (2016). Bulletin #2253, Growing Highbush Blueberries. The University of Maine Cooperative extension. Available at: https://fruit.wisc.edu/wp-content/uploads/sites/36/2016/03/Growing-Highbush-Blueberries.pdf (accessed 7 May 2018).

Hanson, E.J., Berkheimer, S.F. and Hancock, J.F. (2007) Seasonal changes in the cold hardiness of the flower buds of highbush blueberry with varying species ancestry. *Journal of the American Pomology Society* 61, 14–18.

Huang, Y.H., Johnson, C.E. and Sundberg, M.D. (1997) Floral morphology and development of 'Sharpblue' southern blueberry in Louisiana. *Journal of the American Society for Horticultural Science* 122, 630–633.

Ismail, A.A. and Kender, W.J. (1967) Respiratory inhibition of mature detached blueberry fruit (*Vaccinium angustifolium* Ait.) by N^6-benzyladenine, kinetin and N-dimethylaminosuccinamic acid. *Botanical Gazette* 128, 206–208.

Ismail, A.A. and Kender, W.J. (1969) Evidence of a respiratory climacteric in highbush and lowbush blueberry fruit. *HortScience* 4, 342–344.

Jacobs, L.A., Davies, F.S. and Kimbrough, J.M. (1982) Mycorrhizal distribution in Florida rabbiteye blueberries. *HortScience* 17, 951–953.

Janes, H.W., Chin, C.-K. and Frenkel, C. (1978) Respiratory upsurge in blueberries and strawberries as influenced by ethylene and acetaldehyde. *Botanical Gazette* 139, 50–52.

Johnston, S. (1939) The resistance of certain highbush varieties to injury by frost. *Michigan Agricultural Experiment Station Quarterly Bulletin* 22, 10–11.

Kirk, A.K. and Isaacs, R. (2012) Predicting flower phenology and viability of highbush blueberry. *HortScience* 47, 1291–1296.

Knight, R.J. Jr and Scott, D.H. (1964) Effects of temperature on self- and cross-pollination and fruit of four highbush blueberry varieties. *Proceedings of the American Society for Horticultural Science* 85, 302–306.

Kovaleski, A.P., Williamson, J.G., Olmstead, J.W. and Darnell, R.L. (2015) Inflorescence bud initiation, development, and bloom in two southern highbush blueberry cultivars. *Journal of the American Society for Horticultural Science* 140, 38–44.

Krebs, S.L. and Hancock, J.F. (1988) The consequences of inbreeding on fertility in *Vaccinium corymbosum* L. *Journal of the American Society for Horticultural Science* 113, 914–918.

Krebs, S.L. and Hancock, J.F. (1990) Early-acting inbreeding depression and reproductive success in the highbush blueberry, *Vaccinium corymbosum* L. *Theoretical and Applied Genetics* 79, 825–832.

Krebs, S.L. and Hancock, J.F. (1991) Embryonic genetic load in the highbush blueberry, *Vaccinium corymbosum* (*Ericaceae*). *American Journal of Botany* 78, 1427–1437.

Kushima, T. and Austin, M.E. (1979) Seed number and size in rabbiteye blueberry fruit. *HortScience* 14, 721–723.

Kushman, L.J. and Ballinger, W.E. (1963) Influence of season and harvest interval upon quality of Wolcott blueberries grown in eastern North Carolina. *Proceedings of the American Society for Horticultural Science* 83, 395–405.

Kushman, L.J. and Ballinger, W.E. (1968) Acid and sugar changes during ripening in Wolcott blueberries. *Proceedings of the American Society for Horticultural Science* 92, 290–295.

Lang, G.A. and Danka, R.G. (1991) Honey-bee-mediated cross- versus self-pollination of 'Sharpblue' blueberry increases fruit size and hastens ripening. *Journal of the American Society for Horticultural Science* 116, 770–773.

Lin, W. and Pliszka, K. (2003) Comparison of spring frost tolerance among different highbush blueberry (*Vaccinium corymbosum* L.) cultivars. *Acta Horticulturae* 262, 337–341.

Lipe, J.A. (1978) Ethylene in fruits of blackberry and blueberry. *Journal of the American Society for Horticultural Science* 103, 76–77.

Lohachoompol, V., Mulholland, M., Srzednicki, G. and Craske, J. (2008) Determination of anthocyanins in various cultivars of highbush and rabbiteye blueberries. *Food Chemistry* 111, 249–254.

Luby, J.J. and Finn, C.E. (1987) Inheritance of ripening uniformity and relationship to crop load in blueberry progenies. *Journal of the American Society for Horticultural Science* 112, 167–170.

Lyrene, P.M. (1984) Late pruning, twig orientation, and flower bud formation in rabbiteye blueberry. *HortScience* 19, 98–99.

Lyrene, P.M. (1985) Effects of year and genotype on flowering and ripening dates in rabbiteye blueberry. *HortScience* 20, 407–409.

Lyrene, P.M. (2008) Breeding southern highbush blueberries. *Plant Breeding Reviews* 30, 354–414.

Lyrene, P.M. and Goldy, R.G. (1983) Cultivar variation in fruit set and number of flowers per cluster in rabbiteye blueberry. *HortScience* 18, 228–229.

Maas, J.L., Galletta, G.J. and Stoner, G.D. (1991) Ellagic acid, an anticarcinogen in fruits, especially in strawberries: a review. *HortScience* 26, 10–14.

Mainland, C.M. (1985) *Vaccinium*. In: Halevy, A.H. (ed.) *Handbook of Flowering*, Vol. IV. CRC Press, Boca Raton, Florida, pp. 451–455.

Mainland, C.M., Buchanan, D.W. and Bartholic, J.F. (1977) The effects of five chilling regimes on bud break of highbush (*Vaccinium corymbosum* L.) and rabbiteye (*V. ashei* Reade) blueberry hardwood cuttings. *HortScience* 12, 411.

Markakis, P., Jarczyk, A. and Krisha, S.P. (1963) Nonvolatile acids of blueberries. *Agricultural Food Chemistry* 11, 8–11.

Merrill, T.A. (1936) Pollination of the highbush blueberry. *Michigan Agricultural Experiment Station Technical Bulletin No. 151.*

Moore, J.N. (1964) Duration of receptivity to pollinations of flowers of the highbush blueberry and the cultivated strawberry. *Proceedings of the American Society for Horticultural Science* 85, 295–301.

Moore, J.N., Reynolds, B.D. and Brown, G.R. (1972) Effects of seed number, size and development on fruit size of cultivated blueberries. *HortScience* 7, 268–269.

NeSmith, D.S., Krewer, G. and Lindstrom, O.M. (1999) Fruit set of rabbiteye blueberry after subfreezing temperatures. *Journal of the American Society for Horticultural Science* 124, 337–340.

Norvell, D.J. and Moore, J.N. (1982) An evaluation of chilling models for estimating rest requirements of highbush blueberry. *Journal of the American Society for Horticultural Science* 107, 54–56.

Paltineanu, C., Coman, M., Nicolae, S., Ancu, I., Calinescu, M., Sturzeanu, M., Chitu, E., Ciucu, M. and Nicola, C. (2017) Root system distribution of highbush blueberry crops of various ages in medium-textured soils. *Erwerbs-Obstbau* https://doi.org/10.1007/s10341-017-0357-3.

Parliment, T.H. and Kolor, M.G. (1975) Identification of the major volatile compounds of blueberry. *Journal of Food Science* 40, 762–763.

Parrie, E.J. (1990) Pollination of hybrid southern highbush blueberries (*Vaccinium corymbosum* L.). MS thesis, Louisiana State University, Baton Rouge, Louisiana.

Parrie, E.J. and Lang, G.A. (1992) Self- and cross-pollination affect stigmatic pollen saturation in blueberry. *HortScience* 27, 1105–1107.

Proctor, A. and Miesle, T.J. (1991) Polygalacturonase and pectinmethylesterase activities in developing highbush blueberries. *HortScience* 26, 579–581.

Rejman, A. (1977) Frost damage on highbush blueberries in central Poland during the years 1971–1975. *Acta Horticulturae* 61, 163–168.

Rowland, L.J., Panta, G.R., Mehra, S. and Parmentier-Line, C. (2004) Molecular genetic and physiological analysis of the cold-responsive dehydrins of blueberry. *Journal of Crop Improvement* 10, 53–76.

Rowland, L.J., Ogden, E.L., Ehlenfeldt, M.K. and Vinyard, B. (2005) Cold hardiness, deacclimation kinetics and bud development in diverse blueberry (*Vaccinium*) genotypes under field conditions. *Journal of the American Society for Horticultural Science* 130, 508–514.

Rowland, L.J., Ogden, E.L., Takeda, F., Glenn, D.M., Ehlenfeldt, M.K. and Vinyard, B.T. (2013) Variation among highbush blueberry cultivars for frost tolerance of open flowers. *HortScience* 48, 692–695.

Sharpe, R.H. and Sherman, W.B. (1971) Breeding blueberries for low chilling requirement. *HortScience* 6, 145–147.

Shine, J. and Buchanan, D.W. (1982) Chilling requirements of 3 Florida blueberry cultivars. *Proceedings of the Florida State Horticulture Society* 95, 85–87.

Shutak, V.G., Hindle, R. Jr and Christopher, E.P. (1957) Growth studies of the cultivated blueberry. *Rhode Island Agricultural Experiment Station Technical Bulletin No. 339.*

Shutak, V.G., Gough, R.E. and Windus, N.D. (1980) The cultivated highbush blueberry: twenty years of research. *Rhode Island Agricultural Experiment Station Technical Bulletin No. 428.*

Spann, T.M., Williamson, J.G. and Darnell, R.L. (2003) Photoperiodic effects on vegetative and reproductive growth of *Vaccinium darrowii* and *V. corymbosum* interspecific hybrids. *HortScience* 38, 192–195.

Spann, T.M., Williamson, J.G. and Darnell, R.L. (2004) Photoperiod and temperature effects on growth and carbohydrate storage in southern highbush blueberry interspecific hybrids. *Journal of the American Society for Horticultural Science* 129, 294–298.

Spiers, J.M. (1976) Chilling regimes affect budbreak in 'Tifblue' rabbiteye blueberry. *Journal of the American Society for Horticultural Science* 101, 84–90.

Spiers, J.M. (1978) Effect of stage of bud development on cold injury in rabbiteye blueberry. *Journal of the American Society for Horticultural Science* 103, 452–455.

Spiers, J.M. (1981) Freeze damage in six rabbiteye blueberry cultivars. *Fruit Varieties Journal* 35, 68–70.

Spiers, J.M. (1995) Substrate temperatures influence root and shoot growth of southern highbush and rabbiteye blueberries. *HortScience* 30, 1029–1030.

Valenzuela-Estrada, L.R., Vera-Caraballo, V., Ruth, L.E. and Eissenstat, D.M. (2008) Root anatomy, morphology and longevity among root orders in *Vaccinium corymbosum* (Ericaceae). *American Journal of Botany* 95, 1506–1514.

Vander Kloet, S.P. (1991) The consequences of mixed pollination on seed set in *Vaccinium corymbosum. Canadian Journal of Botany* 69, 2448–2454.

Vander Kloet, S.P. and Lyrene, P.M. (1987) Self-incompatibility in diploid, tetraploid and hexaploid *Vaccinium corymbosum. Canadian Journal of Botany* 65, 660–665.

White, E. and Clark, J.H. (1939) Some results of self-pollination of the highbush blueberry at Whitesbog, NJ. *Proceedings of the American Society for Horticultural Science* 36, 305–309.

Williamson, J.F., Darnell, R.L., Krewer, G., Vanderwegen, J. and NeSmith, S. (1995) Gibberellic acid: a management tool for increasing yield of rabbiteye blueberry. *Journal of Small Fruit and Viticulture* 3, 203–218.

Williamson, J.F., Krewer, G. Maust, B.E. and Miller, E.P. (2002) Hydrogen cyanamide accelerates vegetative budbreak and shortens fruit development period of blueberry. *HortScience* 37, 539–542.

Windus, N.D., Shutak, V.G. and Gough, R.G. (1976) CO_2 and C_2H_4 evolution in highbush blueberry fruit. *HortScience* 11, 515–517.

Wood, G.M. (1962) The period of receptivity in flowers of the lowbush blueberry. *Canadian Journal of Botany* 40, 685–686.

Woodruff, R.E., Dewey, D.H. and Sell, H.M. (1960) Chemical changes of Jersey and Rubel blueberry fruit associated with ripening and deterioration. *Proceedings of the American Society for Horticultural Science* 75, 387–401.

Wright, G. (1993) Performance of southern highbush and rabbiteye blueberries on the Corindi Plateau, NSW, Australia. *Acta Horticulturae* 346, 141–148.

Young, M.J. and Sherman, W.B. (1978) Duration of pistil receptivity, fruit set, and seed production in rabbiteye and tetraploid blueberries. *HortScience* 13, 278–279.

LIGHT, PHOTOSYNTHESIS AND YIELD IN BLUEBERRIES

INTRODUCTION

In this chapter, a global vision of the physiology (functioning) of a blueberry plant will be described with regard to the generation and distribution of carbohydrates. We will look at the factors that are involved in dry matter production and its partitioning among the various plant organs, as well as the effect of several environmental variables and management practices on dry matter partitioning to vegetative and reproductive organs.

The plant is a set of organs (roots, shoots, leaves and fruit) that grow and develop in harmony. Most of the dry matter (what is left after removal of water from tissues) is carbohydrates, which are the main product of the photosynthetic process. During photosynthesis, solar energy is converted into chemical energy, which is then transformed into different compounds and stored in various organs within the plant, including the fruit (Fig. 4.1). In order for a crop, such as blueberry, to be productive in the long term, there is a need to

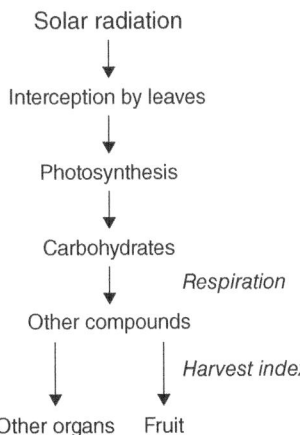

Fig. 4.1 Scheme of the transformation of solar energy into carbohydrates and other compounds, and their distribution among different organs in a fruit crop.

provide the appropriate conditions to maintain high rates of photosynthesis and also to establish an equilibrium in the partitioning of carbohydrates to the various organs of the plant.

FACTORS THAT AFFECT YIELD

For any crop (including blueberries), yield (Y) can be expressed as the product of the amount of photosynthetically active radiation (PAR) that reaches the crop in a given period (I_o), the fraction of PAR intercepted by the crop (F_{PAR}), the efficiency with which the canopy converts that radiation into new biomass (radiation use efficiency, E) and the proportion of the carbohydrates that are apportioned to harvestable organs (harvest index, H):

$$Y = I_o \times F_{PAR} \times E \times H$$

In order to reach high and sustained yields, cultural practices must be focused on maximizing the level of each individual factor.

QUANTITY OF PAR THAT REACHES THE CROP IN A GIVEN PERIOD (I_O)

The quantity of PAR that arrives at a blueberry field defines the amount of energy available for growth and development of the crop. This quantity of radiation is a function of the season of the year and the geographical zone in which the orchard is located; in the short term, this radiation will vary according to cloudiness and competition with other plants in that environment (Table 4.1).

Table 4.1. Summer and winter monthly averages of daily solar radiation (photosynthetically active radiation (PAR) in W/m^2) for various zones where blueberries are cultivated.

Location	Latitude	Summer	Winter	Reference
San Fernando, Chile[a]	34°35'S	244.0	99.1	Sarmiento (1995)
Collipulli, Chile[a]	37°55'S	224.9	85.7	Sarmiento (1995)
Karlsruhe, Germany[b]	49°03'N	385.1	137.2	Häder et al. (2007)
Pisa, Italy[b]	43°43'N	390.3	180.4	Häder et al. (2007)
Malaga, Spain[b]	36°43'N	414.2	219.4	Häder et al. (2007)
Athens, Greece[b]	37°58'N	393.8	214.1	Häder et al. (2007)
Forest Grove, Oregon, USA[a]	45°33'N	226.8	108.0	Bryla et al. (2008)

[a]Seasonal average.
[b]Clear skies.

Many blueberry cultivars are hybrids of different blueberry species (see Chapter 2, this volume). Most of the materials used for generating these cultivars originates from temperate latitudes and the plants often live naturally as an understorey plant (below deciduous forests), which means that in their natural environment they grow under intermediate light intensity and diffuse light (Eck *et al.*, 1990). However, several *Vaccinium* spp. come from open areas.

The amount of radiation that a blueberry plant receives not only determines the potential production of carbohydrates through photosynthesis but also defines fruit quality (colour), plant morphology (e.g. shoot length, leaf size, stomatal density) and the number of flower buds for the next season. This is the reason why plants in the shaded zones within the canopy not only generate a lower amount of carbohydrates but also fewer fruit (because of a lower potential to induce flower buds), and why they are slower in developing fruit colour (due to the need for radiation to form anthocyanins, the pigments responsible for the blue colour of blueberry fruit).

Light intensity and floral induction

For fruit crops in general, the formation of flower buds requires at least 30% full sun (30% of the radiation received by the outermost leaf). The light levels needed to induce flower buds in highbush blueberries have not been published; however, in rabbiteye blueberries, minimum light levels of around 25% full sun have been shown to be required for flower induction (Yáñez *et al.*, 2009). Light levels decrease sharply within the canopy; thus, 60 cm from the periphery, light levels are less than 40% full sun, except near the end of the season (24 weeks after full bloom (WAFB)) when the weight of fruit opens the canopy (Table 4.2). In the field, it is common to observe low numbers of fruit in the centre of the canopy, especially in adult plants of cultivars with dense canopies that have not received adequate pruning.

Another important factor influencing fruit load is that for floral induction to occur, sufficient light must be present during specific periods of development. Gough (1994) speculated that increased canopy density might reduce flower bud initiation during these critical periods. Information from morphological studies of bud development and the effects of the growth regulator gibberellic acid on the magnitude of flower induction (Retamales *et al.*, 2000) indicate that flower induction in highbush blueberries occurs at around 12–17 WAFB.

To test whether floral induction is time specific, an experiment was conducted in south-central Chile in which canopies were opened for 4 weeks at different times during the season for 14-year-old rabbiteye 'Choice' plants (Yáñez *et al.*, 2009). It was found that the timing of canopy opening did indeed have a major impact on flower bud induction and that flower induction

Table 4.2. Light availability (percentage of full sun) at different levels within 15-year-old 'Choice' rabbiteye blueberry plants during the season and the number of flower buds at each level the following winter. Data are from Cato, Chillán, Chile (36°21′S, 71°50′W). (Adapted from Yáñez *et al.*, 2009.)

Canopy level	Height (cm)	Percentage of full sun			Flower buds per cane (winter)
		8 WAFB	13 WAFB	24 WAFB	
1	180–200	100	100	100	7.1
2	160–179	75	69	89	14.9
3	140–159	38	39	65	13.7
4	120–139	21	25	38	10.7
5	100–119	12	16	19	2.6
6	80–99	11	10	12	3.4
7	60–79	4	13	7	1.3

occurred earlier in rabbiteyes compared with highbush blueberries. Opening the canopy in December (which corresponds to 8–12 WAFB) almost doubled the total number of flower buds per cane (compared with canopies that were not open) in this rabbiteye blueberry cultivar, especially in the 100–179 cm height range (Table 4.3). It was also observed that, in all treatments, the highest proportion of flower buds was induced at the top 60 cm of the canopy.

Shading and blueberry performance

As mentioned above, many native blueberries grow in the shade; thus, commercial cultivation of these plants in the open field is likely to subject them to radiation and temperature stress. For this reason, the use of coloured shading nets (also known as photoselective films) in areas of high solar irradiance can reduce stress and allow better growth and higher yields (Lobos *et al.*, 2009, 2012). However, such increases in yield were obtained in experiments done in central Chile (latitudes 35–37°S) (Table 4.4), but not in Michigan (latitude 42°N) (Fig. 4.2).

Shading can alter various characteristics of blueberries, including: (i) radiation availability and quality; (ii) physiological traits such as leaf photosynthesis, stomatal conductance and chlorophyll fluorescence (Fv/Fm); (iii) leaf characteristics such as angle, size, weight, temperature, and chlorophyll and nitrogen (N) contents; and (iv) fruit yields and quality (soluble solids, size, colour, weight loss in postharvest).

Table 4.3. Effect of the time of canopy opening for 4 weeks starting 12 December, 17 January, 21 February, 30 March or 10 May (corresponding to 8, 13, 18, 24 or 29 weeks WAFB, respectively) on the number of flower buds per cane produced the following winter at different heights in 14-year-old rabbiteye 'Choice' blueberries. Data are for Cato, Chillán, Chile (36°21′S, 71°50′W). (Adapted from Yáñez *et al.*, 2009.)

Height (cm)	Flower buds per cane in winter					
	Control	8–12 WAFB	13–17 WAFB	18–23 WAFB	24–28 WAFB	29–33 WAFB
180–200	7.1	11.4	15.6	10.4	5.8	7.8
160–179	14.9	32.2	20.1	23.4	29.0	25.2
140–159	13.2	26.6	11.6	17.7	26.4	18.3
120–139	10.8	16.1	5.8	8.0	11.1	6.3
100–119	2.6	7.3	2.3	2.4	5.8	1.9
80–99	3.4	3.3	0.8	1.3	0.9	0.2
60–79	1.3	0.6	0.9	0.4	0.3	0.0
40–59	0.8	0.1	0.1	0.0	0.0	0.0
20–39	1.0	0.0	0.0	0.0	0.0	0.0
0–19	0.1	0.0	0.0	0.0	0.0	0.0
Total	55.2	97.6	57.2	63.6	79.3	59.7

Radiation availability and quality

Net colour can affect plant responses through changes in light quality and quantity (Trewavas, 2014). Photoselective nets influence the PAR transmitted through the canopy, with white nets allowing significantly higher radiation to penetrate than red and black nets at all (25, 50 and 75%) shade levels (Lobos *et al.*, 2013). It has been found that the percentage shading provided by suppliers of nets does not always correspond to the light availability for the plant, and thus results may vary depending on the actual shade provided. There were significant effects of net colour on light quality within the canopy. Large differences in irradiance were observed under nets: red ones reduced the visible spectrum and increased infrared wavelengths, whereas white nets reduced ultraviolet radiation. Black nets had almost neutral effects on light quality throughout the spectrum (Lobos *et al.*, 2012).

Physiological traits

Research done in South Korea on the effect of shade on leaf photosynthesis rate (A_n) of NHB 'Bluecrop' plants grown in pots showed that the maximum A_n at 31, 60, 73 and 83% shade was 11.8, 11.0, 8.4 and 7.5 mol/m^2/s,

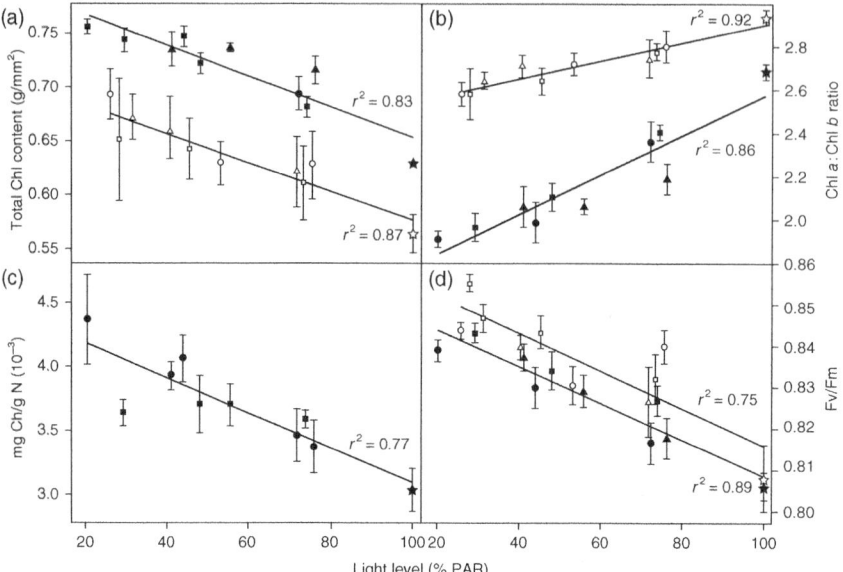

Fig. 4.2. (a) Total chlorophyll (Chl) content, (b) chlorophyll *a*:chlorophyll *b* ratio, (c) total chlorophyll:leaf nitrogen ratio, and (d) chlorophyll fluorescence (maximum photosystem II photochemical activity, Fv/Fm), in relation to light level (%PAR). Data were obtained 7 days before harvest from 'Elliott' plants growing under nets of different colours: black (○, ●), red (□, ■) and white (△, ▲), and control at full sun (☆, ★), in two locations, Chillán, Chile (open symbols), and Gobles, Michigan (closed symbols). Bars show standard error. (Adapted from Lobos *et al.*, 2012.)

respectively (Kim *et al.*, 2011). Studies done in the USA (Lobos *et al.*, 2009) showed that leaf photosynthesis decreased linearly as shading intensity increased. Red and black nets at 70% shading had the lowest rates.

While Lobos *et al.* (2012) reported that chlorophyll fluorescence (measured as Fv/Fm) increased linearly with the percentage of shading (Fig. 4.2d), being significantly lower in plants without shading (control) compared with bushes under 70% shading (white, black or red), Kim *et al.* (2011) found no significant differences in Fv/Fm under black nets providing 31–83% shading. In the case of Lobos *et al.* (2012), no treatment was below 0.8 Fv/Fm (Fig. 4.2d), which is considered a normal level for non-stressed plants, while all treatments were below this value in the trials by Kim *et al.* (2011). Lobos *et al.* (2009) also reported that stomatal conductance (g_s) was not significantly altered by shading treatments, while Kim *et al.* (2011) found that the leaves of plants grown under 83% shade had a significantly lower g_s compared with those grown under 31, 60 and 73% shade. These differences in physiological response found in the two trials might in part be attributed to genetic factors (different cultivars), as well as growing conditions. The trial by Lobos *et al.* (2009) was carried out on mature field-grown plants, while the research of Kim *et al.* (2011) was done using pot plants growing in the greenhouse.

Leaf characteristics

Leaf characteristics (e.g. size, angle, stomatal density, pigment concentrations) are important factors influencing the absorption of light energy (Tsukaya, 2005). In order to absorb sufficient light energy, leaves must be as large as possible and oriented optimally towards the sun. At the same time, to facilitate gas exchange of carbon dioxide (CO_2), oxygen (O_2) and water (H_2O), leaves must be as large and thin as possible. However, if leaves are too large and too thin, they become quickly dehydrated. Thus, leaf area and thickness are mainly restricted by the availability of water. Shade leaves usually invest assimilates in increasing leaf size in order to capture the maximum amount of sunlight available for photosynthesis in sectors of the canopy where light levels are low (Sims *et al.*, 1994; Niinemets *et al.*, 1998).

Shading has a significant impact on leaf characteristics influencing light energy absorption. Experiments carried out on NHB 'Brigitta' blueberries in Chile during the 2008/09 season showed that, as shading increased, the leaf angle increased. Thus, while leaves in the control treatment (no shade) had a 25° angle, leaf angles for white 25 and 70% shade nets were 45° and 63°, respectively (Cobo, 2010). These changes in leaf angles were observed within a week after imposing the shade treatments.

When Kim *et al.* (2011) shaded NHB 'Bluecrop' plants with black nets at various intensities (31, 60, 73 and 83%), they reported that, with increasing shade level, leaf length, width and area increased, but leaf thickness decreased. However, they found no obvious tendency in leaf length:width ratio with increasing shade levels. Leaves growing under more shade had less dense stomata than more sun-exposed leaves, but the stomata were larger in leaves growing under greater shade.

In trials done in Michigan with the NHB 'Elliott' (Lobos *et al.*, 2009), leaf dry weight under the various nets was negatively correlated with the percentage of shade ($r^2 = 0.72$). Although Lobos *et al.* (2009) found that leaf temperature in 'Elliott' plants decreased as shading increased ($r^2 = 0.84$), with ranges between 29.9°C (control) and 26.4°C (black, 75%), Kim *et al.* (2011) reported that 'Bluecrop' diurnal leaf temperatures changed similarly for all black net shadings (31–83%). Kim *et al.* (2011) found that increased shading with black nets (31–83%) significantly increased the soil plant analysis development (SPAD) readings for chlorophyll (the light-collecting pigment within leaves). Lobos *et al.* (2012) reported that no matter what colour was used for shading, leaves under shade increased their chlorophyll concentration as shading intensity increased ($r^2 = 0.83$ and 0.87 in the USA and Chile, respectively) (Fig. 4.2a), presumably to compensate for lower light intensity. The same tendency was observed in the total chlorophyll:leaf N concentration ratio (Fig. 4.2c), with it being significantly ($P < 0.01$) lower in plants without shading (control) compared with plants under 75% shading (white, black or red). The chlorophyll a:chlorophyll b ratio showed a positive correlation with the percentage of PAR (%PAR) (Fig. 4.2b). The increase in both total chlorophyll concentration and total chlorophyll:leaf N ratio, as well as the decrease in chlorophyll a:chlorophyll b ratio as the level of shade increased clearly indicates an acclimatory response of highbush blueberry to the light environment.

Fruit yield

Shading can have a significant impact on yield. Trials done in Michigan over two growing seasons on NHB 'Elliott' blueberries found that fruit yield had a positive, curvilinear relationship with %PAR and reached an asymptote at 50% PAR. The interaction between net colour (black, red or white) and shade intensity (25, 50 or 75%) was not significant (Lobos *et al.*, 2013). Studies done in Chile on the NHB 'Berkeley' showed that with 50% shading, only black nets reduced yield, while the other net colours (grey, white or red) markedly raised yield (Retamales *et al.*, 2008). The highest yield increases were for red or white 50% shading (Table 4.4). In Michigan (Fig. 4.3), it was found that nets began

Table 4.4.   Effect of the use of shading nets (for two seasons on the same plants) on PAR levels and the yield of NHB 'Berkeley' blueberries (planted in 1994). Data are for Miraflores, Linares, Chile (latitude 35°S). PAR was measured from February to March 2004; yield was estimated based on 3300 plants/ha; the effect on yield is shown as a 2-year mean. (Adapted from Retamales *et al.*, 2008.)

Treatment		PAR radiation (W/m^2)	Yield (t/ha)		Effect on yield (% of control)
Colour	Shade (%)		2003/04	2004/05	
Control	0	910	12.5^c	22.1^{cd}	–
Black	50	495	12.1^c	16.0^e	−18.8
Black	35	500	17.2^{abc}	20.4^{de}	+8.7
White	35	710	13.9^{bc}	26.6^{bc}	+17.1
Red	35	695	16.5^{abc}	24.6^{bcd}	+18.8
Grey	50	640	18.0^{abc}	27.6^{abc}	+31.8
Grey	35	685	20.0^{ab}	27.8^{ab}	+38.2
Red	50	625	23.0^a	29.4^{ab}	+51.4
White	50	706	23.9^a	32.2^a	+62.1

[a,b,c,d,e]Mean values within a column with non-identical superscript letters are statistically different at $P < 0.05$ (Duncan's test).

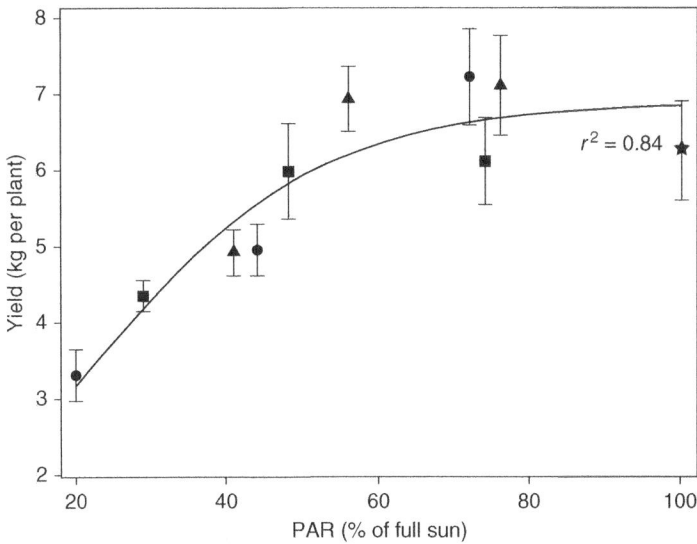

Fig. 4.3. Yield (kg per plant) in relation to the intensity of PAR (percentage of full sun) for 15-year-old 'Elliott' plants growing under nets of different colours (●, black; ■, red; ▲, white) and at full sun (★) in Gobles, Michigan. Bars show standard error. (Adapted from Lobos *et al.*, 2013.)

to have a negative impact on yield at greater than 50% shading levels. In the case of black nets, they only had positive effects on yield (20% more than the control) for 25% shading, while with 75% shading, black nets had the strongest negative impact (47.3% less than the control) among all treatments. Red nets had either no impact on yield (at 25 and 50% shading) or a negative effect (at 75% shading: 28% lower than the control). White nets produced similar yield increases at 25% (16.4% increase) and 50% (14.9% increase) shading, but had the lowest negative impact on yield (19.4% decrease) among the treatments with 75% shading (Lobos *et al.*, 2013).

A number of reproductive growth characteristics are influenced by shade. In their trials on 'Bluecrop', Kim *et al.* (2011) showed that increasing shading (from 31 to 83% black shade) significantly reduced flower number, fruit set per flower bud and yield. In a 2-year trial on 'Elliott', Lobos *et al.* (2013) found that after 1 year of shading, the rate of flower bud development decreased linearly as %PAR increased, but the total number of flower buds per cane and terminal shoots increased gradually as light levels increased, reaching an asymptote at 50% PAR. They concluded that placing red or white nets over mature plants at less than 60% PAR had no detrimental effects on return bloom or yield.

Fruit quality (soluble solids, size, colour and weight loss)

Shading has a significant impact on fruit development and quality. In trials done in Michigan, the proportion of mature 'Elliott' fruit on earlier dates was higher in control plants and decreased as %PAR was reduced. The harvest delay was maximal at the lowest light level (25%) and decreased exponentially with increasing %PAR (Lobos *et al.*, 2013). While they found that, in the USA, fruit quality (weight, soluble solids and firmness) had significant colour × shade and colour × harvest time (fruit weight, soluble solids and titratable acidity) interactions, Retamales *et al.* (2008) found no effects of 35 or 50% shading (white, black, red or grey nets) on fruit quality (fruit size, soluble solids) in Chile. Trials done on the NHB 'Elliott' in Michigan by Lobos *et al.* (2013) showed that fruit soluble solids increased linearly ($r^2 = 0.86$) and water content decreased linearly ($r^2 = 0.88$) as %PAR increased. In Michigan, the various shading treatments increased 'Elliott' fruit weight but reduced soluble solids, with black 50 and 70% having the greatest impact. Black 50 and 70% also reduced soluble solids (Lobos *et al.*, 2009). There were no effects of the colour or degree of shading of nets on fruit weight loss after storing 'Berkeley' fruit for 30 days at 4°C.

These differences between the two sites could in part be explained by higher radiation and ambient temperatures (and higher water stress) in Chile compared with Michigan. The date when the nets were placed over the fields may also have influenced the yield differences between Michigan and Chile, as the nets were installed immediately after petal fall in Chile, while in Michigan they were placed 1 month after petal fall.

When the different effects are analysed in conjunction, it is clear that blueberries are able to adjust their physiological processes and morphology to varying light levels. As fruit quality is either not affected or increased under shading nets, the sustained increases in fruit yields and the delays in fruit maturity obtained with the use of shading nets could prove advantageous and profitable in many blueberry-growing regions. Although selection of the appropriate colour and degree of shading would depend on latitude, the results reported so far indicate that 50% shading using white (grey) or red nets would provide the most benefits. However, as appreciation of fruit colour is greatly impaired under red nets, the use of white or grey nets would be the best alternative.

FRACTION AND AMOUNT OF LIGHT INTERCEPTED BY THE CROP (F_{PAR})

In agricultural crops, most of the radiation is intercepted by the organs that are specialized for that function, the leaves. However, green fruit can intercept radiation and possess anatomical characteristics that allow them to carry out photosynthesis. In rabbiteye blueberries, fruit photosynthesis provided 85 and 50% of the carbohydrates required by this organ at 5 and 10 days after full bloom, respectively. In a whole season, 15% of the carbohydrates required by the rabbiteye blueberry fruit was generated by this organ (Birkhold *et al.*, 1992; Darnell and Birkhold, 1996).

As an important function of the leaves is to intercept and utilize solar radiation, it should not come as a surprise that, for any crop, there is a high association between total foliar area (number of leaves × average leaf size) and the amount of intercepted radiation; the association is somewhat weaker between foliar area and total production of dry matter. In rabbiteye blueberries, a high association ($r^2 = 0.59$) has been established between canopy volume and yield (Haman *et al.*, 1997). A sensitivity analysis in apples revealed that light interception was influenced most by changes in leaf area and leaf optical properties (Green *et al.*, 2003).

The quantity of radiation intercepted by a crop at a given time will depend, among other factors, on: (i) plant density (number of plants/ha); (ii) rate of leaf development; (iii) leaf area duration (period from leaf unfolding until leaf drop); (iv) rate of leaf area removal (the speed at which leaves are detached or removed from the plant); and (v) distribution of leaves within the plant (plant architecture).

Plant density

In all fruit crops, denser plantings have higher early yields per hectare than less dense plantings. This would be a function of the surface area of leaves, but as the plantings fill up the allotted space and the block of plants reaches maximum light interception, the beneficial effect on yield tends to level off later in the life of the field. Strik and Buller (2002) found that cumulative yield of the mid-season NHB 'Bluecrop' from year 3 to year 7 was 104% higher at an in-row spacing of 0.45 m compared with 1.2 m, but to the best of our knowledge, no other research has been done examining the long-term effect of planting density in blueberries. If the crop canopy becomes too dense, several negative side effects can occur: (i) fruit bud induction and colour development can be impaired; (ii) spray coverage becomes more difficult; (iii) pruning needs to be more intense; and (iv) drying of aboveground organs after rain or overhead sprinkler use would be slower.

Early cropping

As leaves have to compete with other organs for carbohydrates, another factor that affects canopy development is the fruit load in the first years of the orchard. Although a high number of fruit buds per plant is desirable to achieve a higher yield per unit area, the greater demand for assimilates by the fruits can reduce whole-canopy leaf area, which leads to reduced fruit quality (Léchaudel *et al.*, 2005). In an experiment to establish the effect of early cropping on the performance of the NHB cultivars 'Duke' (early), 'Bluecrop' (mid-season) and 'Elliott' (late), it was found that plant growth at the start of year 3 was adversely impacted by early cropping in years 1 and 2. Evaluations done on year 3 showed that early cropping reduced the dry weight of the root system, crown, and 1–3-year-old wood in all cultivars. Early-cropped plants had a lower percentage of fruit buds than control plants. In addition, early cropping reduced the yield by 44, 24 and 19% in year 3, compared with control plants, in 'Elliott', 'Duke' and 'Bluecrop', respectively (Strik and Buller, 2005). In their trials, Strik and Buller (2005) found that early cropping did not improve cumulative yield (years 1–4) of 'Bluecrop' and 'Duke', and significantly reduced cumulative yield in 'Elliott'. This supports the hypothesis that early cropping is more stressful on late-season cultivars, as the fruit is competing with vegetative growth for a longer period. Thus, depending on the cultivar, there is a long-term risk associated with early cropping.

Rate of development of leaf area

In perennial crops such as blueberries, the rate of leaf area development is greater than in annual crops. This is because perennial fruit crops already have an existing structure (roots, canes of different ages, laterals and buds), and only require the opening of leaf buds to deploy the foliage that will allow them to capture sunlight. Most temperate zone plants, including blueberries, enter a dormant period during late autumn and winter, which is characterized by lack of growth and greatly reduced metabolic activity of aboveground parts. This dormant condition is a defence mechanism that enables plants to survive the cold. The development of dormancy and cold hardiness is a gradual process, which begins in late autumn or early winter in response to shorter days and lower temperatures during the autumn (see Chapter 3, this volume).

Once fully dormant, a blueberry plant must be exposed to a period of cool temperatures before it will break dormancy and start to grow normally again the following spring. The chilling requirement varies according to the species and cultivar. Temperatures between 0 and 7°C appear to be the most effective at satisfying the chilling requirement of blueberries, while temperatures between 7 and 13°C contribute little to chilling, and temperatures above 21°C in late autumn and winter probably negate some chilling (Lyrene and Williamson, 2004). Darnell and Davies (1990) found that the percentage of vegetative bud break (calculated as the amount of bud break at a given chilling time relative to the maximum amount of bud break that occurred) increased with increasing chilling duration (100–1000 h at 7°C) in all rabbiteye blueberry cultivars studied ('Climax', 'Tifblue' and 'Woodard'). Another factor that affects the accumulation of chilling is the presence of leaves; thus, in regions with mild climates, leaves remain attached to the plants and the bushes will not accumulate chilling hours as quickly as defoliated plants (Lyrene and Williamson, 2004). In highbush and rabbiteye blueberries, the emergence of leaves can occur simultaneously with bloom (as in the rabbiteye 'Bonita'), before bloom (as in the rabbiteye 'Climax') or after bloom (as in most SHB cultivars in mild climates such as Florida) (Lyrene and Williamson, 2004). When the emergence of flower buds occurs before leaf emergence, the development of flowers will depend for several days exclusively on carbohydrate reserves from the previous season.

The import of photosynthates in leaves of dicotyledonous plants (such as blueberries) ends at between 30 and 60% of their maximum growth, but it should be noted that developing leaves can still import photosynthates even though they are also exporting their own organic products (Turgeon, 1989). Leaves of rabbiteye blueberry approach maximum photosynthetic potential at full expansion (Andersen *et al.*, 1979). However, the rate at which leaf gas exchange declines after full expansion is extremely species dependent; ageing is associated with the reduction of leaf osmotic potential, lower responsiveness of stomata to climatic variations, lower N and chlorophyll contents, and reduced maximum rates of leaf gas exchange (Andersen *et al.*, 1979; Andersen, 1989).

Under conditions of high competition for carbohydrates between vegetative and reproductive growth, carbohydrate availability can restrict or delay leaf area expansion (Léchaudel *et al.*, 2005). This condition may be due to: (i) an unusually high number of fruit produced the previous season (this effect is especially deleterious in late cultivars because fruit growth will consume for a longer period the carbohydrates that could otherwise go to storage); (ii) environmental conditions that reduce the photosynthetic rate (cold temperatures, low light intensities); or (iii) an excessively high fruit load at the time of leaf expansion in the spring, especially in those cultivars where leaves and flowers develop simultaneously.

Leaf area duration

If winter temperatures are mild, rabbiteye and SHB blueberries will retain an important proportion of their leaves during winter, while NHB blueberries lose their foliage in autumn as day length gets shorter and minimum temperatures approach freezing. The retention of some leaves by certain blueberry species and cultivars will reduce their accumulation of chilling hours in the autumn and winter, which will delay bud break and canopy development (Lyrene and Williamson, 2004). Trials in rabbiteye blueberries show that the reduction in the final fruit load caused by defoliation was restricted only to the flower buds in the vicinity of the defoliated sector and that a late defoliation (when natural leaf fall was starting) did not alter plant performance (Lyrene, 1992).

At the time buds open in the spring, up to six leaf primordia exist in each vegetative blueberry bud. As the shoot grows after bud opening, a new leaf bud will be formed about every 5 days in the shoot apex. Growth of individual shoots of the blueberry is simpodial (zigzag or irregular form) and episodic, being accompanied by a varying number of apical abortions ('black-tip stage'). Each abortion terminates a 'flush' of growth (Eck *et al.*, 1990). The aborted shoot apex usually remains visible on individual shoots for a short period, after which it falls off, and usually the next bud located to the base of the aborted one resumes growth (next flush). The length of individual shoots and the number of flushes that occur on a single shoot vary and may affect the potential number of flower buds. Bañados (2006) found that the late-season 'Elliott' had more flushes of shoot growth (four to five) than early- or mid-season NHB cultivars such as 'Bluecrop' and 'Duke'.

Nutrition and soil mulching can markedly influence shoot growth. In mature NHB 'Bluecrop' plants, high rates of N (200 kg N/ha compared with 0 or 100 kg N/ha) increased the number of vigorous shoots per bush and the number of flushes of growth (three flushes instead of two; Bañados, 2006). Mulching or amending the soil with sawdust in 2- to 4-year-old 'Bluecrop' plants produced more and longer shoots than when bark was used (Kozinski, 2006). During fruit maturation in mid-summer, the fruit provide a highly competitive sink for nutrients and carbohydrates, considerably reducing the availability of resources to other parts of the plant (Gough, 1994). Consequently, leaf expansion will be reduced near fruit harvest, and another flush of vegetative growth may occur after fruit harvest (if temperatures are adequate). The effect of competition between leaf and fruit tissues is local, as leaves of the current season (without fruit) have greater vigour and larger leaves than those of the previous season (when the fruit are swelling).

The expansion and development of each leaf is controlled by several meristems (tissues with the capacity to undergo cell division). The form and final size of a leaf will depend on the genetic makeup and on environmental/cultural factors. For example, only black (35 or 50% shading) or grey (50% shading) nets significantly increased both leaf length and width compared with control (open field) NHB 'Berkeley' plants, although the leaf length:width ratio was not altered (Retamales *et al.*, 2008).

Rate of removal of leaf area

Besides the more or less predictable effects of environment (photoperiod and temperature) on leaf area duration (leaf emergence and leaf fall), there are unpredictable biotic and abiotic stresses that are faced by the crop during a given season and which cause the removal of a proportion of the leaf area. In this context, various abiotic stresses (temperature, soil moisture, salinity) reduce both the photosynthetic rate and the length of time that the leaf remains active. Water stress (by flooding or by scarcity of water) will reduce leaf size and accelerate leaf senescence and leaf drop. In rabbiteye and high-bush blueberries, the photosynthetic rate is significantly reduced after 1 or 2 days of flooding (Davies and Flore, 1986a,b), while after 10–14 days of flood-ing, the use of carbohydrates by respiration exceeds the formation of these compounds by photosynthesis (i.e. the plant is living on its food reserves). It has been shown that new spring vegetative flushes can be killed by the same tem-peratures (−2.2°C) that kill open flowers and fruit (Lyrene and Williamson, 2004). Young blueberry plants are sometimes damaged in field nurseries during late autumn and winter if they have not been properly hardened.

Pest or disease damage to leaves can also induce their early fall and reduce leaf photosynthetic potential. Experiments in the rabbiteye 'Premier' and SHB 'Bluecrisp' showed that 15–20% of necrotic leaf area caused by the fungus *Septoria albopunctata* reduced the photosynthetic rate by 50% (Roloff and Scherm, 2004). Leaves with higher disease severity abscised significantly earlier than healthy leaves. Lyrene (1992) demonstrated that an early autumn defoliation of rabbiteye blueberries resulted in a significantly lower percentage of nodes that produced flower buds the following spring. This indicates the need for early control of pests and diseases that damage the foliage.

Plant architecture

Plant architecture refers to the natural (genetic) or artificial (training, prun-ing, growth regulators) arrangement of the plant parts. Within each species, architectural differences are found in branching intensity, branch extension and leaf display. Such differences in plant architecture have been interpreted in terms of their potential adaptive value in different light environments (Kawamura and Takeda, 2002). As a rule, the higher the density of leaves on the outside of the canopy, the less light that is available in the more internal layers. Hence, any training methods aiming to improve plant architecture must focus on maximizing light interception and reducing internal shading. For these goals to be reached, new leaves growing within the canopy should be properly distributed spatially to expand light interception and avoid shading within the plant.

Pruning can have a marked effect in changing plant architecture. Base pruning (elimination close to the ground of older, poorly illuminated and less productive canes at the centre of plants) is used in many northern climates to reduce the number of canes per plant in mature bushes. This practice can result in yield gains by increasing light penetration into the bush and enhanc-ing the levels of flower bud formation (Strik *et al.*, 2006). Fine pruning and chemical reduction in the number of reproductive organs (through the application of gibberellic acid at flower induction time) can increase fruit weight in overbearing plants by reducing competition among fruits for limited resources (see Chapters 6 and 7, this volume).

Plant organs have plastic reactions in response to the environment they face. Depending on the amount and quality of available radiation, plants may alter shoot elongation (internode length), foliar angle, leaf size and rate of branching. Leaves fully exposed to the sun are quite different from those that grow under shade; among other variables, leaves from shaded sectors have a greater leaf area, greater leaf angles and reduced thickness and stomatal density, as well as lower rates of photosynthesis and respiration. Experiments on the effect of shading on NHB 'Berkeley' plants carried out in Chile (latitude 35°S) showed that shading had no effect on shoot number; however, internode and shoot length were significantly increased with black nets at both 35 and 50% shading (Retamales *et al.*, 2008).

The ability of a plant to capture solar radiation can be expressed in a number of different ways but the most commonly used are the leaf area index (LAI) and the leaf:fruit (L:F) ratio, expressed as numbers of each organ. LAI is defined as the relationship between foliar area (measured on only one side of the leaf) and the area of soil that a plant occupies (related to planting density). The changes in LAI within the season are important not only for photosynthesis but also with respect to pesticide application, estimation of nutrient status of the plant, and to model leaf shading in different sectors of the plant (and with it the capacity of the various sectors to induce flowers and grow fruit).

L:F ratios have been studied in blueberries. In adult NHB 'Jersey' plants, reductions in fruit weight and accumulation of soluble solids were observed when the L:F ratio was below 0.5. Studies done on potted half-high 'Northland' and NHB 'Bluecrop' plants showed that berries on shoots with five leaves per fruit (L:F ratio 5:1) matured 3 days earlier, were heavier (22% in 'Bluecrop' and 35% in 'Northland') and had similar soluble solid levels and lower titratable acidity (0.58 versus 1.29 in 'Bluecrop' and 0.92 versus 1.00 in 'Northland') than fruits on control shoots with a 1:1 L:F ratio (Suzuki *et al.*, 1998). Compared with the 1:1 ratio, the 5:1 ratio also increased the number of vegetative buds and shoot length in both cultivars, while the number of flower buds was not altered in 'Bluecrop' and was reduced in 'Northland' (Suzuki *et al.*, 1998). In an experiment performed in a 4-year-old commercial NHB 'Brigitta' field in southern Chile (38°29'S, 72°23'W) at 70 days after full bloom (during the second growth stage), plants were subjected to girdling (removal of 1 cm wide band of bark at the shoot base) and different L:F ratios (10:1, 1:1 or 1:10). Girdling and a lower fruit amount (L:F ratios of 1:1 and 10:1) treatments induced a reduction in the maximum photosynthetic rate (P_{max}), g_s, N and total chlorophyll, and increases in intrinsic water-use efficiency, dark respiration, and the chlorophyll *a*:chlorophyll *b* ratio. The impact of girdling was counterbalanced by an L:F ratio of 1:10, with non-girdled shoots with one fruit per leaf and a 1:10 L:F ratio having similar values (Jorquera-Fontena *et al.*,

2016). Blueberry fruit weight has been correlated with shoot vigour (Eck *et al.*, 1990; Suzuki *et al.*, 1998). Changes in the source–sink relationship in 'Brigitta' plants led to a rearrangement of physiological and structural leaf traits leading to an adjustment in the daily balance between carbon (C) assimilation and absorbed light energy. In young commercial blueberry orchards in Chile, L:F ratios of 1:1 to 0.6:1 have been measured (Jorquera-Fontena *et al.*, 2016), but the impact of such relationships on current season fruit quality as well as flower induction for the coming season has not been studied.

Studies done on various wine grape cultivars across years and with a variety of training systems discovered that about 0.8–1.2 m² leaf area per kg fruit was needed for optimum fruit maturity in vines trained to single-canopy trellis systems and 0.5–0.8 m²/kg for vines trained to divided-canopy trellis systems (Kliewer and Dokoozlian, 2005). Maintaining a balance between vegetative and reproductive growth is needed to optimize blueberry yield and quality. There are various means for regulating vegetative/reproductive balance. Although, to the best of our knowledge, no chemical fruit thinning is currently used in blueberries, in the future this may be a viable method for regulating blueberry crop load (Matta *et al.*, 2005). Fruit size of rabbiteye blueberries has been increased by selective cane removal, and mature highbush blueberries should be pruned annually to reduce overbearing (Eck *et al.*, 1990). Perennial fruit plants with a high L:F ratio, as in young plants or those with a low fruit load, often form large fruits with a 'spongy' tissue, which reduces postharvest life and increases susceptibility to disease. As fruit density increases, the L:F ratio decreases, resulting in a lower supply of photosynthate per fruit; fruit size therefore decreases (Fischer *et al.*, 2012).

PHOTOSYNTHETIC EFFICIENCY IN TRANSFORMING SOLAR ENERGY INTO CHEMICAL ENERGY (*E*)

Nearly 90% of the dry matter in plants is derived from photosynthetic activity (DaMatta, 2007). In this process, CO_2 enters through the pores (stomata) located on the surface of leaves and fruits. The energy from light is stored as sugars formed from water (obtained from the soil) and CO_2 in leaf cells that contain pigments such as chlorophyll and carotenes, which are capable of trapping the light. However, through the same stomata that allow CO_2 into the plant, water is lost to the environment through transpiration (a process that is needed to regulate plant temperature and move nutrients from the soil into the plant) (Fig. 4.4). A low resistance for CO_2 capture coincides with high permeability to water losses by the plant.

A large proportion of the sugars (40–50%) formed by photosynthesis are used by plants for respiration. Respiration fulfils two vital roles: it generates energy (in the form of ATP) for various functions (e.g. stomatal control,

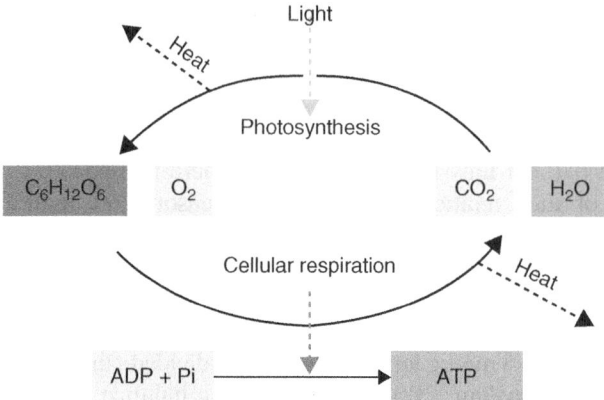

Fig.4.4. General scheme of the interdependency of photosynthesis and respiration. The glucose ($C_6H_{12}O_6$) and O_2 produced by photosynthesis are used up in cellular respiration. Energy is transferred to form ATP (from ADP + Pi); heat, CO_2 and H_2O are released as waste products. The CO_2 and H_2O are then available as raw materials for photosynthesis. (Adapted from Open Learning Initiative, 2015.)

translocation of substances), and produces the building blocks for the various compounds required by the plant (e.g. proteins, hormones, pigments, regulatory and defence compounds).

Two basic types of respiration have been defined: maintenance respiration (R_m: destined to secure the functioning of the existing plant processes and systems) and growth respiration (R_g: linked to an increase in plant size) (Fig. 4.5). A basic axiom is that the plant must always secure enough substrate for R_m: thus, if carbohydrate supply is reduced due to deleterious environmental conditions or management practices, R_g would have a greater chance of being affected compared with R_m (Loomis and Connor, 1992).

The final product of photosynthesis is generally glucose, which through respiration is converted into the other substances needed by the plant. The biochemical efficiency of the transformation of glucose into other compounds is defined as the production value (V_p) or efficiency of conversion. As can be seen in Table 4.5, from the same amount of the initial compound (glucose), almost two times more cellulose or starch will be obtained than protein. Hence, if the plant has to form tissues with an important proportion of compounds with high V_p, its growth potential will be higher than if the tissues to be formed have a low V_p. The global energy efficiency of blueberries is unknown, but it is expected that the efficiency of conversion should be high, as a high proportion of plant tissues contain compounds with a high V_p.

The overall efficiency of photosynthesis for agricultural plants is low, because, out of a total of 100 units of energy that reach an agricultural land in a given period, less than 4 units end up being stored in the different tissues

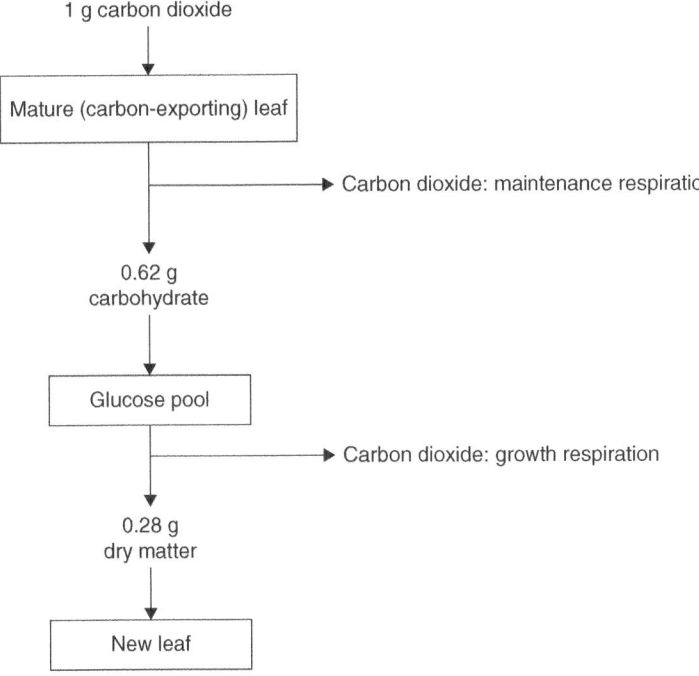

Fig. 4.5. Scheme of growth and maintenance respiration in fruit crops, showing the carbon cost of leaf construction. (Adapted from Weinbaum, 1978.)

Table 4.5. Production values (V_p), defined as the efficiency of converting 1 g of glucose into different plant components. (Adapted from Loomis and Connor, 1992.)

End product	V_p (g/g)
Sucrose	0.92
Cellulose, starch	0.86
Lignine	0.52
Protein	0.45
Lipid	0.36

produced by the plant (Long *et al.*, 2006). Except in tropical climates, a large amount of this energy is lost in winter and early spring because there is no leaf tissue to trap solar radiation (a situation that in blueberries can be partially overcome through the use of tunnels or greenhouses or by evergreen cultivation with low-chill SHB cultivars), another part of the energy is reflected from leaves (because a large portion of the light has a quality that does not interact with blueberry leaf pigments), another proportion is lost due to saturation of the leaf by the radiation (in blueberries, the photosynthetic rate reaches a

maximum at 40–50% full sunlight; Fig. 4.6), and finally an important proportion is 'lost' as heat (due to inefficiency in the conversion of glucose to other
compounds through respiration). Thus, even though growers may optimize
management of blueberries in the field (e.g. by irrigation, nutrition, use of
shading nets, phytosanitary controls), the maximum amount of solar energy
that a blueberry field can capture approaches only 4% of the total incoming
solar energy.

Variables that affect the photosynthetic rate

In any process that requires the participation of different components, the rate
of the process will depend on the level of the variable or component that is
the furthest from its optimum value; this concept is known as the law of the
minimum.

Although photosynthesis cannot occur without light, excess light can
damage plant tissues. The extra light energy that reaches a blueberry leaf must
rapidly be dissipated through transpiration to avoid a reduction in photosynthetic rates. In blueberries, leaf temperatures at full sun can reach up to 15°C
above those of the air. High temperatures and drought have deleterious effects
on blueberries. It has been reported that rabbiteye blueberries tolerate heat and
drought stress better than highbush blueberries (Galletta and Ballington,
1996). In their screening of wild species germplasm, Erb *et al.* (1988a,b) found

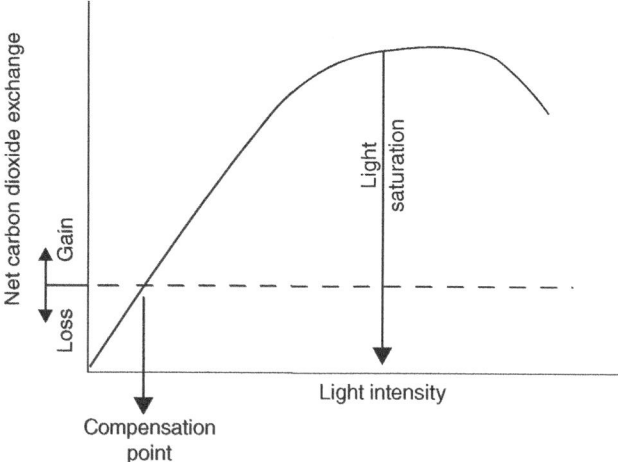

Fig. 4.6. Effect of light intensity on the net exchange of carbon dioxide in
blueberries. The compensation point indicates the light intensity at which losses
of carbon dioxide released by respiration equal carbon dioxide gains as a result of
photosynthesis. (Adapted from Teramura *et al.*, 1979; Davies and Darnell, 1994.)

V. elliottii, V. darrowii and *V. virgatum* to be the most drought tolerant species, and this trait was transmitted to the hybrid progeny.

Successful adaptation to heat and drought in blueberry may depend on how rates of CO_2 assimilation, transpiration and water-use efficiency are affected by changes in leaf temperatures (Hancock *et al.*, 2008). A study on *V. corymbosum* showed that the temperature optimum was 18–26°C for 'Jersey' and 14–22°C for 'Bluecrop', while for *V. darrowii* it was 25–30°C (Moon *et al.*, 1987a). As the temperature was raised from 20 to 30°C, the photosynthetic rate decreased by 30% in 'Bluecrop' but only by 20% in 'Jersey' and 9% in *V. darrowii* (Moon *et al.*, 1987a). The authors concluded that 'Jersey' and *V. darrowii* possess a greater tolerance to high temperature and drought conditions, which operates through restriction of water loss by decreasing stomatal apertures via high epicuticular wax deposition (Moon *et al.*, 1987b). Considering that the condition favouring survival of *V. darrowii* at a higher temperature is heritable, those SHB cultivars that have a *V. darrowii* component in their background (Hancock *et al.*, 2008) would have a greater tolerance to higher temperatures than NHB cultivars. It also indicates that, in certain growing regions, NHB plants could be under temperature stress for a few hours during typical summer days.

As in other fruit crops, photosynthetic rates in blueberries increase as light reaches higher intensities (Fig. 4.6). In dim light, the CO_2 released by respiration exceeds the small amounts of CO_2 fixed by photosynthesis. Further increases in light intensity will eventually allow fixation to begin compensating the loss by respiration. The light intensity at which this occurs is known as the 'light compensation point' (Fig. 4.6), which for *V. darrowii* and *V. corymbosum* is about 50 µmol/m^2/s (Moon *et al.*, 1987a). Photosynthesis is eventually saturated as light intensity increases; in blueberries (rabbiteyes, NHBs and *V. darrowii*), this occurs at around 700–800 µmol/m^2/s (Moon *et al.*, 1987a; Teramura *et al.*, 1979), and corresponds to 40–50% of maximum light intensity. Li *et al.* (2012) found that in the NHB 'Bluecrop', the saturation point was slightly higher in plants growing in the field (770 µmol/m^2/s) than in those growing in the greenhouse (720 µmol/m^2/s).

The maximum net photosynthetic rate measured in NHB leaves (11.5–11.9 µmol/m^2/s) is 25–35% higher than that for *V. darrowii* (Moon *et al.*, 1987a) and double that reported for excised shoots of rabbiteye blueberries (Teramura *et al.*, 1979) and those growing in the field (Andersen, 1989); however, research on potted rabbiteye 'Tifblue' plants showed rates of 9.0 µmol/m^2/s (Wright *et al.*, 1993).

It has been reported that, at various PAR levels (0–1200 µmol/m^2/s) and CO_2 concentrations (0–400 µmol/mol), NHB 'Bluecrop' blueberries growing in the field had nearly 10% higher photosynthetic rates than those in a greenhouse (Li *et al.*, 2012). As reported by Singsaas *et al.* (2001), this difference could be due to genetic variability, different environmental conditions or the type of plant material analysed (excised shoots versus plants). Experiments on the NHB 'Elliott' in Chile (Lobos *et al.*, 2012) showed that: (i) the photosynthetic rates in open-field plants are similar to those published previously for this species; and (ii) the presence of fruit generally increases the rate of photosynthesis of nearby leaves by about 20% (probably due to higher demand for carbohydrates).

Trials on 4-year-old NHB 'Brigitta' plants in which the sink–source relationship was manipulated by leaf and fruit removal as well as through shoot girdling, showed that the maximum rate of photosynthesis (P_{max}) was closely and negatively correlated to the accumulation of carbohydrates in the leaves. This indicates that sugar-sensing mechanisms play an important role in the regulation of blueberry leaf photosynthesis (Jorquera-Fontena *et al.*, 2016). Further experiments from this group in which 4- and 5-year-old 'Brigitta' plants were subjected to various degrees of pruning (50–80% of fruit buds were removed with respect to a mild pruning treatment) showed that the light-saturated photosynthetic rate diminished with increasing pruning severity, indicating a sink limitation of photosynthesis. Furthermore, various fruit parameters such as weight, dry matter, and glucose and fructose levels were negatively correlated with fruit load (Jorquera-Fontena *et al.*, 2014).

As stomatal closure during water stress impedes the diffusion of one of the compounds (CO_2) needed for photosynthesis, water-deficient plants have reduced photosynthetic rates. On sunny days, the photosynthetic rate generally shows a midday depression (Li *et al.*, 2009); however, Li *et al.* (2012) reported that this trend was clearer in potted greenhouse-grown NHB 'Bluecrop' plants than in open-field plants where the photosynthetic rate steadily decreased as the day advanced (from around $10\,\mu mol/m^2/s$ at 6:00 down to nearly $2\,\mu mol/m^2/s$ at 18:00). Similar trends of decreasing photosynthetic rates (around $12\,\mu mol/m^2/s$ at 8:00–10:40 down to nearly $9\,\mu mol/m^2/s$ at 17:00–19:40) were reported by Jorquera-Fontena *et al.* (2016) for sunny days in open-field-grown NHB 'Brigitta' plants. The midday photosynthetic rate depression can be explained by: (i) an accumulation of photosynthates in the leaf that stops the photosynthetic process (feedback inhibition); (ii) stomatal closure because transpiration exceeds water absorption and conduction, and thus the leaves undergo temporary water stress; and (iii) a low CO_2 concentration, as the plant has to process a large air volume to obtain the C (CO_2) needed for photosynthesis, and if the air is still, the CO_2 available within the leaf (intercellular CO_2) can be a limiting factor (Li *et al.*, 2009). In the case of cloudy days, Li *et al.* (2012) working with potted 4-year-old NHB 'Bluecrop' plants detected minimal differences in photosynthetic rates within the day, which varied from 0 to $4\,\mu mol/m^2/s$ and from 0 to $2\,\mu mol/m^2/s$, for open-field and greenhouse-grown plants, respectively. The highest rates were measured at 10:00–12:00.

Variables affecting respiration

Temperature is the most important environmental factor influencing respiration. Respiration is very low at 0°C (a fact that is used to extend the life of the fruit in postharvest) and increases up to a maximum near 38°C. At higher temperatures, the respiratory rate is dramatically reduced due to damage to the enzymes that catalyse the processes. The effect of temperature on the rate of any biological process can be measured through a concept known as Q_{10} (increase in the rate of a process when the temperature is raised by 10°C). The respiratory process has a Q_{10} near 2; this means that the rate of respiration doubles when the temperature is increased by 10°C (Flore and Layne, 1999).

HARVEST INDEX (*H*)

The total production of dry matter in a plant is a function of the total radiation intercepted (mainly by the leaves). In other words, the plant will increase in size as the leaves are able to intercept more radiation during the season (Green *et al.*, 2003). However, the goal in a blueberry planting is to have a high and sustained production of quality fruit; therefore, management practices must be developed to channel a high proportion of the dry matter produced to the fruit, and this has to be done without altering the long-term plant productivity. There are providers or 'sources' of carbohydrates, and receivers or 'sinks'. In the summer, the most important sources for the plant are the leaves, but in the spring during bud opening and early leaf and floral expansion, the most important sources are the reserve tissues (older shoots, buds, crown and structural roots) (Flore and Layne, 1999).

In crop physiology, the proportion of the total amount of dry matter partitioned to the harvested portion (the fruit) is called the harvest index (*H*) (Lakso and Flore, 2003). *H* increases with plant age and depends on various factors such as cultivar type, agroecological conditions and crop management (Fischer *et al.*, 2012). Growers can rely on a number of methods that directly or indirectly influence sink activity (fruit growth) and photosynthesis. Among these, the most important are plant height, planting design, fruit thinning, pruning, fertilization, application of growth regulators, irrigation and phytosanitary control (Flore and Lakso, 1989; Fischer *et al.*, 2012).

The plant distributes resources to the various sinks according to three concepts: (i) priority (which follow the general descending order: fruit > shoots > leaves > roots); (ii) distance between source and sink, with preference to those sinks that are closer to the source or have a direct vascular connection with the source (orthostichy); and (iii) sink strength, which is the product of the size × activity of the sinks (Flore and Lakso, 1989). Thus, if there are two sinks with similar priority and distance to a given source, the one that has the greatest strength will receive a higher proportion of the carbohydrates, and hence will have the highest growth potential.

Temperature plays an important role in carbohydrate partitioning (Fischer *et al.*, 2012). The optimum temperature for transporting carbohydrates in most species is 20–30°C and, according to Guardiola and García-Luis (1993), translocation diminishes with decreasing temperatures (due to the viscosity of the phloem solution). Night temperature is of great importance for carbohydrate translocation, as growth occurs more during the night than the day. Water-stressed plants delay carbohydrate transport due to an increase in the viscosity of the solution translocated. Prolonged water deficits cause the accumulation of abscisic acid, a hormone that inhibits phloem loading in leaves (Fischer *et al.*, 2012). Deficiency or imbalance of mineral nutrients may affect the distribution of assimilates through the initiation and development of sink organs and by restricting or reducing source functioning. Potassium is claimed to be essential in the process of loading and unloading the phloem (due to high concentrations of potassium in companion cells of the sieve elements, which transport carbohydrates) (Taiz and Zeiger, 2006).

Throughout the life of a blueberry plant, the sinks vary in their importance (Fig. 4.7). Initially a greater proportion of carbohydrates is assigned to vegetative organs (roots, canes and leaves), while the fruit gains increasing importance later on until the productive life of the orchard is complete (usually around 20–30 years after planting). From this point of view, the relative importance of the fruit decreases as a sink for carbohydrates and an important part of the resources is assigned to the foliage and, to a lesser extent, to the canes and roots (following the previously mentioned priority). Retamales and Hanson (1989) found in Michigan that field-grown 22-year-old NHB 'Bluecrop' plants had accumulated throughout their lifetime a total of 10.1 kg of dry matter. From this total, 96.3% was partitioned to vegetative tissues and 3.7% to reproductive tissues. The distribution among vegetative tissues was: leaves, 7.3%; 1-year-old shoots, 10.7%; 2–3-year-old shoots, 13.8%; shoots older than 3 years, 17.0%; crown, 36.3%; and roots, 11.2%.

In a trial established in Oregon on the NHBs 'Duke' and 'Liberty' to evaluate management practices for organic production, after two growing seasons it was found that total plant dry weight was generally greater on raised beds than on flat beds, but the difference varied depending on the fertilizer and type of mulch used. Of the total plant biomass, shoots and leaves accounted for 60–77%, whereas roots were 7–19% and fruit accounted for 4–18%. Yields were 33% higher on raised beds than on flat beds and 36% higher with a weed mat than with sawdust mulch. Yield was also higher when the plants were fertilized with a low rate of fish emulsion for 'Duke' but was unaffected by fertilizer source or rate in 'Liberty'. Although raised beds and sawdust or sawdust + compost produced the largest total plant dry weight, the greatest shoot

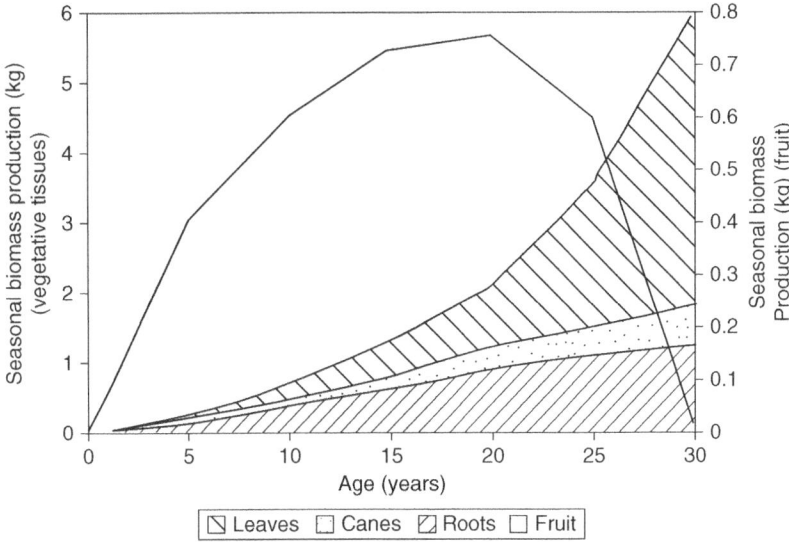

Fig. 4.7. Distribution of dry matter in wild highbush blueberries to various organs throughout the lifetime of the plant. (Adapted from Pritts and Hancock, 1985.)

growth and yield occurred when plants were mulched with a weed mat or with compost + sawdust on raised beds in both cultivars (Larco *et al.*, 2013).

In order to respond to market requirements, breeders have generated a steady increase in fruit size; however, there has not been a concomitant increase in yield, as fruit numbers per plant have simultaneously been reduced (Fig. 4.8). This is known as the compensatory effect, which means that yield cannot expand indefinitely, and that for every increase in one variable (in this case, fruit size), another is reduced (in this case, fruit number per plant). Fruit weight and soluble sugar concentration typically decrease when the ratio between fruit number (fruit load) and leaf area is high. This is because the C source supply (mainly the leaves) is not able to meet the demand of the C sinks (mainly the fruit) (Seehuber *et al.*, 2011).

When there is a need to establish the impact of various cultural practices on yield, it is useful to utilize the concept of yield components. The influence of a cultural practice can be evaluated by examining the effects it has on the component parts of the yield. In the case of blueberries, these components are:

Yield = number of canes per plant × number of fruit per cane × weight per fruit

There are statistical methods (path analysis) that have been used to define the interactions among these components, and between them and yield (Fig. 4.9). In NHB 'Jersey' plants, the results of such an analysis indicated that the number of fruits per cane is the component with the highest effect on yield (the one with the largest positive coefficient assigned in its relationship with

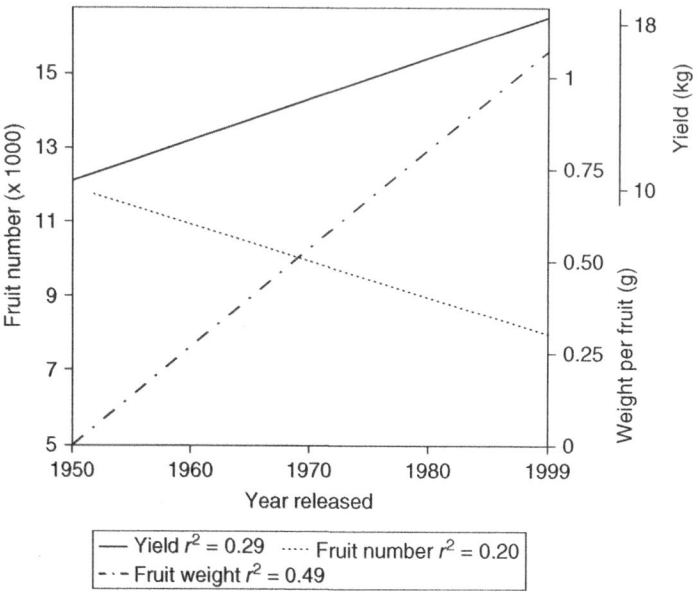

Fig. 4.8. Effect of the date the cultivar was released in NHB blueberries ('Berkeley', 'Bluecrop', 'Blueray', 'Earliblue', 'Elliott', 'Jersey', 'Rubel' and 'Spartan') on the size of fruit, fruit number and yield. (Adapted from Siefker and Hancock, 1986.)

yield), followed in descending order by cane number per plant and weight per fruit. Compensatory effects were apparent, as increased fruit numbers impacted negatively on all other yield components (Siefker and Hancock, 1986).

In the future, market requirements are expected to demand cultural practices focused on greater profitability, including the possibility of producing larger fruit at the expense of yield. If this becomes the case, it will be beneficial to better understand the impact of various management practices on the yield components in different cultivars and productive regions.

CONCLUSIONS

Photosynthesis (or the conversion of light energy into chemical energy stored in plant tissues) sustains life on earth and is fundamental for agricultural productivity. The yield of a crop depends on various factors: the capture of solar energy (mainly by leaves), the transformation of this energy into various compounds required by the plant for its activities, and finally the distribution of these compounds among diverse plant organs.

In the different blueberry regions around the world, the availability of radiation is variable. This has an impact on cultural practices, particularly

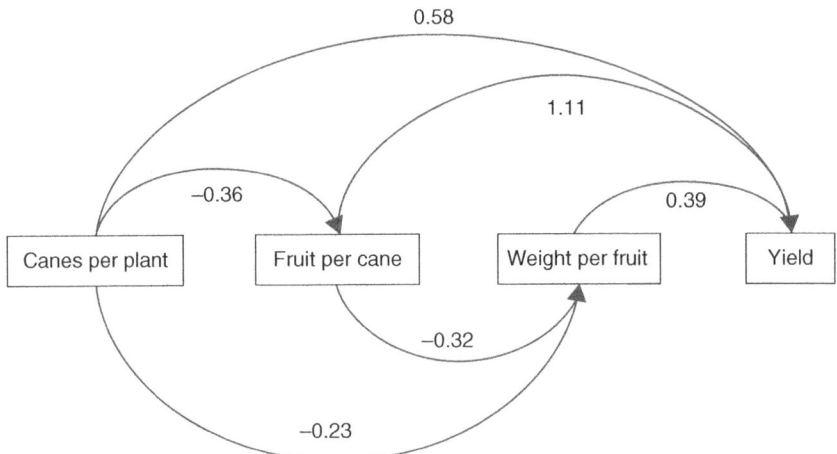

Fig. 4.9. Representation of the yield components in blueberry 'Jersey'. Numbers represent path coefficients. (Adapted from Siefker and Hancock, 1986.)

pruning and the use of shading nets. Judging from the diverse responses to the use of shading nets in highbush blueberries in Chile and the USA, levels of natural light are excessive for blueberries in Chile but not in the USA. The colour of nets is also important, with black nets having (at the same level of shading) more detrimental effects (such as reduced soluble solids and delayed colour development) than white, grey or red nets.

Most of the carbohydrates required for fruit production (around 85%) come from the radiation intercepted by leaves. The amount of light intercepted depends, among other factors, on: (i) plant density; (ii) rate of foliage expansion; (iii) duration of leaf area; (iv) rate of leaf area removal; and (v) plant architecture. The capacity of a plant to capture light can be estimated through the LAI and the relationship between the number of leaves and fruit (L:F ratio). Although LAI is important for evaluating crop development and defining management practices, this variable has not been studied in blueberries. The availability of radiation influences the quality of the current year crop (colour and level of anthocyanins) as well as the flower induction for next year's crop. A value of 25% full sun has been defined as the critical minimum to trigger flower induction in rabbiteye blueberries. There is a need for more studies on this topic in order to define light capture and distribution within the crop throughout the season.

The amount of compounds available for plant growth is a function of the total amount of photosynthesis at a certain period, as well as how much glucose is used through respiration to synthesize those compounds. The rate of photosynthesis depends on various factors, including genotype, light intensity and temperature. The photosynthetic rate becomes saturated (reaches its highest level) at 40–50% of maximum light intensity. The rest of the energy must be dissipated through transpiration in order to maintain leaf temperature and avoid damage to leaf structures. Photosynthetic rates and temperature optima are similar for highbush and rabbiteye blueberries.

Temperature is the factor with the greatest effect on the respiratory rate; the Q_{10} value for respiration is near 2. Two types of respiration have been defined: growth and maintenance respiration. The plant must always secure enough carbohydrate for maintenance respiration. The transformation of glucose into other compounds has specific efficiencies of conversion. As a corollary, because the formation of various compounds has different efficiencies of conversion, the growth capacity of a plant will depend not only on the photosynthetic capacity of the plant but also on the proportion of different compounds in plant tissues.

In order to maintain the productivity of a plant over the long term, carbohydrates must be provided to the various organs of the plant in the required quantity and at the correct time. A relationship is established between sources (leaves and reserve organs) and sinks (e.g. flowers, fruit, roots). In the plant, the carbohydrates are distributed according to: distance, priority (fruits have the highest priority) and sink strength (joint effect of sink activity and sink size). Sinks vary in importance throughout the lifespan of a blueberry planting and according to the management practices used (e.g. pruning, thinning, nutrition, irrigation). Through breeding and management, strong efforts have been made to increase fruit weight. Although yield has also increased, due to compensatory effects it has expanded at a lower intensity than fruit size. The effect of management practices on yield can be analysed through the concept of yield components, which for blueberries are canes per plant, fruits per cane and weight per fruit.

There is a close relationship between solar radiation, photosynthesis and yield, but there is a need to conduct research in order to improve our understanding of the interaction among these factors, as well as to develop a greater knowledge of the effect of various cultural practices performed in different environmental conditions and on different plant materials.

REFERENCES

Andersen, P.C. (1989) Leaf gas exchange characteristics of eleven species of fruit crops in North Florida. *Proceedings of the Florida State Horticultural Society* 102, 229–234.

Andersen, P.C., Buchanan, D.W. and Albrigo, L.G. (1979) Water relations and yields of three rabbiteye blueberry cultivars with and without drip irrigation. *Journal of the American Society for Horticultural Science* 104, 731–736.

Bañados, M.P. (2006) Dry weight and ^{15}N-nitrogen and partitioning, growth, and development of young and mature blueberry plants. PhD thesis, Oregon State University, Oregon.

Birkhold, K.T., Koch, K.E. and Darnell, R.L. (1992) Carbon and nitrogen economy of developing rabbiteye blueberry fruit. *Journal of the American Society for Horticultural Science* 117, 139–145.

Bryla, D.R., Linderman, R.G. and Yang, W.Q. (2008) Incidence of *Phytophthora* and *Pythium* infection and the relation to cultural conditions in commercial blueberry fields. *HortScience* 43, 260–263.

Cobo, N. (2010) Cambios morfológicos y fisiológicos en arándanos (*Vaccinium corymbosum* cv. Brigitta) inducidos por la modificación de su ambiente lumínico. [Morphological and physiological changes in blueberries (*Vaccinium corymbosum* cv. Brigitta) induced by the modification of their light environment.] Undergraduate thesis, Universidad de Talca, Talca, Chile.

DaMatta, F.M. (2007) Ecophysiology of tropical tree crops: an introduction. *Brazilian Journal of Plant Physiology* 19, 239–244.

Darnell, R.L. and Birkhold, K.B. (1996) Carbohydrate contribution to fruit development in two phenologically distinct rabbiteye blueberry cultivars. *Journal of the American Society for Horticultural Science* 121, 1132–1136.

Darnell, R.L. and Davies, F.S. (1990) Chilling accumulation, budbreak, and fruit set of young rabbiteye blueberry plants. *HortScience* 25, 635–638.

Davies, F.S. and Darnell, R.L. (1994) Blueberries, cranberries and red raspberries. In: Schaffer, B. and Andersen, P. (eds) *Handbook of Environmental Physiology of Fruit Crops.* CRC Press, Boca Raton, Florida, pp. 43–84.

Davies, F.S. and Flore, J.A. (1986a) Flooding, gas exchange and hydraulic root conductivity of highbush blueberry. *Physiologia Plantarum* 67, 545–551.

Davies, F.S. and Flore, J.A. (1986b) Short-term flooding effects on gas exchange and quantum yield of rabbiteye blueberry (*Vaccinium ashei* Reade). *Plant Physiology* 81, 289–292.

Eck, P., Gough, R.E., Hall, I.V. and Spiers, J.M. (1990) Blueberry management. In: Galletta, G.J. and Himelrick, D.G. (eds) *Small Fruit Crop Management.* Prentice-Hall, Upper Saddle River, New Jersey, pp. 273–333.

Ehlenfeldt, M.K. and Martin R.B. Jr (2002) A survey of fruit firmness in highbush blueberry and species-introgressed blueberry cultivars. *HortScience* 37, 386–389.

Erb, W., Draper, A.D. and Schwartz, H.J. (1988a) Methods of screening blueberry populations for drought resistance. *HortScience* 25, 312–314.

Erb, W., Draper, A.D. and Schwartz, H.J. (1988b) Screening interspecific blueberry seedling populations for drought resistance. *Journal of the American Society for Horticultural Science* 113, 599–604.

Fischer, G., Almanza-Merchán, P.J. and Ramírez, F. (2012) Source–sink relationships in fruit species: a review. *Revista Colombiana de Ciencias Hortícolas* 6, 238–253.

Flore, J.A. and Lakso, A.N. (1989) Environmental and physiological regulation of photosynthesis in fruit crops. *Horticultural Reviews* 11, 111–157.

Flore, J.A. and Layne, D.R. (1999) Photoassimilate production and distribution in cherry. *HortScience* 34, 1015–1019.

Galletta, G.J. and Ballington, J.R. (1996) Blueberries, cranberries and lingonberries. In: Janick, J. and Moore, J.N. (eds) *Fruit Breeding: Vine and Small Fruit Crops,* Vol. II. Wiley, New York, New York, pp. 1–107.

Gough, R.E. (1994) *The Highbush Blueberry and its Management.* Food Products Press, Binghampton, New York.

Green, S.R., McNaughton, K.G., Wünsche, J.N. and Clothier, B.E. (2003) Modeling light interception and transpiration of apple tree canopies. *Agronomy Journal* 95, 1380–1387.

Guardiola, J.L. and García-Luis, A. (1993) Transporte de azúcares y otros asimilados. [Transport of sugars and other assimilates.] In: Azcón-Bieto, J. and M. Talón (eds). *Fisiología y Bioquímica Vegetal*. McGraw-Hill Interamericana de España, Madrid, Spain, pp. 149–171.

Häder, D.P., Lebert, M., Schuster, M., del Campo, L., Helbling, E.W. and McKenzie, R. (2007) ELDONET – A decade of monitoring solar radiation on five continents. *Photochemistry and Photobiology* 83, 1348–1357.

Haman, D.Z., Smajstrla, A.G., Pritchard, L.T. and Lyrene, P.M. (1997) Response of young blueberry plants to irrigation in Florida. *HortScience* 32, 1194–1196.

Hancock, J.F., Lyrene, P., Finn, C.E., Vorsa, N. and Lobos, G.A. (2008) Blueberries and cranberries. In: J.F. Hancock (ed.). *Temperate Fruit Crop Breeding: Germplasm to Genomics*. Springer, New York, New York, pp 115–150.

Jorquera-Fontena, E., Alberdi, M. and Franck, N. (2014) Pruning severity affects yield, fruit load and fruit and leaf traits of 'Brigitta' blueberry. *Journal of Soil Science and Plant Nutrition* 14, 855–868.

Jorquera-Fontena, E., Alberdi, M., Reyes-Díaz, M. and Franck, N. (2016) Rearrangement of leaf traits with changing source–sink relationship in blueberry (*Vaccinium corymbosum* L.) leaves. *Photosynthetica* 54, 508–516

Kawamura, K. and Takeda, H. (2002) Light environment and plant architecture of two temperate *Vaccinium* species: inherent growth rules versus degree of plasticity in light response. *Canadian Journal of Botany* 80, 1063–1077.

Kim, S.J., Duk, J.Y., Tae-Choon, K. and Hee, J.L. (2011) Growth and photosynthetic characteristics of blueberry (*Vaccinium corymbosum* cv. Bluecrop) under various shade levels. *Scientia Horticulturae* 129, 486–492.

Kliewer, W.M. and Dokoozlian, N.K. (2005) Leaf area/crop weight ratios of grapevines: influence on fruit composition and wine quality. *American Journal of Enology and Viticulture* 56, 170–181.

Kozinski, B. (2006) Influence of mulching and nitrogen fertilization rate on growth and yield of highbush blueberry. *Acta Horticulturae* 715, 231–235.

Lakso, A.N. and Flore, J.A. (2003) Carbohydrate partitioning and plant growth. In: Baugher, T.A. and Singh, S. (eds). *Concise Encyclopedia of Temperate Tree Fruit*. Food Products Press, Binghampton, New York, pp. 21–30.

Larco, H., Strik, B., Bryla D.R. and Sullivan D.M. (2013) Mulch and fertilizer management practices for organic production of highbush blueberry. I: plant growth and allocation of biomass during establishment *HortScience* 48, 1250–1261.

Léchaudel, M., Joas, J., Caro, Y., Genard, M. and Jannoyer, M. (2005) Leaf:fruit ratio and irrigation supply affect seasonal changes in minerals, organic acids and sugars of mango fruit. *Journal of the Science of Food and Agriculture* 8, 251–260.

Li, X., Chen, W. and Li, Y. (2012) Study on photosynthetic characteristics of blueberry in greenhouse. *Acta Horticulturae* 926, 315–319.

Li, Y., Chen, W., Zhang, Z. and Wulin, L. (2009) Study on the influence of three production methods on blueberry photosynthesis. *Acta Horticulturae* 810, 521–525.

Lobos, G.A., Retamales, J.B., del Pozo, A., Hancock, J.F. and Flore, J.A. (2009) Physiological response of *Vaccinium corymbosum* L. cv. Elliott to shading nets in Michigan. *Acta Horticulturae* 810, 465–470.

Lobos, G.A., Retamales, J.B., Hancock, J.F., Flore, J.A., Cobo, N.G. and del Pozo, A. (2012) Spectral irradiance, gas exchange characteristics and leaf traits of *Vaccinium corymbosum* L. 'Elliott' grown under photo-selective nets. *Environmental and Experimental Botany* 75, 142–149.

Lobos, G.A., Retamales, J.B., Hancock, J.F., Flore, J.A., Romero-Bravo, S. and del Pozo, A. (2013) Productivity and fruit quality of *Vaccinium corymbosum* cv. Elliott under photo-selective shading nets. *Scientia Horticulturae* 153, 143–149.

Long, S.P., Zhu, X.G. Zhu, Naidu, S.L. and Ort, D.R. (2006) Can improvement in photosynthesis increase crop yields? *Plant, Cell and Environment* 29, 315–330.

Loomis, R.S. and Connor, D.J. (1992) *Crop ecology: Productivity and Management in Agricultural Systems.* Cambridge University Press, Cambridge.

Lyrene, P.M. (1992) Early defoliation reduces flower bud counts on rabbiteye blueberry. *HortScience* 27, 783–785.

Lyrene, P.M. and Williamson, J.G. (2004) Protecting blueberries from freezes in Florida. Department of Horticultural Sciences, University of Florida, Florida. Available from: http://edis.ifas.ufl.edu/pdffiles/HS/HS21600.pdf (accessed 26 January 2018).

Matta, F.B., Cartagena, J.R. and Spiers, J.M. (2005) Response of rabbiteye blueberries to chemical thinning agents. Research Report of the Mississippi State University Agricultural Research Station, Vol. 23. Available from: http://mafes.msstate.edu/publications/research-reports/rr23-15.pdf (accessed 26 January 2018).

Moon, J.W., Flore, J.A. and Hancock, J.F. (1987a) A comparison of carbon and water vapor gas exchange characteristics between a diploid and highbush blueberry. *Journal of the American Society for Horticultural Science* 112, 134–138.

Moon, J.W., Hancock, J.F., Draper, A.D. and Flore, J.A. (1987b) Genotypic differences in the effect of temperature on CO_2 assimilation and water use efficiency in blueberry. *Journal of the American Society for Horticultural Science* 112, 170–173.

Niinemets, Ü., Kull, O. and Tenhunen, J.D. (1998) An analysis of light effects on foliar morphology, physiology and light interception in temperate deciduous woody species of contrasting shade tolerance. *Tree Physiology* 18, 681–696.

Open Learning Initiative (2015) Introduction to biology (open + Free). Module 22: Photosynthesis/cellular respiration cycle. Available at: https://oli.cmu.edu/jcourse/workbook/activity/page?context=90d4016280020ca600fc1dafce92befd (accessed 27 February 2018).

Pritts, M.P. and Hancock, J.F. (1985) Lifetime biomass partitioning and yield component relationships in the highbush blueberry, *Vaccinium corymbosum* L. (Ericaceae). *American Journal of Botany* 72, 446–452.

Retamales, J.B. and Hanson, E.J. (1989) Fate of ^{15}N-labeled urea applied to mature highbush blueberries. *Journal of the American Society for Horticultural Science* 114, 920–923.

Retamales, J.B., Hanson, E.J. and Bukovac, M.J. (2000) GA_3 as a flowering inhibitor in blueberries. *Acta Horticulturae* 527, 147–151.

Retamales, J.B., Montecino, J.M., Lobos, G.A. and Rojas, L.A. (2008) Colored shading nets increase yields and profitability of highbush blueberries. *Acta Horticulturae* 770, 193–197.

Roloff, I. and Scherm, H. (2004) Photosynthesis of blueberry leaves as affected by *Septoria* leaf spot and abiotic leaf damage. *Plant Disease* 88, 397–401.

Sarmiento, P. (1995) *Energía Solar: Aplicaciones e Ingeniería*, 3rd edn. [Solar Energy: Applications and Engineering.] Ediciones Universitarias de Valparaíso, Valparaíso, Chile.

Seehuber, C., Damerow, L. and Blanke, M. (2011) Regulation of source:sink relationship, fruit set, fruit growth and fruit quality in European plum (*Prunus domestica* L.) using thinning for crop load management. *Plant Growth Regulation* 65, 335–341.

Siefker, J.H. and Hancock, J.F. (1986) Yield component interactions in cultivars of the highbush blueberry. *Journal of the American Society for Horticultural Science* 111, 606–608.

Sims, D.A., Gebauer, R. and Pearcy, R.W. (1994) Scaling sun and shade photosynthetic acclimation of *Alocasia macrorrhiza* to whole plant performance. II. Simulation of carbon balance and growth at different photon flux densities. *Plant, Cell and Environment* 17, 889–900.

Singsaas, E.L., Ort, D.R. and DeLucia, E.H. (2001) Variation in measured values of photosynthetic quantum yield in ecophysiological studies. *Oecologia* 128, 15–23.

Strik, B. and Buller, G. (2002) Improving yield and machine harvest efficiency of 'Bluecrop' through high-density planting and trellising. *Acta Horticulturae* 574, 227–231.

Strik B. and Buller, G. (2005) The impact of early cropping on subsequent growth and yield of highbush blueberry in the establishment years at two planting densities is cultivar dependant. *HortScience* 40, 1998–2001.

Strik, B., Fisher, G., Hart, J., Ingham, R., Kaufman, D., Penhallegon, R., Pscheidt, J., William, R., Brun, C., Ahmedullah, M., Antonelli, A., Askham, L., Bristow, P., Havens, D., Scheer, B., Shanks, C. and Barney, D. (2006) Blueberry pruning. Northwest Berry & Grape Information Network. Available at: http://berrygrape.org/blueberry-pruning/ (accessed 25 August 2010).

Suzuki, A., Shimizu, T. and Aoba, K. (1998) Effects of leaf/fruit ratio and pollen density on highbush blueberry fruit quality and maturation. *Journal of Japanese Society for Horticultural Science* 67, 739–743.

Taiz, L. and Zeiger, E. (2006) *Plant Physiology*, 4th ed. Sinauer Associates, Sunderland, Massachusetts.

Teramura, A.H., Davies, F.S. and Buchanan, D.W. (1979) Comparative photosynthesis and transpiration in excised shoots of rabbiteye blueberry. *HortScience* 14, 723–724.

Trewavas A. (2014) *Plant Behaviour and Intelligence*. Oxford University Press, Oxford.

Tsukaya, H. (2005) Leaf shape: genetic controls and environmental factors. *International Journal of Developmental Biology* 49, 547–555.

Turgeon, R. (1989) The sink–source transition in leaves. *Annual Review of Plant Physiology and Plant Molecular Biology* 40, 119–138.

Weinbaum, S.A. (1978) Carbohydrate and nitrogen assimilation and utilization in the almond. In: Micke, W. and Kester, D. (eds) *Almond Orchard Management*. University of California, Berkeley, California, pp. 97–107.

Wright, G.C., Patten, K.D. and Drew, M.C. (1993) Gas exchange and chlorophyll content of 'Tifblue' rabbiteye and 'Sharpblue' southern highbush blueberry exposed to salinity and supplemental calcium. *Journal of the American Society for Horticultural Science* 118, 456–463.

Yáñez, P., Retamales, J.B., Lobos, G.A. and del Pozo, A. (2009) Light environment within mature rabbiteye blueberry canopies influences flower bud formation. *Acta Horticulturae* 810, 471–474.

NUTRITION IN BLUEBERRIES

INTRODUCTION

The nutrient demand of blueberries is low compared with fruit trees (Table 5.1). Some authors have reported adequate growth and fruiting in rabbiteye blueberries grown for several seasons in low-fertility soils with no fertilizer application (Austin and Brightwell, 1977; Austin and Gaines, 1984). However, in most situations, regular fertilizer applications are usually needed for commercial fields (Hanson and Hancock, 1996; Krewer and NeSmith, 1999).

Table 5.1. Sufficient or normal foliar concentrations of nutrients for NHB blueberries and apple (dry-weight basis).

Nutrient	NHB blueberry[a]	Apple[b]
Macroelement (%)		
N	1.70–2.10	2.20–2.40
P	0.08–0.40	0.13–0.33
K	0.40–0.65	1.35–1.85
C	0.30–0.80	1.30–2.00
Mg	0.15–0.30	0.35–0.50
S	0.12–0.20	–
Microelements(ppm)		
B	25–70	35–50
Cu	5–20	7–12
Fe	60–200	>150
Mn	50–350	50–150
Zn	8–30	35–50

[a]Data from Hanson and Hancock (1996).
[b]Data from Stiles and Reid (1991).

There are various conditions, in both the plant and the soil, that explain the low nutritional requirements of blueberries compared with other fruit crops. Blueberries are said to be calcifuge plants, which means they are adapted to acidic soil conditions. Best growth and productivity is obtained when blueberries grow in soil with a pH in the range of 4.2–5.5. At this pH, the availability of most soil nutrients is limited, and this reduces the amount of mineral elements that are absorbed by the plant (Korcak, 1989; Hanson and Hancock, 1996). Blueberry roots are shallow and devoid of hairs (which limits the surface area in contact with the soil or substrate), and in natural habitats they are colonized by a specialized type of fungus called ericoid mycorrhizae. Studies on NHBs showed that increasing fertilization rates decreased ericoid mycorrhizae colonization of 'Duke' and had little influence on colonization of 'Reka' (Golldack *et al.*, 2001). This type of cultivar-specific response in sensitivity to ericoid mycorrhizae colonization by nutrient availability may be responsible for some of the differences in the frequency and intensity of colonization that were detected among different highbush blueberry cultivars (Scagel, 2005). In addition, the fine root system of blueberries demands a loose soil, which makes sandy loams high in organic matter preferable for their cultivation.

There has been a significant expansion in the amount of land planted to blueberries across the world, and as a result, soils that are less optimal for blueberry production are often being used. In many cases, amendments such as organic mulches and acidification are needed to provide adequate conditions for plant growth and development.

Acidic soils are characterized mainly by low pH (pH 4.5–5.5, strongly acid, or pH <4.5, extremely acid), low cation exchange capacity and low base saturation. The solubility of several potentially toxic metal ions including aluminium (Al^{3+}), manganese (Mn^{2+}) and iron (Fe^{3+}) increases as the pH decreases, whereas several nutrients, such as calcium (Ca), magnesium (Mg) and phosphorus (P) become deficient because of leaching or precipitation as metallic salts (Zhao *et al.*, 2014). In acidic soils, plant growth is usually limited by a combination of toxic concentrations of Al, Mn, Fe and protons (H^+) and deficiencies of P, potassium (K), Ca, Mg and molybdenum (Mo). The relative dominance of these constraints on plant growth depends on the species and genotype of the plant, soil characteristics, ion species and climatic factors (Marschner, 1986). Under field conditions, plant growth is limited by these specific factors individually and collectively by their interactions (Zhao *et al.*, 2014).

In many growing areas, nitrogen (N) is the most frequent, if not the only, mineral nutrient that must be applied to blueberries (Hanson and Hancock, 1996). Soils high in organic matter have a greater N supply and fertilization rates can be lower. However, when organic mulches are added, additional N needs to be provided, as N is used by the microbes to decompose these materials (Eck *et al.*, 1990). Ca is another important nutrient because it impacts on fruit quality. Several characteristics of the blueberry soil and the plant influence the Ca supply to the fruit.

Once the amount of nutrient to be applied is defined, there is a need to establish the method of fertilizer application. Fertilizers can be broadcast, applied through fertigation or sprayed on the leaves. A combination of methods is commonly used. The decision to use a certain type of application will depend on technical and economic factors (Hart *et al.*, 2006).

In this chapter, we will examine some fundamental concepts of plant nutrition and then describe the critical factors influencing blueberry nutrition. These factors include: (i) supply of nutrients from the soil (substrate); (ii) soil pH; (iii) role of ericoid mycorrhizae; (iv) determination of nutrient status; (v) absorption and translocation of nutrients within the plant; and (vi) application of nutrients.

FUNDAMENTAL CONCEPTS OF MINERAL NUTRITION

The goal of fertilization is to remove limitations to yield and quality by supplying the blueberry crop with ample nutrition in advance of demand. A fertilizer application should be based on soil and plant analysis, information on environmental conditions, plant performance and management, and grower's experience. A fertilizer application should produce a measurable change in plant growth, plant performance and/or nutrient status. Results from nutrient applications can vary from year to year and from field to field. The use of fertilizers must be part of a complete management package. If some parts or components of the blueberry-growing system are not working properly, they cannot be substituted by additional fertilization. Soil properties such as pH, moisture and organic matter influence the nutritional status of the orchard. Increasing fertilizer rates will not correct other limiting factors (Hart *et al.*, 2006).

To obtain optimum yields, plants must have sufficient nutrient levels at all times. Nutrient deficiencies (or excesses) will have an impact on both yield and fruit quality. The importance of the impact will depend on the magnitude, opportunity and duration of the deviation from the optimum (Marschner, 1986). In commercial plantings, the high yield and quality required by the market forces the grower to monitor the crop constantly in order to satisfy the nutrient requirements in terms of amount and opportunity needed to avoid nutrient deficiencies or excesses (Hart *et al.*, 2006).

Plants interact with the environment (nutrients, light, water and biotic factors) to generate growth. The amount of growth and the balance between reproductive and vegetative growth determines yield. Adequate nutrition is based on the interaction between the soil and the plant (Marschner, 1986). There are specific conditions that maximize root growth and absorption of water and nutrients. For plant growth to occur, the soil must satisfy certain biological, physical and chemical conditions. Plants contain mineral elements that are 'essential' for their metabolism, growth and development, which includes reproduction (flower formation and fruit development). The definition of 'essentiality' implies that such an element is required by the plant to complete its cycle and that it cannot be replaced by another element. The essential elements are classified as 'macro-' or 'microelements', based on their proportion in plant tissues and not on their relative importance (Lawlor, 1991). Plants are fairly selective in the absorption of mineral elements from the soil, so they usually take what they need and not necessarily what the soil or substrate offers.

Soils have the three states of matter: solid, liquid and gas. The solid phase is constituted by organic and inorganic materials. The organic component consists of the residues of plants and animals in various stages of decomposition, and of a stable part called humus. The inorganic fraction is composed of primary and secondary minerals with different particle sizes. The fractions of the soil that participate in the ionic exchange are those components with particle diameters less than 0.02 mm and are called colloids. Colloids are particles that have a large surface area and if suspended in water will not settle. Particle sizes in the different soil types are: sand, <2.00 to 0.02 mm; silt, <0.02 to 0.002 mm; and clay, <0.002 mm (Loomis and Connor, 1992).

Colloids (organic and inorganic) develop negative charges on their surfaces. The cations (with positive charge: K^+, sodium (Na^+), H^+, Ca^{2+}, zinc (Zn^{2+}), Mg^{2+}, copper (Cu^{2+}) and Al^{3+}) are attracted and retained by these surfaces, while the anions (with negative charge: nitrate (NO_3^-) and hydroxide (OH^-)) are not retained so firmly. The number of cations (expressed as milliequivalents per 100 g dry soil (meq per 100 g)) is called the cation exchange capacity (CEC). CEC is one of the most important chemical properties of the soil. It represents the number of cations that are easily exchangeable with other cations and as a result are available for the plant. Thus, the CEC of a soil represents the total number of exchangeable cations that a soil can retain. Representative CEC values for clay range from 30 to 100 meq per 100 g, while for organic matter CEC may be as great as 300 meq per 100 g (Loomis and Connor, 1992).

The CEC associated with soil organic matter is called pH-dependent CEC. This means that the actual CEC of the soil will depend on the pH of the soil. Given the same amount of organic matter, a neutral soil (~pH 7) will have a higher CEC than an acid soil (e.g. ~pH 5). In other words, the CEC of a soil with a pH-dependent charge will increase with a raise in pH. As blueberries are adapted to acid soils, the pH-dependent CEC in this crop tends to be lower than in most fruit crops. This is one of the reasons why blueberries have low nutrient requirements.

As shown in Table 5.2, the CEC in soils planted to blueberries ranges from 1 to 25 meq per 100 g soil. The higher the CEC, the more clay or organic matter is present in the soil (and the higher the water-holding capacity). The ability to change the soil pH also depends on the CEC. High CEC soils require more elemental sulfur (S) to change their pH. Low CEC soils are more prone to develop cation deficiencies. Thus, for sandy soils, more frequent additions of small

Table 5.2. Characteristics of soils planted to blueberries.

Site	Latitude	Soil depth (cm)	Texture	Organic matter (weight%)	pH	CEC (meq per100g)	Bases (% of CEC)		
							Ca	Mg	K
USA									
Oregon[a]	44°56'N	>200	Clay loam	3–4	5.0–5.6	11–20	57–66	9–11	11–17
Michigan[b]	42°40'N	40–70	Sandy loam	7–10	4.5	22–25	>60	>15	>10
Georgia[c]	31°46'N	60	Loamy	1.0	4.7–5.1	3–10	50–62	9–12	12–20
Europe									
Huelva, Spain[d]	37°28'N	25–250	Sandy	0.4–0.7	4.8–5.4	2–5	37–55	33–48	2–4
South America									
Collipulli, Chile[e]	37°95'S	20	Clay loam	3.3	5.3	9	72	20	3
Osorno, Chile[e]	40°60'S	60–120	Loamy, clayey silt	13.4	5.0	11	67	15	15
Concordia, Argentina[f]	31°22'S	40–60	Sandy loam, sand	0.2–5.2	4.6–5.7	1–22	63	12	1
Buenos Aires, Argentina[f]	34°66'S	40–60	Clay loam	1.5–4.7	5.4–6.1	17–24	62	14	7

[a]Data from W.Q. Yang (2011, Oregon, personal communication).
[b]Data from Hanson (1987).
[c]Data from D.S. NeSmith (2011, Georgia, personal communication).
[d]Data from Vadillo (2006).
[e]Data from Tosso (1985).
[f]Data from R.S. Lavado (2010, Buenos Aires, Argentina, personal communication).

amounts of fertilizer are better. In these soils, a large one-time addition of cations can lead to large leaching losses, because the soil is not able to hold on to the excess cations.

THE NITROGEN CYCLE

As previously described, N limitation reduces plant growth in acidic soils (Zhao *et al.*, 2014). In most blueberry cropping situations, N is the nutrient most commonly applied and whose deficiency is most prevalent (Hanson and Hancock, 1996). N is present in the biosphere in various chemical forms. Molecular nitrogen (N_2) represents 80% of the atmosphere composition. However, blueberry plants cannot directly use this form of N. N_2 enters the biological nitrogen cycle in three main ways: through biological fixation (prokaryotic conversion of N_2 to ammonium (NH_4^+) with the widely known association between leguminous plants and rhizobia), by atmospheric fixation (lightning and photochemical conversion of N_2 to NO_3^-), and by the Haber–Bosch industrial fixation of N_2 to produce NH_4^+. Once N is fixed as NO_3^- or NH_4^+, it has two main fates: (i) NO_3^- and NH_4^+ can undergo biochemical processes that transform them back to N_2 (Marschner, 1986); or (ii) they can be reduced and/or assimilated for the biosynthesis of N-containing metabolites (Kraiser *et al.*, 2011).

In addition to the regulation of inorganic and organic N uptake systems, plants display considerable developmental plasticity in response to variations in the concentration and distribution of external nutrients. One of the most dramatic plant adaptations to ensure adequate N acquisition is modulation of the root system architecture in response to N supply. The proliferation of lateral roots within a localized NO_3^--rich zone is a response that occurs in many crops and represents a common adaptation phenomenon. Additional effects of N supply on root architecture and root developmental plasticity include changes in primary root growth, lateral root initiation and elongation (Kraiser *et al.*, 2011).

Soil N is in a constant state of flux, moving and changing chemical forms (Subbarao *et al.*, 2006). The nitrogen cycle is mediated by microorganisms, whose activity is dependent on chemical and physical soil conditions (Fig. 5.1). There are processes that increase N availability for the plants (nitrification and mineralization), while others reduce the amounts of N available for the crop (immobilization, denitrification, volatilization and leaching). The main characteristics of these processes and their impact on N nutrition in blueberries are addressed below.

Nitrification

Nitrification (NO_3^- assimilation), defined as the biological oxidation of NH_4^+ to NO_3^- (Subbarao *et al.*, 2006), is the process by which decomposing proteins,

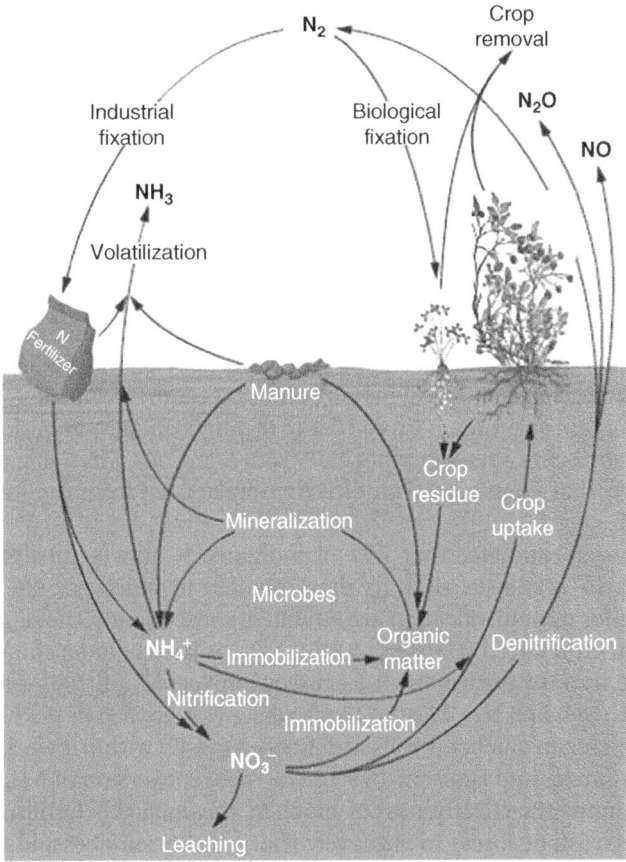

Fig. 5.1. The nitrogen cycle in blueberry. (Adapted from Johnson *et al.*, 2005.)

inorganic N and other nitrogenous substances that originate from organic matter are transformed to NO_3^- by microorganisms. NO_3^- assimilation is a two-step process involving the reduction of NO_3^- to NO_2^- (nitrite) by nitrate reductase (NR) and further to NH_4^+ by nitrite reductase. Reduction of NO_2^- to NH_4^+ and assimilation of NH_4^+ are relatively rapid processes as accumulation of these compounds can be toxic to cells. Therefore, the reduction of NO_3^- to NO_2^- catalysed by NR is the rate-limiting step in NO_3^- assimilation. Consequently, NR gene expression and activity are highly regulated by various factors such as light and photosynthates (sugars), particularly in shoot tissues. Such regulation of NO_3^- assimilation by light and photosynthates may allow plants to coordinate carbon (C) and N metabolism. It has been proposed that whole-plant NO_3^- assimilation capacity in blueberries is generally low and that this may in turn result in the reduced total NO_3^- acquisition capacity by the roots

(Darnell and Hiss, 2006). However, it is also likely that lack of substantial NO_3^- uptake and its translocation to the shoots limits the NO_3^- assimilation capacity of the shoot tissues, as NO_3^- can regulate its own assimilation. Low levels of NO_3^- in the shoot tissues may not be sufficient to induce NR activity and downstream metabolism (Alt *et al.*, 2017).

Plants have evolved inorganic and organic N uptake systems to cope with the heterogeneous N availability in the soil. Soil organic compounds can also contribute to plant N nutrition. Amino acids represent the largest fraction of low-molecular-weight dissolved soil organic N. Amino acids, urea, small polypeptides and other N-containing biomolecules can be released back into the environment by secretion, excretion or the decay of organic matter. These organic forms of N can also be used as N sources by plants and other organisms. The amino acid pool is dynamic because it is quickly taken up by plants and microorganisms. Urea is excreted into the environment by various organisms and represents a readily available N source in soils. Physiological studies have shown that plant roots can directly uptake urea from the soil (Kraiser *et al.*, 2011)

It has been estimated that 75% of inorganic N may be nitrified in cultivated soils. In acid soils, such as those planted to blueberry, the microbes responsible for most nitrification are inhibited; consequently, the rate of nitrification is much reduced (Hanson, 2006). NO_3^- is subject to leaching and can also escape into the environment as gaseous molecules (nitrous oxide (N_2O), nitric oxide (NO) and N_2). Most of the fertilizer N is applied in the NH_4^+ form, which in Michigan blueberry soils was nitrified mostly within 4 weeks of application (Retamales and Hanson, 1990). The rapid conversion of NH_4^+ to NO_3^- in the soil limits the effectiveness of much of the applied N fertilizer. NH_4^+ is held by electrostatic forces to negatively charged clay surfaces and functional groups of organic matter. This binding would be sufficient to limit N losses by leaching (Subbarao *et al.*, 2006). Thus, nitrification of NH_4^+-N results in the transformation of N from a relatively soil-immobile N form (NH_4^+) to a highly mobile form (NO_3^-). Management practices can increase the potential for nitrification. It is generally agreed that the population of nitrifying organisms increases rapidly upon moderate addition of NH_4^+ to the soil. In Michigan, soil-nitrifying populations and nitrification rates were higher in old blueberry fields than in adjacent, undisturbed forest soils (Hanson *et al.*, 2002). This could be due to NH_4^+ applications, improved drainage, or alterations in soil chemistry or structure. Organic matter composition, soil texture, CEC, drainage and pH can also affect the rate of nitrification (Subbarao *et al.*, 2006). NH_4^+ usually predominates in acidic soils because of their low nitrification rates, whereas NO_3^- is the main form of N in neutral and calcareous soils (Zhao *et al.*, 2014). However, research in blueberries has shown that nitrification rates are not related to soil pH (Hanson *et al.*, 2002). Production practices in blueberry (application of NH_4^+-containing fertilizers, addition of P and K fertilizers, cultivation, composition and quality of litter, and drainage) lead to increased populations of nitrifying bacteria and higher nitrification capacity. The practical significance of these findings is that the optimum timing of fertilizer application depends on the specific nitrification capacity of the site. On soils that nitrify readily, NO_3^- is formed rapidly, and multiple applications of lower N rates should be used to reduce leaching losses and increase fertilizer-use efficiency (Subbarao *et al.*, 2006). Growers have observed that plants grow more slowly when replanted on old blueberry sites compared with virgin soils. Perhaps a factor contributing to this slow growth on replanted sites is that they have higher nitrification rates, which increase N loss through leaching and reduce the efficiency of fertilizer use. NO_3^- leaching can be measured in the field by placing porous ceramic capsules beneath the rooting depths of blueberries (i.e. greater than 80 cm deep; Eck *et al.*, 1990). A vacuum pump connected to the samplers via flexible polyethylene tubing can be used to collect the samples.

Blueberries absorb NO_3^- less readily than NH_4^+, and this has an important impact on N nutrition in blueberries (Eck *et al.*, 1990). For this reason, it is desirable to reduce the nitrification rate in order to improve the N-use efficiency of blueberries and limit the potential leaching of NO_3^- into soil water. Nitrification rates and N recovery by 3-year-old 'Bluecrop' were measured after applications of ammonium sulfate $((NH_4)_2SO_4)$ with or without the nitrification inhibitor dicyandiamide (DCD) on sandy loams at pH 4.8. Concentrations of fertilizer-derived NO_3^- were significantly lower in DCD-treated soils 2 weeks following application, but DCD had no effect on total NO_3^- levels or fertilizer-derived NO_3^- later in the season. DCD also had no effect on fertilizer-derived or total N levels in plants. It seems that the effect of DCD on NO_3^- levels is short lived and, when the whole season is considered, the impact would be minimal or negligible (Throop and Hanson, 1998).

Denitrification

Denitrification is the transformation of NO_3^- or nitrite (NO_2^-) to gaseous N either as molecular N (N_2) or an oxide of nitrogen (NO or N_2O). The escape of these gases from the soil represents a net loss of N from the field. The bacteria involved in this transformation use oxygen from NO_3^- for their respiration. The populations of these bacteria increase with the organic matter content of the soil (Loomis and Connor, 1992). Wetting and drying cycles change the O_2 availability in the soil and have major effects on soil microbial processes. Nitrification and denitrification rates increase dramatically after wetting of air-dried soils. In order to reduce the impact of denitrification, growers should avoid marked fluctuations in soil water content, especially when high levels of NO_3^- are present in the soil.

Volatilization

Volatilization is the loss of N through the conversion of NH_4^+ to free ammonia (NH_3) gas, which is then released to the atmosphere. Free NH_3 increases about 10-fold with each unit increase in pH. Thus, about 0.004% of the N is present as free NH_3 at pH 5 and nearly 0.04% at pH 6 (Loomis and Connor, 1992). Hot and windy conditions also favour volatilization losses. Additions of NH_4^+ fertilizers can lower the pH in localized areas and reduce the rate of N loss. When NH_4^+ fertilizers are broadcasted, volatilization losses can be reduced by irrigating shortly after application. Up to 20% of the applied fertilizer can be lost by volatilization from blueberry fields if rain is not received or the field is not irrigated in a timely fashion (Krewer and NeSmith, 1999).

Leaching

Leaching of NO_3^- is often the main cause of N loss and is of high concern in terms of water quality (Subbarao *et al.*, 2006). Environmental agencies in many places have set standards for the maximum amount of NO_3^- permitted in drinking water; the level is 10 ppm in the USA. The rate of leaching depends on soil drainage, rainfall, quantity of NO_3^- present in the soil and rate of crop uptake. Well-drained soils, low crop yield, high N inputs (especially when roots are not active) and high rainfall are all conditions that increase the potential for NO_3^- leaching (Loomis and Connor, 1992; Subbarao *et al.*, 2006). The coarse, porous soils that are common to blueberry croplands may facilitate leaching (Hanson, 2006). In commercial NHB blueberry fields, measurements taken 2–3 months after fertilization have shown a decline in topsoil (0–30 cm) NO_3^- and NH_4^+ and an increase in subsoil (30–60 cm) NO_3^- and NH_4, which was probably due to leaching (Retamales and Hanson, 1990).

Immobilization

Immobilization refers to the incorporation of inorganic N into microbial biomass, and then more permanently into humus (Myrold and Bottomley, 2008). All living organisms require N, and soil microorganisms compete with plants for N. Immobilization occurs through the incorporation of N (NO_3^- or NH_4) into the microbial biomass with resultant resistance to further availability for plant use. Substantial proportions of applied N may be incorporated into the microbial biomass within 24 h of application of $(NH_4)_2SO_4$ and near-complete immobilization has occurred within 1 week of application of N fertilizer (Myrold and Bottomley, 2008). The immobilized N is incorporated into proteins, nucleic acids and other organic N constituents of microbial cells and cell walls; as such, it becomes part of the microbial biomass. As the microbes die and decay, some of the microbial biomass N is released as NH_4^+ through the process of mineralization; the remainder undergoes conversion to more stable organic N compounds, ultimately becoming part of the soil organic matter. The stabilized organic compounds are not readily available to plants; therefore, the net result of immobilization/mineralization (see below) is a decrease in the availability to the crop of the N added to the soil as fertilizer, and also the partial conversion of this N to a form (NH_4^+) that is not subject to loss from most soils (Azam *et al.*, 1993).

Of particular importance for N availability is the C:N ratio: the ratio of available C to mineral N (NH_4^+ and NO_3^-). When the C:N ratio is less than or equal to 20, mineralization exceeds immobilization, whereas at C:N ratios of 30 or more, immobilization exceeds mineralization. The C:N ratio of the materials incorporated into the soil determines whether N will be immobilized (unavailable) or mineralized (available) (Yang *et al.*, 2002). The C:N ratio of the residues declines as they decay (Loomis and Connor, 1992). As explained in Chapter 6 (this volume), the incorporation of materials with a high C:N ratio (e.g. sawdust, straw, bark) will cause greater demand for N, and thus may result in N immobilization. In these cases, extra N should be added to compensate for microbial immobilization (including mycorrhizal fungi) and leave N available for the blueberry crop (Yang *et al.*, 2002; Hanson, 2006). Immobilization locks up N temporarily. When the microorganisms die, the organic N in their cells becomes available for plant uptake through mineralization.

Mineralization

Mineralization is the process by which microbes decompose organic N from manure, dead microbes, organic matter and crop residues to produce NH_4^+. As mineralization is a biological process, its rate varies with factors that affect microbial activity, such as soil temperature, moisture and the amount of O_2 in the soil (aeration). Mineralization occurs readily in warm soil (20–35°C). Between 1 and 10% of the soil organic reserves may be mineralized within a year (Loomis and Connor, 1992). Assuming as an average that each 1% of organic matter content releases about 7 kg N/year, in most blueberry soils this would amount to 30–120 kg N/ha/year (Krewer and NeSmith, 1999). It is important to estimate mineralization and immobilization potentials of a given soil when determining N application rates. In some cases, native inorganic N released from organic matter may be considerable and sufficient to satisfy plant requirements.

SOIL PH REQUIREMENTS OF BLUEBERRIES

The soil pH range recommended for highbush blueberries is 4.5–5.5, and pH 4.2–5.0 for rabbiteye blueberries. The pH of the soil influences the availability of nutrients for plants (Fig. 5.2). High soil pH is a common problem encountered in new blueberry sites. When blueberries are grown in high soil pH, their leaves turn yellow with green veins or are completely yellow. These leaves are small and often turn brown and fall from the plant before the season is over. Little growth occurs, and some plants may die. Plants stunted by high soil pH usually do not recover, even when the soil pH is reduced (Hart *et al.*, 2006). Plants established in high-pH soils often require replanting to obtain a uniform and vigorous stand. Fe, Mn or Cu deficiencies are common in soils with high pH; thus, rather than application of these elements to the soil, correcting the pH will usually be more helpful and effective.

Soils are acidified either with elemental S incorporated before planting or with sulfuric acid (H_2SO_4) applied through the irrigation system. In the case of soil S application, as the transformation of S into acid is a process mediated by microorganisms, it requires time, moisture and warm temperatures for pH change. As a result, soil pH should be corrected at least 1 year before planting. Table 5.3 provides estimates of the amount of elemental S required to shift the pH to a final pH of 5.0.

As shown in Table 5.3, there are two variables that affect the amount of elemental S needed to drop the pH. The first is the initial pH of the soil and the

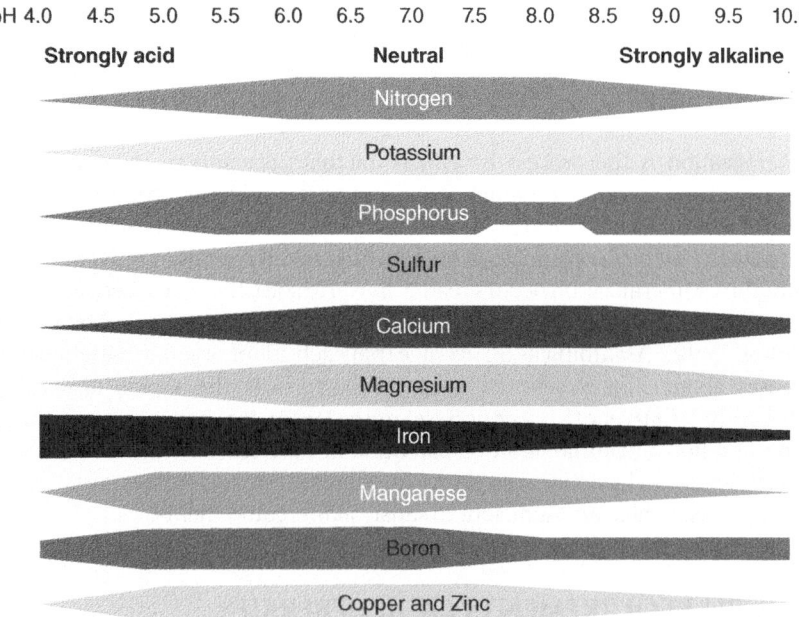

Fig. 5.2. Effect of soil pH on availability of nutrients for crops. (Adapted from Extension, 2017.)

Table 5.3. Estimated levels of elemental S required to change the pH of soil from 6.5, 6.1 or 5.7 to a desired pH of 5.0 according to the soil's CEC ranging from less than 14 to more than 25 meq per 100 g. (Adapted from Horneck *et al.*, 2004.)

Desired soil pH	CEC (meq per 100 g)	Amount of elemental S required (t/ha) at current soil pH of:		
		6.5	6.1	5.7
5.0	<14 (sandy)	2.02–2.35	1.23–1.68	0.56–0.90
5.0	14–25 (silt loam)	2.69–3.14	1.79–2.24	0.90–1.35
5.0	>25 (clay loam)	4.03–4.93	2.80–3.36	1.35–1.68

second is the CEC of the soil. The higher the difference between the initial pH and the desired pH, and the higher the CEC (or the buffer capacity of the soil), the more elemental S will be required to adjust soil acidity. If more than 3.4 t of S is needed per hectare, the dosage should be split. The elemental S needs to be thoroughly mixed and incorporated in the first 15–20 cm of the soil. Studies done over 2 years on half-high 'Northblue' blueberries showed that only when peat was used, either as a soil amendment or as mulch, was the impact on soil pH during the first year significant on the top 5 cm of the soil (pH 4.3–4.5 versus 5.9 for the control). In contrast, sawdust caused little change in pH in the top 5 cm (pH 5.7 versus 5.9 for the control). Strik *et al.* (2017) reported a rise in soil pH from 4.9 to 6.9 when on-farm compost was added as a pre-plant amendment or as part of the mulching programme in various NHB cultivars grown under organic management. Similarly, Karp *et al.* (2006) reported that after 5 years of being used as a mulch or soil amendment, peat altered the pH at a depth of 0–10 cm from 5.9 to 4.3–5, while the pH in deeper layers (10–15 cm) ended up between 5.8 and 6.1 with the use of peat. When the effects of hardwood woodchip fines (2–5 cm in length) were studied for 3 years in SHB 'Star' blueberries planted in three sites, the soil pH dropped in the first season by 0.4–0.7 pH units; however, in the second season this pH level increased an average of 0.2 pH units (Cox, 2009). After one season, experiments on the NHBs 'Concord' and 'Pioneer' by Kramer *et al.* (1940) reported no change in pH with respect to control soil (pH 4.5) with various mulches (peat/sawdust, pine leaves, oak leaves or straw) and *Lespedeza* cover.

The most common forms of N fertilization (urea, $(NH_4)_2SO_4$) acidify the soil. Soil pH is affected by the transformation of inorganic N sources and the uptake of NH_4^+ and NO_3^- by plants. H^+ is released into soils through the nitrification of NH_4^+ to NO_3^- in the soil and the uptake of NH_4^+ by plants; consequently, high NH_4^+ concentrations are associated with a low soil pH. In contrast to the uptake of NH_4^+, the uptake of NO_3^- by plants releases OH^- and increases soil pH (Zhao *et al.*, 2014). $(NH_4)_2SO_4$ is known to be more acidifying than urea (Fig. 5.3), because it produces twice as many H^+ ions from nitrification (Hart *et al.*, 2013), and it is less prone to leaching. A survey done in Oregon blueberry

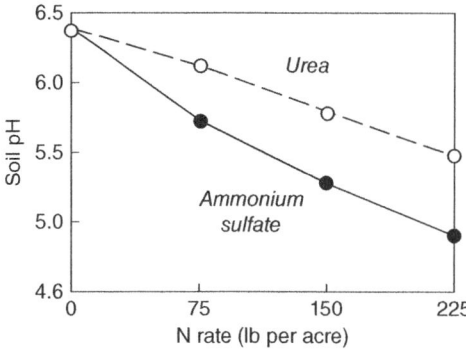

Fig. 5.3. Effect of 5 years of fertigation with different rates of urea or $(NH_4)_2SO_4$ on soil pH of 'Bluecrop' blueberries (1 lb per acre = 1.12 kg/ha). (Adapted from Vargas and Bryla, 2015.)

fields showed that the average soil pH was 5.46 when 45–110 kg N/ha were used, and 4.92 when 340–500 kg N/ha were applied (Scagel and Yang, 2005).

H_2SO_4, incorporated through drip irrigation, acidifies the soil more quickly than elemental S, especially in weakly buffered soils (sandy, low in organic matter). However, this material is hazardous and difficult to use (Horneck *et al.*, 2004). The application of 1.6 litres H_2SO_4 is equivalent to 1 kg elemental S.

ROLE OF MYCORRHIZAL FUNGI IN BLUEBERRY NUTRITION

Fungi interact with plant roots in various ways, from mutualistic symbioses (i.e. when both organisms live in direct contact with each other and establish mutually beneficial relationships) to parasitism. Plants and fungi have co-evolved since the origin of terrestrial plants and the concept of a mutualism–parasitism continuum is useful to describe the extended range of relationships that have developed over evolutionary times. Mycorrhizal fungi are a heterogeneous group of taxa that establish symbioses with over 90% of all plant species (du Jardin, 2015). Ericoid mycorrhizal fungi form symbiotic associations with blueberry roots and help them prosper in soils with low pH, low NO_3, low Ca and high organic matter (Vega *et al.*, 2009). Mycorrhizal inoculation increased plant, root and shoot dry weight without influencing shoot:root ratios, but leaf photosynthetic rate, transpiration and water-use efficiency were not affected by mycorrhizal inoculation in blueberries (Yang *et al.*, 2002). Mycorrhizae increase the uptake of soil nutrients and the efficiency of fertilizer application, improve water use and protect the blueberry plant from toxic levels of elements, such as Al, whose concentration increases as pH decreases (Scagel and Yang, 2005). Mycorrhizal associations also increase the ability of ericaceous plants to tolerate high Cu and Zn concentrations.

There is increasing interest in the use of mycorrhiza to promote sustainable agriculture, considering the widely accepted benefits of such symbioses to nutrition efficiency (for both macronutrients, especially P, and micronutrients), water balance, and biotic and abiotic stress protection of plants. Recent knowledge also indicates the existence of hyphal networks that interconnect not only fungal and plant partners but also individual plants within a plant community. This could have significant ecological and agricultural implications as there is evidence that the fungal conduits allow interplant signalling. In order to reap the benefits of the mycorrhizal associations, crop management practices and plant cultivars should be adapted to the interaction with microorganisms (du Jardin, 2015).

Enhanced N nutrition from ericoid mycorrhizas has been found to be the major benefit for ericaceous plants. The fungus may take up N as NH_4^+ or NO_3^- or in the form of organic N compounds such as amino acids and peptides. Ericoid mycorrhizal fungi can also increase host plant P uptake and may contribute to plant Fe uptake. The ability of ericoid mycorrhizae to detoxify phenolic compounds and their resistance to toxic metals such as Cu is believed to be involved in the ability of their host plants to tolerate environmental stresses (Caspersen *et al.*, 2016). The ability of these fungi to enhance the uptake of soluble inorganic N and P and utilize organic or insoluble N and P substances in the soil can be important in blueberry nutrition. Ericoid mycorrhizae are capable of transferring C and N simultaneously to the host plant when organic N sources are applied, therefore offsetting a portion of the C drain required to sustain fungal growth (Yang *et al.*, 2002). In highbush blueberries, Valenzuela-Estrada *et al.* (2008) found similar colonization by mycorrhizae in the first two root orders (fine: less than 50 μm diameter, absorptive roots) and the decline in colonization in third- and fourth-order root branching.

Mycorrhizal colonization of blueberries varies significantly with cultivar, rate of fertilizer application, and the amount and type of soil organic matter present in the soil. Usually, increasing the amount of fertilizer decreases mycorrhizal colonization, and the effects are cultivar specific (Hanson, 2006). Addition of organic materials (e.g. rotted sawdust, peat), increases in N fertilization (Scagel and Yang, 2005) and high total soil N content (Sadowsky *et al.*, 2012) and NH_4^+ levels (Scagel and Yang, 2005; Sadowsky *et al.*, 2012) can reduce ericoid mycorrhizae root colonization in highbush blueberries. In contrast, the lack of effect of N fertilizer on the occurrence of ericoid mycorrhizae in blueberries in the field or in natural habitats found by some authors might partly be related to N immobilization by organic mulches or litter layers (Caspersen *et al.*, 2016).

Mycorrhizal associations are most prevalent in natural environments but can be important in nursery and commercial plantings. Surveys of highbush blueberry fields in Oregon have shown large variations in mycorrhizal infection levels (0.5–44% of total root length), with most colonization occurring in the top 15 cm of the soil profile (Scagel and Yang, 2005). Levels of root colonization can be doubled if plants are inoculated in the nursery. Inoculation in the nursery of container-grown blueberry plants increased total plant biomass in six of the seven highbush blueberry cultivars studied (Scagel, 2005). The colonization of blueberry roots by ericoid mycorrhizae may have some level of host-fungus specificity, as there have been reports of variation among mycorrhizal isolates in their ability to increase nutrient solubility or plant uptake. Roots on highbush blueberry cultivars that fruit early in the season tended to have higher levels of colonization than cultivars that fruit later in the growing season (Scagel and Yang, 2005).

ESTIMATING THE NUTRIENT NEEDS OF BLUEBERRIES

The total fertilizer needs of a blueberry planting can be determined by calculating the demand as well as the supplies of each element. The relationship between demand, supplies and fertilization needs is established through the following equation:

Fertilizer need = (nutrient demand – nutrient supplies)/efficiency

In the case of young blueberry plantings that have not yet reached maturity, their vegetative mass as well as fruit production is expanding each year. During this period, the balance between vegetative and reproductive growth changes from one season to the next. This fact has to be considered in estimating the amount of fertilizer needed to satisfy nutrient needs. For this reason, age- or size-based fertilizer recommendations that are considered appropriate for the typical growth patterns and yields at each location have been developed in different regions of the world (Tables 5.4 and 5.5). These data reflect the large differences in soil nutrient supply, plant growth and expected yields that exist in different blueberry-producing regions.

Table 5.4. Recommended rates of N for highbush blueberry fields of different ages in Michigan and Oregon.

Years in the field	Michigan (kg/ha)[a]	Oregon[b]	
		g per plant	kg/ha
2	16.8		
3	–	22.7	
4	33.6	25.5	
5	–		112.1
6	50.4		140.1
7	–		162.5
8	72.9		184.9

[a]Data from Hanson and Hancock (1996).
[b]Data from Hart et al. (2006).

In the case of a mature blueberry planting, the fertilizer needs will be determined by the amount of nutrients extracted from the planting, which is a function of the nutrient content in the harvested fruit plus the material removed by pruning (assuming it is not reincorporated into the orchard) (Tables 5.6 and 5.7). The total nutrient supply in a planting is dependent on the natural fertility of the soil, the nutrient content of the irrigation water, and the addition of any nutrient-containing material (organic or inorganic, such as mulch).

The equation above provides the basis for a gross estimation of the total amount of each nutrient that would have to be added as fertilizer. As an example, if for a mature 'Elliott' planting we assume a yield of 10 t, the nutrient extraction by the fruit will be highest for N and K (10.8 and 11.2 kg N and potassium oxide (K_2O), respectively), while the extraction will be lowest for P, Ca and Mg (Table 5.8). However, if we assume that the grower removes 20% of

Table 5.5. Suggested fertigation rates for blueberries. (Adapted from Krewer and NeSmith, 1999.)

Plant diameter (cm)	Fertilizer doses (g per plant per week)		
	N	K_2O	P_2O_5
30	1.0	0.5	0.5
60	1.5	0.75	0.75
90	2.0	1.0	1.0
≥120	2.5	1.2	1.2

Table 5.6. N, K and P concentrations (percentage of dry weight) in different plant parts of mature (8–10-year-old) blueberry cultivars in mid-winter. (Adapted from Bañados *et al.*, 2006a.)

Plant part	'O'Neal'			'Bluejay'			'Brigitta'			'Elliott'		
	N	K	P	N	K	P	N	K	P	N	K	P
Bud, flower	2.7	0.85	0.26	2.6	0.64	0.24	1.8	0.52	0.19	1.7	0.38	0.18
Bud, vegetative	2.0	0.47	0.18	2.0	0.36	0.16	1.5	0.23	0.12	1.7	0.25	0.14
Wood, 1-year-old	0.9	0.36	0.07	1.0	0.30	0.09	0.7	0.29	0.07	0.9	0.37	0.07
Wood, 2-year-old	0.9	0.30	0.06	1.0	0.25	0.07	0.6	0.24	0.05	0.9	0.33	0.06
Wood, 3-year-old	0.7	0.19	0.05	0.7	0.19	0.05	0.6	0.18	0.04	0.7	0.23	0.05
Wood, 4-year-old	0.7	0.15	0.05	0.6	0.19	0.04	0.5	0.17	0.04	0.6	0.18	0.05
Crown	0.7	0.19	0.05	0.7	0.19	0.05	0.6	0.18	0.04	0.7	0.23	0.05
Roots	1.6	0.19	0.20	1.2	0.20	0.12	1.2	0.21	0.09	1.0	0.24	0.12
Significance[a]	***	***	**	***	***	**	***	***	**	***	**	**

[a]Significance of means in each column: ** $P < 0.01$; *** $P < 0.001$.

Table 5.7. Dry matter and dry-weight partitioning (percentage of total) in different plant parts of mature (8–10-year-old) blueberry cultivars during mid-winter. (Adapted from Bañados *et al.*, 2006a.)

	'O'Neal'		'Bluejay'		'Brigitta'		'Elliott'	
Plant part	Dry weight (g)	% of total	Dry weight (g)	% of total	Dry weight (g)	% of total	Dry weight (g)	% of total
Bud, flower	42.5	0.50	10.2	0.15	25.3	0.33	16.4	0.45
Bud, vegetative	9.6	0.11	0.9	0.01	6.0	0.08	2.0	0.05
Wood, 1-year-old	551.2	6.41	127.6	1.89	450.3	5.94	247.6	6.73
Wood, 2-year-old	658.9	7.66	198.4	2.94	722.6	9.53	269.2	7.31
Wood, 3-year-old	882.3	10.26	223.0	3.30	614.0	8.10	537.0	14.59
Wood, 4-year-old	2097.4	24.37	2815.6	41.68	1533.9	20.24	487.9	13.26
Crown	2758.6	32.06	1833.7	27.15	2665.1	35.16	1116.0	30.33
Roots	1602.7	18.63	1545.6	22.88	1562.4	20.62	1004.0	27.28
Total plant	8603.2	100.0	6755.0	100.0	7579.6	100.0	3680.1	100.0

1–4-year-old wood by pruning and this material is taken out from the field, this would amount to an extraction of 2.27 kg N, 0.17 kg phosphorus pentoxide (P_2O_5) and 0.76 kg K_2O (Table 5.8). The fruit would then, in this scenario, represent 82.6, 93.9 and 93.6% of the annual removal of N, P and K from this field, respectively.

The efficiency of N fertilization for mature NHB 'Bluecrop' was estimated at 32% in Michigan (Retamales and Hanson, 1989) and 22–43% in Oregon (Bañados *et al.*, 2006b). The quantity of fertilizer that needs to be applied must be increased in order to compensate for nutrient loss due to runoff, weed uptake, volatilization and immobilization. In order to account for various factors that influence the rate of fertilizers to apply, Hirzel (2008) established a 'dosage factor', which varies according to the nutrient, the supply of the nutrient from the soil and reserves, the efficiency of application, the degree of fixation (for P and K) and presence of weeds (Table 5.8).

The amount of N fertilizer calculated in this exercise is much lower than the 185 kg N/ha/year suggested for Oregon by Hart *et al.* (2006) and lower than the 73 kg N/ha/year recommended for mature blueberry fields (more than 8 years old) by Hanson and Hancock (1996) in Michigan (Table 5.4). Although some fields in Oregon have been reported to reach 44.8 t/ha, which would remove 48.4 kg N/ha/year and according to our calculations would require 103–206 kg N/ha/year to be applied as fertilizer, a survey of 100 fields in Oregon, of which 56% of the plantings were more than 8 years old, showed that highbush blueberry growers apply an average of 193 kg N/ha/year (Scagel and Yang, 2005). Most growers surveyed (96%) used overhead

Table 5.8. Estimation of nutrients extracted in a mature blueberry field cultivar, 'Elliott', by the fruit (yield = 10 t) and wood (20% of aboveground wood removed by pruning), and the amount of nutrients applied through fertigation needed to replace this extraction.

Item	Level of nutrient in fruit or vegetative tissue[a]				
	N	P_2O_5	K_2O	CaO	MgO
Concentration in fruit (mg per 100 g fresh fruit)[b]	108	26	112	13.5	9
Nutrient removed (kg/ha) for 10 t of fruit[b]	10.8	2.6	11.2	1.35	0.9
Removed by pruning: 20% wood over 1 year old (kg)[c]	2.27	0.17	0.76	–	–
Total removed: fruit + wood (kg/ha)	13.07	2.77	11.96	1.35	0.9
	Fertilizer dose (kg/ha) according to soil fertility				
Low-fertility soil					
Dosage factors[d]	3.8–4.6	7.2–9.0	5.9–6.7	11.9–14.8	11.1–13.3
Estimated rate	50–60	20–25	70–80	16–20	10–12
Medium-fertility soil					
Dosage factors[d]	3.1–3.4	5.4–6.5	4.2–5.0	8.9–11.1	6.7–8.9
Estimated rate	40–45	15–18	50–60	12–15	6–8
High-fertility soil					
Dosage factors[d]	2.3–2.7	3.6–5.1	2.9–3.3	5.9–7.4	4.4–5.6
Estimated rate	30–35	10–14	35–40	8–10	4–5

[a]In order to obtain elemental levels (P, K, Ca and Mg), the values of P_2O_5, K_2O, CaO and MgO must be divided by 2.29, 1.2, 1.4 and 1.67, respectively.
[b]Data from Hirzel (2008); low-fertility soils correspond to sandy soils and low organic matter.
[c]Data from Bañados *et al.* (2006a)
[d]Data from Hirzel (2008), who considered the soil supply, supply from reserves, efficiency of application, fixation (for P and K) and presence of weeds.

irrigation, while the calculations in Table 5.8 were based on fertigated plantings, which have been reported to have a more efficient use of the fertilizers. The survey also found that the N fertilization rate increased by 6 kg/ha/year for every year of age (Scagel and Yang, 2005).

In the case of P_2O_5, the estimations from the data provided in Table 5.8 (10–25 kg/ha) cover levels recommended for soil tests in the range of more than 30 ppm for mineral soil, and more than 10 ppm for organic soils (see Table 5.10 in Hanson and Hancock, 1996). For K_2O, only fertilizer rates suggested for the 0–10 ppm range by Hanson and Hancock (1996) would not be covered with rates calculated using Table 5.8.

SOIL AND FOLIAR ANALYSIS

Several tools are used to establish the nutrient status of blueberry fields, including soil and foliar analysis, spectrometry and visual symptoms. Visual symptoms should not be used in commercial operations because, by the time symptoms are visible, some of the negative effects (i.e. lower yields, reduced growth and lower fruit quality) have already taken place. Soil analysis is commonly used about 1 year before planting a new field in order to establish the initial fertility of the soil, the organic matter content and especially the pH. If the pH is >5.8, it should be corrected at least 1 year before planting. Applications of elements such as K and Mg are also more available to roots if applied before planting and mixed with the soil, rather than being broadcast over the surface after planting. Once the blueberries are planted, the use of soil analysis is usually restricted to checking soil pH and salinity (Hart *et al.*, 2006). For this purpose, soil samples should be obtained between the canopy drip line and the plant crown (Hart *et al.*, 2006). The area sampled should be restricted to a uniform soil type and condition within the field (Stiles and Reid, 1991). The area included should not exceed 5 ha. The surface 2–3 cm of soil must be scraped away, and then samples collected from the 2–40 cm depth. Each soil sample submitted for analysis should be a composite of 20–40 subsamples taken throughout the area (Hanson and Hancock, 1996).

Foliar analysis is used to determine the concentration of elements in the plant at a certain moment in the season. They can be used to estimate fertilizer needs, diagnose deficiencies and evaluate the performance of fertilizer programmes. Annual foliar analysis is recommended. In highbush blueberries, Hanson (1987) found that the nutrient levels in soil samples had a weak correlation with leaf nutrient concentrations in NHB blueberry fields in Michigan. In samples from plants of various ages, these correlations were 0.084 for P, 0.239 for K, 0.088 for Ca and 0.132 for Mg. For plants less than 7 years old, the correlations were somewhat higher: 0.333 for K, 0.228 for Ca and 0.294 for Mg.

In order to be useful, tissue collection must follow strict procedures including timing, type of tissue and number of plants to be sampled. Standards have been developed for a period when leaf nutrient levels are stable (first 2 weeks of harvest), which in the case of NHB blueberries is usually late July to mid-August in the northern hemisphere and mid-December to mid-February in the southern hemisphere. Fully expanded leaves from the mid-portion of current-season shoots should be collected. Usually two to five leaves are collected from 10–50 plants distributed randomly in the field (avoiding borders). The plants should represent a uniform condition (e.g. age, cultivar, soil, irrigation system). If a nutritional disorder is suspected, leaves from affected plants should be collected as one sample and compared with samples of 'normal' plants. Only one cultivar should be included in a sample. The sample should not represent a field larger than 5 ha (Krewer and NeSmith, 1999; Hart *et al.*, 2006).

Recommended foliar nutrient concentrations have been developed for different producing regions (Table 5.9). In general, the recommended ranges for the various elements are usually, but not always, similar. For example, recommended levels of P, B and Fe in Missouri are lower than in other regions, and recommended Mn levels are higher in Michigan than elsewhere (Table 5.9). These differences indicate the need to develop and use local standards. Weather (high and low temperatures, rainfall), fruit load, shoot growth, soil data, soil moisture, pruning intensity, yield, and insect and disease load can all affect plant functioning and the nutrient status of the plant (Stiles and Reid, 1991). In a recent 2-year study on the evolution of nutrients in NHB blueberries ('Aurora', 'Bluecrop', 'Draper', 'Duke', 'Legacy' and 'Liberty') over the season, Strik and Vance (2015) found that the pattern of nutrient changes was similar between organic and conventional sites, but they found fewer differences in nutrient concentrations among cultivars at the organic site. In addition, the cultivar had a significant effect on all fruit nutrients except for P at the conventional site.

For managing nutrition, the fruit industry is looking for low-cost, instantaneous and easy-to-implement techniques suited for routine analysis of plant

Table 5.9. Sufficient or normal foliar concentrations of nutrients for highbush and rabbiteye blueberries in different locations.

Nutrient	Highbush blueberry			Rabbiteye	
	Oregon[a]	Michigan[b]	Missouri[c]	South-east USA[d]	Georgia[e]
Macroelements (%)					
N	1.76–2.00	1.70–2.10	1.50–2.10	1.20–1.70	1.20–1.70
P	0.10–0.40	0.08–0.40	0.07–0.12	0.08–0.20	0.08–0.17
K	0.41–0.70	0.40–0.65	0.40–0.80	0.35–0.60	0.28–0.60
Ca	0.41–0.80	0.30–0.80	0.40–0.90	0.25–0.70	0.24–0.70
Mg	0.13–0.25	0.15–0.30	0.10–0.30	0.14–0.20	0.14–0.20
S	0.11–0.16	0.12–0.20	0.10–0.20	0.11–0.25	–
Microelements (ppm)					
B	31–80	25–70	20–50	12–35	12–35
Cu	5–15	5–20	–	2–10	2–10
Fe	61–200	60–200	40–70	25–70	25–70
Mn	30–250	50–350	40–250	25–100	25–100
Zn	8–30	8–30	–	10–25	10–25

[a]Data from Hart *et al.* (2006).
[b]Data from Hanson and Hancock (1996).
[c]Data from Fuqua *et al.* (2005).
[d]Data from Plank and Tucker (2000).
[e]Data from Krewer and NeSmith (1999).

tissues. In this context, attenuated total reflectance Fourier transform infrared spectroscopy (wave number range 375–7500/cm) was carried out on the petioles of the vineyard cultivars 'Chardonnay', 'Semillon' and 'Shiraz' in New South Wales, Australia. Good predictive models were produced for all macronutrients, with values of r^2 (coefficient of determination) between 0.961 and 0.849 and predictive capacities for Mg > K > N > Ca > P > S, while the predictive capacity of the models for Na and micronutrients was lower (0.835–0.612) and in decreasing r^2 was Zn > Na > Fe > Mn > B > Cu (Smith *et al.*, 2014).

The absorption features in visible, near-infrared and short-wave infrared spectral ranges have been used to predict foliar N in various crops. Results from trials in Nova Scotia, Canada, concluded that reflectance spectra may be used to estimate and ultimately map foliar N in wild blueberries (*V. angustifolium*) with r^2 values between 0.71 and 0.78. Wavelengths around 550, 610, 1510, 1690, 1730, 1980 and 2030 nm were used to develop models for *in situ* foliar N estimations (Maqbool *et al.*, 2012). Similar developments are likely to occur in the coming years in cultivated blueberries.

METHODS OF APPLYING FERTILIZERS

Soil application

Until the last decade, most fertilizer applied to blueberry plantings was broadcast on the soil surface. In most instances, the applications were concentrated at the beginning of the season when root growth was scarce and the chances of nutrient losses were high. In the case of N, it was shown that, despite the low pH (which would reduce nitrification rates), NO_3^- formation was high and an important proportion of the N applied in NH_4^+ form was lost through leaching (Retamales and Hanson, 1990). Studies demonstrating the inefficiency of this system induced growers to implement changes to this practice. Among these innovations are: the use of controlled-release fertilizers, the use of split applications during the growing season and fertigation.

In the case of young plantings, these are expanding their vegetation as well as their fruit production. In this period, the balance between vegetative and reproductive growth changes from one season to the next, so is difficult to estimate the amount of fertilizer needed to satisfy these needs. Some guidelines published in the literature show the wide range of recommended rates in different zones (Table 5.4). The data reflect the large differences in soil nutrient supply, plant growth and expected yields.

The use of fertigation (the application of fertilizers through pressure irrigation systems) has increased markedly in blueberry production in recent years. In many producing regions, blueberries are established on ridges with sawdust, plastic or bark chip mulch and drip irrigation. The advantages of fertigation include reduced delivery costs (no tractors or spreaders needed), and the ability to target application of specific nutrients during particular stages of crop development, improved efficiency of fertilization, minimal losses to leaching, optimization of the plant's nutritional balance by supplying nutrients directly to the root zone, reduced potential for fertilizer burn and greater control of nutrient concentration in the soil solution. However, disadvantages include the costs associated with the need for higher quality fertilizers (i.e. purity and solubility), greater technical skills in the personnel and investments in equipment required to inject the fertilizer through the irrigation system. System costs are even higher when injection of corrosive materials such as H_2SO_4 and acidified fertilizers are needed (Bryla and Strik, 2015).

In fruit crops, dramatic increases in N-use efficiency have been measured using fertigation versus surface application. Uptake efficiency was increased two to three times over previous surface fertilizer applications. A study was conducted on NHB ('Blueray' and 'Duke') and half-high ('Northblue') blueberries planted in a silt loam soil where equivalent rates of N, P, K and Mg were applied through fertigation or granular surface application (Finn and Warmund, 1996). After 3 years, greater fruit yield and plant volume were obtained under the fertigation regime, without any difference in N leaf levels between the two treatments. Apparently, the greater plant volume from fertigation resulted in higher yields. As foliar N levels were not affected by the fertilization regime, but plant volume and fruit load were greater with fertigation, N uptake on a per plant basis was probably higher in the fertigated plants. A more consistent availability of N in the root zone would have allowed plants with fertigation to utilize N more efficiently.

Due to low application efficiency, higher N rates are recommended for fertigation in the first 2 years after planting. After this, rates for fertigation and granular fertilizer application are similar (Bryla and Strik, 2015). Bryla and Machado (2011) and Machado *et al.* (2014) found that plant growth declined with higher rates of granular $(NH_4)_2SO_4$ during the first 2 years after planting and suggested that this reduced growth was the result of high salinity from the fertilizer (up to 8 dS/m). They concluded that fertigation was safer in terms of soil salinity (always less than 1 dS/m) but was less efficient at lower N rates than granular fertilizer because at least half of the fertilizer delivered through drip emitters was located outside the root zone.

The effects of N rate (0–150% of current production guide rates) and method of application (fertigation versus broadcast) on the performance of NHB 'Duke' during the first 4 years after planting were studied in south-coastal British Columbia (Ehret *et al.*, 2014). N was applied with three equal applications of broadcast granular $(NH_4)_2SO_4$ each spring or by fertigation (drip irrigation) with ten equal applications of liquid $(NH_4)_2SO_4$ injected every 2 weeks from early spring to late summer each year. Yield increased with increasing N rate during years 2 and 3 of fruit production. The yield response as well as flower number and plant size were greater with fertigation than with broadcast fertilizers. Fruit firmness also improved consistently with increased N rates, while fruit size either increased or decreased, depending on year. There were no effects of N on fruit oxygen radical absorbance capacity (ORAC), titratable acidity or soluble solids. However, the composition of fruit anthocyanins changed, with concentrations of seven anthocyanins decreasing and three others increasing with N rate.

A trial carried in western Oregon, during the first 5 years of fruit production (years 3–8) compared fertigation using liquid N sources, including $(NH_4)_2SO_4$ and urea, with granular fertilizer applications in NHB 'Bluecrop' blueberries (Vargas and Bryla, 2015). Plants were grown on raised beds and mulched every 2 years with sawdust. Liquid fertilizers were injected weekly through a drip system from mid-April (bud break) to early August (60 days prior to the end of the growing season). Granular fertilizers were applied on each side of the plants, in three split applications from mid-April to mid-June, and washed into the soil using microsprinklers. Each fertilizer was applied at three N rates, which were increased yearly as plants matured (63–93, 133–187 and 200–280 kg N/ha) and compared with non-fertilizer control treatments. Canopy cover and fresh pruning weight were greater with fertigation and often increased with N rate when plants were fertigated but decreased at the highest rate when granular fertilizer was applied. Yield increased with N rate and was 12–40% greater with fertigation. Leaf N was greater with fertigation in four out of five years and averaged 1.68% (fertigation) versus 1.61% (granular). Leaf N was often greater with $(NH_4)_2SO_4$ and increased as more N was applied. Soil pH declined as N rates increased and was lower with granular fertilizer than with fertigation during years 1–3 of fruit production and lower with $(NH_4)_2SO_4$ than with urea in every year but one. Soil electrical conductivity was less than 1 dS/m in each treatment but on average was two- to three-fold greater with granular fertilizer and 1.4–1.8 times greater with $(NH_4)_2SO_4$ than with urea. Overall, total yield averaged 32–63 t/ha in each treatment over the first 5 years of fruit production and was greatest when plants were fertigated at rates of 63–93 kg N/ha/year. These results show that fertigation increased plant growth over granular fertilizer. The effect was more probably due to reducing salt stress and maintaining near optimum levels of $NH_4{}^+$–N within the root zone throughout the season.

Suggested fertilizer rates for application through fertigation are shown in Table 5.5. The following fertilizer split within the season has been recommended (Hirzel, 2014): bud break to fruit set, 10%; fruit set to pink fruit, 30%; pink fruit to harvest, 40%; and postharvest, 20%.

The performance of controlled-release N fertilizers was equivalent to regular fertilizers in NHB blueberries (Hanson and Retamales, 1992). S-coated urea had the greatest and $(NH_4)_2SO_4$ the lowest plant growth (urea was intermediate) when applied to 1-year-old 'Tifblue' rabbiteye blueberries (Patten *et al.*, 1988). The main advantages of slow-release fertilizers are that they have reduced risk of fertilizer burn to the plants and that they require less frequent applications to satisfy nutrient needs (Krewer and Ruter, 2009); however, their higher cost may offset these benefits.

Foliar applications

Foliar application of nutrients is a means of rapidly supplying nutrients directly to tissues (foliage, flowers or fruit) at times when a quick response is needed. Although soil treatments last longer, soil-applied micronutrients are unavailable to the plants under some conditions; therefore, foliar sprays may be necessary (Strik *et al.*, 2010). For example, low air temperatures and cold soils in the spring often reduce nutrient availability. In certain seasons and production areas, the demand for certain elements required for the rapid development of leaves and shoots can exceed the supply from roots and reserves. Soil pH can decrease nutrient availability, and foliar nutrient application may ameliorate the deficit faster than correction of soil pH. Foliar sprays may end up being the best way to supply a given element either because the soil supply is insufficient or because there is a need to precisely control the time and rate at which the element is available to the plant. Foliar applications are often timed to coincide with specific vegetative or fruiting stages of growth, and the fertilizer formula is adjusted accordingly. Applications may also be used to aid plants in their recovery from transplant shock, hail damage or the consequences of other weather extremes (Kuepper, 2003). However, foliar applications are more expensive than soil fertilization on a nutrient unit base and usually must be repeated several times to be effective for an entire season (Hart *et al.*, 2006). Therefore, foliar sprays should be aimed at accomplishing a particular outcome in response to a specific need. As a result, foliar sprays should be considered as supplements to, and not replacements for, soil fertilization (Stiles and Reid, 1991).

A detailed discussion on foliar application is presented in Chapter 7 (this volume). In general terms, to get the most efficient results from nutrient sprays and avoid damage to the crop, three considerations must be followed: (i) the proper application rates must be established; (ii) the modes of application must be considered (e.g. adjuvants, volume, time of season, time of day, droplet size); and (iii) the effectiveness of the foliar application should be evaluated based on target tissues and specific environmental conditions at the time of application.

Certain micronutrients such as B and Cu can cause extensive damage to blueberry plants if applied foliarly at higher amounts than necessary for plant growth (Strik *et al.*, 2010). Research has shown that foliar N sprays can damage leaves and are an inefficient way to apply this element to highbush blueberries. No more than 5% w/v urea, equivalent to 16 kg N/ha, can be applied to blueberry plants without burning leaves. Because of the waxy cuticle of blueberry leaves, they are not very effective in taking up N. Less than 50% of the applied N would enter the plant via the leaves (Hart *et al.*, 2006). A study on NHB blueberries in which N-enriched potassium nitrate was applied foliarly to 3-year-old 'Jersey' potted plants and a liquid NPK (10:10:10 liquid formulation; Nachurs Alpine Solutions, Marion, Ohio) was sprayed for four seasons on mature commercial plantings of NHB 'Bluecrop', 'Jersey' and 'Rubel' blueberries showed the following: (i) the foliar N obtained from the spray amounted to only 0.7% (single application) or 1.2% (double application) of the total elemental N in leaves; (ii) the contribution of foliar-derived translocated N to the N status of the new leaf growth was minimal (less than 0.3% of total leaf N); and (iii) foliar sprays increased NPK leaf levels but had no effect on yield, berry weight or soluble solids (Widders and Hancock, 1994).

Studies at full canopy development in various fruit crops have shown that, depending on the volume of water used, 44–58% of the spray ends up on leaves, and only around 2% of the volume sprayed reaches the fruit (Hall, 1991). However, to be effective, the nutrient has to penetrate the leaf (Petit-Jimenez *et al.*, 2009). This reduces the efficiency of foliar sprays even further. If the application aims to deposit nutrients on fruit, as is the case of Ca, then the efficiency of a foliar spray is expected to be very low.

ORGANIC MANAGEMENT OF NUTRITION

According to a recent survey, the proportion of the worldwide highbush blueberry area devoted to organic production was estimated to be 5% in 2010/11, with that from the USA and Chile amounting to almost 80% (Strik, 2014). Although organic blueberry production is expected to have reduced yields (Strik, 2014), there are few reports comparing yields in organic versus conventional blueberry fields. Four years after establishing organic and conventional rabbiteye blueberries in Georgia, plants grown under organic cultivation had 90% of the growth and 70% of the yield of conventionally grown plants (Tertuliano *et al.*, 2012). From a regression analysis of grower survey data for 717 highbush blueberry fields in south-central Chile (Maule region), Retamales *et al.* (2015) concluded that conventional fields had a higher probability of obtaining larger yields than organic ones.

Growers with organic fields utilize a 'feed-the-soil' system that incorporates cover crops, peat, compost, fish meal, humus, residues from agro-industrial processes and manures rich in naturally produced N. The fertilizers may be produced on-farm, acquired from local farms, enterprises or communities, or purchased as commercial products (Caspersen *et al.*, 2016). These natural nutrients promote the growth of beneficial soil microorganisms and supposedly do not harm the symbiotic endomycorrhizal fungi associated with the blueberry root system. These decomposers process biomass materials and indirectly relay N, P, K and other available plant nutrients through the crop rhizosphere (Wang *et al.*, 2008).

In organic production of blueberries, high levels of soil organic matter are especially important not only for their contribution to the soil's ability to retain and supply moisture to the crop, buffer pH and release nutrients through decay, but also because they are a desirable environment for the symbiotic mycorrhizal fungi that assist blueberry roots in absorbing water, N, P and other minerals (Yang *et al.*, 2002). Green manures in advance of planting can play an important role in cycling organic matter into the soil system, as can applications of composts and livestock manures (Kuepper and Diver, 2004). Livestock manure and manure/bedding mixtures can be adequate sources of nutrients for blueberries. The nutrient content of manure needs to be known in order to estimate the application rates; however, there is great variability in their nutrient content depending on animal species and their diet, the presence of bedding or other additions, and handling procedures (Hanson and Hancock, 1996). Strik (2016) found that the addition of yard debris compost to the mulch increased soil and leaf K but had little effect on plant N.

Before the establishment of an organic blueberry culture in soils with a pH above the optimum level for blueberries, elemental S may be added to increase soil acidity. The minimum recommended time interval between S application and planting varies from 6 months to 1 year, the time needed for soil pH reduction depending on the S formulation and amount of S added (Gough, 1994). For organic blueberries in particular, incorporation of peat or pine bark has been recommended to fulfil, at least partially, the pH reduction needed to restrict negative effects of S on soil organisms (Caspersen *et al.*, 2016). In organic blueberry culture, the long-term development of soil pH has been shown to depend on both the type of fertilizer and the type of mulch used. For example, soil pH was lower with fish emulsion than with feather meal as a fertilizer, while compost + sawdust mulch raised the soil pH (Larco *et al.*, 2013). Sullivan *et al.* (2014) concluded that the ability of maintaining a pH <5.5 is the most important characteristic for assessing the suitability of composts for highbush blueberry production.

Once a blueberry planting is established, supplemental N is the greatest concern in organic production, followed by K. Fertilizer recommendations are based on foliar analysis. However, Strik and Vance (2015) speculated that tissue nutrient standards may need to be adjusted for some nutrients when plants are grown in organic systems. Organic fertilizers are usually less soluble than inorganic ones. It has been suggested that these fertilizers should be applied 1–4 weeks ahead of the schedule recommended for soluble fertilizers (Kuepper and Diver, 2004). Generally, all of the NH_4^+-N and 25–50% of the organic N will be available for the blueberry plant in the year of application. Usually, N rates should be increased by 50–100% for organic materials because the microorganisms will tie up the N (Hanson and Hancock, 1996). Kuepper and Diver (2004) stated that, despite the slower release of organic-based N, the carry-over from previous seasons probably results in roughly the same amount of N released each season as is applied.

The application of organic fertilizers in organic blueberry production may cause nutrient imbalances, in particular due to high levels of N, K, Ca, Na or chloride (Cl⁻). The high pH and/or electrical conductivity of most organic amendments is also a challenge, and acidification with elemental S is often necessary, leading to increased electrical conductivity and nutrient solubility (Sullivan *et al.*, 2014). However, both the nature of the ions contributing to a high conductivity as well as the electrical conductivity level per se might be critical, as toxic effects appear to be partly responsible for the negative impact of Na and Cl ions on blueberries. Organic manures or composts have been reported to increase levels of available Zn, Fe, Mn, Cu, B and Mo in the soil. One concern related to the use of organic fertilizers is the sensitivity of blueberries to Na, Cl, K, Ca and, for young plants in particular, high amounts of inorganic N. The use of organic mulches in blueberry production may protect the plants against high salt concentrations and weed problems, but these organic amendments may also allow the build-up of toxic forms of Al and Mn in the soil (Caspersen *et al.*, 2016).

Fertilizers are usually split into two or three applications (Kuepper and Diver, 2004). In a trial to compare organic and conventional nutrient management of highbush and rabbiteye blueberry nursery stock, the authors concluded that it was more difficult to supply enough nutrients for optimal growth using organic formulations, and they recommended a constant and steady input of low concentrations of a balanced nutrient mix (Miller *et al.*, 2006). In another trial on mature NHB 'Duke' and 'Liberty' blueberries, Larco *et al.* (2013) applied feather meal and fish emulsion fertilizers at 29 and 57 kg N/ha. Application of organic fertilizer to blueberries affected the macro- and micronutrients in soil and leaves during the first 2 years after planting, but the results varied depending on the fertilizer source and associated nutrient content. They found that feather meal contained 12 times more Ca and seven times more B than fish emulsion, and resulted in higher levels of soil Ca and soil and leaf B in both cultivars, whereas fish emulsion contained three times more P, 100 times more K and 60 times more Cu, and generated higher levels of soil P, K and Cu, as well as higher levels of leaf P and K. Fish emulsion also reduced soil pH. Organic farmers commonly apply liquid fish emulsion directly to the soil, especially in young plantings. Granular feather meal and grain-based products are also commonly applied directly to the in-row area, after opening the weed mat if present. Organic sources of fertilizer contain high levels of nutrients other than N, particularly P, K and Ca. The addition of these nutrients when using organic fertilizers, even when they are not needed by plants, must be considered in these production systems. The cost of organic fertilizers tested by Strik (2016) ranged from US$5.60 to US$17.95/kg N and varied by method of application.

K for blueberries grown under organic management is often adequately provided by decaying mulches. The need for further supplementation should be determined by soil and/or tissue testing. Where additional K is needed, it can be applied in a number of mineral forms, including sulfate of potash magnesia, granite meal and greensand. Some forms of potassium sulfate are also allowed in organic production (Kuepper and Diver, 2004).

Fertigation – the injection of soluble fertilizers through drip irrigation lines – is a common practice in conventional blueberry production. Some materials accepted in organic production such as spray-dried fish protein and poultry protein, as well as several organic liquid fertilizers derived from fish emulsion, seeds, kelp or seaweed satisfy the requirement of complete water solubility of fertilizers (Kuepper and Diver, 2004). Strik (2016) reported that, during a study over several seasons, liquid sources of fish and grain blends were successfully fertigated through drip irrigation with little impact on emitter performance.

Foliar feeding of blueberries is practised by some organic growers and is especially helpful when plants are stressed. Foliar fertilization programmes usually employ seaweed and fish emulsion.

REQUIREMENTS FOR SPECIFIC ELEMENTS AND THEIR DEFICIENCY SYMPTOMS

Macronutrients

In this section, we analyse the different nutrients needed in blueberry production. For each of them, their soil availability and the symptoms of deficiency and/or toxicity are described and rates of application are suggested, as well as the timing and application methods. Greater coverage is given to N and Ca because of their importance in yield and fruit quality.

Nitrogen

Deficiencies in N are the most frequently reported nutrient deficiency in blueberries worldwide. Plants deficient in N are commonly stunted, with low vigour and pale green to chlorotic (yellow) leaves. The chlorosis is uniform across the leaf, with no mottling or pattern. Fewer canes are initiated. Symptoms appear first on lower (older) leaves and will eventually include the entire plant if N is not applied. Leaves drop early and yields are usually reduced (Hanson and Hancock, 1996; Hart *et al.*, 2006).

Excessive N results in plants with numerous, vigorous shoots and large, dark green leaves. During the season, plants may produce several growth flushes. Growth occurring at the end of the season may not harden properly before winter. The tips of these shoots are often killed by low winter temperatures. Plants with excessive N have reduced yields and smaller berries that ripen later (Hanson and Hancock, 1996; Hart *et al.*, 2006).

As discussed above, proper timing of fertilizer applications may increase N use by plants and avoid waste. Urea is one of the most commonly used fertilizers in blueberries. After urea was broadcast near bloom, N levels in the root zone increased for only 6–8 weeks after application, suggesting that multiple applications may be necessary to maintain sufficient soil N levels throughout the period of high demand (bloom to harvest) (Retamales and Hanson, 1990).

As shown previously (Table 5.4), recommended rates of N application vary greatly among various producing regions. While 73 kg N/ha is recommended for mature plantings (7-year-old fields or older) in Michigan (Hanson and Hancock, 1996), 185 kg N/ha is suggested in Oregon (Hart *et al.*, 2006). For highbush blueberries grown under mulch, application rates equivalent to 158–170 kg N/ha in the establishment year and 238–257 kg N/ha for subsequent years have been estimated in Oregon. However, recent trials in that region comparing N sources (urea versus $(NH_4)_2SO_4$) and mode of application (granular versus fertigation) found that the yield was greatest when plants were fertigated at rates of 63–93 kg N/ha/year (Vargas and Bryla, 2015). These differences reflect variations in crop demand (vegetative and reproductive tissue), soil supply and fertilizer efficiency. It is generally accepted that fertilizers supplying N as NH_4^+ (urea, $(NH_4)_2SO_4$) are preferable for blueberries as this crop is adapted to acidic soils that contain NH_4^+ as the predominant N form.

There has been considerable research on the preference of blueberries for NO_3^- or NH_4^+. While some studies have established the preference of blueberries for NH_4^+, in others no differences in vegetative growth due to N form were observed (Hanson, 2006). In a recent study, Alt *et al.* (2017) investigated potential limitations of NO_3^- assimilation in two blueberry species, the rabbiteye 'Alapaha' and SHB 'Sweetcrisp', by supplying NO_3^- to the roots, leaf surface or through the cut stem of cuttings. They found that both *Vaccinium* spp. acquired both forms of inorganic N and were able to assimilate NO_3^- within the roots. The N form supplied to the roots did not affect the NO_3^- concentration in the xylem sap, did not alter the rate of NO_3^- supply to the shoots, and did not induce NR activity in the shoots. They concluded that NO_3^- acquired by blueberry roots is assimilated within the roots or stored for later assimilation within the plant, limiting its translocation to the shoots. Direct supply of NO_3^- to the shoots enhanced NO_3^- metabolism-related gene expression as well as NR activity, at least transiently. These data suggest that blueberry shoots have the capacity to respond to NO_3^- availability. However, the induced NR activities were still considerably lower than those reported in other woody species. Considering the generally lower capacity of blueberry plants to acquire NO_3^-, it is likely that transport mechanisms involved in the root uptake of NO_3^- and its translocation to the shoots are important limitations to its utilization in blueberries.

Merhaut and Darnell (1996) studied N uptake and N and C partitioning in SHB 'Sharpblue' blueberries grown in acid-washed silica sand and fertilized with NH_4^+ or NO_3^-. The nutrient solution pH was adjusted to 3.0 and 6.5 for the NO_3^-- and NH_4^+-treated plants, respectively. After 12 months of growth, plants were dual-labelled with $^{14}CO_2$ and 10% enriched ^{15}N as either sodium nitrate or $(NH_4)_2SO_4$ and harvested 12 h after labelling. Fertilization with NO_3^- increased leaf, stem and root dry weights compared with NH_4^+ fertilization. Total ^{15}N uptake did not differ between N fertilization treatments; thus, whole-plant and root ^{15}N concentrations were greater in NH_4^+-fertilized versus NO_3^--fertilized plants. Fertilization with NO_3^- increased C partitioning to new shoots compared with NH_4^+-fertilized plants. However, C partitioning to other plant parts was not affected by the N form. Although NO_3^- uptake in blueberries appears to be restricted relative to NH_4^+ uptake, this limitation does not inhibit vegetative growth. Additionally, there appears to be adequate available carbo-hydrate to support concurrent vegetative growth and N assimilation, regard-less of N form.

A trial done in SHB 'Biloxi' blueberries showed increased plant growth when the the NH_4^+:NO_3^- ratio was 0:100 and the pH of the soil was 5.0, while a ratio of 50:50 favoured leaf length and size after 1 week of starting the trial but not at later stages (Crisóstomo *et al.*, 2014). Regarding the foliar nutrient concentrations, a NH_4^+:NO_3^- ratio of 100:0 and a pH of 5 favoured higher Ca and Mg levels (2.02 and 0.21%, respectively) as well as Mn and Zn (300 and 9.75 mg/1, respectively), but this was not the case for K, as a higher K leaf content of (4.97%) was found in plants with an NH_4^+:NO_3^- ratio of 0:100 and a soil pH of 6. These results suggest that, during the vegetative phase, 'Biloxi' plants are capable of taking in N as both NH_4^+ and NO_3^-.

Differences in response to N form may be due to variability in rhizosphere pH (the pH in the immediate vicinity of the roots). Absorption of NO_3^- is accompanied by a net release of excess OH^-, which will increase the rhizos-phere pH, while the uptake of NH_4^+ requires a net release of H^+ with a con-comitant drop in rhizosphere pH (Merhaut and Darnell, 1995). In leguminous crops, it was found that the acidification of the rhizosphere in the presence of NH_4^+ occurred in the presence of light (Rao *et al.*, 2000). When exposed to equal concentrations, blueberries absorb NH_4^+ more rapidly. Less energy is necessary to assimilate NH_4^+ than NO_3^- (Merhaut and Darnell, 1995). NR, the enzyme that mediates NO_3^- transformation in the plant, has been found in stems, leaves and roots of blueberries, but the activity in blueberry is very low compared with other crops (Merhaut, 1993). This in part may be due to the fact that the enzyme contains Fe in one of its subunits, an element that is little available in acid soils (Poonnachit and Darnell, 2004). There is controversy on the effect of N source (NH_4^+ versus NO_3^-) on the level of enzyme activity. How-ever, the lower uptake rate of NO_3^-, as well as the need for its transformation within the plant, may lead to slower growth rates when this compound is the primary source of N (Poonnachit and Darnell, 2004; Hanson, 2006). From these and other studies, it can be concluded that: (i) although some NO_3^- can be metabolized by blueberry plants, for commercial purposes most N should be supplied to blueberries as NH_4^+; (ii) there is some variability in the effects of N source depending on the plant material; (iii) the impact of the NH_4^+:NO_3^- ratio supplied to blueberry plants is affected by the pH of the soil (or substrate); and (iv) the ratio of N forms supplied to blueberries impacts on the metabolism of other plant nutrients.

Blueberries recover relatively low percentages of soil-applied N, depending on environmental conditions and cultural practices. Mature bushes treated with labelled urea (to track N fate) at bud break recovered only 32% of applied N by the end of the season. The remaining N was still in the root zone (15%) or unaccounted for (53%) (Retamales and Hanson, 1989). Lower N rates (100 versus 200 kg N/ha) had greater N recovery (Bañados *et al.*, 2006b).

In fruit crops, the greatest amount of N is absorbed when their leaf mass is highest (Weinbaum *et al.*, 1978). In NHB 'Bluecrop' blueberries, Bañados *et al.* (2006b) found that fertilizer recovery was greater as the plants developed throughout the season and with denser planting within the row (0.45 versus 1.2 m). High N demand and absorption capacity in blueberries occurs from late bloom until the end of fruit harvest. Fertilization practices that maintain sufficient N in the root zone during this 2–3-month period probably optimize N-use efficiency. In 3-year-old NHB 'Bluecrop' blueberries, it was found that application timing greatly affected the amount of fertilizer absorbed by the plants. Plant uptake appeared to be most influenced by plant demand and growth. Efficient uptake occurred only after shoots and leaves had begun growth; absorption decreased as growth ceased late in the season (Throop and Hanson, 1997).

Precise assessment of N demand is complicated in perennial plants such as blueberries, because part of the N absorbed one year is retained in the plant and used in the following season. Increasing plant N reserves late in the season may benefit bushes the following year. In potted 2-year-old rabbiteye blueberries, reserves supplied 90% of the N required by reproductive tissues at bloom and 50% as late as fruit maturity. Under field conditions, only 6% of the N concentration of mature bushes at the end of the season was derived from fertilizer applied that spring (Birkhold and Darnell, 1993). In this context, late-season fertilization can be beneficial. However, late N applications can promote a greater number of actively growing shoots, which may be damaged by cold at the end of the season due to insufficient hardening (Smolarz and Mercik, 1989).

Phosphorus

As in the case of other fruit crops, symptoms of deficiency for this nutrient are rarely seen (Stiles and Reid, 1991); however, there are some conditions where P deficiency occurs. P deficiencies are associated with lower availability of P in very acid soils, possible leaching of P in very sandy soils and the fact that some virgin blueberry sites are naturally very low in P.

Compared with other major nutrients, P is by far the least mobile and least available nutrient to plants in most soil conditions. The supply of P to the roots of perennial crops is particularly constrained in acid, calcareous/alkaline and old, highly weathered soils (Plassard and Dell, 2010). A large fraction of soil P is bound tightly to the surface of soil particles or is tied up (fixed) as organic P compounds, and is therefore relatively unavailable for plant uptake (Kochian, 2012). A distinction can be made between different major soil types based on the soil total P content. The first large group consists of soils with actual low total P content such as acrisols and sandy soils. Other major soil types such as nitisols, acid andosols and calcareous/alkaline soils do contain considerable amounts of P, but a large proportion is bound to different soil constituents, forming complexes of limited bioavailability. This group of soils is commonly referred to as P-fixing soils. In both cases, the concentration of inorganic P in the soil solution is suboptimal for crop production (Ramaekers *et al.*, 2010). In acidic soils, P can be fixed with Al and Fe oxides on the surface of clay minerals, rendering it unavailable for root uptake. Hence, P can be a constraint for plant growth in acidic soils (Zhao *et al.*, 2014). In addition, organic material present in the soil (e.g. from manure or crop debris) can bind phosphate, in particular phytate (inositol compounds). Present estimates are that plants use only 10–25% of applied inorganic P, so there is room for improvement.

For improving P uptake, root exudates are thought to assist in mobilizing P from fixed sources in the soil. The exudates include protons and organic acids, such as citrate, malate and oxalate. In acid soils, organic acids may not necessarily improve P uptake directly, but they could be effective by providing protection from Al toxicity to root growth and in turn indirectly increase P uptake through a better-developed root system. Organic acid exudation will also affect microorganisms involved in nutrient mobilization. Mycorrhizal symbiosis is another important opportunity for improving P uptake (van de Wiel *et al.*, 2015).

The effects of excessive P are rarely seen in blueberry fields (Stiles and Reid, 1991; Krewer and NeSmith, 1999). Leaf P levels are highest at the beginning of the season and lowest at harvest. Tissue concentrations are little affected by variations in crop load and moisture status. Threshold foliar P levels to establish deficiency vary. They were defined as 0.07% in Michigan, 0.05% in Massachusetts and 0.09% in Wisconsin and Minnesota (Hart *et al.*, 2006).

Under P deficiency, plants can be stunted and leaves are unusually small. Another common symptom is a purplish coloration on older leaves and stems, although this symptom may be caused by other factors, such as low soil temperatures and water-saturated soils. Leaves may lie unusually flat against the stems (Hanson and Hancock, 1996; Hart *et al.*, 2006).

Fertilizer P added to soils undergoes various adsorption, precipitation and absorption reactions with soil components. The end result is that most fertilizer P adds to the soil reserves, and solution P levels are increased only slightly (Hanson, 2001). Recommended P applications are based on soil and leaf tests. However, the low pH and high Fe and Al concentrations in most blueberry soils might not allow traditional soil tests to accurately reflect plant-available P

(Roper and Schmitt, 2007). This might be the reason for the low correlation ($r = 0.084$) between soil and leaf P levels that Hanson (1987) found for 539 leaf and soil samples collected from NHB blueberries of various ages in Michigan. In Oregon, applications of P are only recommended if soil test (Bray) readings are below 50 ppm and leaf P is below 0.10%. At 26–50 ppm soil P and 0.08–0.10% leaf P, up to 45 kg P_2O_5/ha is recommended. When soil P is below 25 ppm and leaf P is below 0.07%, 45–67 kg/ha of P_2O_5 is recommended (Hart *et al.*, 2006). In Michigan, although recommendations are based on soil P levels, they suggest P application only when foliar P levels fall below 0.08% (Table 5.10).

Many virgin fields benefit from a pre-plant application of P. It has been estimated that it takes 8 kg phosphate to increase the P level in the soil by 1 kg. In pine bark bed culture, it may be necessary to apply P three or four times a year. This is due to the poor P-holding capacity of pure pine bark (Krewer and NeSmith, 1999).

Potassium

Levels of K are rarely low in blueberries, except on sandy soils. As with other fruit crops, chlorosis of leaf margins in older leaves is the first detectable symptom of K deficiency. This symptom can lead to scorching along the margins, cupping, curling, and necrotic spots and dieback of shoot tips (Hart *et al.*, 2006). Younger leaves (near the shoot tips) may develop interveinal chlorosis similar to symptoms of Fe deficiency (Hanson and Hancock, 1996). Low leaf K can be due to a number of factors: reduced root function, flooding, poor drainage, high N levels, drought and very acid soils (Stiles and Reid, 1991). Root growth is important for K nutrition. In clayey or compacted soils, root proliferation will be reduced; in these conditions, soil nutrient analysis can show high K, but leaf K could be low (Shaw, 2008). K deficiency affects shoot and root development. In roots, it impairs both lateral root initiation and

Table 5.10. Recommended potash and phosphate application rates for highbush blueberries. (Adapted from Hanson and Hancock, 1996.)

	Recommended rate (kg/ha)		
	K_2O	P_2O_5	
Soil test (ppm)	All soils	Mineral soils	Organic soils
0–10	101	168	112
10–20	84	140	84
20–30	67	112	56
30–40	34	56	0
40–50	22	0	0
>50	0	0	0

development. It would have a depressive effect on primary root growth (Chérel *et al.*, 2014). K has recently been associated with an increasing number of plant ecological processes, such as resistance to diseases, herbivory, salinity, cold, frost and waterlogging. Besides the well-known physiological processes of enzymatic function, internal transport systems and cell membrane integrity, K uptake is closely linked to water economy and soil water content; thus, K has a role in increasing water-use efficiency and limiting water loss (Chérel *et al.*, 2014; Sardans and Peñuelas, 2015). K is present in high amounts in plant cells (up to 10% of plant dry weight) and is absolutely required for plant growth. It is the main inorganic cation within plant cells, with concentrations between 100 and 200 mM (Chérel *et al.*, 2014).

Excessive K (leaf K over 0.9%) can result in nutrient imbalances, particularly Mg and Ca deficiencies (Stiles and Reid, 1991). Excess K can interfere with Mg uptake, so it should not be applied unless foliar analysis indicates a deficiency (Pritts, 2000).

K constitutes 2.1–2.3% of the earth's crust. Agricultural soils should possess sufficient levels of K, and application of this element should not be required; however, the constant K extraction by crops and the presence of various factors that reduce plant uptake generate K deficiencies. Roots absorb K as the K^+ ion. Within the plant, it moves as K^+ through the xylem vessels and balances mainly with the NO_3^- anion. K availability for crops depends on soil dynamics of exchange. There is an active fraction of immediate intake and another long-term passive fraction. The passive fraction has no contribution in K nutrition during a growth cycle. The active fraction has three components: K in the soil solution (0.1–0.2% of total soil K but can be much lower at the root surface due to local depletion; Sardans and Peñuelas, 2015), exchangeable K and non-exchangeable K (occluded within phyllosilicate clays). The presence of high levels of other monovalent cations, such as Na^+ and NH_4^+, interferes with K absorption. In contrast, optimum soil K levels improve absorption of Cu, Mn and Zn, which indicates synergism between these elements. Release of exchangeable K is often slower than plant uptake and, as a consequence, the content of K in some soils does not satisfy the need of crops in phenological stages of higher demand. Soil K application efficiency has been estimated as 40–60% and depends on the fertilizer form and dose, as well as the crop's absorption capacity (Guerrero-Polanco *et al.*, 2017). Krewer and NeSmith (1999) reported that in Georgia significant amounts of K can leach from soils planted to blueberries.

As fruit is an important sink for K in the plant, leaf K levels are also greatly influenced by fruit load. Fruit K levels increase strongly as the fruit matures, averaging around 60 mg per berry when the fruit is ripe (Hart *et al.*, 2006). Hirzel (2014) reported that 10 t of fruit at harvest removed 6.5, 7 and 8 kg K/ha in 'Brigitta', 'O'Neal' and 'Duke' highbush blueberries, respectively. When leaf K levels are below normal, fruit yields have been increased with K fertilization on a range of soil types. In Oregon, K fertilizers are not recommended if soil levels are above 150 ppm and leaf K concentration is higher than 0.40%. Up to 84 kg K_2O/ha are suggested if soil test readings are between 101 and 150 ppm and tissue K is 0.21–0.40%. An application of between 84 and 112 kg K_2O/ha is advised when soil levels are 0–100 ppm and leaf K is less than 0.2% (Hart *et al.*, 2006). Similar application levels are recommended in Michigan (Table 5.10), but it is suggested that if crop load is high, leaf K levels between 0.35 and 0.40 would be adequate (Hanson and Hancock, 1996). In general, on light-textured soils where leaching readily occurs, one or two applications of K are needed annually on bearing plants (Krewer and NeSmith, 1999). Growers prefer potassium chloride (KCl; muriate of potash) because it is cheaper, but blueberries are sensitive to the Cl^- ion. Damage can occur when high rates are used or the material is applied to young plants or not spread uniformly (Pritts, 2000). K can be applied any time of the year.

Calcium

Blueberries are said to be calcifuges that thrive in low pH, are efficient in Ca^{2+} uptake and have low Ca requirements relative to other temperate fruit crops. Healthy bushes typically have 0.3–0.8% Ca in leaf tissue (Eck, 1988) compared with 1–3% in temperate tree crops (Shear and Faust, 1980). With regard to foliar levels, blueberries are seldom deficient in Ca (Hanson and Hancock, 1996; Hart *et al.*, 2006); however, Ca nutrition has recognized effects on various fruit quality characteristics (fruit texture, firmness and ripening rate), even when leaf levels indicate that plants are adequately supplied with Ca (Hanson *et al.*, 1993). Ca deficiency in fruit tissues can occur due to insufficient mobilization of Ca from internal stores or a reduced supply of Ca through the xylem (often a result of low transpiration rates; Hocking *et al.*, 2016). Although deficiencies have not been reported in the field, due to the slow translocation of Ca within the plant, deficiency symptoms should tend to occur on the younger plant tissues (Hirschi, 2004). Interveinal chlorosis and/or browning (scorched look) on the edges of newly formed leaves is characteristic of Ca deficiency in other crops (Hanson and Hancock, 1996). A high K and N supply, as well as wide fluctuations in moisture during the season, can accentuate the severity of low plant Ca levels (Hirschi, 2004). Low leaf Ca levels can also occur in heavily fertilized, vigorously growing plants (Stiles and Reid, 1991). Ca deficiency symptoms are most common in plants growing in lower pH soils, which also tend to have low Ca levels. Plants living in calcareous soils generally contain much more Ca than those, such as blueberries, that are adapted to acidic conditions (Demarty *et al.*, 1984). In the presence of high levels of substrate Ca^{2+}, calcifuges such as blueberries cannot regulate Ca^{2+} influx and may accumulate excessive amounts. Plant roots absorb Ca from the soil solution as Ca^{2+} ions; however, the high content of humus leads to a limitation on the quantity of free Ca^{2+} ions, while organic acids form chelate complexes with Ca^{2+} ions. A low pH of the substrate provides conditions for high concentrations of Al^{3+}, Fe^{2+} and Mn^{2+} ions, inhibiting Ca uptake (Ochmian, 2012).

Ca serves various roles within plant cells, including structural, defence and communication among tissues and organs. To accomplish these roles, narrow concentration ranges must be maintained within the cells, but the Ca levels needed for each role are quite different; 10^{-4} M Ca is needed for the structural role and 10^{-7} for communication tasks (a 1000-fold gradient). When Ca uptake exceeds the needs of the plant, some Ca is sequestered to form Ca oxalates, which also helps in defence and detoxification of heavy metals (Franceschi and Nakata, 2005). This can, in part, explain the weak relationships that are usually found between Ca applications and changes in Ca-related processes such as firmness and decay prevention. Wright *et al.* (1995) found that high levels of supplemental soil Ca^{2+} led to greater uptake of Na^+ and concluded that, when the calcifuge blueberry was exposed to salinity, high Ca^{2+} accentuates the detrimental effects of Na^+ on cell metabolism. Low-pH soils also have abnormally high Mn levels. As a benchmark, if Mn leaf levels are above 450 ppm, soils are likely to have low pH and low soil Ca (Hart *et al.*, 2006).

High leaf Ca can be due to low crop load or high soil Ca (Strik *et al.*, 2010). Excessive Ca can reduce Fe uptake by plant roots, as well as interfere with Mg and K metabolism in plants. Ca and Mg should be in balance; thus, a ratio of 8–10 units of soil Ca to 1 unit of Mg is desirable. As most virgin blueberry soils are low in Mg, it is often included in small amounts for balanced fertilization (Krewer and NeSmith, 1999).

Ca is absorbed preferentially by young roots. Soil NH_4^+, K and Mg interfere with Ca absorption by roots; hence, high levels of these nutrients will reduce fruit Ca levels. Ca moves to the fruit through transpiration. Because leaves transpire more than fruit, they accumulate more Ca (White and Broadley, 2003). The fruit tend to accumulate most of their Ca when shoot growth is limited. Low levels of Ca in the fruit are more critical than leaf concentrations. The ratio of leaf Ca:fruit Ca concentrations (dry-weight basis) is indicative of the Ca partitioning among reproductive and vegetative organs. Leaf:fruit Ca ratios in the NHB 'Elliott' were 1.3:1 at 18 days after full bloom, increased to 3.6:1 by 65 days after full bloom and finished at 11.0:1 at 134 days after full bloom (Stückrath *et al.*, 2008). Strik and Vance (2015) speculated that the particularly low fruit Ca concentration in NHB 'Draper' blueberries may be due to the presence of many competing shoot tips during the fruit development period. Seeds in the fruit are involved in fruit Ca accumulation (Buccheri and Di Vaio, 2005). Fruit in cultivars that regularly produce low numbers of seeds or that are grown in seasons with poor environmental conditions at pollination are expected to have lower fruit Ca levels.

Although no studies have been done on *Vaccinium* spp. to establish the functionality of the xylem during the season, in other fruit crops (e.g. apples, grapes, kiwifruit), Ca tends to accumulate in the fruit during the first half of its growth when the xylem (conducting tissue) is functional (Creasy *et al.*, 1993; Dichio *et al.*, 2003; Drazeta *et al.*, 2004). Later in the season, the accumulation of Ca would be limited to the phloem and the build-up of Ca in the fruit should be strongly reduced. Fruit face challenges in terms of Ca nutrition: (i) fruit are architecturally isolated; (ii) their water and nutrient supply changes during development; and (iii) they have low rates of transpiration and xylem transport when compared with other plant organs, which limits fruit Ca delivery. As Ca is a phloem-immobile nutrient, it relies mainly on transpirational water flow for its accumulation within fruit. Fruit transpiration rate is highest at fruit set (e.g. in kiwifruit it can be as high as $2.3\,mmol/m^2/s$), but this quickly declines to as little as 10% of this value later in development, whereas leaf transpiration stays at more than $10\,mmol/m^2/s$ (Montanaro *et al.*, 2015). Most Ca accumulates in the fruit at these early stages of development. As blueberry fruit grows, Ca concentrations decrease by dilution (Fig. 5.4).

The importance of Ca nutrition in determining susceptibility to major horticultural disorders has been established. However, the amelioration of these disorders and improvement in pathogen resistance through Ca fertilization has not delivered reliable results (Hocking *et al.*, 2016). As in other fruit crops, fruit Ca levels in blueberries influence fruit firmness and postharvest life (Hanson *et al.*, 1993). However, these relationships have been difficult to document. In one study, soil applications of lime and gypsum for 5 years to NHB blueberries had a limited impact on fruit Ca levels and no effect on yield, fruit size or firmness (Hanson and Berkheimer, 2004). The highest assimilation of Ca from sprays is achieved by young fruit, as the permeability of cuticle is the highest at

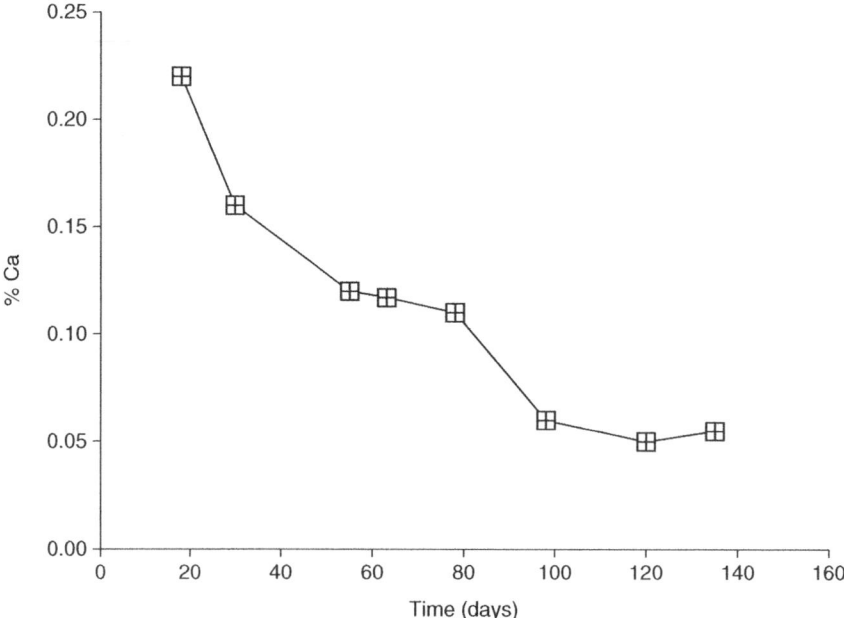

Fig. 5.4.   Changes in Ca concentration during the season (days after bloom) in fruit of NHB 'Elliott' blueberries. (Adapted from Stückrath *et al.*, 2008.)

this stage (Petit-Jimenez *et al.*, 2009), and the properly functioning stomatal apparatus provides an easy passage for Ca^{2+} (Ochmian, 2012). Penetration rates, however, have been shown to vary with cultivar, application method and formulation of Ca used (Saure, 2005). Ca sprays (0–24.2 kg Ca/ha) as calcium chloride ($CaCl_2$) or calcium trihydroxyglutarate (Nutri-Cal) applied for two seasons to NHB 'Bluecrop' blueberries failed to alter the proportion of soft or rotten berries, and had no influence on berry firmness and berry Ca levels (Hanson, 1995). Studies on the effect of three foliar applications at four rates (0, 47.5, 90 and 180 g Ca per 100 l water) of either $CaCl_2$ or calcium nitrate at 15, 30 and 45 days after fruit set on NHB 'Bluecrop', 'Blueray' and 'Ivanhoe' blueberries showed modest impacts on leaf Ca levels. The lowest Ca rate (47.5 g Ca per 100 l water) increased Ca levels in fruit skin and seeds, but 90 g Ca per 100 l water was required to increase Ca levels in the pulp. Both Ca sources (Cl^- and NO_3^-) had similar effects on fruit Ca levels. Sensory evaluation of firmness determined a greater proportion of firm fruit only for 'Bluecrop' and 'Blueray' (Retamales and Arredondo, 1995). Stückrath *et al.* (2008) applied foliar Ca fertilizers (120 g Ca^{2+}/l) 12 times during the season and determined significantly higher Ca fruit levels at harvest of NHB 'Elliott' blueberries at 5 ml/l but not at 0.5 or 3.0 ml/l. Texture and the levels of pectic substances were not affected by Ca sprays.

Recent research on the effect of Ca sprays was carried out in mature highbush blueberry commercial fields in Oregon. For this, two plantings were used for 'Legacy' and 'Spartan' and three fields for 'Draper' and 'Liberty'. Ca formulations tested were $CaCl_2$, $CaCl_2$ + boron, Ca silicate, Ca chelate and calcium acetate, which were compared with a water-only control. The rates used supplied equal amounts of Ca/ha for each treatment. Ca concentration varied from 0.05 to 0.16% applied with a volume of 748 l water/ha. Treatment applications started at the early green fruit stage and were repeated three ('Spartan') or four (all other cultivars) times at 7–16-day interval. Backpack and electrostatic sprayers were compared for 'Draper' and 'Legacy'. Compared with the control, none of the Ca treatments or method of application changed leaf or fruit Ca concentration, fruit quality, firmness or shelf-life in any cultivar tested (Vance *et al.*, 2017).

When this nutrient is sprayed on to the foliage, good coverage is needed, as Ca must be present in the tissue where it is needed because there is little retranslocation of Ca among plant tissues (Hirschi, 2004). There have been reports of scorching of young leaves after Ca sprays in studies done in the USA with pre-harvest applications of Ca chloride at 0.08% w/v (Hanson, 1995), but not in Chile at double the rate (Retamales and Arredondo, 1995). To avoid damage to tissues, foliar Ca spraying should not be carried out in conditions of high humidity or temperatures above 25°C, especially when the leaves are young (Stiles and Reid, 1991).

Hanson *et al.* (1993) showed that fruit firmness was increased using Ca dips carried out for 30–240 s (1–4% w/v $CaCl_2$); however, the fruit had a saline taste that was unacceptable to tasters in a panel. Dipping fruit also removes the desirable waxy bloom coating on blueberry fruit (Vance *et al.*, 2017).

Magnesium

Deficiencies in Mg have been reported in fields in many blueberry-growing areas (Eck *et al.*, 1990). Mg deficiencies are seen occasionally in Georgia (Krewer and NeSmith, 1999) and periodically in Michigan (Hanson and Hancock, 1996).

Mg is the eighth most abundant mineral element in the earth's crust. Soil Mg originates from source rock material containing various types of silicates. The Mg content of different silicate types varies considerably, with muscovite > biotite > hornblende > augite > olivine. Variations in Mg content depend on the degree to which Al^{3+} is substituted for Mg^{2+} within silicates. Due to high variation in Mg content among source materials, the total content of Mg in soils ranges widely (0.05–0.5%). Differences in soil silicate content also explain the higher Mg contents typically found in clay and silty soils compared with sandy soils. Mg availability to plants depends on the distribution and chemical properties of the source rock material and its degree of weathering, site-specific climatic factors and, to a high degree, the horticultural management practices established in a specific field, including the species and cultivar used as well as the organic/mineral fertilization practices (Gransee and Führs, 2013).

Lower Mg levels are common in low-pH fields. Mg deficiencies are most severe in rapidly growing plants or in those with heavier fruit loads. Mg as the central atom is an essential constituent of chlorophyll *a/b*. Due to the complex roles of Mg in chlorophyll and protein biosynthesis, severe Mg deficiency causes interveinal chlorosis of older and fully mature leaves, as Mg is highly mobile within the plant (Gerendás and Führs, 2013). Although interveinal chlorosis is the most characteristic symptom of Mg deficiencies, Krewer and NeSmith (1999) stated that the most common symptom they observed in young rabbiteye blueberries was leaves that were pink on the edges and yellowish between the veins. Eventually, these leaf areas turn yellow to bright red, while the tissue adjacent to the main veins remains green (Hanson and Hancock, 1996). Leaves may eventually turn red, yellow or brown and drop prematurely from the plant as the deficiency becomes more severe. Basal (lower) leaves of shoots and canes are the first to show symptoms. Leaves on shoot tips usually stay symptomless (Hanson and Hancock, 1996).

Although the leaf deficiency level is generally thought to be below 0.1%, there have been reports of Mg deficiency in bushes with leaf levels as high as 0.2% (Hanson and Hancock, 1996). Higher optimum Mg levels occur when leaf K levels are also high. High Ca and/or K reduces Mg absorption and may also indicate a need for Mg. Desirable ranges of the percentage of bases (as a proportion of the CEC) in soil samples are 60–80% Ca, 15–30% Mg and

10–15% K. If, in the soil analysis, Mg is less than 4% of the bases or if K exceeds Mg as a percentage of the bases, Mg should be applied (Hanson and Hancock, 1996). For practical purposes in fruit crops in general, a ratio of the percentages of K:Mg greater than or equal to 4:1 in foliar samples usually indicates that the Mg supply is inadequate (Stiles and Reid, 1991). Excessive tissue Mg (above 0.4%) usually indicates that the soil pH is too high (Hart *et al.*, 2006). Decisive quality parameters important for horticultural crops, such as soluble solids and acidity, are often more closely correlated with cation ratios (e.g. Ca:Mg and K:Mg) than with Mg concentrations alone. For instance, the Ca:Mg ratio mainly determines the functional properties such as product firmness, texture and storability, which are mostly determined by the role of Ca in stabilizing cell walls. As Mg can replace Ca from binding sites, imbalanced Ca:Mg ratios in the tissue often negatively affect product quality (Gerendás and Führs, 2013).

If foliar levels indicate a deficiency, the decision on which fertilizer to use should be based on soil pH. If the soil pH is above 4.5, magnesium sulfate ($MgSO_4$; Epsom salts) or SulPoMag (21–24% K_2O, 21% S, 10–18% magnesium oxide (MgO)) is recommended. Epsom salts can be supplied through the irrigation system. However, if the pH is below 4.5, dolomitic lime at a rate of 1 t/ha should be the choice (Hart *et al.*, 2006). Mg deficiency can be corrected by applying 17–56 kg Mg/ha (Hanson and Hancock, 1996; Krewer and NeSmith, 1999). All soil applications should be done in the autumn (Hart *et al.*, 2006). Water quality should be tested to determine the quantities of Mg that could be added through the water supply.

Sulfur

Plants take up S in the form of sulfate (SO_4^{2-}) from the soil, reduce it and assimilate it into bioorganic compounds, with cysteine being the first product. SO_4^{2-} uptake and assimilation are highly regulated by the demand for reduced S, availability of nutrients and environmental conditions. S compounds are involved in responses to abiotic and biotic stress, in detoxifying reactive oxygen species and as substrate for the synthesis of glucosinolates in defence against herbivores and pathogens. SO_4^{2-} assimilation is also strongly interconnected with the assimilation of NO_3^- and C (Davidian and Kopriva, 2010).

Deficiencies in S are rare in blueberries and are often confused with N deficits. Plants low in S are stunted and their leaves are light green with no pattern or mottling. The first symptoms appear in younger tissues. S is routinely applied when the pH is too high (Table 5.3). As S deficiencies are rare, S is seldom applied to correct a deficiency of this element, although blueberries have responded to S applications (Beaton, 1966). If S is low, S-containing-materials such as $(NH_4)_2SO_4$, ordinary superphosphate, potassium-magnesium sulfate or $MgSO_4$ could be used in the fertilization programme (Owen Plank and Kissel, 2010).

Aluminium

Al is the most abundant metal element in the earth's crust, comprising approximately 7% of its mass (Yang *et al.*, 2013). In soils, Al is bound with minerals and occurs mainly as insoluble aluminosilicates and aluminium oxides that are non-toxic to plants. In mildly acidic or neutral soils, Al occurs primarily as insoluble deposits and is essentially biologically inactive. In acidic solutions (pH <5.0), Al becomes soluble and available to plants as the Al^{3+} and $Al(OH)^{2+}$ forms, when they become potentially toxic to plants. Soil solution Al is generally used as a predictor of Al toxicity to plant roots growing in acidic soils (Zhao *et al.*, 2014). In the soil, low pH desorbs Al from minerals into solution and increases the toxic levels of Al^{3+} ions in soil solutions (Reyes-Díaz *et al.*, 2010). This leads to Al toxicity in acidic soils (Zhao *et al.*, 2014). Organic acids (mainly malate, citrate and oxalate) that commonly chelate Al^{3+} have been reported to play an important role in the Al tolerance of many plant species. However, the presence of P and NH_4^+ has been shown to alleviate Al toxicity and reduce the secretion of organic acids from roots under Al stress. Organic matter mineralization also results in a release of hydrogen ions (H^+), decreasing soil pH (Reyes-Díaz *et al.*, 2009); however, organic amendments and mulches may mediate fluctuations in soil moisture and temperatures and also bind active forms of Al that are released when the mineral soils are acidified before planting (Yang and Goulart, 1997). These ions accumulate in the roots and inhibit root expansion, but also have been found to produce a deleterious effect on top-growth physiology, including chloroplast functioning, total chlorophyll content, photosynthetic rates, canopy development and yields (Yang and Goulart, 1997; Reyes-Díaz *et al.*, 2010). The toxic effects of Al are largely associated with its interference in P and N metabolism, as the symptoms were reversed by foliar P and N application as NH_4^+ polyphosphate to highbush blueberries. Ericoid mycorrhizal infection could ameliorate Al toxicity, but once Al binding and accumulation sites in mycorrhizal-infected roots are fully occupied, excessive Al may be transported into leaves of mycorrhizal plants and cause damage (Yang and Goulart, 1997).

Studies on highbush blueberries found that 'Brigitta' was least affected by high Al^{3+} substrate levels, with 'Legacy' being intermediate and 'Bluegold' most susceptible. 'Brigitta' accumulated more Al in roots and leaves and had a faster recovery of photochemical parameters (Reyes-Díaz *et al.*, 2009).

Micronutrients

Iron

Fe deficiencies are common in blueberries. The margins of young leaves in Fe-deficient plants become chlorotic while the veins remain green. As the deficiency progresses, the leaves become brown or bronze–gold and may drop. Fe deficiency symptoms differ from those caused by Mg deficiency in that the main veins and many minor veins remain green in Fe-deficient leaves. Shoot growth and leaf size are often reduced. Symptoms of Fe deficiency are generally the first indicator of high soil pH (>5.5) (Hanson and Hancock, 1996). Rather than a lack of Fe in the soil, it is the high pH that makes Fe unavailable to the plant (Hart *et al.*, 2006). This deficiency has also been associated with soils that are saturated, poorly drained, or have very high Mn or P levels (Stiles and Reid, 1991).

Leaf Fe levels are not always a reliable indicator of Fe plant status, as symptoms may appear over a wide range of leaf Fe concentrations, but the problem is that part of the Fe is unavailable for metabolism (Krewer and NeSmith, 1999). In fact, plants with leaf levels in what is considered the deficiency range may sometimes exhibit no symptoms of deficiency.

Soil applications of Fe fertilizers can correct plant Fe deficiency. Fe fertilizers belong to three main groups: inorganic Fe compounds, synthetic Fe chelates and natural Fe complexes. Synthetic Fe(III) chelates have often been applied to blueberries. They are derivatives of the ethylenediamine–carboxylic acids. They are expensive, and their use is therefore restricted to soilless horticulture and where its application is economically advisable. Furthermore, these xenobiotics strongly affect metal availability and mobility in the soil because of their extended persistence in the environment, questioning their sustainability in modern agriculture (Briat *et al.*, 2015). Fe chelate at a rate of 17–34 kg/ha can be applied to the soil (Hart *et al.*, 2006). As there are many formulations of Fe chelate, the rate must be adjusted according to product concentration. Iron sulfate can also be used where the pH is high, as it lowers the pH and supplies Fe. A suggested rate is three teaspoons per m of bush height, spread evenly under the bush (Krewer and NeSmith, 1999). Soil applications of Fe sources seldom benefit Fe-deficient bushes (Hanson and Hancock, 1996). Instead, the most effective and efficient means of correcting Fe deficiencies is to adjust the soil pH. Nevertheless, if leaf foliar levels are in the deficiency range, two foliar applications of Fe chelate (10% Fe) at a rate of 1 kg per 400 l water per ha has been recommended (Hart *et al.*, 2006). A surfactant is necessary to enhance penetration through the waxy cuticle of blueberry leaves (Krewer and NeSmith, 1999). Foliar Fe fertilization is usually cheaper and more environmentally friendly, and has traditionally been used in crops for which the application of chelates to the soil is too expensive. The success of foliar fertilization depends on many factors, including the ability to penetrate the cuticle and/or stomata, to be transported through the apoplast and plasma membrane of leaf cells, and ultimately to reach the chloroplasts (Briat *et al.*, 2015).

As an alternative to foliar or soil Fe applications, it is well established that intercropping of grass and non-grass plants, especially on calcareous soils, can lead to a 1.5–2.5-fold increase in the Fe content of shoots and reproductive tissues of the non-grass plant in such an intercropping system. The phytosiderophores (high-affinity Fe(III) chelators) secreted by the grass plant help to solubilize Fe, which is then actively reduced and taken up by the Fe-deficient non-grass partner (blueberry plant) (Briat *et al.*, 2015).

Manganese

Unless the pH exceeds the recommended range, Mn deficiencies are rarely found in most blueberry-growing areas (Hanson and Hancock, 1996; Fuqua *et al.*, 2005; Hart *et al.*, 2006). Mn becomes toxic under acidic and redox soil conditions that favour the Mn^{2+} form (Rojas-Lillo *et al.*, 2014). With Mn deficiency, sectors near the leaf margin may die. There may also be isolated dead spots throughout the leaf. The leaves are smaller, and interveinal chlorosis of young leaves has been found associated with Mn deficit (Fuqua *et al.*, 2005). The main toxic effect of Mn^{2+} on plants is reductions in the biosynthesis of photosynthetic pigments (chlorophyll *a/b* and carotenoids), net photosynthesis rates and growth (Stiles and Reid, 1991; Rojas-Lillo *et al.*, 2014). The damaging effect of high Mn concentrations can be aggravated by greater exposure to ultraviolet (UV) radiation. Rojas-Lillo *et al.* (2014) reported that a toxic level of Mn^{2+} (500 mM $MnCl_2$) + UV-B radiation (a daily dose of $94.4\,kJ/m^2$) caused a more negative effect on net photosynthesis (P_n), stomatal conductance (g_s), the photochemical parameters of photosystem II and the chlorophyll *a*:chlorophyll *b* ratio than treatments with toxic Mn^{2+} or UV-B alone. They also showed that NHB 'Brigitta' blueberries had a better acclimation response in P_n and g_s than 'Bluegold'. An increased concentration of photo-protective compounds and enhanced resistance to oxidative stress in 'Brigitta' could underpin increased resistance to the combined stress. Abnormalities due to high Mn in SHB 'O'Neal' blueberries included multiple flushes of growth over the season, multiple laterals arising from a single position on the shoots (often called 'witches broom') and small, crinkled leaves with a reddish colour in the margins. Leaf levels for mild and severe symptoms are very similar in 'O'Neal', with mild symptoms present when foliar levels were 426 and 111 ppm for first and second flushes of growth, while severe symptoms occurred when 476 and 151 ppm were measured for the first and second flushes, respectively. Mn toxicity symptoms (abnormal shoot growth) in 'O'Neal' were correlated with low pH and high Mn in the sawdust and manure applied as mulch and before planting. It is recommended to measure Mn levels in organic soil amendments before application (Bañados *et al.*, 2009).

Tissue Mn levels increase as pH drops. Thus, tissue Mn levels can serve as an indicator of soil pH levels. Tissue Mn levels above 450 ppm are considered excessive, especially if they are present for long periods. However, there are reports that highbush blueberry plants can grow normally with leaf Mn concentrations as high as 650 ppm, as they have mechanisms for Mn tolerance (Hart *et al.*, 2006). Cultivars may differ in their response to Mn. No detrimental effects were reported by Korcak (1989) with plants of rabbiteye 'Delite' blueberries at a soil pH of 5.1 (1175 ppm foliar Mn) or at pH 6.9 (994 ppm Mn) or

'Tifblue' plants at pH 5.1 (531 ppm Mn) or pH 6.9 (343 ppm Mn). The SHB 'O'Neal' has been found to be especially susceptible to high Mn levels.

High soil water content resulting in low O_2 availability has also been found to induce Mn toxicity in blueberries. Under these conditions (low O_2, high Mn levels, low pH), the more mobile Mn^{+2} form is more available for root uptake than the Mn^{+3} (Bañados *et al.*, 2009). High Mn also induces Ca deficiency, resulting in small, crinkled, malformed leaves. The foliar application of various fungicides increases foliar surface levels of Mn; however most of these materials do not penetrate the leaves and can lead to spurious leaf Mn readings (Stiles and Reid, 1991). If Mn is found to be deficient, two foliar applications of Mn chelate (2–8% Mn) at 7 kg/ha or manganese sulfate (32% Mn) at 2.2 kg/ha during the summer have been recommended (Hanson and Hancock, 1996).

Boron

Tip dieback is the most prevalent symptom of B deficiency in blueberry tissues. Leaves close to aborted shoot tips develop a mottled chlorosis and cupped shape. Internodes tend to be shorter in affected shoots (Hanson and Hancock, 1996). B plays a key role in reproductive processes affecting anther development, pollen germination and pollen-tube growth. For this reason, in B-sensitive crops, abortion of flower initials and poor fruit set are observed when B is deficient (Ganie *et al.*, 2013). Flower and vegetative buds may fail to open or develop in severely affected blueberry plants. Winter injury tends to be greater in B-deficient plants (Hart *et al.*, 2006).

B deficiency is one of the most common among plant micronutrient deficiencies worldwide (Gürel and Başar, 2016). B deficiency is particularly prevalent in light-textured soils, where water-soluble B readily leaches down the soil profile and becomes unavailable to the plants. B deficiency may also occur in soils with adequate B levels if its uptake is impeded by overliming, dry or wet soil conditions, and a low level of soil O_2 (Ganie *et al.*, 2013). Deficiency is aggravated by dry weather and a heavy crop load. Low plant B may accentuate deficiencies of other nutrients because of impaired root function. In fruit crops in general, low B is often associated with Ca deficiency problems (Stiles and Reid, 1991). Its incidence varies across the blueberry-producing regions: while common in Oregon (Hart *et al.*, 2006), it has not been found in Michigan (Hanson and Hancock, 1996).

If B is deficient, 11–22 kg borax/ha (11% B) can be applied in the autumn or early spring prior to rain. Alternatively, 0.9–2.7 kg Solubor (20% B) in 950 l water/ha can be sprayed before bloom, or after harvest and before leaf senescence. A yearly application of 560 g B/ha has been suggested (Hart *et al.*, 2006). Foliar and soil treatments of B (four applications of 0.2 kg B/ha between early bloom and 6 weeks after bloom) to NHB 'Bluecrop' blueberries increased leaf and flower B levels, as well as fruit soluble solids, but failed to alter plant vigour, the number of flowers per cane, fruit set or yield (Wojcik, 2005). In contrast, a 4-year-study on the effect of autumn and spring B foliar applications to NHB 'Blueray' and 'Collins' blueberries grown in Missouri increased yield by an average of 10% for all seasons, which was mostly due to an increased number of berries per plant. B sprays also reduced tip die back symptoms (Blevins *et al.*, 1998). The foliar B levels need to be monitored carefully, as toxic levels of B can be reached rapidly (Hanson and Hancock, 1996).

Zinc

Symptoms of Zn deficiency include short internodes and small leaves. Low Zn causes a uniform yellowing of young leaves early in the season with no interveinal pattern. Affected leaves can fold upwards along the midrib. The importance of Zn deficiency varies across productive regions. Symptoms have been reported in Oregon (Hart *et al.*, 2006) but not in Michigan (Hanson and Hancock, 1996) or Missouri (Fuqua *et al.*, 2005). Zn deficiency is accentuated at high pH (>6.0) and low soil temperature. Excessive use of P may also lead to Zn deficiency (Stiles and Reid, 1991).

The total Zn content in soil depends upon the parent rock, weathering, organic matter, texture and pH. The most quoted normal range of total Zn in soils is 10–300 mg/kg with a mean value of 50 mg/kg. Soils formed from basic rocks (i.e. basalt) are richer in Zn than those from acid rocks (i.e. granite and gneisses). Total Zn content is generally lower in lighter soils. Zn must be present in the soil solution or adsorbed in a labile form in order to be available to plants. Among the factors affecting Zn availability to plants are: total Zn content, clay content, calcium carbonate content, soil moisture status, redox conditions, microbial activity in the rhizosphere, concentration of other trace elements, concentration of macronutrients (especially P) and climate. Thus, Zn deficiency can be observed, even though high amounts of Zn are available in the soil. The root–shoot barrier, a major controller of Zn transport in plants, is highly affected by changes in the anatomical structure of conducting tissue and adverse soil conditions (Sharma *et al.*, 2013).

If blueberries are deficient in Zn, a foliar spray of 454 g Zn chelate (14% Zn) is recommended after harvest and before leaf fall in a volume of 935 l/ha. Another option is a soil application of Zn chelate at a rate of 11–34 kg/ha (Hart *et al.*, 2006).

Copper

Cu deficiencies are rarely seen in either rabbiteye or highbush blueberry fields (Hanson and Hancock, 1996; Krewer and NeSmith, 1999; Hart *et al.*, 2006), although deficiencies have been reported on high-organic-matter blueberry soils in North Carolina (Krewer and NeSmith, 1999). In other areas, symptoms have only been induced by removing this nutrient experimentally from the substrate. Cu deficiency symptoms are similar to those associated with insufficient Mn (Hanson and Hancock, 1996). They include interveinal chlorosis of young leaves and, in severe cases, shoot dieback.

Although Cu concentrations in soils range between 3 and 100 mg/kg, the majority is bound to organic matter, and only about 1–20% is readily bioavailable (Marschner, 1986). In soil, Cu can be found in solid and liquid phases. Solid-phase Cu includes being water soluble, exchangeable and complex in secondary minerals such as clays and Fe and Mn oxyhydroxides, organic matter and primary silicate minerals or co-precipitated with carbonates and phosphates. Cu deficiency may be more severe in soils high in organic matter (over 25%) (Hart *et al.*, 2006). Coarser-texture soils, high soil pH and high P soil levels tend to accentuate low tissue Cu levels (Stiles and Reid, 1991). Cu bioavailability is controlled by the total Cu concentration, CEC, soil organic matter and soil pH. Moreover, in acidic Cu-contaminated soils, Cu bioavailability is more influenced by rhizosphere pH than the bulk soil pH. Cu is an essential element for plants at low level, but in excess it is phytotoxic because it interferes with various metabolic processes that are vital for plant growth and development. When compared with other potentially toxic essential trace elements, such as Mn and Zn and non-essential cadmium, excess Cu is more toxic to plants and less harmful to animals and humans. Root growth is more sensitive to Cu toxicity than shoot growth (Adrees *et al.*, 2015).

In North Carolina, 6 kg elemental Cu is applied before planting on high-organic-matter sites and repeated every 5 years. A trial application is suggested if Cu leaf levels are less than 3 ppm (Krewer and NeSmith, 1999). When needed, a broadcast application of copper sulfate ($CuSO_4$; 25% Cu) has been recommended at a rate of 34–56 kg/ha. Another option is a foliar spray of 0.5 kg $CuSO_4$ in a volume of 900 l/ha any time leaves are present (Hanson and Hancock, 1996; Hart *et al.*, 2006).

Biostimulants

A plant biostimulant is any substance, microorganism or non-essential element applied to plants that is aimed at enhancing nutrition efficiency, abiotic stress tolerance and/or crop quality traits, regardless of its nutrient content. By extension, plant biostimulants also represent commercial products containing mixtures of such substances and/or microorganisms. Many biostimulants improve nutrition regardless of their nutrient content. Biofertilizers, which have been proposed as a subcategory of biostimulants, increase nutrient-use efficiency and open new routes for nutrient acquisition by plants. In this sense, microbial biostimulants include mycorrhizal and non-mycorrhizal fungi, bacterial endosymbionts and plant growth-promoting rhizobacteria. Thus, microorganisms applied to plants can have a dual function of both biocontrol agent and biostimulant. The scientifically demonstrated effects of all biostimulants converge to at least one or several of the following agricultural functions: enhanced nutrition efficiency, abiotic stress tolerance and/or crop quality traits (du Jardin, 2015).

In this chapter, we will cover only those biostimulants that have been tri-alled in *Vaccinium* spp. These include: silicon (Si; inorganic element), humic substances and beneficial bacteria. Additionally, the effect of chitosan and other biopolymers will be covered in Chapter 9 (this volume).

Silicon

Chemical elements that promote plant growth and may be essential to particu-lar botanical groups but are not required by all plants are called beneficial ele-ments. The main beneficial elements are: cobalt (Co), Na, selenium (Se) and Si. They are present in soils and plants as different inorganic salts and as insoluble forms. The beneficial functions can be constitutive, such as strengthening of cell walls by silica (SiO_2) deposits, or expressed in defined environmental condi-tions, such as pathogen attack for Se and osmotic stress for Na. Although the modes of action are not yet fully established, these inorganic compounds influ-ence osmotic, pH and redox homeostasis, hormone signalling and enzymes involved in the stress response (e.g. peroxidases) (du Jardin, 2015).

Although Si is the second most abundant element in the earth's crust, it is not considered an essential element. However, both research results and practi-cal experience show beneficial impacts of Si on the growth and development of many plant species, especially when exposed to abiotic or biotic stress. The Si content amounts to 200–350 g/kg in clay soils and 450–480 g/kg in sandy soils. Although most soil Si is present as insoluble oxides or silicates, some water-soluble Si also occurs. The predominant Si forms in mineral soils include SiO_2, and primary (e.g. quartz, feldspar, mica) or secondary (e.g. clay minerals) silicate minerals, which contain Si, oxygen and metals such as Al (alumina sili-cates) and Mg (talc). SiO_2 comprises up to 45% of the soil mass. Weathering of Si-containing minerals is the major source of chemical elements for terrestrial plants. The mineral breakdown releases soluble Si mainly as silicic acid (H_4SiO_4) into the soil solution (liquid phase). The dissolved Si concentration varies con-siderably depending on the type of minerals and the biotic and abiotic environ-ment (Farooq and Dietz, 2015). H_4SiO_4 does not dissociate at pH <9 and thus plants take up Si in this non-ionic form actively or passively, depending on the external Si concentration and their inherent requirements. The latter vary considerably, as indicated by the large variation in tissue Si concentrations among species, which range from 0.1 to 10% of dry weight (Savvas and Ntatsi, 2015).

It has been shown that Si alleviates salt, drought and nutrient stress as well as stress associated with climatic conditions, minimizes metal and metal-loid toxicities, and may delay plant senescence processes. However, the mecha-nisms underlying Si-mediated alleviation of abiotic stresses remain poorly understood. Limited information is currently available about the effect of Si application as a biostimulant in fruit crops; most trials refer to *Citrus* spp. Si can be applied through foliar spraying, soil incorporation or fertigation. Overall, foliar Si sprays do not seem to be as effective as applications through the root system in mitigating abiotic stress. Given that Si is probably the only element conferring resistance to multiple stresses, and that it is not toxic to humans and the environment, the use of this element as a biostimulant in horticulture is expected to increase considerably in the future (Savvas and Ntatsi, 2015).

Si has generally been a neglected element in blueberry nutrition, so the literature is scarce on this topic. Carnelli *et al.* (2002) conducted paleo-ecological studies on 20 species, including some Ericaceae, and found leaf Si contents of 0.4 and 0.3 mg/g dry weight in *V. myrtillus* L. and *V. vitis-idaea* L., respectively. In blueberry cuttings (*V. corymbosum* L. 'Bluecrop'), Morikawa and Saigusa (2004) found mean leaf concentrations of 32 and 60 mg/g dry weight in young and old leaves, respectively.

The commercial formulation Quick-Sol (Quick-Sol Global, LLC, Comfort, Texas) contains 36% Si (as ionized sodium silicate), humic and fulvic acids (2%) and 1% of each of the following elements: Ca, Fe, Zn, Cu, Mg and Mn. A rate of 100 ml per 100 l of this formulation was sprayed at early bloom, fruit set and onset of shoot growth on to mature SHB 'O'Neal' plants growing in a commercial field in Salto Grande, Entre Ríos Province, Argentina (latitude 30°S). Plants received 400 l/ha of Quick-Sol or water (control). The fruit were harvested at first pick (5% ripe fruit) and at full harvest (50% ripe fruit) and evaluated at harvest and after storage for 24 days at 1°C and 90% relative humidity. At both the 5 and 50% ripe fruit picks, there was no effect of Quick-Sol on fruit texture or soluble solids. The only effect was on acidity in the 5% ripe fruit pick, where it induced lower fruit acidity at harvest but not after cold storage. This change in acidity at harvest in the first pick caused a greater soluble solids:acid ratio at that time, but not after cold storage or at the other harvest date (50% ripe fruit). Except at harvest at the first pick where there were no differences compared with the control in the proportion of rotten fruit, Quick-Sol reduced the proportions of rotten fruit after 24 days of cold storage in fruit collected in the first pick (21 versus 34% in the control), as well as in the 50% ripe fruit pick both at harvest (7 versus 15% in the control) and after 24 days of cold storage (5 versus 13% in the control). The largest proportion of rotten fruit in this trial was due to *Alternaria* spp. infection. The incidence of *Alternaria* spp. was significantly reduced by application of Quick-Sol in both picks and in both postharvest evaluations (A.M. Heredia, INTA-Famaillá, Tucumán, Argentina, personal communication).

Humic substances

Humic substances (HSs) are natural constituents of the soil organic matter, resulting from the decomposition of plant, animal and microbial residues, but also from the metabolic activity of soil microbes using these substrates. They represent the major pool of organic C at the earth's surface and contribute to the regulation of many crucial ecological and environmental processes. HSs are collections of heterogeneous compounds, originally categorized according to their molecular weights and solubility into humins, humic acids and fulvic acids. HSs and their complexes in the soil thus result from the interplay between organic matter, microbes and plant roots. They are extracted from naturally humified organic matter (e.g. peat or volcanic soils), composts and vermicomposts, or mineral deposits (leonardite, an oxidation form of lignite) (du Jardin, 2015).

A meta-analysis of trials performed in crop species estimated shoot dry-weight increases of $22 \pm 4\%$ and root dry-weight increases of $21 \pm 6\%$ in response to the application of HSs (Rose *et al.*, 2014). However, actual responses varied considerably and were influenced mainly by the source of the HS applied, the rate of application and, to a lesser extent, the plant type and growing conditions. HSs from compost sources significantly outperformed lignite and peat-derived ones in terms of growth promotion, while the application rate non-linearly moderated the growth response under different circumstances. Reported benefits of HSs include improved soil properties and structure, greater bioavailability of soil nutrients, increased microbial population sizes and plant hormone-like effects (Bryla and Strik, 2015).

To study the effect of N form during plant establishment in Oregon, a total of 100 kg/ha N was applied through two drip lines per row to 1-year-old NHB 'Draper' blueberry plants (Bryla and Vargas, 2014). The fertilizers tested included: humic acid + NP (10 N:2.3 P), urea–H_2SO_4, NP, liquid urea, granular urea, liquid + granular urea, controlled-release fertilizer and controlled-release fertilizer + liquid urea. The humic acid + NP (Actagro products, Fresno, California) was injected weekly from early bud break (mid-April) to fruit bud set (late July). Total plant dry weight was greatest in plants fertilized with humic acid + NP or urea–H_2SO_4. The authors recommend the use of urea–H_2SO_4 or humic acids during plant establishment in high-pH soils. Root growth was particularly enhanced by humic acids during years 1 and 2 after planting a new field of NHB 'Draper' blueberry in western Oregon. The use of humic and fumic acids appears promising for blueberry, but the mechanism of their effect on root characteristics and whether they increase fruit production is unclear (Bryla and Strik, 2015).

Beneficial bacteria

Bacteria interact with plants in different ways: (i) as with fungi, there is a continuum between mutualism and parasitism; (ii) bacterial niches extend from the soil to the interior of cells, with intermediate locations in the rhizosphere; (iii) associations may be transient or permanent; and (iv) functions influencing plant life cover involvement in the biogeochemical cycles, supply of nutrients, increases in nutrient-use efficiency, induction of disease resistance, enhancement of abiotic stress tolerance and modulation of morphogenesis by plant growth regulators (du Jardin, 2015).

Eighteen-month-old SHB 'Legacy' blueberries were planted in a volcanic ash soil (Andosol) in south-central Chile (Chillán; latitude 36°36'S) (Schoebitz *et al.*, 2016). At planting, each plant received 15 g of the commercial product

Biosolve (Oikos Chile Ltda, Cauquenes, Chile) containing humic acid (derived from leonardite shale), which was added in powder form and manually mixed into the plantation holes (Biosolve contains 70% humic acid, 15% fulvic acid and 10% K_2O (w/w)). At planting and every 2 months from August (bud break) to March (onset of leaf fall), the microbial consortium Oiko Bac 174 (Oikos Chile Ltda) containing *Bacillus subtilis, Bacillus licheniformis, Bacillus megaterium, Bacillus polymyxa, Bacillus macerans, Pseudomonas fluorescens, Pseudomonas putida, Nocardia corallina, Saccharomyces cerevisiae* and *Trichoderma viride* was applied at cell concentrations of 10^8 colony-forming units/g. One year after planting, it was found that the combined application of microbial consortium and HSs was the most effective treatment with 50% higher shoot dry weight and 43% greater root dry weight compared with control plants. The microbial inoculant treatment also increased shoot (32%) and root (31%) dry weights with respect to control plants. The combined microbial consortium and HS treatment improved N and K uptake. The soil NO_3^- content was higher when HSs were applied alone, while changes were also observed in the rhizobacterial community composition. No significant effects were observed on aggregate stability, bulk density, total porosity, soil available water and shoot P levels with any of the treatments studied (Schoebitz *et al.*, 2016).

CONCLUSION

Blueberries grow on low-pH soil and require less mineral fertilizer than most fruit crops. The CEC is strongly associated with the fertility of the soil and with the buffer capacity for changes in soil pH. Soil pH must be adjusted to 4.5–5.5 in order to avoid nutrient imbalances and deficiencies. Usually, S applied one year before planting is used to adjust the soil pH. Most blueberry roots are confined to the top 30 cm of soil. They are colonized by a special type of mycorrhizae. These ericoid mycorrhizae reduce their level of infection with the application of mineral fertilizers and with root age. They protect the plant from Al toxicity, which is prevalent in low-pH soils.

Soil analysis is recommended before planting to determine the nutritional status and pH. Once the field is planted, leaf analysis should be the basis for nutrient management, along with soil pH monitoring. In order to be useful, strict procedures must be followed for leaf sample collection. Normal foliar levels vary somewhat in P, B, Fe and Mn levels across production regions, which may reflect environmental conditions for best performance.

Fertilizer need is based on nutrient demand and supply, but the efficiency of fertilizer use has to be considered. The most common method for fertilizer application has been broadcast; however, fertigation has increased in acceptance due to higher efficiency and ease of application. Organic nutrient management is also expanding for blueberries. Whatever the method, the efficiency usually increases as the number of applications increases and the rate of each is reduced. Except for some micronutrients and on specific occasions, foliar fertilization is not very efficient and has proven costly. After correction of pH and P levels at planting, N and sometimes K are the nutrients that usually require yearly application in blueberry fields.

Deficiency symptoms have been described for each of the major blueberry nutrients, along with specific fertilization practices to restore sufficiency. Biostimulants (HSs, Si and microbial consortia) can alleviate some biotic and abiotic stresses, but there is a need for more research on this topic.

REFERENCES

Adrees, M., Ali, S. Rizwan, M., Ibrahim, M. Abbas, I., Farid, M., Zia-ur-Rehman, M., Irshad, M.K. and Bharwana, S.A. (2015) The effect of excess copper on growth and physiology of important food crops: a review. *Environmental Science Pollution Research* 22, 8148–8162.

Alt, D.S., Doyle, J.W. and Malladi, A. (2017) Nitrogen-source preference in blueberry (*Vaccinium* sp.): enhanced shoot nitrogen assimilation in response to direct supply of nitrate. *Journal of Plant Physiology* 216, 79–87.

Austin, M.E. and Brightwell, W.T. (1977) Effect of fertilizer applications on yield of rabbiteye blueberries. *Journal of the American Society for Horticultural Science* 102, 36–39.

Austin, M.E. and Gaines, T.P. (1984) An observation of nutrient levels in old, unfertilized rabbiteye blueberry plants. *HortScience* 19, 417–418.

Azam, F., Simmons, F.W. and Mulvaney, R. L. (1993) Immobilization of ammonium and nitrate and their interaction with native N in three Illinois Mollisols. *Biology and Fertility of Soils* 15, 50–54.

Bañados, M.P., Bonomelli, C. González, J. and Juillerat F. (2006a) Dry matter, nitrogen, potassium and phosphorus partitioning in blueberry plants during winter. *Acta Horticulturae* 715, 443–448.

Bañados, M.P., Strik, B.C. and Righetti, T. (2006b) The uptake and use of ^{15}N-nitrogen in young and mature field-grown highbush blueberries. *Acta Horticulturae* 715, 357–364.

Bañados, M.P., Ibáñez, F. and Toso, A.M. (2009) Manganese toxicity induces abnormal shoot growth in 'O'Neal' blueberry. *Acta Horticulturae* 810, 509–512.

Beaton, J.D. (1966) Sulfur requirements of cereals, tree fruits, vegetables, and other crops. *Soil Science* 101, 267–282.

Birkhold, K.T. and Darnell, R.L. (1993) Contribution of storage and currently assimilated nitrogen to vegetative and reproductive growth of rabbiteye blueberry fruit. *Journal of the American Society for Horticultural Science* 118, 101–108.

Blevins, D.G., Scrivner, C.L., Reinbott, T.M. and Schon, M.K. (1998) Foliar boron increases berry number and yield of two highbush blueberry cultivars in Missouri. *Journal of Plant Nutrition* 19, 99–113.

Briat, J.F., Dubos, C. and Gaymard, F. (2015) Iron nutrition, biomass production, and plant product quality. *Trends in Plant Science* 20, 33–40.

Bryla, D.R. and Machado, R.M.A. (2011) Comparative effects of nitrogen fertigation and granular fertilizer application on growth and availability of soil nitrogen during establishment of highbush blueberry. *Frontiers in Plant Science* 2, 1–8.

Bryla, D.R. and Strik, B.C. (2015) Nutrient requirements, leaf tissue standards and new options for fertigation of northern highbush blueberry. *HortTechnology* 25, 464–470.

Bryla, D.R. and Vargas, O. (2014) Nitrogen fertilizer practices for rapid establishment of highbush blueberry: a review of six years of research. *Acta Horticulturae* 1017, 415–421

Buccheri, M. and Di Vaio, C. (2005) Relationship among seed number, quality, and calcium content in apple fruits. *Journal of Plant Nutrition* 27, 1735–1746.

Carnelli, A.L., Madella, M., Theurillat, J.P. and Ammann, B. (2002) Aluminum in the opal silica reticule of phytoliths: a new tool in palaeoecological studies. *American Journal of Botany* 89, 346–351.

Caspersen, S., Svensson, B., Håkansson, T., Winter, C., Khalil, S. and Asp, H. (2016) Blueberry–soil interactions from an organic perspective. *Scientia Horticulturae* 208, 78–91.

Chérel, I., Lefoulon, C., Boeglin, M. and Sentenac, H. (2014) Molecular mechanisms involved in plant adaptation to low K^+ availability. *Journal of Experimental Botany* 65, 833–848.

Cox, J. (2009) Comparison of plastic weedmat and woodchip mulch on low chill blueberry soil in New South Wales, Australia. *Acta Horticulturae* 810, 475–482.

Creasy G.L., Price, S.F. and Lombard, P.B. (1993) Evidence for xylem discontinuity in Pinot noir and Merlot grapes: dye uptake and mineral composition during berry maturation. *American Journal of Enology and Viticulture* 44, 187–192.

Crisóstomo, M.N., Hernández, O.A., López, J., Manjarrez-Domínguez and C. y Pinedo-Alvarez, A. (2014) Relaciones amonio/nitrato en soluciones nutritivas ácidas y alcalinas para arándano. [Ammonium/nitrate relations in acid and alkaline nutrient solutions for blueberry.)] *Revista Mexicana de Ciencias Agrícolas* 5, 525–532.

Darnell, R.L. and Hiss, S.A. (2006) Uptake and assimilation of NO^{3-} and iron in two *Vaccinium* species as affected by external nitrate concentration. *Journal of the American Society for Horticultural Science* 131, 5–10.

Davidian, J.C and Kopriva, S. (2010) Regulation of sulfate uptake and assimilation – the same or not the same? *Molecular Plant* 3, 314–325.

Demarty, M., Morvan, C. and Thellier, M. (1984) Calcium and the cell wall. *Plant, Cell and Environment* 7, 441–448.

Dichio, B., Remorini, D. and Lang, S. (2003) Developmental changes in xylem functionality in kiwifruit fruit: implications for fruit calcium accumulation. *Acta Horticulturae* 610, 191–195.

Drazeta, L., Lang, A., Hall, A.J., Volz, R.K. and Jameson, P.E. (2004) Causes and effects of changes in xylem functionality in apple fruit. *Annals of Botany* 93, 275–282.

du Jardin, P. (2015) Plant biostimulants: definition, concept, main categories and regulation. *Scientia Horticulturae* 196, 3–14.

Eck, P. (1988) *Blueberry Science*. Rutgers University Press. New Brunswick, New Jersey, pp. 106–109.

Eck, P., Gough, R.E., Hall, I.V. and Spiers, J.M. (1990) Blueberry management. In: Galletta, G.J. and Himelrick, D.G. (eds) *Small Fruit Crop Management*. Prentice Hall, New Jersey, pp. 273–333.

Ehret, D.L., Frey, B., Forge, T., Helmer, T., Bryla, D.R. and Zebarth, B.J. (2014) Effects of nitrogen rate and application method on early production and fruit quality in highbush blueberry. *Canadian Journal of Plant Science* 94, 1165–1179.

Extension (2017) pH and availability. USDA National Institute of Food and Agriculture, New Technologies for Ag Extension project. Available at: http://articles.extension.org/sites/default/files/w/c/c6/PHandavailability.jpg (accessed 30 August 2017).

Farooq, M.A. and Dietz, K.J. (2015) Silicon as versatile player in plant and human biology: overlooked and poorly understood. *Frontiers in Plant Science* 6, 1–14

Finn, C.E. and Warmund, M.R. (1996) Fertigation vs. surface application of nitrogen during blueberry plant establishment. *Acta Horticulturae* 446, 397–402.

Franceschi, V.R. and Nakata, P.A. (2005) Calcium oxalate in plants: formation and function. *Annual Review of Plant Biology* 56, 41–71.

Fuqua, B., Byers, P., Kaps, M., Kovacs, L., and Waldstein, D. (2005) *Growing Blueberries in Missouri*. Bulletin 44, State Fruit Experiment Station, Mountain Grove, Missouri.

Ganie, M.A., Akhter, F., Bhat, M.A., Malik, A.R., Junaid, J.M., Shah, M.A., Bhat, A.H. and Bhat T.A. (2013) Boron – a critical nutrient element for plant growth and productivity with reference to temperate fruits. *Current Science* 104, 76–85.

Gerendás, J. and Führs, H. (2013) The significance of magnesium for crop quality. *Plant Soil* 368, 101–128.

Golldack, J., Schubert, R., Tauschke, M., Schwarzel, H., Hofflich, G., Lentzsch, P. and Munzenberger, B. (2001) Mycorrhization and plant growth of highbush blueberry (*Vaccinium corymbosum* L.) on arable land in Germany. In: Smith, S.E. (ed.) *Diversity and Integration in Mycorrhizas: Proceedings of the 3rd International Conference on Mycorrhiza*, 8–13 July 2001, Adelaide, Australia. Springer, Dordrecht, The Netherlands.

Gough, R.E. (1994) *The Highbush Blueberry and its Management*. Food Products Press, Binghampton, New York..

Gransee, A. and Führs, H. (2013) Magnesium mobility in soils as a challenge for soil and plant analysis, magnesium fertilization and root uptake under adverse growth conditions. *Plant Soil* 368, 5–21.

Guerrero-Polanco, F., Alejo-Santiago, G. and Luna-Esquivel, G. (2017) Potassium fertilization in fruit trees. *Revista Bio Ciencias* 4, 143–152.

Gürel, S. and Başar, H. (2016) Effects of applications of boron with iron and zinc on the contents of pear trees. *Notulae Botanicae Horti Agrobotanici* 44, 125–132.

Hall, F.R. (1991) Influence of canopy geometry in spray deposition and IPM. *HortScience* 26, 1012–1017.

Hanson, E.J. (1987) Integrating soil tests and tissue analysis to manage the nutrition of highbush blueberries. *Journal of Plant Nutrition* 10, 1419–1427.

Hanson, E.J. (1995) Preharvest calcium sprays do not improve highbush blueberry (*Vaccinium corymbosum* L.) quality. *HortScience* 30, 977–978.

Hanson, E.J. (2001) Phosphorus management in Michigan fruit crops. *Massachusetts Berry Notes* 13, 12.

Hanson, E.J. (2006) Nitrogen fertilization of highbush blueberry. *Acta Horticulturae* 715: 347–351.

Hanson E.J. and Berkheimer, S.F. (2004) Effect of soil calcium applications on blueberry yield and quality. *Small Fruits Review* 3, 133–139.

Hanson, E.J. and Hancock, J.F. (1996) *Managing the Nutrition of Highbush Blueberries*. Extension Bulletin E-2011, Michigan State University Extension, East Lansing, Michigan.

Hanson, E.J. and Retamales, J.B. (1992) Effect of nitrogen source and timing on highbush blueberry performance. *HortScience* 27, 1265–1267.

Hanson, E.J., Beggs, J.L. and Beaudry, R.M. (1993) Applying calcium chloride postharvest to improve highbush blueberry firmness. *HortScience* 28, 1033–1034.

Hanson, E.J., Throop, P.A., Serce, S., Ravenscroft, J. and Paul, E.A. (2002) Comparison of nitrification rates in blueberry and forest soils. *Journal of the American Society for Horticultural Science* 127, 136–142.

Hart, J., Strik, B., White, L. and Yang, W. (2006) *Nutrient Management for Blueberries in Oregon*. EM 8918, Oregon State University Extension Service, Corvallis, Oregon.

Hart, J.M., Sullivan, D.M., Anderson, N.P., Hulting, A.G., Horneck, D.A. and Christensen, N.W. (2013) *Soil Acidity in Oregon: Understanding and Using Concepts for Crop Production*. EM 9061, Oregon State University Extension Service, Corvallis, Oregon.

Hirschi, K.D. (2004) The calcium conundrum. Both versatile nutrient and specific signal. *Plant Physiology* 136, 2438–2448.

Hirzel, J. (2008) Principios de fertilización en frutales y vides. [Fertilization principles in fruits and grapes.] In: J. Hirzel (ed.) *Diagnóstico nutricional y principios de fertilización en frutales y vides*. (Nutritional diagnostic and fertilization principles in fruits and grapes). Colección Libros INIA-24. Instituto de Investigaciones Agropecuarias, Chillán, Chile, pp. 219–251.

Hirzel, J. (2014) Principios de fertilización en frutales y vides (Fertilization principles in fruits and grapes). In: Hirzel, J. (ed.) *Diagnóstico Nutricional y Principios de Fertilización en Frutales y Vides*. [Nutritional diagnostic and fertilization principles in fruits and grapes.] Colección Libros INIA-31, 2nd edn. Instituto de Investigaciones Agropecuarias, Chillán, Chile, pp. 225–293.

Hocking, B., Tyerman, S.D., Burton, R.A. and Gilliham, M. (2016) Fruit calcium: transport and physiology. *Frontiers in Plant Science* 7, 569.

Horneck, D., Hart, J., Stevens, R., Petrie S. and Altland J. (2004) *Acidifying Soil for Crop Production West of the Cascade Mountains* (Western Oregon and Washington). EM 8857-E, Oregon State University Extension Service, Corvallis, Oregon.

Johnson, C., Albrecht, J., Ketterings, Q., Beckman, J. and Stocking, K. (2005) *Nitrogen Basics: the Nitrogen Cycle*. Fact Sheet 2, Agronomy Fact Sheet Series. Cornell Cooperative Extension, Ithaca, New York.

Karp, K., Noormets, M., Starast, M. and Paal, T. (2006) The influence of mulching on nutrition and yield of 'Northblue' blueberry. *Acta Horticulturae* 715, 301–305.

Kochian, L.V. (2012) Rooting for more phosphorus. *Nature* 488, 466–467.

Korcak, R.F. (1989) Variation in nutrient requirements of blueberries and other calcifuges. *HortScience* 24, 573–578.

Kraiser, T., Gras, D.E., Gutiérrez, A.G., González, B. and Gutiérrez, R.A. (2011) A holistic view of nitrogen acquisition in plants. *Journal of Experimental Botany* 62, 1455–1466.

Kramer, A., Evinger, E.L. and Schrader, A.L. (1940) Effect of mulches and fertilizers on yield and survival of the dryland and highbush blueberries. *Proceedings of the American Society for Horticultural Science* 38, 455–461.

Krewer G. and NeSmith, D.S. (1999) *Blueberry Fertilization in Soil.* No. 01-1, University of Georgia Cooperative Extension, Athens, Georgia.

Krewer G. and Ruter, J. (2009) Fertilizing highbush blueberries in pine bark beds. Available at: https://secure.caes.uga.edu/extension/publications/files/pdf/B%20 1291_3.pdf (accessed 30 January 2018).

Kuepper G.L. (2003) Foliar fertilization. ATTRA – National Sustainable Agriculture Information Service, Butte, Montana. Available at: https://attra.ncat.org/attra-pub/download.php?id=286 (accessed 30 January 2018).

Kuepper G.L. and Diver, S. (2004) Blueberries: organic production. ATTRA – National Sustainable Agriculture Information Service, Butte, Montana. Available at: https://www.attra.ncat.org/attra-pub/blueberry.html (accessed 2 October 2010).

Larco, H., Strik, B.C., Bryla, D.R. and Sullivan, D.M. (2013) Mulch and fertilizer management practices for organic production of highbush blueberry. II. Impact on plant and soil nutrients during establishment. *HortScience* 48, 1484–1495.

Lawlor, D.W. (1991) Concepts of nutrition in relation to cellular processes and environment. In: Porter, J.E. and Lawlor, D.W. (eds). *Plant Growth: Interactions with Nutrition and Environment.* Seminar Series 43, Society for Experimental Biology. Cambridge University Press, Cambridge, pp. 1–32.

Loomis R.S. and Connor, D.J. (1992) *Crop Ecology: Productivity and Management in Agricultural Systems.* Cambridge University Press, Cambridge.

Machado, R.M.A., Bryla, D.R. and Vargas, O. (2014) Effects of salinity induced by ammonium sulfate fertilizer on root and shoot growth of highbush blueberry. *Acta Horticulturae* 1017, 407–414.

Maqbool, R., Percival, D.C., Adl, M.S., Zaman, Q.U. and Buszard, D. (2012) *In situ* estimation of foliar nitrogen in wild blueberry using reflectance spectra. *Canadian Journal of Plant Science* 92, 1155–1161.

Marschner, H. (1986) *Mineral Nutrition of Higher Plants.* Academic Press, London.

Merhaut D.J. (1993) Effects of nitrogen form on vegetative growth, and carbon/nitrogen assimilation, metabolism, and partitioning in blueberry. PhD thesis, University of Florida, Gainesville, Florida.

Merhaut, D.J. and Darnell, R.L. (1995) Ammonium and nitrate accumulation in containerized southern highbush blueberry plants. *HortScience* 30, 1378–1381.

Merhaut, D.J. and Darnell, R.L. (1996) Vegetative growth and nitrogen/carbon partitioning in blueberry as influenced by nitrogen fertilization. *Journal of the American Society for Horticultural Science* 30, 1378–1381.

Miller, S.A., Patel, N., Muller, A., Edwards, C.M. and Solomona, S.T. (2006) A comparison of organic and conventional nutrient management protocols for young blueberry nursery stock. *Acta Horticulturae* 715, 427–432.

Montanaro, G., Dichio, B., Lang, A., Mininni, A.N. and Xiloyannis, C. (2015) Fruit calcium accumulation coupled and uncoupled from its transpiration in kiwifruit. *Journal of Plant Physiology* 181, 67–74.

Morikawa, C.K. and Saigusa, M. (2004) Mineral composition and accumulation of silicon in tissues of blueberry (*Vaccinium corymbosum* cv. Bluecrop) cuttings. *Plant and Soil* 258, 1–8.

Myrold D.D. and Bottomley, P.J. (2008) Nitrogen mineralization and immobilization. In: Schepers, J.S. and Raun, W.R. (eds) *Nitrogen in Agricultural Systems.* Agronomy Monograph No. 49, American Society of Agronomy, Madison, Wisconsin, p. 157–172.

Ochmian, I. (2012) The impact of foliar application of calcium fertilizers on the quality of highbush blueberry fruits belonging to the 'Duke' cultivar. *Notulae Botanicae Horti Agrobotanici* 40, 163–169.

Patten, K.D., Haby, V.A., Leonard, A.T., Neuendorff, E.W. and Davis, J.V. (1988) Nitrogen source effects on rabbiteye blueberry plant–soil interactions. *Communications in Soil Science and Plant Analysis* 19, 1065–1074.

Petit-Jimenez, D., Gonzalez-Leon, A., Gonzalez-Aguilar, G., Sotelo-Mundo, R. and Baez-Sanudo, R. (2009) Permeability of cuticular membrane during the ontogeny of *Mangifera indica* L. *Acta Horticulturae* 820, 213–220.

Plank, C.O. and Kissel, D.E. (2010) *Plant Analysis Handbook for Georgia.* University of Georgia College of Agricultural and Environmental Sciences, Athens, Georgia. Available at: http//aesl.ces.uga.edu/publications/plant/Intro.htm (accessed 2 October, 2010).

Plank, C.O. and Tucker, M.R. (2000) Blueberry, rabbiteye. In: Campbell, C.R. (ed.) *Reference Sufficiency Ranges for Plant Analysis in the Southern Region of the United States.* Southern Cooperative Series Bulletin No. 394, North Carolina Department of Agriculture and Consumer Services Agronomic Division, Raleigh, North Carolina, pp. 99–100. Available at: http://www.ncagr.gov/agronomi/saaesd/scsb394.pdf (accessed 30 January 2018).

Plassard, C. and Dell, B. (2010) Phosphorus nutrition of mycorrhizal trees. *Tree Physiology* 30, 1129–1139

Poonnachit, U. and Darnell, R. (2004) Effect of ammonium and nitrate on ferric chelate reductase and nitrate reductase in *Vaccinium* species. *Annals of Botany* 93, 399–405.

Pritts, M. (2000) Blueberry nutrition on upland soils. *New York State Horticultural Society* 8, 15–16.

Ramaekers, L., Remans, R., Rao, I.M., Blair, M.W. and Vanderleyden J. (2010) Strategies for improving phosphorus acquisition efficiency of crop plants. *Field Crops Research* 117, 169–176

Rao, T.P., Yano, K., Yamauchi, A. and Tatsumi, J. (2000) Rhizosphere pH changes induced by exposure of shoots to light. *Plant Production Science* 3, 101–107.

Retamales, J.B. and Arredondo, G. (1995) Calcio en arándano. [Calcium in blueberry.] In: Yuri, J.A. and Retamales, J.B. (eds) *Calcio en Fruticultura.* [Calcium in Fruit Culture.] Universidad de Talca, Escuela de Agronomía, Talca, Chile, pp. 129–141.

Retamales, J.B. and Hanson, E.J. (1989) Fate of ^{15}N-labeled urea applied to mature highbush blueberries. *Journal of the American Society for Horticultural Science* 114, 920–923.

Retamales, J.B. and Hanson, E.J. (1990) Effect of nitrogen fertilizers on leaf and soil nitrogen levels in highbush blueberries. *Communications in Soil Science and Plant Analysis* 21, 2067–2078.

Retamales, J.B., Mena, C., Lobos, G. and Morales, Y. (2015) A regression analysis on factors affecting yield of highbush blueberries. *Scientia Horticulturae* 186, 7–14.

Reyes-Díaz, M., Alberdi, M. and Mora, M.L. (2009) Short-term aluminum stress differentially affects the photochemical efficiency of photosystem II in highbush blueberry genotypes. *Journal of the American Society for Horticultural Science* 134, 14–21.

Reyes-Díaz, M., Inostroza-Blancheteu, C., Millaleo, R., Cruces, E., Wulff-Zottele, C., Alberdi, M. and Mora, M.L. (2010) Long-term aluminum exposure effects on physiological and biochemical features of highbush blueberry cultivars. *Journal of the American Society for Horticultural Science* 135, 212–222.

Rojas-Lillo, Y., Alberdi, M., Acevedo, P., Inostroza-Blancheteau, C., Rengel, Z., Mora, M. and Reyes-Díaz, M. (2014) Manganese toxicity and UV-B radiation differentially influence the physiology and biochemistry of highbush blueberry (*Vaccinium corymbosum*) cultivars. *Functional Plant Biology* 41, 156–167.

Roper, T. and Schmitt, W. (2007) Using anion exchange membranes to monitor soil-available phosphorus for cranberries. *HortScience* 42, 897.

Rose, M.T., Patti, A.F., Little, K.R., Brown, A.L., Jackson, W.R. and Cavagnaro, T.R. (2014) A meta-analysis and review of plant-growth response to humic substances: practical implications for agriculture. *Advances in Agronomy* 124, 37–89.

Sadowsky, J.J., Hanson, E.J. and Schilder, A.M.C. (2012) Root colonization by ericoid mycorrhizae and dark septate endophytes in organic and conventional blueberry fields in Michigan. *International Journal of Fruit Science* 12, 169–187.

Sardans, J. and Peñuelas, J. (2015) Potassium: a neglected nutrient in global change. *Global Ecology and Biogeography* 24, 261–275.

Saure, M.C. (2005) Calcium translocation to fleshy fruit: its mechanism and endogenous control. *Scientia Horticulturae* 105, 65–89.

Savvas, D. and Ntatsi, G. (2015) Biostimulant activity of silicon in horticulture. *Scientia Horticulturae* 196, 66–81.

Scagel, C.F. (2005) Inoculation with ericoid mycorrhizal fungi alters fertilizer use of highbush blueberry cultivars. *HortScience* 40, 786–794.

Scagel, C.F. and Yang, W.Q. (2005) Cultural variation and mycorrhizal status of blueberry plants in NW Oregon commercial production fields. *International Journal of Fruit Science* 5, 85–111.

Schoebitz, M., López, M.D., Serri, H., Martínez, O. and Zagal, E. (2016) Combined application of microbial consortium and humic substances to improve the growth performance of blueberry seedlings. *Journal of Soil Science and Plant Nutrition* 16, 1010–1023.

Sharma, A., Patni, B., Shankhdhar, D. and Shankhdhar S.C. (2013) Zinc – an indispensable micronutrient. *Physiological and Molecular Biology of Plants* 19, 11–20.

Shaw, M. (2008) Soil pH is more important than fertilizer for blueberries. *New York Fruit Quarterly* 16, 25–28.

Shear, C.B. and Faust, M. (1980) Nutritional ranges in deciduous tree fruits and nuts. *Horticultural Reviews* 2, 142–163.

Smith, J.P., Schmidtke, L.M., Müller, M.C. and Holzapfel, B.P. (2014) Measurement of the concentration of nutrients in grapevine petioles by attenuated total reflectance Fourier transform infrared spectroscopy and chemometrics. *Australian Journal of Grape and Wine Research* 20, 299–309.

Smolarz, K. and Mercik, S. (1989) Growth and yield of highbush blueberry Bluecrop cv. (*Vaccinium corymbosum* L.) in relation to the level of nitrogen fertilizer. *Acta Horticulturae* 241, 171–174.

Stiles, W.C. and Reid, W.S. (1991) *Orchard Nutrition Management*. Information Bulletin 219, Cornell Cooperative Extension, Ithaca, New York.

Strik, B.C. (2014) Organic blueberry production systems – advances in research and industry. *Acta Horticulturae* 1017, 257–267.

Strik, B.C. (2016). A review of optimal systems for organic production of blueberry and blackberry for fresh and processed markets in the northwestern United States. *Scientia Horticulturae* 208, 92–103.

Strik, B.C. and Vance, A.J. (2015) Seasonal variation in leaf nutrient concentration of northern highbush blueberry cultivars grown in conventional and organic production systems. *HortScience* 50, 1453–1466.

Strik, B.C, Vance, A.J. and Finn, C.E. (2017) Northern highbush blueberry cultivars differed in yield and fruit quality in two organic production systems from planting to maturity. *HortScience* 52, 844–851.

Strik, N., Fisher, G., Hart, J., Ingham, R., Kaufman, D., Penhallegon, R., Pscheidt, J., William, R., Brun, C., Ahmedullah, M., Antonelli, A., Askham, L., Bristow, P., Havens, D., Scheer, B., Shanks, C. and Barney, D. (2010) *Highbush Blueberry Production*. Publication PNW215, Oregon State University Extension Service, Corvallis, Oregon.

Stückrath, R., Quevedo, R., de la Fuente, L., Hernández, A. and Sepúlveda, V. (2008) Effect of foliar application of calcium on the quality of blueberry fruits. *Journal of Plant Nutrition* 31, 1299–1312.

Subbarao, G.V., Ito, O., Sahrawat, K.L., Berry, W.L., Nakahara, K., Ishikawa, T., Watanabe, T., Suenaga, K., Rondon, M. and Rao, I.M. (2006) Scope and strategies for regulation of nitrification in agricultural systems: challenges and opportunities. *Critical Reviews in Plant Science* 25, 303–335.

Sullivan, D.M., Bryla, D.R. and Costello, R.C. (2014) Chemical characteristics of custom compost for highbush blueberry. In: He, Z. and Zhang, H. (eds) *Applied Manure and Nutrient Chemistry for Sustainable Agriculture and Environment*. Springer Science+Business Media, Dordrecht, The Netherlands, pp. 293–311.

Tertuliano, M., Krewer, G., Smith, J.E., Plattner, K., Clark, J., Jacobs, J., Andrews, E., Stanaland, D., Andersen, P., Liburd, O., Fonsah, E.G. and Scherm, H. (2012) Growing organic rabbiteye blueberries in Georgia, USA: results of two multi-year field studies. *International Journal of Fruit Science* 12, 205–215.

Throop, P.A. and Hanson, E.J. (1997) Effect of application date on absorption of ^{15}N by highbush blueberry. *Journal of the American Society for Horticultural Science* 114, 728–732.

Throop, P.A. and Hanson, E.J. (1998) Nitrification and utilization of fertilizer nitrogen by highbush blueberry. *Journal of Plant Nutrition* 21, 1731–1742.

Tosso, J. (1985) *Suelos Volcánicos de Chile.* [*Chilean Volcanic Soils.*] INIA, Santiago, Chile.

Vadillo, J. (2006) *Plan General de Transformación Zona Regable Andévalo Occidental Fronterizo (Huelva)*. Available at: http://www.juntadeandalucia.es/medioambiente/web/Bloques_Tematicos/Calidad_Ambiental/Prevencion_Ambiental/evaluacion_ambiental_planes_y_programas/anejo02edafologiayygeologia.pdf (accessed 23 February 2011).

Valenzuela-Estrada, L.R., Vera-Caraballo, V., Ruth, L.E. and Eissenstat, D.M. (2008) Root anatomy, morphology, and longevity orders in *Vaccinium corymbosum* (Ericaceae). *American Journal of Botany* 95, 1506–1514.

van de Wiel, C.C.M., van der Linden, C.G. and Scholten, O.E. (2015) Improving phosphorus use efficiency in agriculture: opportunities for breeding. *Euphytica* 207, 1–22.

Vance, A.J., Jones, P. and Strik, B.C. (2017) Foliar calcium applications do not improve quality or shelf life of strawberry, raspberry, blackberry, or blueberry fruit. *HortScience* 52, 382–387.

Vargas, O.L. and Bryla, D.R. (2015) Growth and fruit production of highbush blueberry fertilized with ammonium sulfate and urea applied by fertigation or as granular fertilizer. *HortScience* 50, 479–485.

Vega, A.R., Garciga, M., Rodriguez, A., Prat, L. and Mella, J. (2009) Blueberries mycorrhizal symbiosis outside the boundaries of natural dispersion for Ericaceous plants in Chile. *Acta Horticulturae* 810, 665–671.

Wang, S.Y., Chen, C., Sciarappa, W., Wang, C.Y. and Camp, M.J. (2008) Fruit quality, antioxidant capacity, and flavonoid content of organically and conventionally grown blueberries. *Journal of Agricultural and Food Chemistry* 56, 5788–5794.

Weinbaum, S.A., Merwin, M.L. and Muraoka, T.T. (1978) Seasonal variation in nitrate uptake efficiency and distribution of absorbed nitrogen in no-bearing prune trees. *Journal of the American Society for Horticultural Science* 103, 516–519.

White, P.J. and Broadley, M.R. (2003) Calcium in plants. *Annals of Botany* 92, 487–511.

Widders, I.E. and Hancock, J.F. (1994) Effects of foliar nutrient application on highbush blueberries. *Journal of Small Fruit and Viticulture* 2, 51–62.

Wojcik, P. (2005) Response of 'Bluecrop' highbush blueberry to boron fertilization. *Journal of Plant Nutrition* 28, 1897–1906.

Wright, G.C., Patten, K.D. and Drew, M.C. (1995) Labeled sodium ($^{22}Na^+$) uptake and translocation in rabbiteye blueberry exposed to sodium chloride and supplemental calcium. *Journal of the American Society for Horticultural Science* 120, 177–182.

Yang, L.T., Qi, Y.P., Jiang, H.X. and Chen, L.S. (2013) Roles of organic acid anion secretion in aluminium tolerance of higher plants. *BioMed Research International* 2013, 173682.

Yang, W.Q. and Goulart, B.L. (1997) Aluminum and phosphorus interactions in mycorrhizal and nonmycorrhizal highbush blueberry plantlets. *Journal of the American Society for Horticultural Science* 122, 24–30.

Yang, W.Q., Goulart, B.L., Demchak, K. and Li, Y. (2002) Interactive effects of mycorrhizal inoculation and organic soil amendments on nitrogen acquisition and growth of highbush blueberry. *Journal of the American Society for Horticultural Science* 127, 742–748.

Zhao, X.Q., Chen, R.F. and Shen, R.F. (2014) Coadaptation of plants to multiple stresses in acidic soils. *Soil Science* 179, 503–513.

6

BLUEBERRY FIELD MANAGEMENT AND HARVESTING

INTRODUCTION

Once a grower has selected both the site and the plant material (cultivar), it is time to get involved in the management of the crop. The goals of blueberry management are: (i) to avoid or minimize stressful conditions to the plant; (iii) to optimize plant function in relation to the environmental conditions; (iii) to properly balance growth among the various parts of the plant; and (iv) to harvest the fruit at a low cost and without reducing its quality and post-harvest life potential. To be cost-effective, these interventions must increase plant growth and yield sufficiently to justify the cost involved. As blueberry cultivars have different genetic backgrounds and are planted in widely different environments, their response to stress and to horticultural practices varies widely. Because environmental stress is not constant throughout the season, it is important to consider not only the degree of intervention but also the appropriate timing.

In this chapter, we cover several management practices of utmost importance in blueberry cultivation, including mulching, protected cultivation, soil-less culture, irrigation, pruning, grafting, pollination and harvesting. The impact of insufficient or excess water is explained. The methods to determine water status in blueberry fields are discussed and the most important irrigation systems compared. In the case of mulching, the characteristics, considerations and effects of both organic and plastic materials are presented. Soil-less and protected cultivation techniques and their effects on plant behaviour are addressed. The timing, intensity and effects of pruning are presented, along with the reproductive biology of blueberries and the most effective planting designs to maximize yield and fruit quality. We include a section on pollination and its impact on fruit size, fruit set and yield. The section on grafting presents the developments that have occurred in this area. We also discuss the different types of harvesting and their effect on fruit quality at both harvest and postharvest.

FUNDAMENTAL CONCEPTS OF IRRIGATION

Water has several functions within the plant: (i) it is the milieu for transporting nutrients and growth substances to the various parts of the plant; (ii) it is necessary for the plant to regulate the temperature of its tissues, especially when plants are grown in hot environments; and (iii) it is needed to maintain normal physiological activity, as water is the driving force for growth. In fruit crops, such as blueberry, there is often a premium price for fresh fruit where water content is maximized.

Water stress can be due to either excess or deficit. Excesses normally occur in blueberries when the plant is dormant or near bud break. Deficits usually happen during the middle of the growing season, when temperatures are high and demand for water is at its maximum (Darnell, 2006). There are various techniques to alleviate water-related stress in perennial crops. They can be classified as: (i) strategic, which are undertaken before planting and comprise cultivar selection, plant density, use of mulch, trellising, irrigation and drainage system selection and design; or (ii) tactical, which are the opportunistic implementation of strategies once the crop is planted, and include fertilizers (dose, method, timing), pruning (timing and intensity), weed control, management of irrigation and drainage systems (Daebeke and Aboudrare, 2004).

To provide an adequate water supply to plants, it is important that: (i) the soil root zone is large enough to supply the plant; (ii) the soil water supply is replenished frequently enough to avoid water stress; (iii) the water freely infiltrates the soil without leaching nutrients to depths below the root zone; and (iv) an effective water-absorbing root system is maintained throughout the plant's life cycle.

The low root density of blueberries leads to slow water movement within the soil profile and consequently, under moderate to high evaporative demand on the canopy, water stress develops within the plant. As a result, the capacity of the water to move within the plant is controlled more by the transpiration induced by the environment than by soil moisture. This places great emphasis on understanding canopy microclimate and the behaviour of stomata (the pores in leaves where plants exchange water and gases with the environment) as a means to control not only water loss but also water potential.

Before discussing the irrigation of blueberries, it is important to understand some basic concepts regarding water relationships in fruit crops.

FUNDAMENTAL CONCEPTS IN WATER RELATIONSHIPS

Water is essential for plant growth and development. The soil must supply large quantities of water to meet the transpiration demands of growing plants. The behaviour of water in the soil and the plant has been unified in a single energy concept: the water potential or Ψ_w. This concept considers the soil–plant–atmosphere chain as a continuum. Although Ψ_w is usually expressed in units of pressure (megapascals, MPa), it indicates the energy of the water in a given part of the system.

The energy concept is used to explain why water enters the soil, is absorbed and transported through plants, and then evaporates into the atmosphere through transpiration. Under most conditions (relative humidity less than 100%), water potential (Ψ_w) is highest in the soil and lowest in the atmosphere, with intermediate values in the plant. There is thus a gradient from the soil to the plant to the atmosphere. Water potentials in the soil and the plant are related (although not linearly), as can be seen in water-stressed blueberries (Table 6.1).

The capacity of a soil to store water is the result of the attraction of the soil matrix (solids) for water. Water interacting with the soil loses energy, so the water potential (or energy) of the water retained by the soil is lower than that of pure water. The interaction of ions (e.g. salts, fertilizers) and other solutes with water (called osmotic forces) further reduces the energy level (or potential) of water in the soil. Gravitational forces also affect the energy or potential of soil water, particularly under saturated moisture conditions. A major consequence of this low energy status (or potential) of soil water is that the removal of water from the soil matrix by a plant root requires the expenditure of energy, which is ultimately supplied by sunlight through photosynthesis. Various forces act on the water in the system. The relationship among the various energies required to move water from the soil and within the plant is expressed by the water potential and is the result of the gravitational potential (Ψ_g), the matric potential (Ψ_p) and the osmotic potential (Ψ_s):

$$\Psi_w = \Psi_g + \Psi_p + \Psi_s$$

Table 6.1. Soil and leaf water potential in well-watered (daily irrigated) and water-stressed (not irrigated for the designated number of days) 2-year-old 'Rancocas' highbush blueberry plants. Values are means of four replicates ± standard error. (Adapted from Lee *et al.*, 2006.)

| Treatment | Day | Water potential (MPa) | |
		Soil	Leaf
Well watered	1	-0.20 ± 0.013	-1.01 ± 0.033
	3	-0.18 ± 0.021	-0.99 ± 0.024
	5	-0.20 ± 0.010	-1.00 ± 0.054
	7	-0.22 ± 0.031	-1.07 ± 0.023
Water stressed	1	-0.21 ± 0.027	-1.02 ± 0.056
	3	-0.33 ± 0.017	-1.42 ± 0.049
	5	-0.83 ± 0.068	-1.58 ± 0.075
	7	-0.99 ± 0.045	-1.79 ± 0.038

As this equation shows, the three components are additive. Ψ has a negative sign, representing the energy difference between different components of the system. Under normal field conditions, after free drainage of water has occurred (24–72 h after rain or irrigation), the gravitational potential is considered insignificant relative to the other components.

The central importance of Ψ in water relationships arises because differences in total Ψ provide the driving force for water movement and therefore determine the direction of water movement in any system. Water moves freely into the root. Water then passes from an area of high water potential (the soil) into the root, which is an area of low water potential. The water then crosses through various tissues until it reaches the vascular cells (xylem). Transpiration provides the pulling force for drawing water up into the shoot of the plant. In this scheme, it is assumed that the gaseous areas of plants including the stomatal cavity are saturated with water vapour and that the relative humidity is 100% (Salisbury and Ross, 1991).

The matric potential not only gives the soil its water-storing ability but also functions to determine water movement. In agricultural situations, the matric potential is measured with a soil tensiometer and is always negative. Water always moves down its water potential gradient from areas of higher water potential (higher water concentration) to areas of lower water potential. In the soil, the osmotic potential, which is also negative, originates from the presence of solutes. The most obvious plant osmotic adjustment in fruit crops is by the fruit, which accumulate large quantities of soluble solids (mostly sugars) over the season as a result of photosynthesis (the transformation of solar energy into carbohydrates; see Chapter 4, this volume).

Unlike the matric potential, the osmotic potential value of a soil does not affect the movement of soil water, nor does it influence the retention of water by the soil matrix. Its main effect is its influence on water uptake by plant roots. A soil solution with a high solute content (such as when high dosages of fertilizer are applied) effectively restricts water movement through the root cells because the presence of salts lowers the energy level in the plant root.

As water is removed from the soil, the remaining water is held at a more negative potential, indicating that the water is under tension and that work must be done to extract water from the soil (Fig. 6.1). The relationship between the matric potential (related to water-storing ability and water movement within the soil) and the moisture content for soils of different textures is called the soil moisture characteristic curve. Under normal (non-saline) conditions, such as those that commonly occur in blueberry fields, the matric potential dominates the availability and movement of soil water. At a given matric potential, the heavier-textured soils (with finer particles, such as clay) hold more water than the lighter-textured (sandy) soils. After free drainage has occurred in a saturated soil ($\Psi_g = 0$), the moisture content of the soil is said to be at 'field capacity'. Depending on the soil texture, this free drainage would occur in 24 h (sandy soil) to 72 h (clayey soil). At this point, the pores are filled

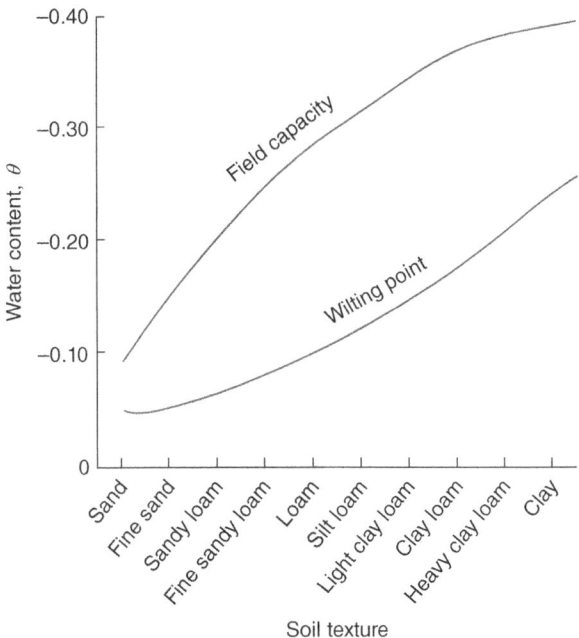

Fig. 6.1. Water-holding properties for various soil textures. For each soil, the available water supply is the difference between the field capacity and the wilting point; retained water volumes correspond to the range between field capacity and wilting point. (Adapted from Shan, 2011.)

with water. Plants remaining in this condition may have insufficient O_2 for normal respiration. For most soils, this point is at a soil matric potential of about -0.033 MPa or -0.33 bar (Fig. 6.1).

The matric potential decreases (becomes more negative) as water is removed from the soil, until this potential is equal to that of the plant. At this point, the plant can no longer remove water from the soil, and the turgor or water pressure in the plant cells (responsible for keeping the cells rigid) drops to zero and the plant wilts. If the water is not replenished, the plant will wilt permanently and will not recover, even if placed in an atmosphere of 100% relative humidity.

Plant species have different water potentials before permanent wilting occurs. For most species, permanent wilting would occur at a water potential of -1.5 MPa or -15 bar. The amount of water retained in a given soil at -1.5 MPa is referred to as the 'permanent wilting point'. The amount of water held between field capacity and the permanent wilting point is known as 'available water' (Fig. 6.1).

The water potential in a non-transpiring plant approaches the water potential in the soil, but as a result of transpiration, a gradient in water potential develops from the soil, to and through the plant; thus, water moves from the soil, through the plant and out to the external atmosphere. The water potential can thus be assessed in different aerial parts of the plant (e.g. leaves, canes). The plant water status depends on a combination of soil, atmospheric and crop factors (Jones and Tardieu, 1998).

Evapotranspiration is defined as crop water losses by both evaporation from the soil and transpiration from the plant. Evaporation is the term used to describe the change from a liquid state to a gaseous state and is purely a physical process. The level below that of soil water potential to which the plant water potential falls during the day depends on the magnitude and duration of the lag of absorption behind transpiration. Transpiration is governed by atmospheric factors, such as radiation, air temperature, relative humidity and wind speed, while water absorption is governed partly by soil factors such as water content and unsaturated conductivity (the capacity of water to move within the soil between the field capacity and the permanent wilting point). Water absorption is also a function of plant factors, such as the amount of absorbing root surface and the permeability of the roots (Salisbury and Ross, 1991).

Transpiration is a biophysical process involving soil moisture content, the passage of water throughout the plant, movement of water from leaf stomata and transport of water into the atmosphere by processes of diffusion (movement from higher to lower solute concentration) and turbulence (air movement that cannot normally be seen). Transpiration involves a physical change of state of water (liquid to gas). The propensity of plants to lose water through the leaves (transpiration) depends on the activity of the stomata. These pores allow photosynthesis and transpiration to occur simultaneously. The degree of stomatal opening is measured as the stomatal conductance (g_s). The degree of stomatal opening is to a large part related to the gradient of vapour pressure between the plant and the environment. Increased transpiration rates are generally associated with higher photosynthetic and dry-weight accumulation rates (Loomis and Connor, 1992).

The amount of fruit harvested can strongly influence the water status of the plant if it results in a reduction in root growth. This relationship is especially important in recently established plants, where high crop loads can severely reduce root growth. A reduced root volume lessens the total soil water availability and results in low water potential over the entire season. It can also impact on plant size and reduce future yield potentials.

WATER DEFICIT IN BLUEBERRIES

Water is the most abundant component in all plants. Depending on the tissue, water encompasses as much as 70% (canes) to 85% (fruit) of the weight of organs in blueberries (Retamales and Hanson, 1989). As a result, all plant processes are eventually affected in diverse magnitudes by water stress. In general, water deficits reduce transpiration the most, followed by photosynthesis and, to a much more limited extent, respiration. A strong reduction in leaf area coupled with reduced photosynthesis results in major reductions in dry-weight accumulation. Heavy cropping accentuates leaf area reduction but increases transpiration rates per leaf area. Water stress reduces mineral availability and water uptake, but specific elements can be affected differently.

In order to select the best drought stress markers in peach trees, Jiménez *et al.* (2013) studied the physiological, biochemical and molecular drought responses of four *Prunus* spp. rootstocks (GF 677, Cadaman, ROOTPAC-20 and ROOTPAC-R) budded with the peach cultivar 'Catherina'. Trees were grown in 15 l containers and subjected to a progressive water stress for 26 days, monitoring soil moisture content by time domain reflectometry. The authors concluded that accumulation of sorbitol, raffinose and proline in different tissues could be used as markers of drought tolerance in peach cultivars grafted on *Prunus* sp. rootstocks. It should be possible in the coming years to develop these kinds of stress indicators for different blueberry genotypes.

Cameron *et al.* (1989) found that water-stressed 2-year-old NHB 'Bluecrop' and 'Jersey' plants allocated a higher percentage of dry matter to their root system at the expense of aboveground organs. Under field conditions and without rain or irrigation during summer, water stress often develops in blueberries within 3–7 days, varying depending on plant age, cultural practices, phenological development, soil texture and weather conditions. Stress symptoms included reduced shoot growth, increased root growth, lower water use and reduced photosynthesis. Young, succulent shoots and leaves wilted readily under dry conditions, and when drought persisted, leaf margins and tips became necrotic and scorched. This scorching is similar in appearance to the salt injury often associated with overfertilization (Bryla, 2011).

Worldwide water shortages are predicted to become more frequent in the near future with global warming. Boland *et al.* (2006) studied the motivations of stone and pome fruit growers in south-eastern Australia for adopting improved management practices. They concluded that adoption of sustainable irrigation practices was generally not limited by lack of knowledge and that water savings are not the key incentive for growers to improve irrigation management (micro-irrigation and soil moisture monitoring), but that other potential benefits (e.g. fruit quality, postharvest life) must also be demonstrated before increased adoption of regulated water deficit occurs.

There is a differential tolerance to water stress among blueberry species. Rabbiteye blueberries are generally more tolerant than highbush (Eck, 1988). Highbush blueberry is a shallow-rooted crop that is highly sensitive to soil water deficit. However, the impact of water stress on plant functioning depends on the level of stress, the time of the season, the extension of the stress period and the cultivar. When highbush blueberries are exposed to even mild episodes of drought, vegetative growth is rapidly reduced and fruit development is often diminished. When Améglio *et al.* (2000) subjected 9-year-old NHB 'Bluecrop' plants in containers to 10 days of water deficit in the middle of the growing season, they found that water potential, embolism, transpiration (leaf and plant) and g_s were highly sensitive to drought (as judged by rapid stomatal closure and reduced transpiration). The rapid drop in g_s effectively restricted water loss and prevented tissue death. However, 5 days after irrigation was initiated, the water-stressed plants had similar transpiration levels to the control plants.

In another experiment, 7-year-old NHB 'Bluecrop' blueberries were grown in peat and pine bark (50:50) substrate in containers and were subjected during the growing season to mild (replacement of 65% of the water transpired by well-watered plants) or severe (replacement of 35% of the water transpired by well-watered plants) water stress. This water stress was imposed for 3 weeks in four different periods: fruit growth (weeks 7–4 before peak harvest), ripening (weeks 4–1 before peak harvest), harvest (with week 2 defined as peak harvest) and postharvest (weeks 4–7 after peak harvest). NHB blueberries exhibited a marked sensitivity to water stress, in particular between fruit formation and maturation (Mingeau *et al.*, 2001). The magnitude of the plant response was related to phenological stage. Plants were able to recover the transpiration levels of control faster under mild stress than under severe stress. The periods of maximum water requirement were the first 2 weeks after petal fall and the 2 weeks before and after harvest. In control plants (no water stress), most shoot elongation occurred during the green fruit stage, but stressed plants had negligible shoot elongation during this period. After rehydration, their shoot elongation rate was greater than control plants, particularly in those that had been under severe stress. For all water stress periods during the vegetative season, stem diameter did not increase while under water stress. Severe water restriction reduced yields by 31% in plants stressed during initial fruit growth and by 49% in plants stressed near harvest. The effect of stress on yield was mainly due to a reduction in fruit size and not fruit number (Table 6.2).

Mingeau *et al.* (2001) also studied the effects of water stress during the postharvest period, at which time NHB blueberries would be expected to initiate flower buds. Severe stress restricted to this period did not have a significant impact on yield the following year. Although severe stress reduced fruit number per bush through a lower number of flower buds per bush (53% of control plants), there was a compensatory effect due to larger fruit size (59% greater than control). This suggests that water stress during flower induction might have a similar effect to winter pruning on subsequent fruit production. However, the authors cautioned that water stress imposed at this time to obtain large-sized fruit could jeopardize plant architecture by encouraging the formation of saplings from the crown. Water stress at this period could also cause early defoliation, which would reduce the level of reserves and, with this, the development of leaf and flower buds in the following season.

Drought and mandatory water restrictions are limiting the availability of irrigation water in many important blueberry-growing regions, such as the Pacific coast (USA), Chile and Australia. New strategies are needed to maintain

Table 6.2. Effects of time and intensity of water stress on yield and its components in 7-year-old 'Bluecrop' NHB blueberries. The stress period indicates dates for the northern hemispheres (Adapted from Mingeau *et al.*, 2001.)

Stress period	Stress intensity (% of transpiration)	Yield per bush (g)	No. of fruit per bush	Fruit weight (g)
Control	100	2850[a]	3150[a]	0.92[a]
Fruit growth (28 May–16 June)	65	2225[b]	2961[a]	0.76[a,b]
	35	1965[b]	3181[a]	0.64[b,c]
Ripening (18 June–7 July)	65	2250[b]	2992[a]	0.77[a,b]
	35	1450[c]	2667[a]	0.56[c]
Harvest (10–29 July)	65	2100[b]	2835[a]	0.76[a,b]

[a,b,c]Mean values within a column with non-identical superscript letters were significantly different at $P < 0.05$ (Newman–Keuls test).

yield and fruit quality using less water. There are two main approaches to implementing deficit irrigation: regulated deficit irrigation (RDI) and partial root-zone drying (PRD). Deficit irrigation occurs when the water applied to a crop is below evapotranspiration requirements. RDI is implemented by imposing prescribed limits on soil moisture depletion and limits to irrigation inputs at specific phenological stages of the crop cycle. PRD involves irrigation inputs that are below evapotranspiration requirements, but irrigation is alternated between plant roots separated into drying and wetted zones. When successfully applied, plants respond to PRD by reducing transpiration by partially closing their stomata without loss of plant turgor (Keen and Slavich, 2012).

To address deficit irrigation, researchers in Chile, Oregon and Australia developed different approaches. Lobos *et al.* (2016) and Lobos (2016) studied the effect of RDI on yield, physiological parameters and fruit quality of NHB 'Brigitta' blueberries. The trials were carried out on 6-year-old plants in Colbún, Chile (latitude 35°41′S) during two seasons, and on 26-year-old bushes in South Haven, Michigan (latitude 42°21′N) for one season. Plants were subjected to irrigation treatments that replaced 50, 75 or 100% (control) of actual evapotranspiration (ET_a). Severe water deficit (50% ET_a) decreased photosynthetic rate, vegetative growth (year 2 in the Colbún and South Haven trials) and fruit quality (berry size, titratable acidity, soluble solids and berry weight), and increased fruit oxidative stress during both seasons in Colbún. The 50% ET_a treatment also had the highest yield reduction during year 2 in Colbún. In contrast, mild water stress (75% ET_a) produced similar fruit yields and quality (firmness, fruit size, titratable acidity, soluble solids and berry weight) as the 100% ET_a treatment but with higher water productivity and intermediate antioxidant capacity. The authors concluded that the application of 25% less water

(75% ET_a) to NHB blueberries produced similar yields as the fully irrigated plants and did not alter fruit quality or levels of antioxidants.

In Oregon, Almutairi *et al.* (2017) evaluated RDI for 2 years in a 7-year-old planting of the NHB 'Elliott' with three potential options for reducing water use: deficit irrigation, irrigation cut-offs and crop thinning. The treatments were: (i) no thinning and 50% crop removal in combination with either full irrigation at 100% of estimated crop evapotranspiration (ET_c); (ii) deficit irrigation at 50% ET_c (applied throughout the growing season); (iii) or full irrigation with irrigation cut-off for 4–6 weeks during early (early to late green fruit) or late (fruit colouring to harvest) stages of fruit development. The authors concluded that deficit irrigation and early irrigation cut-offs would be two possible options for reducing irrigation water use in NHBs, but each had positive and negative points. Deficit irrigation used half as much water as full irrigation but had little to no effect on yield or fruit quality. However, deficit irrigation resulted in less vegetative growth than full irrigation, which reduced pruning labour each year but, if not managed properly, could diminish fruit production. Deficit irrigation also hastened fruit ripening in one year, which, depending on the cultivar, labour availability and the market, could be an advantage or disadvantage. Cutting off irrigation early had no effect on yield in year 1 and delayed fruit ripening in year 2. However, it decreased yield in year 2 when the plants were not thinned, and produced smaller berries with less soluble solids content than either full or deficit irrigation. Judicious use of early cut-off irrigation may therefore be warranted at times but should probably be restricted to years of water scarcity. Cutting off irrigation late also reduced water use, but yields dropped by 35%, and it resulted in smaller but firmer berries than full or deficit irrigation. However, the fruit had higher soluble solids and acids, and lasted longer in cold storage. Thus, while late cut-offs reduced yield, deficit irrigation could be used during late stages of fruit development to increase fruit quality and storage life. More research is needed to find a good balance between late-season water restrictions and fruit yield/quality in NHBs. Fruit removal (crop thinning) was laborious and showed little promise for reducing water stress during moderate or severe soil water deficits. Its only advantage was greater vegetative growth, which was important when irrigation was cut off early to increase berry weight. Fruit bud thinning through proper pruning is essential for maintaining production and quality in NHBs. However, it does not appear that overthinning through severe pruning is an effective tool for mitigating water restrictions.

Keen and Slavich (2012) carried out a glasshouse experiment in New South Wales, Australia (latitude 29°S), to assess the feasibility of applying PRD to 1-year-old SHB 'Star' blueberry plants. A subsequent field experiment was established on 5-year-old field-grown 'Star' plants on a basaltic clay loam ferrosol soil to assess four irrigation strategies aimed at improving water-use efficiency. Applying PRD to plants during the glasshouse experiment reduced g_s without reducing plant water potential. Hindered by high rainfall, a physiological response to PRD was not observed in field-grown plants. However, irrigation scheduling using a single crop coefficient (K_c) curve constructed from FAO 56 guidelines (http://www.fao.org/docrep/X0490E/X0490E00.htm) and postharvest RDI delivered annual water savings of 0.8 and 1.3 million l/ha, respectively, compared with a total 3.6 million litres/ha applied using a 'rule-of-thumb' approach commonly adopted by Australian blueberry growers. These savings were achieved without reducing berry yield or quality.

Sensitivity to short-term water deficits varies among NHB cultivars. As soil water was depleted, 'Duke' maintained, on average, significantly higher stem water potentials and greater g_s than 'Elliott', while 'Bluecrop' appeared to be less tolerant to short-term water deficits (Bryla and Strik, 2006).

Spermidine (SPD) plays an important role in plant defence responses to drought stress. Chen *et al.* (2017) investigated the effects of exogenously applied SPD on plant growth, net photosynthetic rate (P_n), antioxidant enzyme activities and chlorophyll, malondialdehyde and phytohormone concentrations in leaves of SHB 'Misty' blueberry plantlets under drought stress. Drought stress severely reduced the relative water content (RWC), chlorophyll concentration, P_n, specific leaf weight and indole-3-acetic acid, gibberellic acid and polyamine concentrations, but increased relative electrolyte conductivity and malondialdehyde, total soluble sugar and abscisic acid concentrations, as well as antioxidant enzyme activities. Compared with no SPD treatment (control), SPD increased the RWC, chlorophyll content, P_n, specific leaf weight and antioxidant enzyme activities, and decreased relative electrolyte conductivity and malondialdehyde, total soluble sugar and abscisic acid contents in plantlets. These results suggest that SPD could be used as a growth regulator for improving blueberry plant growth under drought stress.

WATER EXCESS (FLOODING)

Although wild highbush blueberries are found growing on hummocks in wetlands, flooded areas are not recommended for blueberries, as these crops perform much better on dry land. It has been suggested that the fibrous root system of blueberries may aid the plant in surviving flooding, as higher O_2 levels necessary for root growth frequently occur close to the soil surface in poorly aerated soils. The O_2 concentration adequate for optimum plant growth has been established at $0.01\,kg/m^3$; these levels were detected at a 30 cm depth (Topp *et al.*, 2000).

Flooding stress is primarily due to a deficiency in soil O_2, as O_2 in soil pores is depleted by microbial and root respiration (Darnell, 2006). Soil O_2 levels can drop from 20% to less than 5% within 2 days of flooding (Crane and Davies, 1987). The effects of flooding on cultivated blueberries vary with the duration, time of the season and sometimes the species. Rabbiteye blueberries are reported to be more flood tolerant than highbush blueberries (Davies and Flore,

1986a; Eck, 1988), but these differences appear slight. It is possible that the perceived differences in flooding tolerance might be due to the increased sensitivity of most highbush blueberry cultivars to *Phytophthora* root rot, rather than to physiological factors (Davies and Flore, 1986a; Crane and Davies, 1988).

Bryla and Linderman (2007) performed a 2-year study in Oregon to determine the effects of irrigation method (overhead sprinklers, microsprays or drip irrigation) and level of water application (50, 100 or 150% of the estimated ET_c requirement) on the development of *Phytophthora* root rot. NHB 'Duke' plants were growing on mulched, raised beds in a silty clay loam field. Less plant growth with drip irrigation was associated with higher levels of *Phytophthora* infection. Infection by the pathogen increased with water application, regardless of irrigation method, but averaged 14% with drip irrigation and only 7% with sprinklers and microsprays. Roots were also infected by *Pythium* spp., whose infection also increased with the total amount of water applied but, unlike *Phytophthora*, was similar among irrigation methods.

Blueberry plants respond to flooding by stomatal closure, which reduces transpiration and slows damaging reductions in water potentials. However, stomatal closure limits gas exchange, which decreases photosynthesis and eventually may lead to growth cessation and death. The levels of g_s and transpiration decreased significantly after 4–5 days of flooding during the growing season, and photosynthetic rates in NHB blueberries decreased to 60% of those of the non-flooded control within 2 days of flooding (Davies and Flore, 1986a). After 1 day of flooding, rabbiteye blueberries in containers had net photosynthetic rates that were 64% lower than non-flooded control plants (Davies and Flore, 1986b). C assimilation (photosynthesis minus respiration) in highbush blueberries became negative after 11–19 days of flooding due to decreased photosynthesis, reduced stomatal conductance to CO_2 and high leaf temperatures, which increased respiration (Darnell, 2006).

It has been reported that the cultivated highbush blueberry can survive extended periods of flooding, if flooding occurs at times other than the spring period of active growth; however, growth and plant development can be severely impacted at any time during the season (Darnell, 2006). Flooding of highbush blueberries reduces water and nutrient uptake, suppresses plant growth and reduces yield and fruit quality (Abbott and Gough, 1987a). The reduction in water uptake under flooded conditions has been attributed to the adverse effects of high CO_2 and low O_2 concentrations on the permeability of root cells to water.

Flooding highbush blueberries for 4 months at different stages reduced both vegetative and reproductive growth (Abbott and Gough, 1987b). The flooding reduced shoot and internode length, leaf size and root dry weight, and plants had fewer flower buds, fewer flowers per bud, delayed bloom, reduced fruit set and weight, and less soluble solids in the fruit (Table 6.3). The highest amount of damage occurred when flooding started at bud break. The negative

Table 6.3. Effect of 4 months of root-zone flooding on reproductive growth in 2-year-old container-grown 'Bluecrop' NHB blueberry. Data on flowers per bud and date of full bloom are means of 1 year, while data on flower buds per shoot, fruit set, fruit weight and soluble solids are means of 2 years. (Adapted from Abbott and Gough, 1987b.)

Treatment	Flower buds per shoot	Flowers per bud	Full bloom	Fruit set (%)	Fruit weight (g)	Soluble solids (%)
Control	4.3[a]	4.2[a]	17 May	87.0[a]	1.40[a]	11.0[a]
Flooded/planted	1.7[a]	1.9[b]	21 May	55.3[b]	1.10[b]	9.9[b]
Continuous flooding	1.1[b]	1.4[b]	17 May	52.2[b]	0.78[c]	8.8[c]

[a,b,c,d]Mean values within columns with non-identical superscript letters were significantly different at $P < 0.05$ (Duncan's test).

effects on reproductive development could have been partially due to a reduction in the production and/or translocation of hormones such as cytokinins and gibberellins, which promote bud activity.

Flooding duration also affects the responsiveness of blueberry plants to environmental stimuli. Davies and Flore (1986a) found that g_s and C assimilation declined during the first 4 or 5 days of flooding, although the stomata were still responsive to changes in vapour pressure gradient. As flooding was prolonged, g_s declined and the C balance became negative. After 24 days of flooding, g_s was near zero and the stomata did not respond to changes in the environment. The C balance continued with negative values and, depending on the cultivar, the leaves became red or chlorotic. Recovery after 24 days of flooding to pre-flood g_s and transpiration values required 16–18 days for rabbiteye blueberries, while highbush blueberries had recovered only 64% of the levels of control plants by that time.

IRRIGATION

In many blueberry-producing regions, rain and/or the water table do not provide all the water requirements of the plants (Williamson *et al.*, 2006). Blueberries are shallow-rooted plants that are rapidly subject to drought injury (Spiers, 1986). The cultivated highbush blueberry root and conduction systems have minimal lateral transport. It has been demonstrated that providing irrigation to one side of the plant (as in drip irrigation) can result in extreme differential growth and disrupted fruit production (Abbott and Gough, 1986). Care should be taken in the design, operation and maintenance of irrigation equipment to ensure that it provides uniform distribution of water for plant use.

Under most commercial production conditions, irrigation is economically justified because of its positive impact on plant growth, yield and fruit quality. The demand for irrigation is greatest and most critical when full foliage is present, maximum berry growth is occurring, and rain is scarce or non-existent (Williamson *et al.*, 2006). The main issue with irrigation management is to determine the frequency, quantity and timing of irrigation in order to optimize both water-use efficiency and crop growth and productivity. Once the grower has decided that irrigation is profitable, it is important to consider that there are some plant characteristics that impact on water application.

Irrigation scheduling

Proper irrigation management is critical in blueberries for producing high yields and good fruit quality. Even after only a few days without rain or irrigation, water stress develops quickly in blueberry, reducing photosynthesis and leading to less growth and fruit production. Over-irrigation, however, reduces blueberry root function, increases soil erosion and nutrient leaching, and enhances the probability of developing crown and root rot infection by soil pathogens such as *Phytophthora* and *Pythium* spp. (Bryla and Linderman, 2007). Developing accurate irrigation regimes requires knowledge of both the timing and amount of water needed to replenish any loss by crop transpiration and soil evaporation.

The timing or frequency of water applications will depend on soil texture (e.g. sand versus clay), the irrigation system used (e.g. drip versus sprinkler), the rate at which the plant is using water and the overall development of the plant's root system. As blueberry is a shallow-rooted plant compared with many perennial fruit crops, when water demands are high, blueberry plants quickly deplete the water from their root zone and require frequent applications of water to avoid water stress (Bryla, 2011).

To avoid drought stress, irrigation, rain and/or the water table must be adequate to provide sufficient water for plant transpiration and evaporation. There are various ways to schedule irrigation. The most commonly used are: (i) determination of orchard water status; and (ii) moisture accounting.

In the water status method, tensiometers or other devices are used to measure the availability of water in the soil. A certain criterion or threshold is defined to start irrigation (Williamson *et al.*, 2006). When tensiometers are used, readings are provided in centibars (cbar). A reading of 10–20 cbar reflects a soil that is at field capacity (i.e. water availability is at its maximum). Readings of 20–85 cbar indicate the need for irrigation, with 20–40 cbar and 60–80 cbar usually indicating the need to irrigate in light- and heavy-textured soils, respectively. However, reports from Florida indicate that, in sandy soils, irrigation should be scheduled when soil water tension reaches 10–20 cbar (Smajstrla and Harrison, 2008). In research done on SHB blueberries grown in sandy soils in Florida, in the first 3 years after establishment, it was found that the highest plant volume was obtained when a 10 cbar threshold was used for scheduling drip irrigation in comparison with 15 or 20 cbar. With 10 cbar scheduling, the average fruit yield for years 2 and 3 was 68 and 394% higher than scheduling with 15 or 20 cbar, respectively (Haman *et al.*, 2005).

The moisture accounting method balances soil moisture gains from rainfall and irrigation against soil moisture losses from evaporation and transpiration. The moisture accounting method has been reported to work well on soils that do not have a high water table. When a water table is present, it can be measured simply by digging a post hole in the row between bushes and installing a piece of drainage tubing (Williamson *et al.*, 2006). The key to successful use of moisture accounting is the estimation of crop water productivity (CWP) (Byers and Moore, 1987). A widely accepted method for estimating CWP is the use of evaporation pans. In this method, evaporation from a standardized evaporation pan (class A, Weather Bureau) is related to CWP. The pan is 1.2 m in diameter and 250 mm deep and should be installed 150 mm off the ground. The normal operating water level is specified as 175–200 mm of water depth. In many blueberry-producing regions, evapotranspiration data can be obtained through the Internet from meteorological websites. CWP and evaporation from the class A pan are related by the crop coefficient (K_c). In other words, to assess the water that a blueberry field has consumed, the water evaporated from the pan has to be multiplied by K_c.

Reported values for K_c vary widely according to plant age (canopy development), plant species and methodology used to estimate it. Williamson *et al.* (2015) argued that autumn and winter K_c values of SHBs in subtropical climates are subject to many factors that potentially influence defoliation, including weather, leaf disease pressure and cultural practices such as irrigation, fertilization and hydrogen cyanamide use (to enhance leaf development). They determined K_c values for Florida during three seasons for 4–7-year-old SHB 'Emerald' blueberries growing in lysimeters (a lysimeter is a container of soil in which measurements of gains and losses of soil water can be made, for example by weighing). The plants were established on a sandy soil and had 50–60% canopy coverage. K_c values ranged from around 0.45 in January–February (when plants were mostly defoliated and initiating bloom) to a maximum of 0.86 in September when maximum canopy development occurred. Values for the rest of the year were near 0.7. These authors made the point than in some subtropical climates with very low chill accumulation (e.g. Peru, Mexico), SHB blueberries are commonly grown under an 'evergreen' or 'non-dormant' system without winter defoliation. Under these conditions, winter and early spring K_c values would probably be higher than reported in their study. Working also with 5–6-year-old NHB 'Bluecrop' plants growing on lysimeters, it was determined that K_c values increased during leaf expansion to 0.19 for 5-year-old plants and 0.27 for 6-year-old plants; these values remained at these levels until leaf senescence. Assuming a cylindrical bush shape, the maximum K_c

value was equal to 1.5 times the measured canopy coverage (in m^3). Canopy coverage was equal to 18% of the total cultivated area (Storlie and Eck, 1996).

Bryla (2011) reported that, in fully mature plants, K_c values increased as the canopy developed from bud break (0.2) to the beginning of fruit ripening (1.0) and then gradually declined until leaf senescence and dormancy when they reached about 0.8. Blueberries would reach full effective canopy cover when the first blue fruit appear, and it is at this stage that water use by blueberry is equal to that of lucerne and $K_c = 1$. Once a planting has 70% cover or more, it reaches an adult condition where ET_c is no longer a function of plant size.

Research in Florida established that K_c values varied from 0.10 to 0.24 for 1-year-old rabbiteye blueberry plants (Haman et al., 1997b). In the case of 2- and 3-year-old plants, K_c ranged from 0.10 to 0.49. In both cases, the highest values were for the summer months. Research in Arkansas estimated the K_c value for young highbush blueberries to be 0.75, although there were no significant differences in total vegetative growth, yield or quality with K_c values of 0.5 and 1.0 (Byers and Moore, 1987). Bryla (2011) reported that, in fruit crops, a correction factor is used for young plantings, which considers a shadow factor adjusted according to the irrigation method.

A problem with the moisture accounting method is that precise K_c values are often difficult to establish due to regional and site-specific variability in climate, soil characteristics, crop physiology and cultural practices. A grower using the moisture accounting method for scheduling irrigation should employ a conservative K_c value, frequently check soil water availability with soil probes and look for stress symptoms in plants (reduced shoot and leaf growth, alteration of leaf angles and wilting). K_c values should then be adjusted until adequate moisture levels are reached.

Martins et al. (2016) developed an automatic system of irrigation capable of controlling water and nutrient supply for blueberries. By acquiring humidity and temperature data, the system can optimize the water supply by delivering the proper quantity of water at the most advisable period, through the use of a programmable logic controller. The system is also accessible through the Internet and includes wireless communication between the sensors boards and the main control board, which allows the user to easily place the sensor wherever they are needed. Equipped with sensors, infrared cameras and intelligence systems, they can target patches of soil requiring irrigation. Gonzalez-Dugo et al. (2013) used drones – unmanned aerial vehicles (UAVs) – to assess the heterogeneity in water status in a commercial drip-irrigated Prunus and Citrus orchard located in south-western Spain, as a prerequisite for precision irrigation management. Recent studies have demonstrated that high-resolution airborne thermal imagery enables the assessment of discontinuous canopies as pure tree crowns, thus eliminating the background effects. A UAV equipped with a thermal camera on board was flown three times during the day at 9:00, 11:00 and 13:00 (local time). Stem water potential was measured at the same time as the flights. In some irrigation units, irrigation was stopped prior to the measurement date to induce water deficits for comparative purposes. Several approaches for using the thermal data were proposed. Daily evolution of the differential between canopy and air temperature ($T_c - T_a$) was compared with tree water status. The slope of the evolution of $T_c - T_a$ with time was well correlated with water status and is proposed as a novel indicator linked to stomatal behaviour. The crop water stress index (CWSI) was calculated with the temperature data from the 13:00 flight using an empirical approach for defining the upper and lower limits of $T_c - T_a$. The assessment of variability in water status was also performed using differences in relative canopy temperatures. Ample variability was detected among and within irrigation units, demonstrating that the approach proposed was viable for precision irrigation management. The assessment led to the identification of water-stressed areas, and to the definition of threshold CWSI values and associated risks. Such thresholds may be used by growers for irrigation management based on crop developmental stages and economic considerations. UAV systems are being developed for wild blueberry production (Percival *et al.*, 2017); a similar methodology could be implemented in cultivated blueberries to monitor water status and allow precision irrigation of commercial fields.

Determining soil moisture

There is a wide range in the characteristics of the methods used to determine water status in agricultural situations. Some establish the availability of water in the soil, while others measure the status of water in the plant.

In recent years, there has been increasing availability of devices and sensors for automatic measurement of soil and crop water. Measurement of soil water potential and soil water content provide an index of the rate at which water is taken up by the plant or lost from the root zone. Data on soil water content and potential are therefore most useful in conjunction with information about the soil–plant–atmosphere system. Although climate- and soil-based methods provide a means for estimating irrigation amount and timing, they do not take into account the variability among cultivars, growth stages and the response of plants to soil moisture deficit (Jones and Tardieu, 1998). The water refill point, which is the lowest possible soil water content with no decrease in yield or fruit quality, varies among cultivars, soils, management practices and seasons. An integrated approach utilizing both soil and plant factors for irrigation scheduling is often beneficial (Al-Yahyai, 2006). Physiological variables (plant water potential and gas exchange), as well as plant growth and fruit production and quality, should be correlated with soil water content prior to determining the appropriate amount and timing of water application to a blueberry field.

Among the methods used to determine soil water, the simplest one is 'feel and appearance'. Field samples are taken, which the operator feels by hand.

The major advantages of this approach are its low cost, rapidity and the possibility of assessing multiple locations. Among the disadvantages are that considerable experience is required and that it has low accuracy (Williamson *et al.*, 2006).

The gravimetric method measures mass water content. Field samples are collected, weighed, oven dried and weighed again. The advantages of this approach are its accuracy and that multiple locations can be measured. However, the process is labour consuming, and there is a time delay between sampling and results.

Electrical resistance blocks (gypsum blocks) and granular matrix sensors (GMSs) measure soil water potential. They tend to work better at higher tensions (lower water contents). They can be set to turn irrigation on automatically when a certain level is reached. GMS technology reduces the problems inherent in gypsum blocks (slow response time and dissolution of the block) by using a mostly insoluble granular fill material. Like gypsum blocks, GMSs operate on the principle of variable electrical resistance (Shock, 2008). A major difficulty is the representativeness of site sensor placement (Améglio *et al.*, 1999). To date, gypsum blocks and GMSs have rarely been employed in blueberry fields.

Tensiometers have been widely used in blueberry management. A tensiometer consists of a porous cup, connected through a rigid body tube to a vacuum gauge, with all components filled with water. They measure soil water potential (tension), which is directly related to the ability of plants to extract water from soil. This reading is a measure of the energy that would need to be exerted by the plant to extract water from the soil (Smajstrla and Harrison, 2008).

The way tensiometers work is that a partial vacuum is created as water moves from the sealed tensiometer tube to the surrounding soil. The change in vacuum levels is translated into a reading, which is a direct indication of the attractive forces between the water and soil particles. As the soil dries, the water potential decreases (tension increases) and the tensiometer vacuum gauge reading increases. Conversely, an increase in soil water content (from irrigation, the water table or rainfall) decreases tension and lowers the vacuum gauge reading. In this way, a tensiometer continuously records fluctuations in soil water potential under field conditions. A tensiometer indicates only when irrigation should be scheduled, and not how much water should be applied. Digital tensiometers can be set up to turn on irrigation systems when a previously defined threshold is reached. Tensiometers are placed below the plant canopy in positions where they will receive typical amounts of rainfall and irrigation. The porous ceramic of the device should be set in the blueberry root zone (usually 30 cm deep) with the ceramic cup firmly in contact with the soil.

Growers often use tensiometers for irrigation scheduling because of a number of advantages: (i) they provide direct measurements of soil moisture status; (ii) they are easily managed; and (iii) they can be automated to control water applications when the soil water potential decreases to a predetermined critical value. Among the drawbacks are that: (i) they need careful placement and constant maintenance; (ii) they are useful only under mostly uniform soil texture; (iii) their practical operating range is about 0–0.75 bar (Fig. 6.2), a range that usually excludes its use for medium- and fine-textured soils; and (iv) they are not appropriate to use if soils are saline, or if saline irrigation water is being used, because in these conditions the osmotic potential will be a large portion of the total soil potential (Wilk *et al.*, 2009).

Determining plant water status

In general, the establishment of water status through plant measurements is more frequently used for research purposes than to schedule irrigation in commercial fields. One exception is the visible symptoms of water stress (Loomis and Connor, 1992; Darnell, 2006), but in this case the assessment is too late, as the yield potential of the plant has already been reduced by the stress experienced (Mingeau *et al.*, 2001). Bryla (2011) stated that the plant-based approach may be the most accurate method to schedule irrigations and avoid water stress during critical stages of growth, but is probably also the most complex and labour intensive.

The methods for measuring water status in plants can be classified as direct and indirect. The measurement of RWC is one of the direct methods and estimates the current water content of the sampled leaf tissue relative to the maximal water content it can hold at full turgidity. Normal values of RWC range from 98% in turgid and transpiring leaves to about 40% in severely

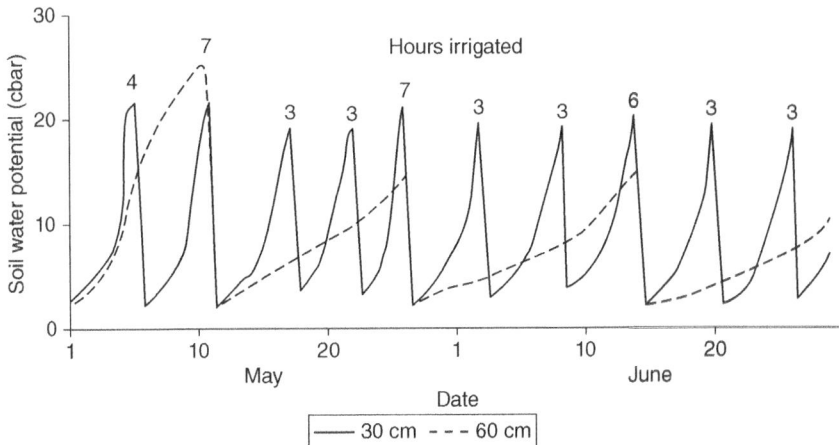

Fig. 6.2.   Effect of irrigation on soil water potential as measured by tensiometers placed at 30 or 60 cm deep. (Adapted from Smajstrla and Harrison, 2008.)

desiccated and dying leaves. In most crop species, the typical RWC around wilting is about 60–70%. To the best of our knowledge, this method has not been used in commercial blueberries to schedule irrigation, but Davies and Johnson (1982) reported that, for rabbiteye blueberries, the RWC changed by 6.4% per 1.0 MPa change in water potential.

Another direct method of water status measurement is the water potential, which measures the energy (tension) status in either the leaf or the stem. To the best of our knowledge, this method has not been used to schedule irrigation in blueberries but has been practised by some elite tree fruit growers. A fully mature leaf or a stem is enclosed in a reflective plastic bag for 1 h to suppress transpiration and allow stem water potential to equilibrate with leaf water potential. Measurements are taken near noon with a pressure chamber (Al-Yahyai, 2006). Stem water potential, which corresponds to the tension of the xylem vessels in the trunk, is representative of the whole plant and is a reliable plant-based water status indicator for irrigation scheduling in fruit crops (Fereres and Evans, 2006; Moriana *et al.*, 2010). The microclimate is not uniform within the plant, and this will generate a range of water potentials reached by organs of different types (leaves, fruits, shoots and canes of different ages).

In fruit trees, it has been reported that, within the canopy, the leaf water potential of exposed leaves was 32% lower than that for shaded leaves (Oyarzún *et al.*, 2010). No data on the variability of this characteristic are available for blueberries. In fruit trees, the stem water potential has been found to be a good indicator of water stress for crops in conditions of heterogeneous soil humidity, particularly when only a small part of the soil contains easily available water (e.g. limited drip or mini-sprinkler irrigation, a patchy root system). In such cases, the use of complementary stress indicators such as sap flow, which are not biased by the spatial distribution of soil water and which therefore are more specific to the actual water stress, would overcome the uncertainty that can arise from the use of stem water potential values alone (Améglio *et al.*, 1999). Another negative issue associated with water potential measurements in plant tissues is that these measurements do not consider internal osmotic adjustment, which corresponds to the active accumulation of solutes in the cell sap. However, as blueberries are not drought-tolerant plants, their ability for osmotic adjustment is minimal (Muralitharan *et al.*, 1992).

Interpretation of water potential for irrigation scheduling is complicated by the fact that values are influenced by weather conditions. For example, due to increasing evaporative demand, leaf water potential tends to decrease with time over the season, regardless of adequacy of irrigation. To overcome this problem, a fully irrigated baseline (reference) value of stem water potentials must be calculated for any given value of midday air vapour pressure deficit. A baseline value is applicable to a wide variety of soil and irrigation conditions and has provided stem water potential guidelines for fully irrigated fruit trees using relative humidity and air temperature. Once developed, data collected from weather stations can be used for baseline estimates in commercial fields throughout a region. Irrigation scheduling is accomplished by comparing actual water potentials to reference values; when actual values fall below reference values, irrigation is increased. Typically, irrigation is increased by 5–10% above the previous week's rate when mean weekly stem water potentials are lower than reference values, and decreased by 5–10% when actual and reference values are equal for two consecutive weeks. To ensure plants are not over- or under-irrigated, soil water content should also be monitored at least monthly (Bryla, 2011).

Sadras and Trentacoste (2011) compiled data sets of stem water potentials for contrasting crop loads of apple, olive, peach, pear and plum. Pooling all the data revealed a unique linear association between plasticity of stem water potential and crop load, irrespective of species and growing conditions. They concluded that this represents a significant shift in perspective, as the effect of crop load on stem water potential is highly contingent, but the effect of crop load on the plasticity of the trait is not.

Indirect methods of measuring plant water status include: (i) crop canopy temperature; (ii) changes in trunk or stem diameter; (iii) g_s; and (iv) sap flow. An infrared thermometer is used to determine crop canopy temperature. A general problem often encountered in assessing plant water stress by canopy temperature is the representativeness of the target area. The inclusion of non-transpiring surfaces (e.g. soil, branches) inside the infrared thermometer field of view generates unwanted shifts in temperature readings. At present, affordable and portable thermal imaging devices with high resolution are available; these have solved the problem of discriminating between foliage and non-transpiring surfaces (Testi *et al.*, 2008).

A trunk diameter fluctuation (TDF) sensor measures the daily cycle of shrinking and swelling in the trunk/stem of crops. This cycle is produced owing to the lag between transpiration and root uptake that is partially compensated for by the water of the trunk. Therefore, the trunk is a water reservoir in the soil–plant–air continuum. The approach has been used successfully in almond and lemon trees but not in olive trees (Moriana *et al.*, 2010). TDF was not useful as a permanent system in plums due to temporal changes as the trees aged (Bonet *et al.*, 2010). To the best of our knowledge, TDF has not been tried in blueberries, but it can be anticipated that the growth pattern of blueberries with multiple shoots (canes) would complicate these measurements.

The basic principle of sap flow is that transpiration occurs as a continuum from soil to plant to atmosphere, and it may be measured or estimated as moisture loss from the soil, liquid flow through the plant stem (xylem sap flow) or vapour loss to the surrounding atmosphere. The stem of a woody plant is a convenient place to measure xylem sap flow and, ultimately, transpiration if measured over a sufficiently long period to negate changes in stem storage. The speed of transport in the xylem sapwood of a woody plant stem can be established if a heat pulse is applied to the trunk and the change in temperature is measured at a given distance from that heat source (Swanson, 1994). Sap flow measurements give reliable, direct estimates of plant or shoot water loss without disturbing the conditions of the leaf environment. In irrigated grapevines, sap flow measurements have been shown to be good estimators of canopy transpiration (Cifre *et al.*, 2005); to the best of our knowledge, these measurements have not been made in blueberry.

The value of g_s (i.e. the ability of these pores to open or close in response to environmental conditions) is used to quantify gas diffusion processes, such as transpiration, between plants and the atmosphere (Byers *et al.*, 1988). Stomatal closure is among the first processes occurring in the leaves in response to drought. Diurnal changes in g_s, leaf water potential and transpiration have been shown to be closely related in blueberries (Fig. 6.3 and Table 6.4) (Bryla and Strik, 2006), and thus it is difficult to isolate one factor from another. A moderate correlation has also been found in both highbush and rabbiteye blueberries between g_s and leaf water potential (Davies and Darnell, 1994), and the various plant parameters of water status are related (Fig. 6.3). Under field conditions, g_s declined rapidly as leaf water potential approached values as high as −0.6 to −0.8 MPa, indicating that highbush blueberry is quite sensitive to even moderate levels of water stress (Bryla, 2011). Regardless of cultivar or in-row spacing, g_s decreases rapidly as the stem water potential approaches −0.6 MPa (Fig. 6.3).

There is variability among NHB cultivars in their response to soil water loss (Bryla and Strik, 2006). 'Duke' maintained, on average, significantly higher stem water potentials (less negative) and greater g_s than 'Bluecrop' and 'Elliott' as soil water was depleted (Table 6.4), which may indicate that this cultivar has the highest tolerance to short-term soil water deficits. 'Bluecrop', on the other hand, had the lowest stem water potentials and g_s, and thus may be more sensitive to water deficits than the other two cultivars. The authors speculated that 'Duke' may require less frequent irrigation than the others because it produced the deepest root system and extracted more water at depths below 0.6 m.

Work by Byers *et al.* (1988) on young NHB 'Bluecrop' blueberries showed that leaf g_s values were high in the early morning, remained high throughout the day and decreased in the late evening. Stomata in highbush blueberries were not as sensitive to water deficits as those of rabbiteye blueberries (Davies and Johnson, 1982). Byers *et al.* (1988) concluded that the root system of highbush blueberries is inefficient in water uptake; thus, even if soil water levels are adequate, temporary drought stress around midday is likely. Considering that young, fully expanded leaves generally have lower stomatal conductivity than mature ones (Davies and Darnell, 1994), if this method is going to be used to monitor water status there is a need to carefully select leaves that will adequately represent the whole plant.

Assessing the water status of a blueberry field is very important to maximize yield and fruit quality. A combination of methods is likely to provide more

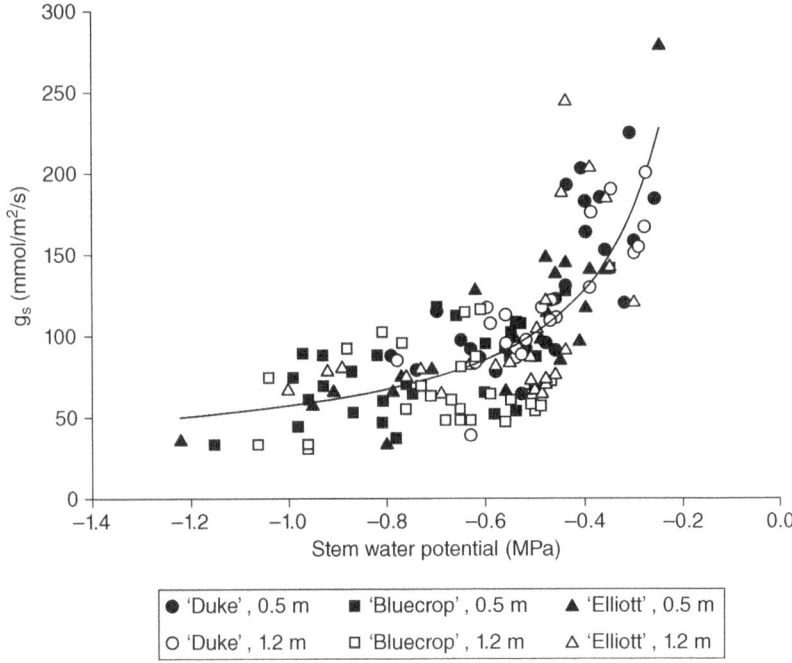

Fig. 6.3. Leaf g_s as a function of stem water potential for highbush blueberry cultivars ('Duke', 'Bluecrop' and 'Elliott') spaced 0.5 or 1.2 m apart within rows. Each symbol represents one measurement. The relationship was fitted with an inverse second-order polynomial ($y = 6.10/x^2 - 25.82/x + 25.38$, with $r^2 = 0.57$ and $P \leq 0.001$). (Adapted from Bryla and Strik, 2006.)

reliable information than the use of individual techniques. Digital tensiometers or electrical resistance blocks (gypsum or GMSs) combined with sap flow meters and TDF sensors appear to be the most promising techniques.

Calculating water needs

Once a method to determine the water status of the crop has been selected, there is a need to calculate the amount of water to be applied. Blueberry cultivars vary in the size and shape of their plant canopy and root system, as well as the timing of harvest, which influence biomass production/partitioning and water requirements. There is considerable variation in the morphological and physiological adaptations of cultivars to tolerate short-term episodes of water deficits, such as deeper root systems, greater water-use efficiency (relationship between net photosynthesis and transpiration) and the ability to maintain higher plant water status (Erb *et al.*, 1991). As most water is lost through the

Table 6.4. Mean g_s and stem water potential of NHB blueberry cultivars ('Duke', 'Bluecrop' and 'Elliott') spaced 0.5 or 1.2 m apart within rows. Values represent the seasonal average (May–September) of seven sets of measurements. (Adapted from Bryla and Strik, 2006.)

Cultivar	In-row spacing (m)	g_s (mmol/m²/s)	Stem water potential (MPa)
'Duke'	0.5	129	−0.55[b]
'Bluecrop'	0.5	83	−0.72[c]
'Elliott'	0.5	96	−0.62[b]
'Duke'	1.2	115	−0.46[a]
'Bluecrop'	1.2	78	−0.72[c]
'Elliott'	1.2	109	−0.58[b]
Analysis of variance			
Cultivar (C)		<0.001	<0.001
Spacing (V)		NS	<0.05
C×V		NS	NS

NS, not significant.
[a,b,c]Mean values within columns with non-identical superscript letters were significantly different at $P < 0.05$ (least significant difference test).

leaves, some authors found that crop water was strongly correlated with canopy size (Wang *et al.*, 2007), while Bryla and Strik (2007) reported that, although the percentage canopy cover in highbush blueberries was up to 246% greater at 0.45 m planting distance within the row than at 1.2 m, water use increased by no more than 10%.

It is important to understand that a crop's irrigation requirements are much greater than its water requirements. Crop water requirements indicate the total amount of water directly used by a crop but do not account for any extra water needed to compensate for non-beneficial water use or loss, such as runoff, deep percolation, evaporation, wind drift, ground cover and weeds. Additionally, irrigation systems do not apply water with 100% uniformity (Bryla, 2011).

Various cultural practices affect crop water use, such as mulching, type of irrigation system, ground cover, cultivation practices and planting density (Allen *et al.*, 1998). In highbush blueberries, close spacings (0.5 m within-row spacing) had significantly higher water uptake per hectare at 0–0.6 m soil depth than wider spacings (1.2 m) (Bryla and Strik, 2006, 2007). Bryla and Strik (2007) found that water use in NHB blueberries was related to ripening period, with the highest water use for 'Duke', which ripened first, and the lowest in 'Elliott', which ripened last. Mulching has been shown to conserve soil moisture and prevent weed growth, which can alter the water availability for blueberry plants (Eck *et al.*, 1990; Haman *et al.*, 2005). The effect of mulching can be greater in soils of low water-holding capacity. Part of the effects of mulching on soil moisture could be attributed to reduced soil temperature in the top 10 cm (Haman *et al.*, 1988).

Williamson *et al.* (2015) used lysimeters to calculate average daily water use in mature SHB 'Emerald' blueberries in monthly intervals beginning in April 2010 and ending in September 2012. Water use increased rapidly during spring through the final stages of fruit ripening and harvest (May), with peak water use occurring during mid- to late-summer (July–September). Plants grown in pine bark beds used more water than plants in pine bark-amended soil during April, May and December 2010, February 2011 and March 2012, but there were no differences during the periods of highest water use (May–September in the northern hemisphere). No differences in water use were observed between single or split-application irrigation treatments. Monthly averages for daily water use during the 30-month period ranged from 1.75 l per plant in January to 8.0 l per plant in mid- to late summer. Monthly water use for SHBs in their first 3 years after establishment are shown in Table 6.5 (Haman *et al.*, 2005).

Average irrigation application efficiencies for well-maintained solid set sprinkler systems generally range from 65 to 75%, and largely depend on the quality of sprinkler overlap. Close spacing and newer sprinkler heads help improve sprinkler water application efficiency. Brand-new drip systems, on the other hand, can generally be designed with 85–93% efficiency, except in cases with major elevation changes. Neglected drip systems have been shown to have actual efficiencies closer to 60–80%. Primary causes for low efficiencies include flow variation due to poor system design, emitter plugging and pressure differences within the field (Bryla, 2011).

Table 6.5. Water use (l per plant) during the first 3 years of SHB blueberry establishment in Florida. (Adapted from Haman *et al.*, 2005.)

Month	Year 1	Year 2	Year 3
January	–	64.4	26.5
February	–	60.6	34.1
March	–	71.9	56.8
April	53.0	87.1	177.9
May (harvest)	56.8	117.3	310.4
June	124.9	147.6	344.5
July	128.7	196.8	393.7
August	121.1	193.1	367.2
September	109.8	166.6	336.9
October	87.1	143.8	200.6
November	75.7	90.8	174.1
December	53.0	53.0	174.1

In the literature, there is large variability in the recommended amounts of water that should be applied to blueberries. The water requirement of an adult blueberry field in New South Wales in Australia has been calculated to be about 25 mm per week during the growth period and up to 40 mm per week in the final 2 weeks of fruit growth (Ireland and Wilk, 2006). Brightwell and Austin (1980) indicated that water requirements for rabbiteye blueberries in Georgia are in the range of 25.5–44.5 mm per week to obtain a large root development during the growing season. In north-eastern USA, 5 l per bush per day is recommended for 3–4-year-old highbush blueberry plants, and 14–27 l per bush per day for mature plants (Kender and Brightwell, 1966). In New Jersey, water use in sunny days during June–August was calculated to be 3.5–4 l per bush per day for 5-year-old plants and 4–4.5 l per bush per day on 6-year-old ones (Storlie and Eck, 1996). In Arkansas, the general recommendation is to apply 3.8 l per bush per day in young plants and 7.6 l per bush per day in adult plants. However, in an experiment on 3-year-old 'Bluecrop', Byers and Moore (1987) used a K_c of 0.74 and tensiometer readings at 15 cm depth (indicating that average soil matric potentials were maintained at levels higher than 0.012 MPa in the intervals between irrigations), and were able to reduce water applied by 68% to 1.3 l per bush per day.

Water stress can be alleviated by increasing the amount of water retained in each portion of the soil profile through added irrigation and also by soil modification. Research done in rabbiteye blueberries showed that plants receiving more than one water-supplementing treatment (irrigation, peat moss incorporation and mulch) had greater root weight than those receiving only one (Patten *et al.*, 1989). Total root weight correlated strongly and positively with plant height and yields. In well-aerated sandy soils, where moisture can be limiting, roots are concentrated in areas where, through various cultural practices, soil moisture is most prevalent. For plants having greater root depths, growers should avoid concentrating soil moisture near the soil surface.

Water quality

Low water quality can have short- and long-term effects on crop performance and the operation of irrigation systems (Ayers and Westcot, 1985), particularly in blueberries. Some blueberry-growing areas (e.g. Texas, Alabama, Mississippi, southern Peru and northern Chile) have particularly low water quality. The constituents of major importance to quality of irrigation waters are the cations Ca^{2+}, Mg^{2+} and Na^+, and the anions Cl^-, SO_4^{2-}, bicarbonate (HCO_3^-) and carbonate (CO_3^{2-}). The relative proportions of these ions in irrigation water determines their hazard to plant growth. Good-quality water should have low salts: total Na^+ <2 mM, and total HCO_3^- and total Cl^- <4.0 mM (Haby and Pennington, 1988). Combinations of Na^+ and Cl^- and HCO_3^- are considered the most harmful salts for plant growth. The sodium adsorption ratio

(SAR) of the irrigation water is used as a measure of the sodicity hazard of the water (Haby *et al.*, 1986). SAR is an indicator of the suitability of water for use in agricultural irrigation. SAR equals the Na^+ concentration (meq/L) divided by the square root of the half sum of Ca^{2+} plus Mg^{2+} concentrations (meq/L). Haby *et al.* (1986) established that growth of rabbiteye 'Tifblue' blueberry plants was more severely decreased by increasing SAR in sandier soils than in the clay loam. Water with a high SAR reduces soil structural stability by clay dispersion, swelling and reorientation. This soil hardening restricts root growth (Smith *et al.*, 2016).

When water contains appreciable amounts of carbonates, Ca^{2+} and to a lesser extent Mg^{2+}, they may precipitate as $CaCO_3$ or $MgCO_3$, respectively. As Na^+ remains soluble, precipitation of divalent cations increases the SAR and potential Na^+ hazard. The use of water with high HCO_3^- levels to irrigate azaleas (a member of the Ericaceae family) has been shown to markedly increase the absorption of Na^+ when Ca^{2+} supplies were low, and decreased the absorption of Ca^{2+} and Fe^{3+} (Haby *et al.*, 1986).

Salinity problems related to water quality can also occur if the total quantity of salts in the irrigation water is high enough that salts accumulate in the crop root zone to the extent that yields are affected. If excessive quantities of soluble salts accumulate in the root zone, the crop has extra difficulty in extracting enough water from the salty soil solution (Ayers and Westcot, 1985). Salt accumulation in the root zone causes the development of osmotic stress, inhibits the uptake of essential nutrients such as K^+, Ca^{2+} and NO_3^-, and generates accumulation of toxic levels of Na^+ and Cl^-. These stresses cause hormonal changes, alter carbohydrate metabolism, reduce the activity of certain enzymes and impair photosynthesis. As a consequence of these metabolic modifications and dysfunctions, cell division and elongation decline or may be completely inhibited while cell death is accelerated. At a whole-plant level, the impacts of salinity are reflected by declines in growth and yield, and in more acute cases may cause leaf injuries, which can lead to complete defoliation of plants and their subsequent desiccation (Paranychianakis and Chartzoulakis, 2005). The damaging effects of poor-quality water on growth of rabbiteye blueberries have been reported to be due to high concentrations of Na^+. Field studies showed that weight gain of rabbiteye blueberry plants irrigated with well water containing 7.83 mM Na^+ was only 65% of that of plants irrigated with surface water containing 0.23 mM Na^+ (Haby *et al.*, 1986).The maximum salt content tolerated in water by blueberries is in the range of 250–300 ppm (Bell, 1982; Freeman, 1983). The most helpful salinity hazard indicator is electrical conductivity (EC). At high EC, the infiltration rate of water in the profile is affected (Loomis and Connor, 1992). Water for blueberry irrigation is generally recommended to have an EC below 0.45 millimho (mmho)/cm or 0.45 dS/m (Ireland and Wilk, 2006), although other authors have established a 1.0 dS/m threshold (Himelrick and Curtis, 1999).

Because both SAR and EC affect water infiltration, both must be considered in estimating water infiltration hazard. In general, the Na^+ hazard increases as SAR increases and EC decreases (Hopkins *et al.*, 2007). For example, in the case of SAR values between 0 and 3, the water infiltration hazard would be low for EC values above 0.7 dS/m, moderate for EC in the range of 0.2–0.7 dS/m, and high for EC values below 0.2 dS/m (Smith *et al.*, 2016). In practice, the severity of water infiltration problems depends partly on soil texture. At a given EC and SAR, water infiltration problems are greater with higher soil clay content. The type of clay also is important. Soils that contain shrink:swell clays at a 2:1 ratio have greater water infiltration problems than 1:1 clays (Hopkins *et al.*, 2007).

To some extent, modification of irrigation geometry can mitigate the effects of salty water. Research on rabbiteye blueberries showed that root-zone salinity was greatest and plant growth least when the wetting front of the emitter focused salt directly under the plant (Patten *et al.*, 1989). To reduce salt build-up near the root system, water should be applied in smaller amounts to a greater volume of soil and more frequently. Mulch can also reduce root-zone salinity by decreasing surface evaporation and improving infiltration.

Water with a pH <6.5–7.0 is also desirable. Alkaline (high pH) irrigation water will eventually raise the soil pH to a level harmful to blueberries, and high-pH water is more likely to contain potentially harmful levels of salts, Na^+ and carbonates. The pH of irrigation water can be adjusted with phosphoric, hydrochloric or sulfuric acid. Sulfuric acid is usually cheaper. Addition of 1.66 l sulfuric acid is equivalent to 1 kg of elemental sulfur. Well water treated at a rate of 21 ml sulfuric acid/l of water changed its pH from 8.7 to 5.0–5.4 (Smith *et al.*, 2016). An injector pump is used to force the acid into the main irrigation line to be thoroughly mixed with the water (Williamson *et al.*, 2006). These acids should be handled with extreme care, as they are very toxic and can irritate the respiratory and digestive tract, as well as eyes and skin.

Low-volume irrigation systems such as microsprinklers and drip irrigation are prone to clogging due to several water quality characteristics. The levels of some of the variables that can cause moderate plugging hazards are: (i) pH: 7.0–7.5; (ii) dissolved solids: 500–2000 mg/l; (iii) manganese and iron: 0.1–0.5 mg/l; (iv) hydrogen sulfide: 0.5–2.0 mg/l; and (v) hardness ($CaCO_3$): 150–300 mg/l. Above and below these levels, severe and slight clogging hazards would occur, respectively (Smith *et al.*, 2016).

Irrigation systems

Various irrigation systems have been used in blueberry production including furrow, sprinkler, microjet and drip (surface or buried emitters). Each of the different irrigation systems has advantages and disadvantages. When selecting an irrigation system, economic and technical parameters such as field size, need to control frosts, topography, availability and quality of water, type of soil, human resources and costs should be considered (Loomis and Connor, 1992; Holzapfel, 2009). The decision as to which irrigation system to use should be made before the field is planted.

Over-the-canopy sprinkler systems are relatively simple to install and maintain, and have been widely used to irrigate blueberries. In a survey carried out in Oregon, 96% of the blueberry fields had overhead irrigation and 4% used drip irrigation (Scagel and Yang, 2005). Overhead irrigation is also commonly used in Florida and Michigan. Overhead sprinklers can be used for frost control if enough water can be delivered during the event (Haman *et al.*, 1997b). Unlike surface irrigation, sprinklers require moderately clean water so that the sprinkler nozzles are neither blocked nor damaged by suspended sediment. Sprinklers provide even wetting, and hence water moves uniformly through the soil profile (Loomis and Connor, 1992). However, sprinklers deliver water on the canopy top and this can increase disease problems. In addition, a portion of the water is deposited between rows where it is unavailable to the crop (Bryla, 2008).

Furrow irrigation is a type of surface irrigation that requires flat land or gentle slopes. Soil physical properties determine infiltration rate and the slope defines the period that water remains on the surface for irrigation. Furrow irrigation is not suitable for soils with high infiltration rates (sand and sandy loam). Moisture levels are quite variable within the field and between periods of irrigation. Water-use efficiency is lower in furrow irrigation than in pressurized systems. Erosion can be a problem if slopes are too steep or the volumes of water are excessive (Loomis and Connor, 1992). Furrow irrigation has been used with good results in some areas (Lyrene and Muñoz, 1997). A study done almost 40 years ago to compare the costs of various irrigation systems established that the total cost of furrow irrigation was 29% cheaper than drip and overhead irrigation (Fereres *et al.*, 1981), but the figures may now be different.

Drip irrigation is used widely in many growing regions. Drip systems are somewhat more expensive to install and more difficult to maintain than furrow and sprinkler systems, but offer better water control and higher distribution uniformity. Drip irrigation has been used most commonly in soils with a high water-holding capacity (Holzapfel *et al.*, 2004). In some areas (e.g. Florida), the use of drip irrigation has resulted in salinity problems around the superficial root system and crown due to a high concentration of calcium and magnesium carbonate in the irrigation water. If rainfall has been sufficient to permit roots to extend beyond the soil volume typically wetted by the emitter, water stress occurs more often during dry periods with drip irrigation than with overhead or microsprinkler irrigation (Patten *et al.*, 1989).

Frequent water applications are especially important when using drip irrigation, which tends to restrict soil wetting and thus produces a smaller root system. High-frequency irrigation may be especially beneficial and perhaps even required when organic matter is incorporated into the planting bed. Organic matter often reduces the water-holding capacity of the soil and can lead to problems with hydrophobicity. Soil hydrophobicity is the lack of affinity of soil to water and is thought to be caused primarily by a coating of long-chained hydrophobic organic molecules, such as those released from decaying organic matter, on individual soil particles. Hydrophobic soils often become very difficult to re-wet once they dry out. Even with drip irrigation, sawdust incorporated into raised planting beds has made it difficult to retain adequate moisture in the upper portions of the soil where many of the blueberry roots are located. To compensate, much longer and more frequent irrigation is required in beds with incorporated sawdust than in those without. Even after 50 mm of rainfall, dry beds with incorporated sawdust tend to remain dry and do not become fully saturated until the following season (Bryla, 2011).

Drip emitters are best suited to young plants with limited root systems, giving better water-use efficiency. If drip emitters are used for mature plants, the wetted area should cover at least 50% of the root zone. When drip irrigation is used in lighter-texture soils, two lines of emitters, one on each side of the plant, are probably needed to provide adequate coverage. The effects of the number of drip laterals (wetted area) and irrigation frequency on yield were evaluated for two seasons in a field of 7-year-old NHB 'Brigitta' plants. The field was located in south-central Chile (latitude $37°19'S$) and had a sandy soil. Seven-year-old bushes were planted on 0.25 m high raised beds mulched with pine bark and irrigated from September (bud break) to April (leaf fall) with two, four or six drip laterals per row, either 4 or 6 days per week. All treatments received the same amount of water per week with a total of 532 mm per season. The volume of water applied through irrigation plus precipitation amounted to 90 and 122% of the theoretical ET_c in years 1 and 2, respectively. Total yield and the proportion of fruit greater than 10 mm in size (exportable fruit) was greatest in plants irrigated with four drip laterals per row. The two irrigation frequencies evaluated did not affect yield. There was a significant interaction between irrigation frequency and number of laterals in terms of the percentage of exportable fruit. When plants were irrigated 6 days per week, the higher number of drip laterals per row increased the percentage of exportable fruit (Holzapfel et al., 2015).

A trial was carried out for 2 years on newly planted NHB 'Duke' blueberries in Oregon comparing drip emitters that were: (i) buried 0.1 m deep on each side of the plants; (ii) on one line suspended at 1.2 m above the plants; or (iii) placed on the soil surface at each side of the plants (Bryla, 2006). During the first 2 years after planting, plants irrigated with a buried drip were larger and produced significantly more whips than the other systems. Subsurface drip had the extra advantage of eliminating water runoff and bed erosion that were observed with both surface drip treatments. It also maintained lower soil water content near the crown, which may have reduced rot due to *Phytophthora* and *Pythium* (Bryla, 2006). However, research on lucerne in Australia has demonstrated that whitefringed weevils (*Naupactus leucoloma*) can damage subsurface drip irrigation lines (Nicholas, 2010).

Ehret *et al.* (2012, 2015) conducted a study on NHB 'Duke' plants growing on a silt loam soil in Agassiz, British Columbia, Canada (latitude 49°14′N), to determine the effects of drip configuration (one or two lines with emitters spaced every 0.3 or 0.45 m) and irrigation at moderate or heavy rates (5 or 10 l per plant). In stage 1 of measurements (from plant establishment to year 3), plant growth and yield were unaffected by irrigation rates of 5 and 10 l per plant applied several times per week compared with zero irrigation during the first 3 years after planting, and were only greater with drip irrigation during the fourth year. However, several fruit quality characteristics such as size, firmness, and soluble solids were affected by irrigation a year or two earlier, but neither yield nor fruit quality was altered by the configurations of the drip system in any year (Ehret *et al.*, 2012). In stage 2 of this trial (years 5 and 6), the researchers found that the plants became more sensitive to soil water deficits with age and, therefore, unlike when they were younger, had greater yields when more water was applied (Ehret *et al.*, 2015). Berry size and firmness were little affected by irrigation in older plants, but fruit oxygen radical absorbance capacity (ORAC) was higher with than without irrigation. As found in stage 1 of this study, growth, yield and fruit quality were unaffected by drip configuration in these last two seasons. Overall, the results revealed that the response of highbush blueberry to drip irrigation changed over time and indicated that irrigation management should be adjusted as a planting matures. Plants become more sensitive to water deficit with age, as shown by a more pronounced relationship between yield and soil moisture. The factors that contribute to yield, such as water supply, plant size, flower count and berry size, also change over time.

Microspray irrigation is a low-pressure irrigation system that is used only rarely in blueberries but offers advantages similar to drip irrigation. As with drip irrigation, water application can be directed to areas were blueberry roots are growing, which will save water compared with overhead sprinklers (Holzapfel *et al.*, 2004). Microspray irrigation is preferred on sandy soils because its greater wetting pattern reaches a larger percentage of the root system. Because microsprays wet more soil volume than drip irrigation, plants tend to produce a larger root system, which may provide an advantage in a shallow, densely rooted crop such as blueberry. Both microspray and drip irrigation allow fertigation, but microspray irrigation is not compatible with plastic mulch. Growers who use microsprinklers in Florida have reported better fruit quality and fewer disease problems than with overhead sprinklers (Haman *et al.*, 1997a).

A system called subirrigation has been widely used for irrigation of field crops and vegetables over the last 20 years (Qiaosheng *et al.*, 2007), and has been adapted to blueberries (Hanson, 2006). Although many fields are suited to the system (which requires specific soil and topography characteristics), it currently amounts to only 4% of the acreage in Michigan. In subirrigation, water is pumped into a tile drain system to elevate the water table. In blueberries, this can be economical if the field needs tiling anyway. The system that several growers have been operating in Michigan includes the typical tile drain system, with water table management boxes at various locations. The boxes contain sliding gates that allow the grower to back the water table up behind the box. Fields are 'zoned' based on elevation differences, and a control box is positioned between the zones.

Among the advantages of subirrigation for crops in general are labour, water and nutrient savings, more uniform plant growth, lower air humidity, fewer foliar diseases and fewer environmental problems from nutrients and chemical leaching (Qiaosheng *et al.*, 2007). Some specific advantages to this system in blueberry fields are that: (i) the drainage can be managed so there can be irrigation advantages even if no water is pumped into the system (controlled drainage); (ii) the plants and sometimes the soil surface stay dry, which will reduce disease and the growth of some weeds; and (iii) there is potential to reduce pollutant movement out of fields, because the water is retained (Hanson, 2006).

A comparative study was established in a silty clay loam on 'Elliott' blueberries in Oregon to evaluate irrigation systems (sprinkler, microspray and drip) and water application levels (50, 100 or 150% of estimated ET_c). During the first year after planting, it was found that soil water content was significantly higher when the plants were drip irrigated (29.7%) and lowest when they were irrigated by sprinklers (24.9%). In the second year, microspray irrigation had the lowest water content (20.4%) and drip irrigation had the highest (31.6%). Soil water content, however, did not differ significantly among the different irrigation levels until the second year after planting, with 150% ET_c having the highest (28.4%) and 50% ET_c having the lowest (22.0%) values. Overall, shoot dry weight was highest in plants irrigated at 100% ET_c by drip or at 150% by microspray irrigation. The authors attributed the benefit of these two treatments to higher soil water content and/or higher irrigation frequency, which probably enhanced plant water status over the other treatments (Bryla, 2008). Other work done in Oregon showed that young plants under drip irrigation had longer roots and greater colonization of mycorrhizae in the upper 15 cm than plants under overhead irrigation (Scagel and Yang, 2005).

In Oregon, Bryla (2011) compared the water requirements for growing blueberries with sprinkler, drip and microspray irrigation to determine the effect of irrigation method on growth after planting. Two NHB blueberry cultivars, 'Duke' and 'Elliott', were evaluated. By the end of the second growing season, drip irrigation produced the largest 'Elliott' plants among the irrigation methods, with 42% less water than microsprays and 56% less water than sprinklers. The benefit of drip irrigation in 'Elliott' was probably a result of superior plant water status due to higher soil water content in the vicinity of the roots. Drip irrigation, however, was not beneficial in 'Duke', as the plants irrigated using a drip were only half the size of those irrigated by sprinklers or microsprays. Root sampling revealed that 'Duke' plants were infected by *P. cinnamomi* (which causes root rot in blueberry), and the wetter soil conditions with drip irrigation were more favourable to the disease. During the first 4 years of production, yields were similar in 'Duke' plants whether they were irrigated by sprinklers or microsprays but lower with drip irrigation, again due to a higher incidence of root rot. In 'Elliott', yields were slightly higher with drip irrigation than with sprinklers and microsprays during the first year of production, and still higher than sprinklers in the second year. However, by year 3, the yield was similar between drip and sprinkler irrigation but higher when plants were irrigated by microsprays. The author concluded that, in terms of early plant growth and water use efficiency, drip irrigation was the best method of the three to establish healthy blueberry plants. However, sprinklers and microsprays may be better alternatives for cultivars such as 'Duke' that are highly susceptible to root rot, especially at sites with heavy soils or a history of the disease.

Another comparison of irrigation methods was carried out for 6 years (in 2–7-year-old plants) in south-central Chile (latitude 36°30′S) on NHB 'Bluetta' blueberries planted in a loamy-clay soil with good internal drainage (Holzapfel *et al.*, 2004). Levels of water application, from 20 to 133% of reference evapotranspiration, under microjet and drip irrigation were evaluated. With drip and microjet irrigation, fruit yield increased with higher amounts of water. During the first 2 years of harvest, at all levels of water application, plants under drip irrigation produced higher yields compared with those that were microjet irrigated. However, in year 4 and subsequent seasons, plants irrigated with microjets surpassed those that were drip irrigated. In the last season, a 7-year-old blueberry had the highest yield of 10,300 kg/ha with microjet irrigation and a level of water replacement of 6200 m³/ha, compared with 6800 kg/ha for drip irrigation with the same amount of water applied (Fig. 6.4).

MULCH

The increasing popularity of blueberries has brought about attempts all across the world to grow the crop outside its natural habitat of the native lowland (acid soils, high organic matter and loose soil) in a range of different soil conditions. In addition to modifying soil acidity, the maintenance of moisture near the soil surface is of great importance because of the extreme shallow rooting of blueberries. Among the various amendments and practices that can enhance soil moisture and reduce weed infestation, the most successful and widespread has been the use of mulch. A mulch is defined as any form of covering applied to the soil surface. This broad definition includes crop residues, weeds and other plant material cut and carried in from elsewhere, as well as artificial materials such as paper and plastic (Kumar *et al.*, 2013). Both natural

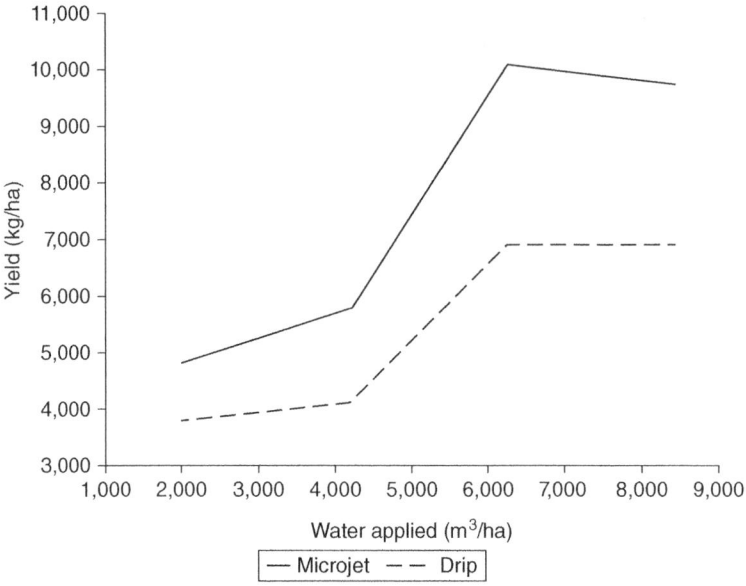

Fig. 6.4. Yield (kg/ha) of 7-year-old NHB 'Bluetta' blueberries with varying amounts of water applied by drip and microjet irrigation. (Adapted from Holzapfel *et al.*, 2004.)

(organic or inorganic) and synthetic non-living mulches are used in agricultural systems. Depending on their characteristics, these mulches can be divided into two main categories: (i) sheet mulches, such as black, clear or coloured polythene, geotextiles (e.g. weed mat), biodegradable films, sprayable substances (Adhikari *et al.*, 2016), paper, needle-punched fabrics and carpets; and (ii) particulate mulches, such as straw and hay, grass clippings, industrial crop waste, coffee grounds, dry fruit shells, shredded and chipped bark or wood, sawdust, crushed rock and gravel (Upadhyaya and Blackshaw, 2007).

The use of mulches is common across the blueberry industry. Mulching is a prevalent approach to weed management in organic blueberry production and offers several other horticultural benefits (DeVetter *et al.*, 2015). A survey carried out in Oregon showed that 58% of blueberry growers used some type of mulch (Scagel and Yang, 2005). From a regression analysis of data from grower's survey of 717 highbush blueberry fields in the Maule region (Chile), Retamales *et al.* (2015) concluded that mulching and weed control with mixed methods had a higher probability of obtaining a high yield than manual weed control, mulch or herbicides as single measures. Mulching is generally more effective against annual weeds than perennial ones (Pannacci *et al.*, 2017).

The types of mulches used cover a wide range, depending on the particular needs of the grower, economic considerations and availability of materials within the area of the blueberry field. The types of mulch most commonly used in blueberry cultivation include peat moss, pine bark, sawdust, straw, hay, manure, leaf litter, plant residues, compost and plastic films (Himelrick *et al.*, 1995; Hart *et al.*, 2006; Cox, 2009). The performance of blueberry plants under these mulches is site dependent, as it varies according to various factors, including soil type (Cox *et al.*, 2014), amount applied, ageing of the material, content of toxic substances, C:N ratio of the materials, mulch placement (on the surface or incorporated), colour and thickness. Each type of mulching material has a particular set of characteristics. The choice of an appropriate mulching material depends on local climate, cost-effectiveness and feasibility for the crop (Kader *et al.*, 2017).

Although the primary objective of mulches is weed control, depending on the material used, mulches may also protect soil from wind and water erosion, add organic matter to the soil, help conserve soil moisture, increase or decrease rainfall penetration, and reduce nutrient leaching and soil O_2 content (Upadhyaya and Blackshaw, 2007). Mulches can improve crop growth and productivity due to their ability to modify and mitigate soil temperature fluctuations, reduce water loss from the soil surface and provide plant mineral nutrients (when they are derived from organic materials), including composts (Clark and Moore, 1991; Burkhard *et al.*, 2009; Cox, 2009). Mulching materials significantly affect the microclimate around the crop canopy by changing the radiation budget of the soil top, soil water dynamics, aerodynamic properties and soil temperature, thus influencing crop yield, evapotranspiration and water-use efficiency (Yang *et al.*, 2015). Magee and Spiers (1995) registered soil temperatures at 10 cm below the soil surface in plants with mulches and under bare soil in southern Mississippi; they reported the following average temperatures between 13 April and 15 May and between 15 July and 15 August, respectively: bark, 21.5 and 31.2°C; white/black plastic, 22.5 and 38.4°C; weed mat, 24.9 and 40.8°C; black plastic, 25.8 and 44.2°C; bare soil, 27.6 and 37.0°C; and air, 21.4 and 33.8°C. Mulching materials, depending on their characteristics, can suppress or reduce infestation by weeds, diseases and insects (Adhikari *et al.*, 2016).

In the case of plant-derived materials, mulches supply organic matter (Clark and Moore, 1991; Himelrick *et al.*, 1995). Sawdust or bark derived from Douglas fir applied to a depth of up to 15 cm is commonly used for mulching throughout Washington. Mulches usually aid soil aeration, but some materials and some conditions can limit it. A 7.5 cm layer of particle mulch is a sufficient depth for weed suppression, but overmulching can reduce O_2 reaching the soil. This is a particular problem in wet or waterlogged conditions (Grundy and Bond, 2007). Use of compost is generally not recommended or widely practised for blueberry production, as composts typically have high pH, EC and K^+ content, all of which are undesirable for blueberry growth and development (Sullivan *et al.*, 2014; DeVetter *et al.*, 2015). Paper mulches were considered initially as an alternative to plastic, but they have been found to rapidly degrade and begin to break apart within weeks of being exposed to soil, rain and wind.

Today, paper-based mulch films are considered to be commercially unviable (Adhikari *et al.*, 2016). Some thermoplastic films that incorporate maize starch have been used for vegetables (Pannacci *et al.*, 2017), but because of their short durable effect (2–4 months), they would not be suitable for blueberries.

Plastic mulches

The use of plastics in agriculture dates back to the post-World War II era. Plastic mulches are used in many blueberry-growing regions. Most benefits from plastic mulches occur in the first years after planting, as this is the period when competition for water, light and nutrients is strongest. Depending on the type and quality of the mulch, they can last from 2 to 7 years, with weed mats (black landscape fabric made from woven polypropylene or polyethylene) having the longest active life (Cox, 2009). Plastic mulching has become a globally applied agricultural practice for its instant economic benefits, such as higher yields, earlier harvests, weed control, increased soil temperature, increased root growth, improved fruit quality, and increased water-use and fertilizer-use efficiency (Adhikari *et al.*, 2016; Steinmetz *et al.*, 2016; Kader *et al.*, 2017). Plastic film mulching is useful to overcome abiotic or water stresses while forming a barrier to restrict soil water evaporation and increase crop transpiration; increasing transpiration can enhance water productivity. Plastic mulching typically increases the water-use efficiency by 20–60% due to reduced evaporation (Steinmetz *et al.*, 2016). The physical features of the mulching film keep the crop root zone moist and protect the soil from wind and water erosion (Kader *et al.*, 2017). The most important aspect of promoting a crop's water-use efficiency derives from changing the balance between evaporation and transpiration under conditions of limited water (Yang *et al.*, 2015). In blueberries, plastic mulches should be used in combination with fertigation, as the fertilizer placed under the plastic is often depleted after 1 or 2 years (Williamson *et al.*, 2006). Mulching with plastic also improves soil temperature and reduces the fluctuation of soil moisture and soil temperature in the 0–25 cm soil layer where most blueberry roots are found (Kader *et al.*, 2017). On heavy, wet, clay soils, plastic mulches lead to anaerobic soil conditions and restrict soil microbial activity. The extent of mulch–soil contact (Tarara, 2000) can influence the aboveground and belowground environment (Cox, 2009). Black, transparent and white mulches are the colours used most commonly. However, colour selection strongly depends on the crop type and the crop's environment, as well as the temperature that can be tolerated by plant roots (Steinmetz *et al.*, 2016).

Polyethylene has become by far the most frequently used base material in agricultural systems. Plastic mulch films are most commonly made of low-density polyethylene and linear low-density polyethylene, and are thin (0.015–0.025 mm) and lightweight ($20\,g/m^2$ for 0.025 mm thickness) (Brodhagen *et al.*, 2015). Polyethylene properties are usually modified by additives such as plasticizers, coloured pigments, UV stabilizers or other polymers. In order to achieve longer life cycles of 3 years or more (which is recommended in the case of blueberries), ethylene vinyl acetate and ethylene butyl acrylate are added to polyethylene mulches as co-polymers (Steinmetz *et al.*, 2016). The polyethylene structural design usually consists of polyolefin, which is non-susceptible to degradation by microorganisms under natural conditions (Yang *et al.*, 2015; Adhikari *et al.*, 2016).

Regarding the effects on soil variables, plastic mulching provides mechanical protection of the surface soil, enhanced root development, stabilization of soil aggregates, increased mucilage production and promotion of soil fauna activity. The accelerated soil processes under plastic mulch can thus alter soil organic matter composition and quality (Steinmetz *et al.*, 2016). Along with alterations in physical properties, soil nutrient availability and microbial activity, plastic mulching can induce shifts in the composition of the soil microbial community. Soil microorganisms are widely used as bioindicators of soil quality (Chen *et al.*, 2014). Mulching has led to slight increases in total microbial diversity compared with non-mulched soil. In this respect, organic mulches alone or in combination with plastic mulches were shown to perform considerably better than plastic mulches. The enhanced productivity under plastic mulches has often been reported to result in lower soil contents of Mg, K, P and N when compared with bare soil (Steinmetz *et al.*, 2016). However, the effects of soil mulch on microbial activity and soil nutrient contents probably depend on both the type of mulch and the environmental conditions of the research site. Accordingly, Strik (2016) reported that, when trialled in Oregon, a semi-permeable weed mat increased soil NO_3^--N, NH_4^+-N, Ca and Mg in autumn, compared with sawdust mulch alone. These higher nutrient levels were attributed to a reduction in nutrient leaching with rainfall. Cox *et al.* (2014) working on SHB 'Star' in New South Wales, Australia, found that soil microbial activity (at 0–10 cm depth) was similar under woodchip and weed mat mulches.

Bacteria are the most abundant and diverse group of soil microorganisms, and are responsible for the vast majority of biogeochemical processes in soils. Microbial communities help maintain healthy soils as a result of their relationship with plant growth, nutrient cycling and suppression of diseases (Chen *et al.*, 2014). Farmer *et al.* (2017) studied the soil microbial population and biological activity in a Hapli-Udic Cambisol (silty loam texture) in Shenyang, Liaoning province, China (41°49′N, 123°34′E), which was continuously planted (28 years) with maize and covered with plastic film. This long-term mulching increased topsoil (0–20 cm depth) populations of bacteria, actinomycetes and fungi by 22.6, 29.3 and 19.7%, respectively, compared with bare soil treatment. Plastic mulching played a major role in shaping the bacterial community structure through significant alteration of soil moisture, pH, total N and soil organic C. Recent research on the effects of white plastic mulching in apple orchards ('Chang Fu #2' grafted on *Malus prunifolia*) in Luochuan, Yan'An, China (35°33′N, 109°47′E) discovered that 5 years of mulching induced significant changes in soil physicochemical properties and bacterial communities compared with a bare soil control (Chen *et al.*, 2014). Thus, populations of *Alphaproteobacteria* and *Actinobacteria* groups were higher under plastic, while numbers of *Deltaproteobacteria* were higher in bare soil. Soil physicochemical properties (temperature, available K, soil organic matter, total N) were significantly higher under the plastic mulch. Mulching with polyethylene is known to alter the absorption and transformation of radiation as well as heat conduction and water dynamics in the soil. Accordingly, the shift in microbial populations may be due in part to the fact that, in this trial, plastic mulching produced the hottest soil as well as the highest soil water content (Chen *et al.*, 2014).

If the mulch has been installed tightly and is in direct contact with the soil, the layer of air between the plastic and the soil is minimized and heat will be transferred readily by conduction (movement of energy by molecular vibrations in a solid or between a solid and a motionless fluid), leading to a rise in soil temperature. Alternatively, if the plastic mulch is laid loosely, leaving an air gap between the plastic and the soil, then heat must first be conducted from the plastic to the still air layer before diffusing through the air gap and being transferred to the soil. Air has a much lower thermal diffusivity than soil, and heat transfer from the mulch in this case is slowed down. If the plastic is not in contact with the soil, most energy at the hot plastic surface will then be transferred by convection (vertical transfer of energy to or from a surface by a moving fluid) to the atmosphere (Tarara, 2000). The increase in soil temperature associated with plastic mulches also brings higher water demands (Larco *et al.*, 2009; Strik, 2016). Black plastic mulch can be sprayed with a mixture of water and brightly coloured latex paint to reduce soil temperatures.

The extent of soil warming is affected by the colour of the plastic. Plastic mulches with high shortwave absorbance (black) or high shortwave transmittance (clear) are expected to generate the highest soil temperatures and have the greatest impact on root growth in the top layers of the soil. As woven weed mats allow some air movement, they have less impact on soil temperature than plastic films. Maximum temperatures are higher but minimum temperatures are lower under woven plastic mulch, compared with wood chips, in blueberries (Table 6.6).

Woven black weed mats have been studied intensively in the last few years and are used in many blueberry production regions. Reports from Australia state that weed mats are successful in weed control, and for protection of soil from erosion and hand-picking disturbance. In contrast to polyethylene films, mats allow rainfall and irrigation water to permeate once the material has aged (Ireland and Wilk, 2006). Weed mats have become very common in organic as well as conventional blueberry fields in the Pacific north-western USA, thus reducing the costs of herbicides and hand weeding (DeVetter *et al.*, 2015; Strik, 2016). Weed mats offered the most economical method of weed control for ten highbush blueberry cultivars in Oregon compared with sawdust alone or a

Table 6.6. Maximum and minimum soil temperatures at 2 cm depth over 2 years in a site planted to SHB 'Star' (loam soil) under woven plastic (weed mat) or wood chip mulch at Corindi, Australia (latitude 30°1′51″S). (Adapted from Cox, 2009.)

Season	Mulch treatment	Maximum daily range (°C)	Maximum (°C)	Minimum (°C)
2006	Weed mat	13.2	36.5	6.2
	Wood chip	5.5	30.7	9.9
2007	Weed mat	13.5	33.3	5.5
	Wood chip	5.7	28.4	8.4

mulch of municipal yard debris compost topped with sawdust. The various mulches tested for weed control in blueberry resulted in relatively little difference in yield or fruit quality over the 8-year study; however, plants grown with the weed mat mulch produced firmer and larger berries than those grown with the compost + sawdust-amendment mulch. Working on NHB blueberries in Oregon, Larco *et al.* (2013a,b) found reduced root:shoot ratios under weed mat compared with Douglas fir sawdust and sawdust + composted yard debris. They also reported reduced root growth under the weed mat in 'Duke' but not in 'Liberty', which could be the differential response of plant material to elevated soil temperature, reduced soil porosity, restricted water infiltration or changes in nutrient availability under the plastic weed mat. Weed mat mulching increased cumulative yield compared with organic mulches (Strik, 2016). In Georgia, a study on rabbiteye blueberry by Krewer *et al.* (2009) established that organic mulches had a similar yield to those with a weed mat in the first 2 years of establishment but higher yields in years 3–5. Pasa *et al.* (2014) after evaluating weed mat versus pine needles (10 cm thick) for two seasons in Pelotas, southern Brazil, reported that the weed mat increased fruit numbers per plant and yield of rabbiteye 'Climax', 'Delite' and 'Powderblue' but not of SHB cultivars 'Georgia Gem', 'Misty' and 'O'Neal'. A weed mat established in plants of all cultivars resulted in a higher cane diameter, cane number and plant height. Trials in NHBs by Strik (2016) showed that while a weed mat reduced average berry weight over seven fruiting seasons, the range of 2.19 g (weed mat) to 2.23 g (sawdust) was not commercially relevant. Plants grown with a weed mat had higher shoot:root ratios during establishment, registered higher soil temperatures and required more irrigation water. Magee and Spiers (1995) found that white-on-black polyethylene-based mulches produced greater plant growth and yield than black plastic or black woven fabric mulches in SHB cultivars as a result of decreased soil temperature under the more reflective mulches.

Plastic mulches also affect the light environment of the plants whose soil is covered by them. No reports of studies on blueberries have been found, but in pepper, twice as much reflected PAR was measured above clear plastic mulch than above black plastic and bare soil. Both red and black plastic reflect about the same amount of PAR, but red plastic increases the ratio of red:far-red in the reflected light. The red:far-red ratio is critical for various characteristics of the plant, such as leaf size, root:shoot ratio and cuticle thickness (Liu *et al.*, 2009). In bell peppers, it was found that the percentage of PAR reflected from the mulches was highest for silver mulches and lowest for black mulches (Diaz-Perez, 2010). Additionally, it has been shown in watermelons that planting holes cut through plastic mulches potentially direct CO_2 (a gas needed for photosynthesis) towards the canopy of plants, the so-called 'chimney effect'. As much as double ambient CO_2 concentrations have been measured above holes cut for transplants (Soltani *et al.*, 1995). As most of the canopy of blueberries develops at greater distances than those of vegetables, most of the effect of this reflected light and CO_2 levels would be expected to occur in young plants, in cultivars with low stature or in lower portions of the canopy.

Although plastic mulching has many advantages, it has some negative effects. The major negative consequence deals with handling the plastic waste and the associated environmental impacts. If not removed, the residual mulch may damage the soil physical structure and can block the infiltration of capillary water. This may affect root growth and can produce secondary salinization of topsoil (Liu *et al.*, 2014). When the remains of the film are less than $16 \, cm^2$, it will not cause crop reduction, but yields have been reduced when the residual plastic exceeds $58.5 \, kg$ per $100 \, m^2$ (Yang *et al.*, 2015). Plastic mulches create difficulties in dumping and emit harmful substances during burning (Kader *et al.*, 2017). Recycling of used mulch is only possible if contaminants (e.g. soil, vegetation, pesticides and fertilizer) make up less than 5% of the total weight of the mulch. Studies have shown that this threshold is exceeded dramatically, with the actual contaminant weight being up to 40–50% (Steinmetz *et al.*, 2016). Many growers in the USA dispose of plastic mulch film in landfills, which is costly (US\$359–584/ha) and labour intensive. Degradation of polyethylene residual mulch film is negligible (e.g. 0.35% in 2.5 years) with the possible formation of environmentally harmful chemicals such as phthalate esters, aldehydes and ketones (Brodhagen *et al.*, 2015). The thickness of the mulch film varies among countries, and this has an impact on its fate, as the thicker mulch film remains mostly intact after use, and almost no residual plastic film mulch is left in farmland soil after mechanized recovery. In China, the film is less than $0.008 \, mm$ thick; in contrast, those used in the USA and Europe are generally $0.02 \, mm$ and in Japan $0.015 \, mm$ (Liu *et al.*, 2014).

While in the short-term plastic mulches usually improve soil conditions (temperature and moisture control) and generate positive effects on yield and quality, the long-term effects on sustainability need to be studied in more detail (Steinmetz *et al.*, 2016). Due to increasingly stringent regulations regarding use of non-degradable plastic in agriculture, they are likely to be phased out in the near future (Adhikari *et al.*, 2016). Due to the negative effects of mulching on ecological variables, there have been advances in the development of products that will degrade naturally and which also have the potential to be used in the field for more than one season. In this context, there has been increasing interest in biodegradable mulches (Grundy and Bond, 2007). Two types of bio-degradable plastic mulch have been developed: bioplastics (or biobased plastics), which are made from renewable resources such as maize starch or cellulose, and oxo-degradable plastics, which are made from petrochemical substances (i.e. fossil fuels). For a truly biodegradable plastic, there must be a reduction in the molar mass (length) of polymers, followed by bioassimilation and/or biological conversion (through microorganisms, such as bacteria and fungi) of the polymer breakdown products and, ultimately, biological conversion to CO_2 and/or methane, and water (Brodhagen *et al.*, 2015).

Any plastic breakdown involves a complex synergy of abiotic (e.g. photo-oxidation, erosion, fragmentation) and biological processes. Polymers based on renewable sources are more ecologically desirable than conventional plastic films (Yang *et al.*, 2015). Although photodegradable plastics are reported to degrade by photoinitiated chemical reactions, their ability to decompose completely (to CO_2 and water) in the soil without light is questionable (Adhikari *et al.*, 2016). Biodegradable mulch films are not pure polymers, and in an agricultural setting, these products need to degrade under variable environmental conditions via the native microorganisms, which may not include those able to break down the polymer (Brodhagen *et al.*, 2015). The rate of breakdown of films made from biodegradable polymers will depend on soil and weather conditions. Most of these films are reported to be relatively weak in mechanical properties, not efficiently degradable and cost prohibitive, as their cost is reported to be two to three times higher than conventional plastics. However, biodegradable plastics should save manpower dedicated to film collecting, as well as the long-term benefits to the environment (Adhikari *et al.*, 2016). Currently, there are few biodegradable polymers available commercially. Preliminary trials on the effects of a biodegradable and compostable mulch (produced using plant starch and oils) on NHB 'Lateblue' blueberries were carried out over two seasons in Turin, Italy, by Girgenti *et al.* (2012). Compared with plants growing in bare soil, plants grown with biodegradable mulch had higher soil moisture, but there were no differences in plant height when comparing the two treatments.

In the search for truly biodegradable plastic mulches, two developments have occurred in recent years: photoselective films and sprayable biodegradable coatings. In contrast to traditional black plastic mulches, photoselective films block only the visible component of solar radiation. The optical properties of these films select both incoming (solar) and outgoing (emitted from the ground) electromagnetic radiations. This avoids thermal damage to plant tissues (Mormile *et al.*, 2017). A 2-year project was carried out in three greenhouses in southern Italy (40°40′N; 14°46′E). Soil temperature for different vegetables at 10 cm depth under 30 μm-thick mulching films was consistently lower under yellow films than under black ones. Increased warming of black mulches is considered beneficial for colder times of the year, but in summer, temperatures exceeding 34°C have been registered under black mulch, which is the threshold value above which root and shoot growth decline for most horticultural species. Root system exposure to high temperatures for several consecutive days can impair root growth and functionality, with negative consequences for crop yield (Bonanomi *et al.*, 2017). Spiers (1995) conducted a greenhouse study to evaluate the influence of substrate temperatures (16, 27 and 38°C) on root and shoot growth of six blueberry clones (three clones each of two types: SHB and rabbiteye). Each clone had a negative linear response to substrate temperatures. Root and shoot growth was best at 16°C, which indicates that both rabbiteye and SHB blueberries would respond favourably to cultural practices that lower soil temperatures during the summer growing season. An added benefit of photoselective mulches is that they combine the effect of transparent and black films, and may also have a repulsive effect on aphid populations (Pannacci *et al.*, 2017).

Sprayable biodegradable polymer mulches, which are easy to apply and versatile, are still in the early stages of development. The ability to mix natural additives, plasticizers and fillers to control the mechanical and biodegradation properties of the core polymeric mulch film has been the driving force behind these innovations. A key challenge for these mulches is to understand how the mulch film forms on different soils and how the condition of the soil surface at the time of application affects the properties of the formulation as it dries to form the mulch film. Sprayable biodegradable mulches are expected to cost less as demand increases due to reduced labour requirements and the cost of raw materials. Any such product needs to be subjected to extensive field trials in order to study the effects of extrinsic factors such as oxidative stress, pH, temperature, UV exposure, and retention of mechanical and radiometric properties. There will also be opportunities to encapsulate other additives (e.g. nutrients and/or agrochemicals) (Adhikari *et al.*, 2016).

Organic mulches

In areas where blueberries have not been planted in naturally acidic highly organic soils (more than 3%), mulching with organic mulches provides many benefits; among them, organic mulches have been reported to increase plant size, berry yield, soil moisture, root weight, organic matter and soil structure, and to reduce frost damage and weed growth, delay vegetative bud opening and leaf drop in autumn, reduce soil temperature in the summer and minimize soil temperature fluctuations (Patten *et al.*, 1989; Clark and Moore, 1991; Gough, 1994; Williamson *et al.*, 2006; Strik, 2016). Part of the effect of organic mulches on soil temperature is due to increased soil moisture (Haman *et al.*, 1988; Ireland and Wilk, 2006); however, factors other than water relationships contribute significantly to the enhanced performance of highbush blueberries under mulch (Clark and Moore, 1991). Sawdust or bark mulch increased photosynthetic and respiration rates of 'Bluecrop' plants by 32–59%, which was partly explained by a 60–69% rise in chlorophyll levels (Wu *et al.*, 2006). However, research has found little or no effect of mulch on fruit quality (Strik, 2016).

Rabbiteye blueberries appear to respond less to mulch than NHBs and SHBs (Haman *et al.*, 1988; Clark and Moore, 1991). Mulches are not commonly used in commercial rabbiteye plantings (Himelrick *et al.*, 1995). However, research on rabbiteye (Patten *et al.*, 1989) reported a reduction in frost damage in mulched plants, which contradicts previous research on fruit trees where it was reported that mulch increased frost damage by lessening the transfer of heat from the soil during the frost (Hogue and Neilsen, 1987). The decrease in frost damage in the blueberry study was attributed both to a delay by the mulch in dormancy release and a retardation in the rate of spring bud opening, and to reduced N levels in plants under mulch (Patten *et al.*, 1989).

The ideal mulching material is one that disintegrates slowly, thus placing less strain on the N supply of the soil and requiring less frequent reapplication (Eck *et al.*, 1990). Sawdust and pine bark are the most widely used mulching materials. A 50:50 mixture of these materials provides most benefits (Himelrick *et al.*, 1995). Aged or rotted sawdust has been commonly shown to increase growth and yield of blueberries. This effect is thought to be due mainly to physical improvement of the soil, as evidenced by greater root growth under decomposed mulch in highbush (Gough, 1980; Odneal and Kaps, 1990) and rabbiteye (Spiers, 1986) blueberries, and greater numbers of fine roots in the top 15 cm (Scagel and Yang, 2005). Application for 11 years of wheat straw to silt loam soils without cultivation in Ohio at 0, 2, 4, 8 or 16 t/ha/year increased available water capacity by 18–35%, total porosity by 35–46% and soil moisture retention by 29–70% (Mulumba and Lal, 2008). Sawdust mulch increased O_2 diffusion rates and porosity, and reduced soil bulk density at the end of the second season in groundnut (Khan *et al.*, 2000). Kozinski (2006), after performing trials on NHB 'Bluecrop' blueberries in Poland, concluded that bark did not benefit the yield as much as sawdust mulch. However, when the material was both incorporated into the soil and used as ground cover, the yield difference between bark and sawdust was smaller. The highest yields in organically grown rabbiteye blueberries in Georgia were obtained with pine bark or wheat straw as a mulch (Tertuliano *et al.*, 2012). Pine-needle mulch was more effective in suppressing weeds than manure + sawdust compost or seafood waste compost; seafood waste compost had high nutrient availability, which induced prolific growth of germinating weed seeds (Burkhard *et al.*, 2009). Williamson *et al.* (2015) reported that 4–7-year-old SHB 'Emerald' blueberry plants grown in pine bark-amended soil (using half the amount of pine bark as beds) did not differ in canopy volume, berry yield or mean berry weight compared with plants grown on pine bark beds.

Fresh sawdust should never be used because it can release toxic compounds and will tie up more N. Red maple and beech have had negative impacts, and use of cedar, oak and walnut has been associated with chlorotic leaves and poor growth. Compost should be applied early in the season and not in the autumn, as it tends to promote vigour and increase the chances of winter injury. Strik *et al.* (2017) after performing trials on a number of highbush blueberry cultivars ('Aurora', 'Bluecrop', 'Bluegold', 'Bluejay', 'Draper', 'Duke', 'Legacy', 'Liberty', 'Ozarkblue' and 'Reka') in a certified organic research site in Oregon reported that on-farm compost as a pre-plant amendment and as part of the mulching programme increased soil pH from 4.9 to 5.9; in turn, this increased organic matter content, Ca, Mg and K levels compared with the weed mat treatment. Straw can be used if locally available but needs to be reapplied more frequently as it decomposes faster. Peat moss can also be used but is expensive, tends to dry out on the surface and is difficult to re-wet (Demchak, 2003).

Within the soil profile, roots will grow primarily where organic matter is present. Mulching has been reported to improve root growth in the upper soil layer (Scagel and Yang, 2005) and a large part of the root system is developed at the soil–mulch interface (Cox *et al.*, 2014). Among the amendments used for increasing organic matter in the planting hole, peat moss has been shown to provide the most consistent benefits. Aged sawdust (either hardwood or softwood) or compost can also be used in the planting hole (Williamson *et al.*, 2006).

Incorporation of organic substances can more frequently cause detrimental impacts or have little effect compared with surface placement of these materials. Sawdust has decreased root length when incorporated in highbush blueberries at planting (Yang *et al.*, 2002). In Oregon, sawdust incorporated prior to planting did not improve plant growth in well-drained sandy loam, loam or silt loam soils compared with heavier soils (Hart *et al.*, 2006).

The effectiveness of organic mulch depends on its thickness (Pannacci *et al.*, 2017). The depth of the organic mulch should be 10–20 cm. In the Pacific Northwest, it is recommended that 9 cm of Douglas fir sawdust be applied initially and 3–8 cm added every other year (Himelrick *et al.*, 1995; Williamson *et al.*, 2006). Strik (2008) recommends increasing the depth of the mulch to 15 cm over a period of years. With greater depths, roots will tend to grow higher in the crown and only within the mulch or in the interface with the soil. A build-up of mulch or mulch applied too deep can restrict root gas exchange (Hanslin *et al.*, 2005). Plantings perform better when mulching is continuous versus only 1, 2 or 3 years after planting (Spiers, 1982, 1986). The rate of deterioration of pine bark in Georgia was estimated to be 2.5 cm/year (Krewer *et al.*, 1997). Once the decision to use mulches has been taken they should not be discontinued. As degradation of the organic mulch proceeds, weed control declines, and some blueberry roots become exposed and can dehydrate, and are subject to herbicide toxicity or physical damage if weeds are controlled with hand tools (Krewer *et al.*, 1997). The C:N ratio, chemical composition, particle size and degree of decomposition of organic materials affect their performance when used as mulches. There is considerable variability in these factors among mulches, and unfortunately few studies have described these characteristics, making it difficult to provide precise directions and specify the impact of organic mulches on blueberry performance. Before its use, the C:N ratio of organic mulches and compost must be measured to determine if there is enough N available for the crop (see below) and whether adjustments are necessary (Demchak, 2003). If N is needed, it should not be supplied directly to the planting hole as young roots are very susceptible to salt damage (Himelrick *et al.*, 1995). The C:N ratio in the organic matter added will determine N availability. Sources with a C:N ratio greater than 30:1 will tie up (immobilize) N (Yang *et al.*, 2002). If the C:N ratio is less than 20:1, the organic matter will mineralize and supply N to the blueberry. Residues with C:N ratios between these values will neither tie up nor release N (Table 6.7). However, when various organic mulches were studied in NHB 'Duke' blueberries in Nova Scotia, Canada, Burkhard *et al.* (2010) found no evidence of net N immobilization despite the high C:N ratios of the pine needles (approximately 73:1) and the manure + sawdust compost (approximately 46:1). The C:N ratio for any organic material decreases as it decomposes. For blueberries, Sullivan *et al.* (2014) suggested that composts with a C:N ratio of 12–25:1 are the most suitable for incorporation in beds or for application as a mulch below a weed mat.

Recommendations have been provided on how much extra N should be applied when mulches are used. As a general rule, it is suggested that when mulches are used, the N rate should be increased by 30%. The rate can then be adjusted according to leaf analysis and subsequent plant growth and

Table 6.7. C:N ratios for various organic materials used for mulching and soil amendments in blueberries, and their propensity to supply or immobilize N. (Data from Demchak, 2003; Whatcom, 2005.)

Organic material	C:N ratio	N supplied (+) or immobilized (−)
Legumes	9–19:1	+
Peat moss	45–58:1	−
Farm manure	90:1	−
Rotted manure	20:1	+
Grass clippings	19:1	+
Straw	20–80:1	−
Lucerne hay	18:1	+
Douglas fir bark	491:1	−
Aged hardwood sawdust	60: 1	−
Fresh sawdust	300–700:1	−
Sawdust weathered for 2 months	625:1	−
Sawdust weathered for 3 years	142:1	−
Douglas fir sawdust	800:1	−

appearance (Williamson *et al.*, 2006). In Oregon, where it is recommended to incorporate 9 cm of Douglas fir sawdust in 30 cm of soil within the row prior to planting, the addition of 107 kg N/ha is suggested to avoid immobilization of N for blueberry plants (Hart *et al.*, 2006). Yard waste compost could be added together with conifer sawdust mulch to counteract N immobilization by the sawdust (Forge *et al.*, 2013). In addition, as composts often have a high pH and EC, dilution with conifer sawdust may be beneficial (Caspersen *et al.*, 2016). Mulch can neutralize poor-quality water better than peat moss applied at the planting hole. Pine bark mulch at the base of each plant (28 dm^3 of medium-sized bark chips) buffered the soil pH, EC and Na$^+$ in the soil, but no effect was found with incorporation of 18.7 dm^3 of peat moss per planting hole. Mulch may reduce soil salinity by decreasing surface evaporation and allowing greater water infiltration (Patten *et al.*, 1989). Large differences in mean pH were found in a trial comparing various organic mulches in NHB 'Duke' blueberries, with pine needles producing the lowest (pH 3.8), and manure compost (pH 6.7) and seafood compost (pH 6.9) the highest (Burkhard *et al.*, 2010).

In the lower south-eastern USA, some SHB blueberry growers use a system called 'pine bark culture'. The pine bark is used to construct beds 20–30 cm deep. Bushes are planted with most of their root system in this bark bed. Provided plants receive adequate water and nutrients, they are vigorous and productive, as the decomposing bark generates a proper environment for root growth. Although the cost of implementation is high, the premium price growers receive for the early fruit justifies the investment (Williamson *et al.*, 2006).

Manipulation of the habitat of predatory arthropods, in this case by installing a cover, may alter both the abundance and diversity of species present in the system, which could have an impact on the efficiency of biological control by these organisms (Ganter *et al.*, 2013). Ground beetles (Carabidae) and rove beetles (Staphylinidae) are generalist predatory species common in temperate agroecosystems. Mulches can affect pupation depth and adult emergence of blueberry maggot, *Rhagoletis mendax*, a major blueberry pest in eastern North America. To learn how predatory beetles that are active when the maggot is pupating may be affected by ground cover, a trial was done for 2 years comparing two organic mulches (hardwood compost and pine needles), with continuous weeding. A 20-year-old NHB 'Bluecrop' and 'Bluejay' blueberry field in Nova Scotia, Canada, was used. Compost mulch and weeding significantly affected carabid populations, while the staphylinid community responded to compost and pine-needle mulches. Effects due to mulch tended to intensify in the year after mulch application for both families. Species richness and diversity for Carabidae and Staphylinidae were similar, but there were indications of higher Carabidae richness in unmulched plots despite fewer individuals being captured. Carnivorous Carabidae were most frequently captured in compost plots in both years, and omnivores were most frequently captured in unweeded compost (Renkema *et al.*, 2012). To study the effect of ground cover on the diversity and abundance of carabid beetles, a trial was carried out on 8-year-old NHB 'Elliott' plants growing in an organic commercial field in Villarrica, Chile. Treatments included woven weed mats, pine needles, weed mats combined with pine needles and bare soil as a control. The number of individuals collected was greater for the weed mat plus pine needles and in bare soil, whereas weed mat treatment alone presented the lowest values. Species richness showed no differences among the treatments (Ganter *et al.*, 2013).

Yeo *et al.* (2017) evaluated alternative, non-chemical, cultural management strategies to reduce *Phytophthora* root rot in a field of NHB 'Draper' plants. The soil was either amended or not with gypsum before planting. The plants were irrigated using narrow (adjacent to the plant crown) or widely spaced (20 cm on either side of the plant crown) drip lines and mulched with Douglas fir sawdust or a black weed mat. Initially, root infection by this pathogen was lower with the combination of gypsum, wide drip lines and sawdust mulch than with any other treatment, except for the fungicide control. The soil under the weed mat accumulated more heat units than under the sawdust and resulted in faster hyphal growth by the pathogen. However, plant growth was similar for both mulch types. Although plant growth was not adversely affected after 2 years with the weed mat, caution is warranted when using weed mats with susceptible cultivars and at sites conducive to *Phytophthora* root rot (e.g. heavy soils and poor drainage).

Pseudosclerotia of *M. vaccinii-corymbosi* overwinter on the soil surface and develop apothecia in early spring, supplying primary inoculum for mummy berry disease of blueberry. To determine whether mulches could suppress apothecia development, mulches of Douglas fir sawdust at 2.5 or 5 cm depth, blueberry leaves at 2.5 cm depth and a bare ground (no mulch) control were assessed for two seasons in a 17-year-old NHB 'Bluetta' field in Corvallis, Oregon. A 5 cm depth of Douglas fir sawdust was associated with greater apothecial suppression in comparison with bare ground. Douglas fir sawdust at a 2.5 cm depth varied in effectiveness, while a 2.5 cm mulch of blueberry leaves was similar in apothecial development to the bare ground treatment. Application timing did not affect apothecial development, but mulches lost significantly more depth when applied at the beginning of the overwintering season compared with late winter mulches (Florence and Pscheidt, 2017).

PRODUCTION UNDER HIGH TUNNELS

The use of protective structures is now widespread in the warmer production regions where SHB blueberries are grown. The most commonly employed structures are tunnels that are plastic covered, unheated and have passive ventilation (Lamont, 2005). Tunnels are used primarily to accelerate fruit ripening but can also serve to protect against freezes and rain, and allow for more efficient fertilizer and water use (Gaskell, 2004; Demchak, 2009; Strik, 2012). High tunnels raise air temperatures by 10–20°C (Kadir *et al.*, 2006) and soil temperatures by 4–8°C (Reiss *et al.*, 2004). Tunnels are now widely used to advance the harvest of many other fruit crops, including raspberry (Dijkstra and Scholtens, 1993; Oliveira *et al.*, 1996), strawberry (Kadir *et al.*, 2006; Salamé-Donoso *et al.*, 2010) and cherry (Lang, 2009). Studies at several locations across the world have shown that SHB production can be advanced by over 1 month under high tunnels compared with open-field production (Ciordia *et al.*, 2002; Baptista *et al.*, 2006; Ozeki and Tamada, 2006; Renquist, 2008).

In a comparison of the performance of the SHB cultivars 'Snowchaser' and 'Springhigh' in tunnels versus open fields in Florida, Santos and Salamé-Donoso (2012) found fruit harvests of both cultivars to be significantly earlier under tunnels. Harvesting of both cultivars began in late February under the tunnels but not until mid-April in the open field. By April 15 the cumulative yields of tunnel-grown 'Snowchaser' were over 22.5 t/ha and 11.2 t/ha for 'Springhigh'. The high tunnels also effectively protected the plants from frost. The minimum air temperature in the open field was -7 to -8°C in the two seasons (61 freezing or near-freezing events) and a little above 0°C inside the tunnels (only 3 days with freezing temperatures).

In a study designed to determine the optimal date (15 December, 2 January and 16 January) to close high tunnels of SHB 'Emerald' and 'Jewel' in Georgia, Ogden and van Iersel (2009) determined the earliest closure date that advanced the harvest date the most. A closure date of 15 December advanced flower initiation by 38 days for 'Emerald' and 39 days for 'Jewel' compared with outdoor control plants. 'Jewel' subsequently developed ripe fruit 80 days after anthesis and 'Emerald' 105 days after anthesis. There was not a significant difference in yield between the various closure dates in the first year of the study, although variability was very high between replicates. No fruit were harvested in the second year due to multiple freezes during flower and fruit development.

In the only study measuring the microclimate and physiological response of highbush blueberries grown in tunnels, Retamal-Salgado *et al.* (2015) found that the temperature under tunnels in south-central Chile was on average 10–12°C higher than in the open field, and the minimum temperature averaged 2–5°C higher. Total PAR decreased an average of 25% under the tunnels, while levels of diffuse PAR increased by more than 150%. The value of g_s ranged from 42 to 99% higher in the high tunnel compared with the control, and was significantly correlated with diffuse PAR. SHB 'O'Neal' blueberries under the high tunnel had 44% higher yields than the controls and were harvested 14 days earlier.

Soil-less culture

Blueberries have traditionally been grown in the open field in soil, but growing them in pots under tunnels is increasing dramatically in popularity where early harvests are desired. Typically, the plants are grown in 15–25 l polyethylene containers on ground that is completely covered with weed mat fabric. Water and nutrients are delivered by drip irrigation.

Kingston (2017) conducted two studies in Oregon with NHB and SHB blueberries to evaluate how medium and fertilizer composition influenced growth and nutrient uptake. In the first study, the performance of SHB 'Snowchaser' was evaluated in ten different combinations of sphagnum peat moss, coconut coir and Douglas fir bark with 10% perlite and one commercially available mix. Four months after transplanting, the total dry weight of the plants had nearly doubled in the media with 60% or more peat or coir compared with those with at least 60% bark. Increasing bark in the medium also reduced nutrient uptake efficiency of N, P, K, S, Ca, Mg, Fe, Mn, B, Cu and Zn relative to peat or coir.

In the second study of Kingston (2017), NHB 'Liberty' and SHB 'Jewel' blueberries were grown in four levels of perlite (0, 10, 20 and 30% by volume) and in four ratios of peat:coir (1:0, 2:1, 1:2 and 0:1). The amount of perlite in the medium was negatively correlated with dry weight in 'Jewel' but had no effect on dry weight in 'Liberty'. Among the media with perlite, the proportion of peat and coir had no effect on the dry weight of either cultivar. However, among the media without perlite, the amount of peat was positively correlated with dry weight in 'Liberty'.

Voogt *et al.* (2014) focused on developing a standard nutrient solution for blueberry in a soil-less growing system in the Netherlands. They studied the NHB 'Draper' planted in a peat:perlite 1:1 v/v/ mix and found that the total nutrient demand by blueberry was low (especially for K) compared with other substrate-grown crops. A high $NH_4^+:NO_3^-$ ratio was needed to produce a sufficiently low pH in the root environment, which led to high SO_4^{2-} concentrations, but this did not appear to be a problem. The uptake of Na^+ and Cl^- was very limited, suggesting that high-quality irrigation water is needed for profitable soil-less production of blueberry.

Xie and Wu (2009) evaluated the physical and chemical properties of several substrates and their effect on the growth and development of half-high 'Northblue' blueberries in China. The substrates included pure and combinations of pine sawdust, peat, garden soil, perlite, sand and gravel. The addition of perlite increased air-filled porosity but lowered the water- and nutrient-holding capacity. The addition of peat increased the organic matter content, lowered the pH, improved air-filled porosity, and increased the water- and nutrient-holding capacity. The addition of peat and sawdust increased the number of plant leaves, average leaf area, shoot length and shoot thickness. Gravel had a negative impact on plant growth.

PRUNING

Highbush blueberry bushes need regular pruning for sustained productivity. Most pruning in NHBs is done during the winter when canes are dormant, while SHBs are often pruned both in summer after harvest and during the dormant season (see below). Pruning can reduce plant size and crop yield in the following season, but if properly conducted, the overall effects of pruning are larger fruit, earlier ripening and greater stability of yields. Pruning to regulate crop load can ultimately lengthen the life of bushes and increase the number of productive harvests. Regular pruning allows light to travel deeper into the canopy and stimulate the formation of more flower buds (Yáñez *et al.*, 2009). Pruning also reduces the conditions favourable for disease development by increasing air circulation, removing diseased canes and providing conditions for improved spray coverage.

Yield in blueberries is a complex interaction among several different yield components including canes per bush, flower buds per cane and fruit weight. When Siefker and Hancock (1987) studied yield component variation in nine NHB cultivars, they found that canes per bush and berries per cane were generally stronger determinants of yield than berry weight. However, a strong negative relationship was found between berry numbers and berry weight, which could partially compensate for yield losses due to reduced berry numbers. This is called 'component compensation' and is why berry size is responsive to pruning intensity. In the different cultivars, overall component interactions ranged from slightly additive in 'Bluecrop' and 'Spartan' to highly compensatory in 'Berkeley' and 'Rubel'. 'Blueray', 'Earliblue', 'Elliott', 'Jersey' and 'Northland' showed intermediate compensatory responses.

Pruning young plants

Fruit yield in young plants needs to be reduced to encourage vegetative growth (including root growth). It is recommended that flower buds be removed on newly set highbush blueberries by rubbing them off by hand or pruning off the tips of shoots. Canes on NHB plants in the second and third year are also commonly prevented from bearing more than two clusters of fruit, although this practice is dependent on the overall vigour of the plants. Reducing yield in young plants allows a mature bearing surface to develop as quickly as possible. In addition, as roots have the lowest priority in carbohydrate allocation by the plant, the removal of fruit in the first seasons of the plant in the field allows adequate development of the underground organ.

Strik and Buller (2005) measured the effect of removing flower buds for 2 years on the growth and yield of young NHB 'Bluecrop', 'Duke' and 'Elliott' bushes over a 4-year period. They found that early cropping significantly reduced the dry weight of roots, crowns and young shoots in all cultivars. Early cropping reduced yield in the fourth year by 44% in 'Elliott', by 24% in 'Duke' and by 19% in 'Bluecrop'. Cumulative yields were similar in the early cropped versus de-budded 'Bluecrop' and 'Duke', while in 'Elliott' early cropping reduced cumulative yield by 20–40%.

Highbush blueberries grown in warm climates can reach mature size in as little as 3–4 years, while those grown in colder climates take as long as 6–8 years to reach full maturity. As a result, overall pruning strategies differ somewhat for SHBs and NHBs. In NHBs, it is generally recommended that growers keep only about two of the new canes produced each year until the bushes reach maturity. In SHBs, young vigorous bushes are commonly left unpruned for the first 2–4 years or thinned to the strongest three or four new canes each year. Well-pruned mature bushes of both NHBs and SHBs should have ten to 20 canes of varying ages, depending on the cultivar's ability to produce renewal canes.

Pruning mature bushes

Annual pruning is recommended for long-term stability of yield. If bushes are pruned only occasionally, an uneven balance of very old and young canes is produced. It has been proposed that the highest-yielding bushes need to have about 15–20% young canes (less than 2.5 cm diameter), 15–20% old canes (over 3.5 cm) and 50–70% middle-aged canes. Palma and Retamales (2017) studied the productivity of canes in 12–14-year-old NHB 'Brigitta' and 'O'Neal' blueberries growing in a commercial field in south-central Chile (latitude 35°S). In this study, canes were grouped into three basal diameter ranges: 0–0.9, 1–1.9 and 2–2.9 cm (which for 'Brigitta' and 'Duke' in Michigan corresponded to 0–2-, 2–3- and 3–5-year-old canes, respectively) and at two locations within the bush: external (growing within 25 cm of the canopy periphery) or internal (in the canopy centre). After two seasons, they established that external canes had higher yields (67–69%) and fruit numbers (54–62%) than internal canes, and this was related to the higher availability of radiation for external (42% full sun) than for internal (27% full sun) canes. They also found that yield increased with greater cane diameter, and this was due mainly to larger fruit number per cane.

During each dormant season, the largest canes should be removed at their base to let as much light as possible into the centre of the bush. The overall condition of canes should be considered when deciding which to remove: weak or diseased canes should receive the highest priority to be removed along with those that are low spreading or mechanically damaged. Many growers in the coldest regions wait until late winter so they can remove canes that have been damaged by extreme cold. The cuts should be as close to the main cane or crown as possible, so that short stubs are not left. In Michigan, most pruning is focused on whole-cane removal, while in Chile and the Pacific Northwest, more effort is focused on the top of bushes to balance floral and vegetative growth.

Siefker and Hancock (1987) compared berry weight, berry number and yield per bush in mature NHB 'Jersey' bushes for 3 years after removing 20–40% of the total base area. Pruning significantly reduced berry number in the first year but not in the second and third, while berry weight was significantly increased by pruning in years 1 and 3. Pruning intensity had a negative impact on yield per plant in the first year but no effect in years 2 and 3. Pruning also significantly increased the number of renewal canes formed. They concluded that moderate pruning may reduce yields in the first year, but that it generally increases fruit weight and may act to prevent an eventual decline in productivity by stimulating the production of new vigorous canes.

Strik *et al.* (2003) compared the effect of pruning intensity on berry weight, yield and harvest efficiency in mature NHB 'Berkeley' and 'Bluecrop' plants in Oregon over a 5-year period. They compared three treatments: (i) conventional pruning with the removal of the most unproductive canes, thinning of 1-year-old shoots, and removal of weak and excessively fruiting shoots from the top of bushes; (ii) 'speed pruning' where only one or two of the most unproductive canes were removed at their base; and (iii) no pruning. Yields were highest in the unpruned controls, but the conventionally pruned bushes had 27% larger fruit and could be harvested in about half the time. Fruit began to ripen on the conventionally pruned bushes about 5 days earlier than on the unpruned ones. The speedily pruned bushes were intermediate for all of these characteristics.

The blueberry physiology research group at the Universidad de Talca, Chile, has been searching for new winter pruning strategies for NHBs that would: (i) reduce the amount of labour needed for this task; (ii) simplify the instructions in order to be able to hire unskilled workers; (iii) improve the light availability in the centre of the canopy in order to increase yield and quality; and (iv) reduce canopy size to reduce the labour needed to harvest the crop. In this context, the idea of sectorial pruning was developed. Under this scheme, one-quarter of the canopy is removed by basal pruning every 1–2 years. In this manner, depending on the growth potential of each site, each quarter of the canopy is pruned every 4–8 years. In the first season of this trial, Espíndola (2018) compared sectorial pruning (basal pruning of all canes in the northeastern quarter of the canopy) with light pruning (removing the oldest cane in each of the four quarters of the canopy). The trial was done in the winter in two mature fields of NHB 'Brigitta' and one of NHB 'Duke' blueberries in south-central Chile (latitude 35°S). Sectorial pruning removed 7.2–10.1% while light pruning removed 1.5–4.7% of the total basal area of each plant. In the following harvest season, fruit weight was on average 66.3% higher for sectorial pruning compared with light pruning in 'Brigitta' and 11% in 'Duke', while the yield for sectorial pruning was 32.4% higher in 'Brigitta' and 87.5% higher in 'Duke'. These results were attributed to increased light availability in the centre and lower parts of the canopy, which would have induced higher numbers of flower buds and allowed leaves to form greater amounts of carbohydrates to feed the growing fruit during the season. On average, the speed of labour for regular pruning of the field was 18 plants/h while for sectorial pruning it was 80 plants/h.

Summer pruning in SHBs

In addition to winter pruning, SHBs are also commonly hedged at 100–122 cm after the fruiting season using a sickle mower or hedge trimmer. This practice is much faster than detailed hand pruning and is done to maintain bush size, reduce disease and pest pressure, prevent overbearing and achieve some drought tolerance. In a study on a mature planting of 'O'Neal' in North Carolina, Mainland (1993) compared seven different pruning treatments conducted in mid-June: (i) no pruning; (ii) removal of weak and damaged canes and bush shaping (lop); (iii) removal of weak and damaged canes, bush shaping and removal of twigs that were damaged or flowering excessively; (iv) hedged flat at 100 cm with no cane pruning; (v) similar to (iv) but with removal of weak and damaged canes and bush shaping; (vi) hedged at an angle with the peak at 122 cm and the edges at 61 cm with no cane pruning; and (vii) similar to (vi) but with removal of weak and damaged canes and bush shaping. Yield was highest in those bushes that were not pruned; however, fruit weight was highest in all treatments that were hedged and they could be stored for longer (Table 6.8). The various treatments had little influence on harvest date. However, it is known that hedging interrupts apical dominance in each shoot. Thus, the nearest vegetative buds underneath the hedging cut are activated to develop. In this scenario, the extra leaves near the tips of the shoots will rapidly and significantly reduce light availability in the canopy centre. As demonstrated by Yáñez *et al.* (2009), this lower light availability in the canopy centre would reduce the induction of flower buds in this sector for the next season.

Experiments have also been conducted on cane removal during the summer in both SHBs and NHBs. The primary objective of this practice is to stimulate laterals to break, while reducing excess plant vigour. The time when this is

Table 6.8. Yield, berry weight and percentage of firm and non-diseased berries after 7 weeks of storage in the first season after harvest in 1989, 1990 and 1991 from various pruning treatments on mature 'O'Neal' plants in North Carolina. See text for details on pruning treatments. (Adapted from Mainland, 1993.)

Pruning treatment	Yield (kg per bush)	Weight per berry (g)	Good berries (%)
None	4.8[a]	1.69[b]	22.4[c]
Lop	3.3[b]	1.80[b]	23.1[b,c]
Lop and detail	3.0[b]	1.87[a,b]	27.2[b]
Top (flat)	3.6[a,b]	2.02[a]	36.0[a]
Top (flat) and lop	3.2[b]	2.05[a]	40.0[a]
Top angle	3.9[a,b]	1.96[a]	33.1[a,b]
Top angle and lop	3.4[b]	2.05[a]	38.2[a]

[a,b,c]Mean values within columns with non-identical superscript letters were significantly different at $P < 0.05$ (Waller–Duncan test).

done is critical, as the number and length of laterals is dependent on how early pruning is done. The goal is to prune on the date that allows good laterals to break and harden off before the winter, while inducing sufficient flower bud development for good yields the following season. While summer pruning is typically done soon after harvest, vegetative bud development should be evaluated, as advanced shoot and bud development will reduce the response. Cultivars differ greatly in their response to summer pruning date.

Bañados *et al.* (2009) studied the effect of summer pruning on lateral shoot growth, flower bud formation, harvest date and fruit weight in the SHBs 'O'Neal' and 'Star' and NHB 'Elliott'. Shoots were cut back to 20–30 cm at monthly intervals during the growing season from 15 December to 15 March (dates for the southern hemisphere). Pruning in mid-December resulted in the highest number and longest shoots in 'O'Neal' and 'Star', and the highest number of flower buds per new shoot in these cultivars (Table 6.9). The largest fruit were produced in 'O'Neal' and 'Star' after pruning in December or January compared with the plants that received no pruning (2.0 versus 1.5 g) or were pruned too late. The harvest season the following year was also delayed by 14 days in these two cultivars after all pruning treatments. The NHB 'Elliott' was less responsive to pruning than the SHBs 'O'Neal' and 'Star'. 'Elliott' plants pruned after December did not produce any laterals. However, fruit were significantly larger in the pruned 'Elliott' plants (2.0 versus 1.7 g) and pruning delayed harvest by 7 days. The authors concluded that summer pruning can increase yield and fruit quality in 'O'Neal' and 'Star' if they are pruned early in

Table 6.9. Effect of summer pruning on lateral length, number of laterals per shoot and number of flower buds per shoot on 'O'Neal' and 'Star' after different monthly pruning dates (southern hemisphere). (Adapted from Bañados *et al.*, 2009.)

Cultivar	Pruning date	Laterals per shoot	Lateral length (cm)	Flower buds per lateral
'O'Neal'	15 December	3[a]	19.0[a]	19.4[a]
	15 January	3[a]	20.4[a]	17.4[a,b]
	28 February	1[a]	9.5[b]	4.1[d]
	15 March	0[b]	3.7[b]	5.2[b,c]
	None	2[a]	10.3[b]	11.4[b,c]
'Star'	15 December	3[a]	15.5[a]	11.5[a]
	15 January	3[a]	17.4[a]	12.7[a]
	28 February	2[a,b]	12.5[a,b]	7.8[a,b]
	15 March	0[b]	1.2[b]	5.2[b]
	None	0[b]	13.5[a,b]	7.6[a,b]

[a,b,c,d]Mean values within columns with non-identical superscript letters were significantly different at $P < 0.05$ (Waller–Duncan test).

the season; otherwise, the effect can be the opposite. If summer pruning is done in 'Elliott', it should be restricted to the most vigorous canes and carried out very early. They also speculated that the dormancy status of the vegetative buds greatly affected the response.

The effects of summer pruning on blueberry yield are dependent on cultivar and location. No pruning and early summer pruning in SHB 'Sharpblue' increased yield compared with late summer pruning in Florida (Williamson and Darnell, 1996). In contrast, Pescie *et al.* (2011) found that summer pruning alone reduced the yield of SHB 'O'Neal' compared with summer followed by winter pruning or with non-pruned plants in Argentina.

Kovaleski *et al.* (2015a,b) determined the effects of summer pruning timing and intensity on vegetative and reproductive traits of mature SHB 'Emerald' and 'Jewel' plants. They measured the effects of pruning date by removing 30% of the canopy in June or July, and measured the effect of pruning intensity by removing 30 or 60% of the canopy in June (followed by shoot tipping in July). The effects of these treatments were compared with unpruned controls over 3 years. Summer pruning, regardless of timing or intensity, generally increased the vigour of vegetative growth for both cultivars and decreased the incidence of leaf diseases such as *Septoria* leaf spot (*S. albopunctata*) and blueberry leaf rust (*Pucciniastrum vaccinii*), in 'Jewel'. The yield of 'Jewel' was unaffected in the first year by pruning treatment but was increased by 48 and 65% in years 2 and 3 with the 30% pruning treatment. None of the pruning treatments affected the reproductive traits of 'Emerald'. The higher level of disease incidence in unpruned 'Jewel' and subsequent defoliation probably affected its yield.

When SHBs are grown in tropical and subtropical climates with little or no chilling such as Trujillo, Peru (latitude 8°S) or Jalisco and Michoacán, Mexico (latitude 20°N), plants can grow and produce fruit continuously. However, marketing fruit from these areas is not profitable at certain times of the year. In this period, plants are forced to have a rest period through heavy pruning. Thus, they are pruned in May–July (northern hemisphere) or January–March (southern hemisphere). In trials done in Michoacán in SHB 'Biloxi', it was found that heading back 30% of the length of each cane (compared with 10, 20 or 50%, 10% + light thinning of laterals, or no pruning) produced the greatest number of shoots and fruits, and the highest fruit weight and yields (Gómez-Martínez, 2010). Pruning on 7 June produced the highest yield (highest fruit number and weight), compared with 23 May, 20 June, 4 July or 18 July.

Renewal pruning and mechanical harvesting

If an old planting has not been pruned in years, a drastic amount of pruning may be necessary to recover high yields. One strategy that has been employed is to remove all canes except the most productive few. This drastically reduces yields, but at least some fruit is produced on the remaining canes. A large flush of new canes appears in the following year, which must be thinned annually until a productive ratio of young to old canes is produced. Another strategy that has been effective is to cut all of the canes to ground level and sacrifice almost the entire yield the following year. Howell *et al.* (1975) found that when large, unthrifty NHB 'Jersey' bushes were sawed off at ground level, the average yield for four seasons after they were pruned to the ground reached 8456 kg/ha/year while those of unpruned controls only had an average yield of 3667 kg/ha/year.

To obtain high yields through mechanical harvesting, it is critical to maintain narrow crown widths so that fruit do not miss the harvester collection plates. This is particularly important in old fields where crown width gradually expands over the years. In the study of Howell *et al.* (1975), they maintained 20, 25 and 30 cm crown widths by controlling suckers with dinitrophenol, paraquat and flames, and found that ground losses of fruit after mechanical harvesting decreased as the crown width was narrowed. The paraquat treatment at the rate of 1.13 kg/ha was most effective at sucker control.

GRAFTING

Grafting on to rootstocks is popular with many woody fruit crops to regulate plant size and precocity and help mitigate biotic and abiotic stresses. Rootstocks have not yet been commercially exploited in blueberry. However, several studies have shown that highbush and rabbiteye blueberry can be grafted successfully on to relatives and provide broadened soil adaptations. Galletta and Fish (1971) evaluated the performance of several cultivars of highbush grafted on to rabbiteye (*V. virgatum*) rootstocks and found them to have broader adaptation to non-traditional blueberry soils than own-rooted cultivars, as well as higher vigour, greater flower bud numbers per shoot and better survival. Ballington (1996) compared the yield and fruit characteristics of own-rooted 'Premier' rabbiteye blueberry (*V. virgatum* Reade) to 'Premier' grafted on the single-axis (tree-like) *V. arboreum* Marsh. (sparkleberry) planted on an upland soil for 3 years in North Carolina. The grafted plants had significantly higher yields and average fruit size for all three harvest seasons. There were no differences in fruit colour, picking scar, firmness or flavour in years 2 and 3, and only a small difference in year 1. In China, Xu *et al.* (2014) found that SHB 'Sharpblue' grafted on to the sea bilberry (*V. bracteatum*) had greater plant height and yield than non-grafted plants, without affecting fruit quality. *V. bracteatum* is known to have broader ecological adaptations than highbush blueberries.

Casamali *et al.* (2016a,b) grafted SHB 'Farthing' and 'Meadowlark' on to *V. arboreum* and compared the yield, berry quality and postharvest fruit life with own-rooted plants grown on amended versus non-amended soil. As

V. arboreum has one axis, the idea of grafting SHBs on to this species was to reduce the proportion of fruit falling to the ground when plants were mechanically harvested. The fruit were harvested both by hand and by machine. In the first fruiting year, yields of hand-harvested plants were generally greater for own-rooted plants grown on amended soil compared with own-rooted plants on non-amended soil or grafted plants on either soil treatment. However, by the second year, hand-harvested yields of grafted plants were substantially greater than own-rooted plants on non-amended soil. Yields of mechanically harvested grafted plants in either soil treatment were similar to yields of mechanically harvested own-rooted plants in amended soil in the second fruiting year and greater than yields of own-rooted plants in non-amended soil. Mechanical harvesting significantly reduced marketable yield compared with hand harvesting; however, grafted plants had significantly reduced ground losses during harvest compared with own-rooted plants for both cultivars. Grafted plants had greater mean berry weight but lower berry firmness at harvest; otherwise, fruit quality at harvest and during postharvest storage was unaffected by the *V. arboreum* rootstocks.

Darnell *et al.* (2015) suggested that *V. arboreum* is adapted to high pH and low-organic-matter soils because both NO_3^- and Fe uptake and assimilation are greater in sparkleberry than in SHB cultivars. This is correlated with the higher activity of nitrate reductase and iron chelate reductase, which are the rate-limiting enzymes for NO_3^- and Fe acquisition, respectively. Rhizosphere acidification does not play a role in the Fe-deficiency response of *V. arboreum* (Nunez *et al.*, 2015).

POLLINATION

Reproductive biology

Because blueberries are not completely self-fertile, cross-pollination generally results in higher seed set and fruit set and larger fruit (Morrow, 1943; Darnell and Lyrene, 1989). Overall, rabbiteye cultivars are less self-fertile than highbush ones, and cross-pollination between cultivars with an overlapping bloom is critical for adequate fruit set (Brevis and NeSmith, 2005). El-Agamy *et al.* (1981) found fruit set in selfed versus outcrossed cultivars to be reduced by an average of 30% in rabbiteye and 15% in SHB blueberries. Seed number per fruit in selfed cultivars was reduced by about 97% in rabbiteyes and 66% in SHBs. In NHBs, selfing reduced the fruit set by up to 133% and the number of seeds per fruit by between 36 and 1469% across several studies (Table 6.10). Selfing also reduced fruit weight in NHBs by up to 35%.

The level of self-fertility is highly variable across rabbiteye and highbush cultivars. In SHBs, El-Agamy *et al.* (1981) found that reductions in fruit set due to selfing ranged from 11% ('Sharpblue') to 50% ('Avonblue'), and

Table 6.10. Effect of self-pollination on seed number per fruit, fruit weight and fruit set in cultivars of highbush blueberry. Data from Ehlenfeldt (2001) (study 1), Krebs and Hancock (1988) (study 2) and J.F. Hancock and G.A. Lobos (unpublished results) (study 3). Studies 2 and 3 were conducted in the field, while study 1 was carried out in the greenhouse.

Cultivar	Study	Seeds per fruit			Fruit weight (g)			Fruit set (%)		
		Selfed	Outcrossed	Difference (%)	Selfed	Outcrossed	Difference (%)	Selfed	Outcrossed	Difference (%)
'Aurora'	3	4.7	43.4	823	1.35	1.89	40	86	90	5
'Bluecrop'	2	10.7	26.7	150	1.87	2.36	26	77	67	−13
	1	9.5	36.5	284	1.55	1.93	25	50	79	58
'Bluegold'	1	10.7	41.8	291	1.15	1.31	14			
'Bluejay'	2	6.2	9.8	58	1.09	1.14	5			
'Brigitta'	3	1.3	20.4	1469	1.69	2.06	22	30	70	133
'Draper'	3	22.1	43.4	96	1.77	1.80	2	90	92	2
'Duke'	3	15.3	40.7	166	1.70	1.80	6	76	85	12
	1	14.7	20.5	39	1.42	1.55	9	40	51	28
'Elliott'	2	7.7	43.7	468	1.60	2.03	27			
'Jersey'	2	15.1	48.4	221	1.16	1.64	41			
'Legacy'	1	13.5	30.1	123	1.59	1.89	19	78	90	15
'Liberty'	3	6.2	14.7	137	1.53	2.06	35	78	78	0
'Nelson'	1	12.6	17.1	36	1.49	1.71	15	64	69	8
'Ozarkblue'	3	10.1	41.5	311	1.64	2.10	28	90	91	0
'Rubel'	1	15.3	24.7	61	0.87	0.85	−2	49	69	41
	2	11.8	22.7	92	0.82	0.96	17			
'Sierra'	1	4.9	29.1	494	1.10	1.69	54	41	69	68
'Spartan'	2	1.3	9.5	631	1.91	2.51	31			
'Sunrise'	1	9.9	27.1	174	1.31	1.72	31	61	89	46
'Toro'	1	4.9	37.2	659	1.29	1.93	50	53	70	32
Average	1	9.9	31.8	221	1.31	1.62	24	57	72	26
	2	7.3	26.8	267	1.41	1.77	26			
	3	9.9	34.0	243	1.61	1.95	21	75	84	12

reductions in seed number per fruit ranged from 65% ('Flordablue') to 70% ('Sharpblue'). In rabbiteyes, they found reductions in fruit set due to selfing varied from 24% ('Climax') to 60% ('Beckyblue') and reductions in seed number per fruit ranged from 77% ('Climax') to 100% ('Aliceblue'). In NHBs, reductions in seed number per fruit due to selfing varied from 36% ('Nelson') to 468% ('Aurora'), and changes in fruit set due to selfing varied from a 58% reduction in 'Bluegold' to a 15% increase in 'Sierra'. Selfing decreased berry weight up to 54% ('Sierra') and increased this parameter in only one cultivar ('Rubel') by 2% (Table 6.10).

Most of the NHB cultivars that had limited fruit size reductions due to self-pollination ('Draper', 'Duke', 'Nelson' and 'Rubel') also had high fruit set (Table 6.10). Likewise, all those that had the most dramatic reductions in fruit weight due to selfing also had low fruit set ('Sierra', 'Sunrise' and 'Toro'). However, several with average to high reductions in fruit weight after selfing still had excellent fruit set ('Aurora', 'Bluecrop', 'Liberty' and 'Ozarkblue'). 'Bluegold' and 'Brigitta' were unusual in that they had a minimal fruit reduction due to selfing, but poor fruit set.

Similar effects of self-pollination have been observed in SHBs. Lang and Danka (1991) compared bee-mediated cross-pollination versus self-pollination in 'Sharpblue' and 'Gulfcoast', and found that cross-pollination increased fruit size by about 14% but did not influence fruit set. The seed count was 58% lower in selfed plants and the harvest percentage of early-ripening fruit was significantly reduced. In another study, Huang *et al.* (1997) found that significantly more ovules aborted after self-pollination of 'Gulfcoast', 'O'Neal' and 'Sharpblue' than after cross-pollination. El-Agamy *et al.* (1981) discovered that fruit set in selfed SHB cultivars averaged 67%, while cross-pollinated ones averaged 82%. The number of seeds per berry averaged 3.9 in self-pollinated cultivars versus 11.2 in outcrossed ones. The relative ranking in self-fertility among cultivars (low to high) was 'Avonblue', 'Sharpblue' and 'Flordablue'.

Meader and Darrow (1947) found that self-pollination also reduced seed set, fruit set and fruit weight in potted rabbiteye blueberries. In a comparison of ten old cultivars that are no longer grown, they discovered that the cultivars fell into three groups: (i) those that were mostly self-sterile; (ii) those that set about one-third of their fruit when self-pollinated; and (iii) one ('Blueboy') that set fruit equally well whether self- or outcrossed. Berry weights after self-pollination versus cross-pollination were reduced from 29 to 82%, and seed numbers per berry were reduced from 10 to 75%. Mainland *et al.* (1979) found that fruit set in plants of the rabbiteye 'Tifblue' averaged only 16% when caged versus 78% when open pollinated. El-Agamy *et al.* (1981) found fruit set averaged 18% in selfed rabbiteye cultivars versus 47% in outcrossed ones, while seed numbers per fruit averaged 1.5 in selfed plants versus 8.5 in outcrossed ones. The relative ranking in self-fertility among cultivars (lowest to highest) was 'Aliceblue', 'Tifblue', 'Beckyblue', 'Climax' and 'Bluebelle'.

Planting designs

For those cultivars that will benefit from cross-pollination, it is recommended that two different cultivars with similar flowering dates are planted in alternate rows. In those fields where machine harvesting prohibits alternate row spacings, our experience suggests that the number of rows of interspersed cultivars should not exceed ten, with closer proximities being better (Hancock *et al.*, 1989). Several growers have indicated that four-row blocks is the smallest workable size that can be machine harvested.

Pollination requirements

All blueberry cultivars require a pollinator to undergo fertilization. Growers commonly place hives of honeybees (*Apis mellifera* L.) in the field to ensure adequate pollination. Bee density is a critical factor in determining fruit set, berry weight and rate of ripening (Dogterom and Winston, 1999; Dedej and Delaplane, 2003), with the goal being to have four to eight bees working each blueberry plant during the hottest part of the day (Pritts and Hancock, 1992). Different densities of beehives are recommended, depending on the attractiveness of different blueberry cultivars to honeybees. Recommendations for NHBs range from 2.5 beehives/ha for 'Bluetta', 'Rubel' and 'Weymouth' to 4.0 for 'Bluecrop', 5.0 for 'Coville' and 'Elliott', and 6.0 for 'Jersey'. The hives should contain at least 45,000 honeybees.

It has not been determined why blueberry cultivars vary in their attractiveness to bees (Eck, 1986). Rodriguez-Saona *et al.* (2011) showed that variation in production of volatiles could have an impact. Marucci (1965) suggested that the difference in attractiveness was more related to differences in the volume of nectar production than to sugar content. However, Brewer and Dobson (1969) found no difference in the nectar production of unattractive NHB 'Jersey' versus attractive 'Rubel' plants, and the nectar of 'Jersey' had a higher sugar concentration than 'Rubel'. Vincent (1971) found that it took bees much longer to fill their honey stomach with nectars at higher sugar content, reducing the time for floral visitations. Flowers of NHB cultivars with short, broad corollas such as 'Bluecrop' are thought to be more easily pollinated by honeybees than those with longer, narrow corollas such as 'Berkeley', 'Coville', 'Earliblue' and 'Jersey' (Eck and Mainland, 1971; Pritts and Hancock, 1992). Corollas of rabbiteye 'Tifblue' flowers are too long to make nectar readily accessible to short-tongued honeybees (Dedej and Delaplane, 2003).

Nectar robbery has been a concern among blueberry growers, where honeybees obtain nectar through slits cut into blueberry blossoms by carpenter bees (*Xylocopa virginica* L.). Although nectar robbers are less effective at pollination than legitimate bee visitors, the impact of corolla slitting by carpenter bees does not appear to be significant on overall fruit set in rabbiteye fields (Sampson *et al.*, 2004).

It is recommended that bees should be placed in the field at the very beginning of bloom (5–25% of flowers open) and kept in the field until petal drop. The hives should be spread out across the farm, with no more than 300 m between hives. Maximum pollination activity is ensured by: (i) removing competing flowers in the field; (ii) placing bees in fields at 5% bloom; (iii) spreading the hives evenly throughout the field; (iv) facing hives towards the east; and (v) avoiding the application of toxic pesticides while bees are in the field (Pritts and Hancock, 1992). To protect bee populations, broad-spectrum insecticides should not be applied when blueberry flowers are open. A late-evening application is more desirable than a morning application, so that the insecticide residue can dry before the bees are active. More information on how to protect bees from toxic chemicals is given by Riedl *et al.* (2006).

Weather conditions can play an important role in the success of bee pollination. In temperate climates, each cultivar typically blooms for 2–3 weeks, and if conditions are unseasonably cold and windy, yields can be diminished. Tuell and Isaacs (2010) compared pollinator activity during conditions they defined as being good for bee activity when temperatures were around 15°C with moderate to low winds and poor when temperatures were between 10 and 15°C and it was windy. Significantly fewer bees were observed foraging during poor weather than during good weather. Fruiting clusters that were exposed during good weather had about five times as many mature seeds, weighed twice as much and had double the fruit set of those not exposed when conditions were good.

In general, bumble bees (*Bombus* spp.) are more efficient at pollination than honeybees (Javorek *et al.*, 2002; Sampson and Spiers, 2002; Heinrich, 2004). Bumble bees and other wild solitary bees 'sonicate' pollen from the anther, which honeybees cannot do (Delaplane and Mayer, 2000). Bumble bees also tend to work under cooler, wetter and windier conditions than honeybees (Tuell and Isaacs, 2010) and they carry more pollen. Javorek *et al.* (2002) estimated that bumble bees deposit 43.1 pollen grains per visit, while honeybees deposit only 11.5, and they also visit more flowers per minute.

Numerous other wild bee species are important pollinators, and growers should make every effort to protect them (Tuell *et al.*, 2009; Isaacs and Kirk, 2010). Production regions vary greatly in the importance of native bees and how relevant they are to fruit size and yield (DeVetter *et al.*, 2016; Gibbs *et al.*, 2016). Over 150 native (wild) bee species have been found in Michigan blueberry fields, and ten of these are considered to have a significant impact on the pollination of blueberries. The majority of these bees are solitary ground-nesting digger bees (*Andrena* spp.), which need undisturbed soil for nesting. In rabbiteye blueberries in Georgia, the most numerous bee visitors are honeybees (*A. mellifera* L.), bumble bee queens (*Bombus* spp.), bumble bee workers, carpenter bees and south-eastern blueberry bees (*Habropoda laboriosa* (F.)) (Delaplane, 1995). The south-eastern blueberry bee and bumble bee queens were found to be the most efficient pollinators on 'Tifblue' (Cane and Payne, 1990). Sampson and Cane (2000) have suggested that *Osmia ribifloris* Cockerell, which is native to the western USA, could also be used as an effective pollinator of rabbiteye blueberries.

In large plantings of blueberries, too few native bees generally exist for full pollination, so the addition of honey and/or bumble bee hives is necessary; however, it is still beneficial to create natural bee habitats with a mix of plants that bloom before and after blueberries (see Fiedler *et al.*, 2008, for more details).

HARVESTING

The development of blueberry harvesters began in the USA soon after World War II when increased industrialization sharply reduced the amount of labour available for blueberry picking (Dale *et al.*, 1994). The first harvesters were hand-held vibrators. A gasoline-powered air compressor was developed in 1957 consisting of a small cart and up to eight vibrators. This allowed harvesting of four rows at a time, with two pickers per row. The system was noisy and not very practical. It was promptly replaced by rechargeable battery units on small pull carts. The vibrators contained individual electrical motors, and a fabric on a metal structure was used to collect the berries.

It has been estimated than more than 2600 electrical shakers were sold up to the 1990s (Brown *et al.*, 1996), after which their production was discontinued. In 1963, up to 35% of Michigan and 20% of Jersey berries were harvested with these devices. Electrical shakers are still being used in small operations across the USA, and new improved models have been evaluated for the fresh market in Chile and Argentina, due to rising costs of labour for picking fruit. Adult NHB 'Brigitta' and 'O'Neal' cultivars from a commercial planting in Linares, Chile (latitude 35°52′S) were used in a trial comparing the effects of hand picking, pneumatic shakers (Campagnola-Sobitec; Sobitec Ltda., Santiago, Chile) and a mechanical over-the-row harvester (Korvan, model 7420; Oxbo, New York). Firmness at harvest, and after 60 days at 0°C plus 1 day at 18°C, was lowest after mechanical harvest in both cultivars. In 'O'Neal', hand- and shaker-harvested fruit had equivalent firmness, while in 'Brigitta', shaker-harvested fruit was softer than hand-harvested fruit. The labour required for harvesting was reduced by half when comparing hand picking versus shaker picking. Mechanical damage was greater for fruit picked with mechanical equipment, and also in pickings done in the morning. Averaging the results for both cultivars, the proportion of fruit for the fresh market was 72, 76 and 83% for mechanical, shaker and hand harvesting, respectively. The authors concluded that there is a potential for harvesting with shakers, but its effects on different cultivars and the cost:benefit ratio need to be studied (Lobos *et al.*, 2014).

Recently, Takeda *et al.* (2017) performed studies in commercial fields of SHB ('Chicadee', 'Farthing', 'Flicker', 'Kestrel', 'Legacy', 'Meadowlark' and 'Springhigh') and NHB ('Draper' and 'Liberty') blueberries to evaluate semi-mechanical harvesting systems consisting of a harvest-aid platform with soft fruit-catching surfaces. The fruit were detached by portable, hand-held pneumatic shakers. The pneumatic shakers removed 3.5–15 times more fruit (g/min) than by hand, and the soft fruit-catching surfaces reduced the impact force and bruise damage. In some cultivars, fruit firmness was higher in fruit harvested by hand compared with that harvested by pneumatic shakers. The bruise area was less than 8% in fruit harvested by hand or with semi-mechanical harvesting system. The percentage of blue packable fruit harvested by pneumatic shakers comprised as much as 90% of the total, but was lower than that of hand-harvested fruit. An ergonomic analysis by electromyography showed that muscle strain in the back, shoulders and forearms was low in workers operating the lightweight pneumatic shakers, which were tethered to the platform with a tool balancer. The harvest efficiency of the semi-mechanical harvesting was improved 10–20-fold compared with hand harvesting.

In 1959, agricultural engineers of the USDA at Michigan State University started to develop the first over-the-row harvesters. The goal was to produce a machine that removed mature blue fruit selectively and had an efficient collection system. Over-the-row harvesters were soon being developed by several other companies and institutions. During the 1960s, several models emerged that were self-propelled or tractor pulled. Some of the tractor-pulled harvesters were used to harvest bushes of small stature.

All current commercial models straddle the row, but several different picking mechanisms are used (Dale *et al.*, 1994). The 'slapper' mechanism consists of bars mounted on a vertical plane. Two sets of bars are staggered on either side of the equipment and swing like gates, independently 'slapping' the bushes as the harvester moves down the row. In some harvesters, there is a 'sway' mechanism, which is a variation of a slapper. In this case, 'swinging gates' are also located on each side of the harvester, but the gates are opposite to one another and work in tandem. The bush is then 'swayed' from one side to the other. A horizontally vibrating 'finger' mechanism is available in some machines. A variation of this is the so-called 'vertirotor', which consists of numerous horizontal 'fingers' arranged around two vertical axes on either side of the equipment. The fingers 'roll through' the plants as the harvester advances through the row. As the fingers in the vertirotor vibrate in a vertical direction, a similar displacement along the fingers is obtained (Dale *et al.*, 1994).

These different mechanisms have their positive and negative points. The 'slapper' usually removes large numbers of fruit and is particularly effective for late-season 'remove all' tasks. However, the vigorous action of this system usually causes greater damage to the bushes and removes a larger proportion of immature fruit. The 'sway' mechanism is generally gentler, removing a high proportion of ripe fruit with less damage to plant structures. The finger system and its variation (the vertirotor) are probably the most delicate, but are not appropriate for clean-up harvests (Dale *et al.*, 1994). The mechanical impacts generated by fruit-catching surfaces have been measured with the berry impact recording device (BIRD) sensor (Yu *et al.*, 2011, 2012). The BIRD sensor is a microcontroller-based data logger that measures and records mechanical impacts in three orthogonal directions using a triaxial accelerometer. The sensor has been miniaturized to the size and weight of a large blueberry fruit. It has enough sensing range and precision to measure impacts generated on the packing lines and in harvesting operations. It has a built-in memory and power supply, enabling it to be operated independently to quantify mechanical impacts (Yu *et al.*, 2011).

In the 1990s, the USDA developed the V45 harvester, which has 45°-angled, spiked-drum shakers, a cane dividing and positioning system, and cushioned catching surfaces (Peterson and Brown, 1996). The cane dividing system bends the canes away from the crown and over the elevating catching surface. The cane dividing system greatly reduces the amount of fruit dropping on the crown rather than on the catch frames. However, the V45 was not widely accepted by growers due to: (i) limited operating speed; (ii) the non-durable fruit-catching surface on the harvester; and (iii) the special cane training and pruning required to minimize plant damage (Takeda *et al.*, 2017). Strik and Buller (2002), in Oregon, showed that supporting NHB 'Bluecrop' plants with a simple two-wire trellis system increased harvest efficiency significantly, mainly by reducing the number of mature fruit missed by the machine. The maximum proportion of fruit remaining after the last machine harvest was reduced from 30.8% in non-trellised plots to 15.5% in trellised plots with canes kept more upright by trellis wires. Most recently, a forced-air concept has been developed that may further reduce damage by better directing and cushioning the falling berries.

The growth of the blueberry industry over the past three decades has been remarkably robust. However, labour shortage for hand harvesting, increasingly higher labour costs and low harvest efficiencies are becoming bottlenecks for sustainable development of fresh market blueberry production (Takeda *et al.*, 2017). Hand harvesting of highbush blueberries is labour intensive and requires as many as 1150 h of labour per hectare (Brown *et al.*, 1983). Comparatively, the use of mechanical vibrators nearly tripled worker productivity and reduced harvesting cost by 55%. The over-the-row harvesters have cut harvest labour to 22 worker-hours per hectare for berries used in processing (Gough, 1994). Other researchers have reported that over-the-row harvesters increased worker productivity by nearly 60 times and reduced harvesting costs by 85% (Brown *et al.*, 1996). Estimates done in the 1970s established that the cost for mechanically harvesting of 1 kg of blueberries was US$0.164

compared with US$0.215 for hand picking (Dale *et al.*, 1994). Estimates by Safley *et al.* (2012) showed that hand harvesting raised the cost of harvesting blueberries from US$1.3/kg to US$2/kg. Julian (2012) estimated the cost of hand picking as US$1.2/kg in Oregon, US$1.58/kg in Alabama and Florida and US$2.93/kg in California. Although over-the-row automotive harvesters can significantly cut harvest labour, they currently cost as much as US$240,000 per unit, making them unaffordable for small- and medium-sized blueberry farms (Takeda *et al.*, 2017).

Although the picking cost is lower for mechanically harvested fruit, more berries are picked per hectare by hand and they have a higher value. An evaluation of mechanical harvesting of highbush blueberries in British Columbia, Canada, established a 14–16% yield reduction compared with hand harvesting (van Dalfsen and Gaye, 1999). Part of this lower yield was due to fruit falling to the ground. In addition, the green fruit harvested was 4.0% of the mechanically harvested yield and 0.35% of the hand-harvested yield over the 3 years of study. Bruising increases with longer falling distance on harder surfaces of the harvester and reduces both external and internal fruit quality (Mainland *et al.*, 1975; Dale *et al.*, 1994). Internal damage to the fruit results in cellular water leakage and development of water-soaked areas in the flesh (Labavitch *et al.*, 1998). Recent studies with a miniaturized instrumented sphere (BIRD) have shown that dropping on the catch pan accounted for over 30% of all impacts in a mechanical blueberry harvester, and that these drop impacts can be reduced by applying cellular silicone as a padding material (Yu *et al.*, 2012).

Postharvest decay caused by various fungal pathogens is a major concern in most blueberry production areas. The incidence of postharvest decay is usually augmented by machine harvesting, storage at high temperatures, infestation of handling surfaces, wetness of the stem scar or the presence of moisture (Cline, 1996). Because the risk of infection is increased by fruit bruising, which in turn is increased by machine harvesting, it has been difficult to mechanically harvest fruit from the early-maturing but soft-textured SHB blueberries for the fresh market. This could change fundamentally with the recent development of SHB genotypes with crisp-textured ('crispy') berries, i.e. fruit with qualitatively firmer flesh and/or more resistant skin. After a 2-year study comparing the effects of hand versus machine harvesting on firmness and postharvest decay of crispy ('Farthing', FL 98-325 and 'Sweetcrisp') and conventional (FL 01-248, 'Scintilla' and 'Star') SHB cultivars, Mehra *et al.* (2013) reported that disease incidence after cold storage was lowest for hand-harvested crispy fruit and highest for machine-harvested conventional fruit. Interestingly, machine-harvested crispy fruit had equivalent or lower disease incidence than hand-harvested conventional fruit. Across all treatments, postharvest disease incidence was inversely related to fruit firmness, with firmness values greater than $220\,g/mm^2$ associated with low disease levels. This study suggested that mechanical harvesting of crisp-textured SHBs is feasible from a postharvest pathology perspective, as the fruit of the crispy genotypes are firmer and more 'springy' than those of their conventional counterparts. The lower postharvest disease levels in crispy cultivars could be explained by two different but related mechanisms: (i) because of their firmer texture, crispy berries are inherently more resistant to direct penetration by fungal pathogens; and (ii) the greater firmness of crispy berries results in lower levels of bruising during harvest, thereby reducing the level of wound-associated infections.

Harvesting machines not only bruise the fruit but also cause visible damage to plant structures, making the plants more susceptible to diseases. Mainland *et al.* (1975) found that only 0.04 canes per bush were damaged by hand pickers compared with 2.2 canes per bush by over-the-row harvesters. The harvester broke nine times as many canes compared with the commercial hand pickers.

In the 1980s, it was estimated that 60% of the highbush blueberries of the east coast of the USA were mechanically harvested, while by the 1990s, machines were used to harvest 70% of Michigan's berries and about half of the New Jersey crop (Dale *et al.*, 1994). Over the years, mechanical harvesters have been developed in attempts to harvest blueberries for the fresh market but with limited success (Peterson and Brown, 1996; Takeda *et al.*, 2008). As the supply of labour for hand picking fresh fruit continues to decrease and the market demand for fresh blueberries increases, there is the requirement for: (i) fruit harvested with mechanical harvesting systems with postharvest quality that is similar to hand-harvested fruit; (ii) a major improvement in mechanical harvesting systems to reduce postharvest handling and sorting operations to separate defective fruit; (iii) development of new cultural practices that increase the yield of high-quality fruit (Takeda *et al.*, 2008), and (iv) development of new cultivars that are amenable to machine harvesting. For the latter point, breeders are considering bush architecture (upright habit and narrow crowns), ease of detachment of mature fruit but not immature fruit, loose fruit clusters, small dry-stem scars, firm fruit and a concentrated ripening period (Olmstead *et al.*, 2013; Takeda *et al.*, 2017). Major changes in bush architecture could enhance harvest efficiency and fruit quality (Dale *et al.*, 1994). In Florida, sparkleberry (*V. arboreum*) has been used in wide crosses with SHB clones in an attempt to introgress traits that may be valuable for machine harvesting. Two eras of sparkleberry hybridization experiments have occurred since the early 1980s. The first era used evergreen blueberry (*V. darrowii*) as a bridge between sparkleberry and tetraploid SHBs, with 'Meadowlark' as an example of the end product. The second era has used chromosome doubling to develop polyploid sparkleberry selections that were directly crossed with tetraploid SHBs. After 1 year of evaluation, an SHB × sparkleberry population showed evidence of introgression and provides an initial step towards improved cultivars for mechanical harvesting (Olmstead *et al.*, 2013).

CONCLUSIONS

A number of management practices are of utmost importance in blueberry cultivation, including irrigation, mulching, protected cultivation, soil-less culture, pruning, grafting, pollination and harvesting. Irrigation is economically justified in most situations, as rain and/or the water table are usually not able to supply the water needs of blueberries. Blueberries have a shallow root system (20–30 cm deep) and are susceptible to water stress due to either excess or deficit. Rabbiteye blueberries are generally more tolerant to water stress than highbush blueberries. The behaviour of water in the soil and the plant has been unified in a single energy concept: the water potential (Ψ), which considers the soil–plant–atmosphere as a continuum. Although wild highbush blueberries are found growing on hummocks in swamps, flooded areas should not be used to grow blueberries. Flooding stress is caused mainly by lack of soil O_2, which closes stomata and reduces transpiration. To determine adequate amounts and timing of water supplies to a blueberry field, physiological variables (plant water potential, gas exchange) should be determined, as well as plant growth, fruit production and quality. To monitor water status in blueberry fields, digital tensiometers or electrical resistance blocks (gypsum or GMSs) combined with sap flow meters and TDF sensors appear the most promising techniques. Various cultural practices (mulching, type of irrigation system, ground cover, cultivation practices and planting density), as well as characteristics of blueberry cultivars (canopy size and shape, root system, timing of harvest) affect their water use and needs. Good-quality water (low salts, EC less than 0.45 mmho/cm) will allow maximum yield under adequate soil and water management practices. The most common irrigation systems used in blueberries are drip and overhead sprinklers.

The use of mulches is widespread among blueberry growers. This practice brings about many benefits, among which weed control and moisture retention are paramount. There are many different materials that can be used, depending on availability and specific needs. Plastic mulches increase soil temperature, and the effect varies according to the type of material (film or woven mat), colour and degree of contact with the soil. Plants under mulch show greater root growth, which may be a consequence of improved soil structure as determined by improved porosity and O_2 availability. Usually 10–20 cm of organic mulch is applied at planting. Considering degradation of the material, 3–8 cm should be added annually to maintain the benefits. Once the practice of organic mulch has been initiated, it should not be interrupted in order to avoid damage to the root system. If the C:N ratio of the organic material used as mulch is greater than 30:1, it will tie up N and leave it unavailable for blueberry plants. To avoid this, it is recommended that N rates are increased by 30% when organic mulching is used.

In warmer production areas, blueberries are increasingly being grown under tunnels to accelerate ripening and reach more profitable markets. Harvest dates can be advanced by as much as a month with tunnels. Most of the tunnel hectarage is planted in soil, but pot culture is becoming more prevalent. Typically, the plants are grown in 15–25 l polyethylene containers on ground that is completely covered with woven polypropylene fabric. The best plant growth is obtained in soil-less medium containing at least 60% peat or coir.

Highbush blueberry bushes need regular pruning for sustained productivity. Most pruning in NHBs is done during the winter when the canes are dormant, while SHBs are often pruned both in summer after harvest and during the dormant season. Fruit yield in young NHB plants needs to be reduced to encourage vegetative growth. Annual pruning is recommended in mature plantings for long-term stability of yield. External canes are higher yielding than internal ones, and cane productivity increases with age. Some younger canes are needed for renewal and a few older canes for support. In Michigan, most pruning is focused on whole-cane removal, while in Chile and the Pacific Northwest, more effort is focused on the top of bushes to balance floral and vegetative growth. In addition to winter pruning, SHBs are also commonly hedged at 100–122 cm after the fruiting season using a sickle mower or hedge trimmer.

Grafting blueberry cultivars on to single-axis sparkleberry can increase yields, improve adaptability to various soil types and reduce losses when plants are machine harvested. This technique could gain commercial interest in coming years as more trials show the benefits for different growing areas and cultural conditions.

Because blueberries are not completely self-fertile, cross-pollination generally results in higher seed set and fruit set and larger fruit. Overall, rabbiteye cultivars are less self-fertile than highbush cultivars, and cross-pollination between cultivars with an overlapping bloom is critical for adequate fruit set. For those cultivars that will benefit from cross-pollination, it is recommended that two different cultivars with similar flowering dates are planted in alternate rows. Growers commonly place hives of honeybees in the field to ensure adequate pollination.

The high cost of labour and the inability to obtain sufficient numbers of pickers have forced many growers to mechanize their harvest operation. This has some trade-offs, as machine-harvested blueberries generally are softer, and have a higher incidence of decay, a greater rate of weight loss and lower post-harvest shelf-life than hand-harvested berries. However, new machines and devices are being developed that are much gentler on the fruit and these could be used for the fresh market.

REFERENCES

Abbott, J.D. and Gough, R.E. (1986) Split-root water application to highbush blueberry plants. *HortScience* 21, 997–998.

Abbott, J.D. and Gough, R.E. (1987a) Growth and survival of the highbush blueberry in response to root zone flooding. *Journal of the American Society for Horticultural Science* 112, 603–608.

Abbott, J.D. and Gough, R.E. (1987b) Reproductive response of the highbush blueberry to root-zone flooding. *HortScience* 22, 40–42.

Adhikari, R., Bristow, K.L., Casey, P.S., Freischmidt, G., Hornbuckle, J.W. and Adhikari, B. (2016) Preformed and sprayable polymeric mulch film to improve agricultural water use efficiency. *Agricultural Water Management* 169, 1–13.

Allen, R.G., Pereira, L.S., Raes, D. and Smith, M. (1998) *Crop Evapotranspiration. Guidelines for Computing Crop Water Requirements.* FAO Irrigation and Drainage Paper No. 56. Food and Agriculture Organization of the United Nations, Rome.

Almutairi, K.F., Bryla, D.R. and Strik, B.C. (2017) Potential of deficit irrigation, irrigation cutoffs, and crop thinning to maintain yield and fruit quality with less water in northern highbush blueberry. *HortScience* 52, 625–633.

Al-Yahyai, R. (2006) Tree water potential for irrigation management of fruit crops. In: *Proceedings of the International Conference on Economic Incentives and Water Demand Management*, 18–22 March 2006, Muscat, Oman. Available at: http://www.nizwa.net/rashid/?download=Alyahyai_EIDM_06pwaj-s.pdf (accessed 8 June 2011).

Améglio, T., Archer, P., Cohen, M., Valancogne, C., Daudet, F.A., Dayau, S. and Cruiziat, P. (1999) Significance and limits in the use of predawn leaf water potential for tree irrigation. *Plant and Soil* 207, 155–167.

Améglio, T., Le Roux, X., Mingeau, M. and Perrier, C. (2000) Water relations of highbush blueberry under drought conditions. *Acta Horticulturae* 537, 273–278.

Ayers, R.S. and Westcot, D.W. (1985) *Water Quality for Agriculture.* FAO Irrigation and Drainage Paper No. 29. Food and Agriculture Organization of the United Nations, Rome.

Ballington, J.R. (1996) Performance of own-rooted 'Premier' rabbiteye blueberry vs. 'Premier' grafted on *V. arboreum* through three harvest seasons on a Fuquay soil. *HortScience* 31, 749–750.

Bañados, P., Uribe, P. and Donnay, D. (2009) The effect of summer pruning date on 'Star', 'O'Neal' and 'Elliott'. *Acta Horticulturae* 810, 501–507.

Baptista, M.C., Oliveira, P.B., Lopes da Fonseca, L. and Oliveira, C.M. (2006) Early ripening of southern highbush blueberries under mild winter conditions. *Acta Horticulturae* 715, 191–196.

Bell, R. (1982) *The Blueberry.* Department of Agriculture NSW, Sydney, Australia.

Boland, A.M., Bewsell, D. and Kaine, G. (2006) Adoption of sustainable irrigation management practices by stone and pome fruit growers in the Goulburn/Murray Valleys, Australia. *Irrigation Science* 24, 137–145.

Bonanomi, G., Chirico, G.B., Palladino, M., Gaglione, S.A, Crispo, D.G., Lazzaro, U., Sica, B., Cesarano, G., Ippolito, F., Sarker, T.C., Rippa, M. and Scala, F. (2017)

Combined application of photo-selective mulching films and beneficial microbes affects crop yield and irrigation water productivity in intensive farming systems. *Agricultural Water Management* 184, 104–113.

Bonet, L., Ferrer, P., Castel, J.R. and Intrigiolo, D.S. (2010) Soil capacitance sensors and stem dendrometers. Useful tools for irrigation scheduling of commercial orchards? *Spanish Journal of Agricultural Research* 8, 52–65.

Brevis, P.A. and NeSmith, D.S. (2005) Transport of cross-pollen by bumblebees in a rabbiteye blueberry planting. *HortScience* 40, 2007–2010.

Brewer, J.W. and Dobson, R.C. (1969) Nectar studies of the highbush blueberry *Vaccinium corymbosum* L. cv. 'Rubel' and 'Jersey'. *HortScience* 4, 332–334.

Brightwell, W.T. and Austin, M.E. (1980) *Rabbiteye Blueberries*. Research Bulletin No. 259. University of Georgia, College of Agriculture Experiment Station, Athens, Georgia.

Brodhagen, M., Peyron, M., Miles, C. and Inglis, D.A. (2015) Biodegradable plastic agricultural mulches and key features of microbial degradation. *Applied Microbiology and Biotechnology* 99, 1039–1056.

Brown, G.K., Marshall, D.E., Tennes, B.R., Booster, D.E., Chen, P., Garrett, R.E., O'Brien, M., Studer, H.E., Kepner, R.A., Hedden, S.L., Hood, C.E., Lenker, D.H., Millier, W.F., Rehkugler, G.E., Peterson, D.L. and Shaw, L.N. (1983) *Status of Harvest Mechanization of Horticultural Crops*. Paper No. 83-3. American Society of Agricultural Engineers, St Joseph, Michigan.

Brown, G.K., Schulte, N.L., Timm, E.J., Beaudry, R.M., Peterson, D.L., Hancock, J.F. and Takeda, F. (1996) Estimates of mechanization effects on fresh blueberry quality. *Applied Engineering in Agriculture* 12, 21–26.

Bryla, D.R. (2006) Drip irrigation configuration influences growth in young highbush blueberries. *HortScience* 41, 1012.

Bryla, D.R. (2008) Water requirements of young blueberry plants irrigated by sprinklers, microsprays and drip. *Acta Horticulturae* 792, 135–139.

Bryla, D.R. (2011) Crop evapotranspiration and irrigation scheduling in blueberry. In: Gerosa, G. (ed.) *Evapotranspiration – From Measurements to Agricultural and Environmental Applications*. InTech Europe, Rijeka, Croatia, pp. 167–186

Bryla, D.R. and Linderman, R.G. (2007) Implications of irrigation method and amount of water application on *Phytophthora* and *Pythium* infection and severity of root rot in highbush blueberry. *HortScience* 42, 1463–1467.

Bryla, D.R. and Strik, B.C. (2006) Variation in plant and soil water relations among irrigated blueberry cultivars planted at two distinct in-row spacings. *Acta Horticulturae* 715, 295–300.

Bryla, D.R. and Strik, B.C. (2007) Effect of cultivar and plant spacing on the seasonal water requirements of highbush blueberry. *Journal of the American Society for Horticultural Science* 132, 270–277.

Burkhard, N., Lynch, D., Percival, D. and Sharifi, M. (2009) Organic mulch impact on vegetation dynamics and productivity of highbush blueberry under organic production. *HortScience* 44, 688–696.

Burkhard, N., Lynch, D., and Percival, D. (2010) Effects of pine-needle and compost mulches and weeds on nitrogen dynamics in an organically-managed highbush blueberry field. *Acta Horticulturae* 873, 253–259.

Byers, P.L. and Moore, J.N. (1987) Irrigation scheduling for young highbush blueberry plants in Arkansas. *HortScience* 22, 52–54.

Byers, P.L., Moore, J.N. and Scott, H.D. (1988) Plant–water relations of young highbush blueberry plants. *HortScience* 23, 870–873.

Cameron, J.S., Brun, C.A. and Hartley, C.A. (1989) The influence of soil moisture stress on the growth and gas exchange characteristics of young highbush blueberry plants (*Vaccinium corymbosum* L.). *Acta Horticulturae* 241, 254–259.

Cane, J.H. and Payne, J.A. (1990) *Native Bees Pollinate Rabbiteye Blueberries*. Bulletin No. 37. Alabama Agricultural Experiment Station, Auburn, Alabama.

Casamali, B., Darnell, R.L., Kovaleski, A.P., Olmstead, J.W. and Williamson, J.G. (2016a) Vegetative and reproductive traits of two southern highbush blueberry cultivars grafted onto *Vaccinium arboreum* rootstocks. *HortScience* 51, 880–886.

Casamali, B., Williamson, J.G., Kovaleski, A.P., Sargent, S.A. and Darnell, R.L. (2016b) Mechanical harvesting and postharvest storage of two southern highbush blueberry cultivars grafted onto *Vaccinium arboreum* rootstocks. *HortScience* 51, 1503–1510.

Caspersen, S., Svensson, B., Håkansson, T., Winter, C., Khalil, S. and Asp, H. (2016) Blueberry-Soil interactions from an organic perspective. *Scientia Horticulturae* 208, 78–91.

Chen, X., Qiu, L., Guo, H., Wang, Y., Yuan, H., Yan, D. and Zheng, B. (2017) Spermidine induces physiological and biochemical changes in southern highbush blueberry under drought stress. *Brazilian Journal of Botany* 40, 841–851.

Chen, Y., Wen, X., Sun, Y., Zhang, J., Wu, W. and Liao, Y. (2014) Mulching practices altered soil bacterial community structure and improved orchard productivity and apple quality after five growing seasons. *Scientia Horticulturae* 172, 248–257.

Cifre, J., Bota, J., Escalona, J.M., Medrano, H. and Flexas, J. (2005) Physiological tools for irrigation scheduling in grapevine (*Vitis vinifera* L.) An open gate to improve water-use efficiency? *Agriculture, Ecosystems and Environment* 106, 159–170.

Ciordia, M., Diaz, M. and Garcia, J. (2002) Blueberry culture both in pots and under Italian-type tunnels. *Acta Horticulturae* 574, 123–127.

Clark, J.R. and Moore, J.N. (1991) Southern highbush blueberry response to mulch. *HortTechnology* 1, 52–54.

Cline, W.O. (1996) Postharvest infection of highbush blueberries following contact with infested surfaces. *HortScience* 31, 981–983.

Cox, J. (2009) Comparison of plastic weedmat and woodchip mulch on low chill blueberry soil in New South Wales, Australia. *Acta Horticulturae* 810, 475–482.

Cox, J.A., Morris, S. and Dalby, T. (2014) Woodchip or weedmat? A comparative study on the effects of mulch on soil properties and blueberry yield. *Acta Horticulturae* 1018, 369–374.

Crane, J.H. and Davies, F.S. (1987) Flooding, hydraulic conductivity, and root electrolyte leakage of rabbiteye blueberry plants. *HortScience* 22, 1249–1252.

Crane, J.H. and Davies, F.S. (1988) Flooding duration and seasonal effects on growth and development of young rabbiteye blueberry plants. *Journal of the American Society for Horticultural Science* 113, 180–184.

Daebeke, P. and Aboudrare, A. (2004) Adaptation of crop management to water-limited environments. *European Journal of Agronomy* 21, 433–446.

Dale, A., Hanson, E.J., Yarborough, D.E., McNicol, R.J., Stang, E.J., Brennan, R., Morris, J.R. and Hergert, G.B. (1994) Mechanical harvesting of berry crops. *HortReviews* 16, 255–382.

Darnell, R.L. (2006) Blueberry botany/environmental physiology. In: Childers, N.F. and Lyrene, P.M. (eds) *Blueberries for Growers, Gardeners and Promoters*. Dr Norman F. Childers Publications, Gainesville, Florida, pp. 5–13.

Darnell, R.L. and Lyrene, P.M. (1989) Cross-incompatibility of two related rabbiteye blueberry cultivars. *HortScience* 24, 1017–1018.

Darnell, R.L., Casamali, B. and Williamson, J.G. (2015) Nutrient assimilation in southern highbush blueberry and implications for the field. *HortTechnology* 25, 460–463.

Davies, F.S. and Darnell, R.L. (1994) Blueberries, cranberries, and red raspberries. In: Schaffer, B. and Andersen, P. (eds) *Handbook of Environmental Physiology of Fruit Crops*. CRC Press, Boca Raton, Florida, pp. 43–84.

Davies, F.S. and Flore, J.F. (1986a) Gas exchange and flooding stress of highbush and rabbiteye blueberries. *Journal of the American Society for Horticultural Science* 111, 565–571.

Davies, F.S. and Flore, J.F. (1986b) Short-term flooding effects on gas exchange and quantum yield of rabbiteye blueberry. *Plant Physiology* 81, 289–292.

Davies, F.S. and Johnson, C.R. (1982) Water stress, growth, and critical water potentials of rabbiteye blueberry (*Vaccinium ashei* Reade). *Journal of the American Society for Horticultural Science* 107, 6–8.

Dedej, S. and Delaplane, K.S. (2003) Honey bee (Hymenoptera: Apidae) pollination of rabbiteye blueberry *Vaccinium ashei* var. 'Climax' is pollinator density-dependent. *Journal of Economic Entomology* 96, 1215–1220.

Delaplane, K.S. (1995) Bee foragers and their pollen loads in South Georgia rabbiteye blueberry. *American Bee Journal* 135, 825–826.

Delaplane, K.S. and Mayer, D.F. (2000) *Crop Pollination by Bees*. CABI Publishing, New York, New York.

Demchak, K. (2003) Mulching and organic matter – keeping your plants happy. In: New England Vegetable & Fruit Conference, 16–18 December 2003. Available at: https://newenglandvfc.org/sites/newenglandvfc.org/files/content/proceedings2003/blueberry1/mulching_organic_matter_keeping_plants_happy.pdf (accessed 5 February 2018).

Demchak, K. (2009) Small fruit production in high tunnels. *HortTechnology* 19, 44–49.

DeVetter, L., Granatstein, D. Kirby, E. and Brady, M. (2015) Opportunities and challenges of organic highbush blueberry production in Washington State. *HortTechnology* 25, 796–804.

DeVetter, L., Watkinson, S., Sagili, R. and Lawrence, T. (2016) Honey bee activity in northern highbush blueberry differs across growing regions in Washington State. *HortScience* 51, 1228–1232.

Diaz-Perez, J.C. (2010) Bell pepper (*Capsicum annum* L.) grown on plastic film mulches: effects on crop microenvironment, physiological attributes, and fruit yield. *HortScience* 45, 1196–1204.

Dijkstra, J. and Scholtens, A. (1993) Growing early and late raspberries in containers. *Acta Horticulturae* 352, 49–54.

Dogterom, M.H. and Winston, M.L. (1999) Pollen storage and foraging by honey bees (Hymenoptera: Apidae) in highbush blueberries (Ericaceae), cultivar Bluecrop. *Canadian Journal of Entomology* 131, 737–768.

Eck, P. (1986) Blueberry. In: Monselise, S.P. (ed.) *CRC Handbook of Fruit Set and Development*. CRC Press, Boca Raton, Florida, pp. 75–85.

Eck, P. (1988) *Blueberry Science*, 1st edn. Rutgers University Press, New Brunswick, New Jersey.

Eck, P. and Mainland, C.M. (1971) Highbush blueberry fruit set in relation to flower morphology. *HortScience* 6, 494–495.

Eck, P., Gough, R.E., Hall, I.V. and Spiers, J.M. (1990) Blueberry management. In: Galletta, G.J. and Himelrick, D.G. (eds) *Small Fruit Crop Management*. Prentice Hall, Englewood Cliffs, New Jersey, pp. 273–333.

Ehlenfeldt, M.K. (2001) Self and cross-fertility in recently released highbush blueberry cultivars. *HortScience* 36, 133–135.

Ehret, D.L., Frey, B., Forge, T., Helmer, T. and Bryla, D.R. (2012) Effects of drip irrigation configuration and rate on yield and fruit quality of young highbush blueberry plants. *HortScience* 47, 414–421.

Ehret, D.L., Frey, B., Forge, T., Helmer, T. and Bryla, D.R. (2015) Age-related changes in response to drip irrigation in highbush blueberry. *HortScience* 50, 486–490.

El-Agamy, S.Z.A., Sherman, W.B. and Lyrene, P.M. (1981) Fruit set and seed number from self- and cross-pollinated highbush (4X) and rabbiteye (6X) blueberries. *Journal of the American Society for Horticultural Science* 106, 443–445.

Erb, W.A., Draper, A.D. and Swartz, H.J. (1991) Combining ability for canopy growth and gas exchange of interspecific blueberries under moderate water deficit. *Journal of the American Society for Horticultural Science* 116, 569–573.

Espíndola, G. (2018) *Efecto de Poda Sectorial en Rendimiento y Calidad de Frutos en Arándano de Arbusto Alto cv. Brigitta y Duke*. [Effect of Sectorial Pruning on Yield and Fruit Quality of Highbush Blueberry cv. Brigitta and Duke.] Memoria de Ingeniero Agrónomo. Facultad de Ciencias Agrarias. Universidad de Talca, Talca, Chile.

Farmer, J., Zhang, B., Jin, X., Zhang, P. and Wang, J. (2017) Long-term effect of plastic film mulching and fertilization on bacterial communities in a brown soil revealed by high through-put sequencing. *Archives of Agronomy and Soil Science* 63, 230–241.

Fereres, E. and Evans, R.G. (2006) Irrigation of fruit trees and vines: an introduction. *Irrigation Science* 24, 55–57.

Fereres, E., Meyer, J.L., Alijibury, F.K., Schulbach, H. and Marsch, A.W. (1981) *Irrigation Costs*. Division of Agricultural Sciences, University of California, Berkeley, California.

Fiedler, A., Tuell, J., Isaacs, R. and Landis, D. (2008) Attracting beneficial insects with native flowering plants. Extension Bulletin E-2973. Michigan State University, East Lansing, Michigan. Available at: http://www.canr.msu.edu/nativeplants/uploads/files/E2973.pdf (accessed 5 February 2018).

Florence, J. and Pscheidt, J. (2017) *Monilinia vaccinii-corymbosi* apothecial development associated with mulch depth and timing of application. *Plant Disease* 101, 807–814.

Forge, T., Temple, W. and Bomke, A. (2013) Using compost as mulch for highbush blueberry. *Acta Horticulturae* 1001, 369–374.

Freeman, B. (1983) *Blueberry Production*. Agfact H3 1.4. Department of Agriculture NSW, Sydney, Australia.

Galletta, G.J. and Fish, A.S. Jr (1971) Interspecific blueberry grafting, a way to extend *Vaccinium* culture to different soils. *Journal of the American Society for Horticultural Science* 96, 294–298.

Ganter, A., Montalba, R., Rebolledo, R. and Vieli, L. (2013) Plastic mulch effects on ground beetle communities (Coleoptera: Carabidae) in an organic blueberry field. *Idesia* 31, 61–66.

Gaskell, M. (2004) Field tunnels permit extended season harvest of small fruits in California. *Acta Horticulturae* 659, 425–430.

Gibbs, J., Elle, E., Bobiwash, K., Haapalainen, T. and Isaacs, R. (2016) Contrasting pollinators and pollination in native and non-native regions of highbush blueberry production. *PLOS ONE* 11, e0158937.

Girgenti, V., Peano, C., Giuggioli, N.R., Giraudo, E. and Guerrini, S. (2012) First results of biodegradable mulching on small berry fruits. *Acta Horticulturae* 926, 571–576.

Gómez-Martínez, M.G. (2010) *La Poda en la Productividad de Arándanos (Vaccinium spp.) en Michoacán.* [Pruning and blueberry (*Vaccinium* spp.) productivity in Michoacán.] Maestría en Ciencias en Horticultura, Departamento de Fitotecnia, Universidad Autónoma de Chapingo, Texcoco, Mexico.

Gonzalez-Dugo, V., Zarco-Tejada, P., Nicolás, E., Nortes, P.A., Alarcón, J.J., Intrigliolo, D.S. and Fereres, E. (2013) Using high resolution UAV thermal imagery to assess the variability in the water status of five fruit tree species within a commercial orchard. *Precision Agriculture* 14, 660–678.

Gough, R.E. (1980) Root distribution of 'Coville' and 'Lateblue' highbush blueberry under sawdust mulch. *Journal of the American Society for Horticultural Science* 105, 576–578.

Gough, R.E. (1994) *The Highbush Blueberry and Its Management.* Food Products Press, Binghampton, New York.

Grundy, A.C. and Bond, B. (2007) Use of non-living mulches for weed control. In: Upadhyaya, M.K. and Blackshaw, R.E. (eds) *Non-chemical Weed Management Principles, Concepts and Technology.* CAB International, Wallingford, UK, pp. 135–153.

Haby, V.A. and Pennington, H.D. (1988) Irrigation water source and quality. In: Baker, M.L., Patten, K.D., Neuendorff, E.W. and Lyons, C. (eds) *Texas Blueberry Handbook.* Texas Agricultural Extension Service, College Station, Texas.

Haby, V.A., Patten, K.D., Cawthon, D.L., Kresja, B.B., Neuendorff, E.W., Davis, J.V. and Peters, S.C. (1986) Response of container-grown rabbiteye blueberry plants to irrigation water and soil type. *Journal of the American Society for Horticultural Science* 111, 332–337.

Haman, D.Z., Smajstrla, A.G. and Lyrene, P.M. (1988) Blueberry response to irrigation and ground cover. *Proceedings of the Florida State Horticultural Society* 101, 235–238.

Haman, D.Z., Smajstrla, A.G., Pritchard, R.T. and Lyrene, P.M. (1997a) Response of young blueberry plants to irrigation in Florida. *HortScience* 32, 1194–1196.

Haman, D.Z., Smajstrla, A.G., Pritchard, R.T., Zazueta, F.S. and Lyrene, P.M. (1997b) Evapotranspiration and crop coefficients for young blueberries in Florida. *Applied Engineering in Agriculture* 13, 209–216.

Haman, D.Z., Smajstrla, A.G., Pritchard, R.T., Zazueta, F.S. and Lyrene, P.M. (2005) *Water Use in Establishment of Young Blueberry Plants.* Bulletin No. 296. University of Florida, Institute of Food and Agricultural Sciences, Gainesville, Florida.

Hancock, J.F., Krebs, S.L., Sakin, M. and Holtsford, T.P. (1989) Increasing blueberry yields through mixed variety plantings. *Michigan State Horticulture Society Annual Report* 119, 130–133.

Hanslin, H.M. Sæbø, A. and Bergersen, O. (2005) Estimation of oxygen concentration in the soil gas phase beneath compost mulch by means of a simple method. *Urban Forestry and Urban Greening* 4, 37–40.

Hanson, E.J. (2006) Irrigation options for blueberries. Department of Horticulture, Michigan State University Extension, Lansing, Michigan. Available at: http://msue.anr.msu.edu/news/irrigation_options_for_blueberries (accessed 5 February 2018).

Hart, J., Strik, B., White, L. and Yang, W. (2006) *Nutrient Management for Blueberries in Oregon.* Publication No. EM 8918. Oregon State University Extension Service, Corvallis, Oregon.

Heinrich, B. (2004) *Bumblebee Economics.* Harvard University Press, Cambridge, Massachusetts.

Himelrick, D. and Curtis, L.M. (1999) Commercial blueberries. Micro-Irrigation Handbook ANR-663. Alabama Cooperative Extension Service, Centreville, Alabama. Available at http://www.aces.edu/anr/irrigation/ANR-663.php (accessed 6 February 2018).

Himelrick, D.G., Powell, A.A. and Dozier, W.A. (1995) Commercial blueberry production guide for Alabama. Available at: http://www.aces.edu/pubs/docs/A/ANR-0904/ANR-0904.pdf (accessed 6 February 2018).

Hogue, E.J. and Neilsen, G.H. (1987) Orchard floor vegetation management. *Horticultural Reviews* 9, 377–430.

Holzapfel, E.A. (2009) Selection and management of irrigation systems for blueberry. *Acta Horticulturae* 810, 641–648.

Holzapfel, E., Jara, J. and Coronata, A.M. (2015) Number of drip laterals and irrigation frequency on yield and exportable fruit size of highbush blueberry grown in a sandy soil. *Agricultural Water Management* 148, 207–212.

Holzapfel, E.A., Hepp, R.F. and Mariño, M.A. (2004) Effect of irrigation on fruit production in blueberry. *Agricultural Water Management* 67, 173–184.

Hopkins, B.G., Horneck, D.A., Stevens, R.G., Ellsworth, J.W. and Sullivan, D.M. (2007) Managing irrigation water quality for crop production in the Pacific Northwest. PNW 597-E. Pacific Northwest Extension, Oregon State University, University of Idaho, Washington State University. Available at: http://ir.library.oregonstate.edu/xmlui/bitstream/handle/1957/20786/pnw597-e.pdf?sequence=3 (accessed 27 September 2017).

Howell, G.S., Hanson, C.M., Bittenbender, H.C. and Stackhouse, S.S. (1975) Rejuvenating highbush blueberries. *Journal of the American Society for Horticultural Science* 100, 455–457.

Huang, Y.H., Johnson, C.E., Lang, G.A. and Sundberg, M.D. (1997) Pollen sources influence early fruit growth of southern highbush blueberry. *Journal of the American Society for Horticultural Science* 122, 625–629.

Ireland, G. and Wilk, P. (2006) Blueberry production in northern NSW. Primefact 195. NSW Department of Primary Industries, Orange, New South Wales, Australia. Available at: https://www.scribd.com/document/71768573/Blueberry-Production-in-Northern-NSW (accessed 6 February 2018).

Isaacs, R. and Kirk, A.K. (2010) Pollination services provided to small and large highbush blueberry fields by wild and managed bees. *Journal of Applied Ecology* 47, 841–849.

Javorek, S.K., Mackenzie, K.E. and Vander Kloet, S.P. (2002) Comparative pollination effectiveness among bees (Hymenoptera: Apoidea) on lowbush blueberry (Ericaceae: *Vaccinium angustifolium*). *Annals of the Entomological Society of America* 95, 345–351.

Jiménez, S., Dridi, J., Gutiérrez, D., Moret, D., Irigoyen, J.J., Moreno, M.A. and Gogorcena, Y. (2013) Physiological, biochemical and molecular responses in four *Prunus* rootstocks submitted to drought stress. *Tree Physiology* 33, 1061–1075.

Jones, H.G. and Tardieu, F. (1998) Modelling water relations of horticultural crops: a review. *Scientia Horticulturae* 74, 21–46.

Julian, J. (2012) Mechanical Harvest of Fresh Blueberry Why? Can we make money at it? Oregon State University, Corvallis, Oregon. Available at: http://cetulare.ucanr.edu/files/168378.pdf (accessed 31 October 2017).

Kader, M.A., Senge, M., Mojid, M.A. and Ito, K. (2017) Recent advances in mulching materials and methods for modifying soil environment. *Soil and Tillage Research* 168, 155–166.

Kadir, S., Carey, E. and Ennahli, S. (2006) Influence of high tunnel and field conditions on strawberry growth and development. *HortScience* 41, 329–335.

Keen, B. and Slavich, P. (2012) Comparison of irrigation scheduling strategies for achieving water use efficiency in highbush blueberry. *New Zealand Journal of Crop and Horticultural Science* 40, 3–20.

Kender, W.J. and Brightwell, W.T. (1966) Environmental relationships. In: Eck, P. and Childers, N.F. (eds) *Blueberry Culture*. Rutgers University Press, New Brunswick, New Jersey, pp. 75–93.

Khan, A.R., Chandra, D., Quraishi, S. and Sinha, R.K. (2000) Soil aeration under different soil surface conditions. *Journal of Agronomy and Crop Science* 185, 105–112.

Kingston, P. (2017) Substrate production of blueberry: evaluation of soilless media and potassium, nitrogen fertility on growth and nutrition. MS thesis, Oregon State University, Corvallis, Oregon.

Kovaleski, A.P., Williamson, J.G., Casamali, B. and Darnell, R.L. (2015a) Effects of timing and intensity of summer pruning on vegetative traits of two southern highbush blueberry cultivars. *HortScience* 50, 68–73.

Kovaleski, A.P., Darnell, R.L., Casamali, B. and Williamson, J.G. (2015b) Effects of timing and intensity of summer pruning on reproductive traits of two southern highbush blueberry cultivars. *HortScience* 50, 1486–1491.

Kozinski, B. (2006) Influence of mulching and nitrogen fertilization rate on growth and yield of highbush blueberry. *Acta Horticulturae* 715, 231–235.

Krebs, S.L. and Hancock, J.F. (1988) The consequences of inbreeding on fertility in *Vaccinium corymbosum* L. *Journal of the American Society for Horticultural Science* 113, 914–918.

Krewer, G., Ruter, J.M., NeSmith, D.S., Thomas, D., Sumner, P., Harrison, K., Westberry, G., Mullinix, B. and Knox, D. (1997) Preliminary report on the effect of tire chips as a mulch and substrate component of blueberry. *Acta Horticulturae* 446, 309–318.

Krewer, G., Tertuliano, M., Andersen, P., Liburd, O., Fonsah, G., Serri, H. and Mullinix, B. (2009) Effect of mulches on the establishment of organically grown blueberries in Georgia. *Acta Horticulturae* 810, 483–488.

Kumar, D., Shivay, Y.S., Dhar, S., Kumar, C. and Prasad, R. (2013) Rhizospheric flora and the influence of agronomic practices on them: a review. *Proceedings of the National Academy of Sciences, India, Section B: Biological Sciences* 83, 1–14.

Labavitch, J.M., Greve, L.C. and Mitcham, E. (1998) Fruit bruising: it's more than skin deep. *Perishables Handling Quarterly* 95, 7–9.

Lamont, W.J. (2005) Plastics: modifying the microclimate for the production of vegetable crops. *HortTechnology* 15, 447–481.

Lang, G.A. (2009) High tunnel tree fruit production: the final frontier? *HortTechnology* 19, 50–55.

Lang, G.A. and Danka, R.G. (1991) Honey-bee-mediated cross- versus self-pollination of 'Sharpblue' blueberry increases fruit size and hastens ripening. *Journal of the American Society for Horticultural Science* 116, 770–773.

Larco, H., Strik, B., Bryla, D. and Sullivan, D. (2009) Establishing organic highbush blueberry production systems – the effect of raised beds, weed management, fertility, and cultivar. *HortScience* 44, 1120–1121.

Larco, H., Strik, B.C., Bryla, D.R. and Sullivan, D.M. (2013a) Mulch and fertilizer management practices for organic production of highbush blueberries. I: Early plant growth and biomass allocation. *HortScience* 48, 1250–1261.

Larco, H., Bryla, D.R., Strik, B.C. and Sullivan, D.M. (2013b) Mulch and fertilizer management practices for organic production of highbush blueberries. II. Impact on plant and soil nutrients, yield, and fruit quality during establishment. *HortScience* 48, 1484–1495.

Lee, H.J., Kim, S.J., Yu, D.J., Lee, B.Y. and Kim, T.C. (2006) Changes of photosynthetic characteristics in water-stressed 'Rancocas' blueberry leaves. *Acta Horticulturae* 715, 111–118.

Liu, E.K., He, W.Q. and Yan, C.R. (2014) 'White revolution' to 'white pollution'– agricultural plastic film mulch in China. *Environmental Research Letters* 9, 091001.

Liu, J.G., Mahoney, K.J., Sikkema, P.H. and Swanton, C.J. (2009) The importance of light quality in crop–weed competition. *Weed Research* 49, 217–224.

Lobos, G.A., Moggia, C., Retamales, J.B. and Sánchez, C. (2014) Postharvest effects of mechanized (automotive or shaker) vs. hand harvest on fruit quality of blueberries (*Vaccinium corymbosum* L). *Acta Horticulturae* 1017, 135–140.

Lobos, T.E. (2016) Regulated deficit irrigation to maintain yield and improve quality and shelf life of highbush blueberry (*V. corymbosum*) fruits. PhD, Universidad de La Frontera, Temuco, Chile.

Lobos, T.E., Retamales, J.B., Ortega-Farías, S., Hanson, E.J., López-Olivari, R. and Mora, M.L. (2016) Pre-harvest regulated deficit irrigation management effects on post-harvest quality and condition of *V. corymbosum* fruits cv. Brigitta. *Scientia Horticulturae* 207, 152–159.

Loomis, R.S. and Connor, D.J. (1992) *Crop Ecology: Productivity and Management in Agricultural Systems*, 1st edn. Cambridge University Press, Cambridge, UK.

Lyrene, P.M. and Muñoz, C. (1997) Blueberry production in Chile. *Journal of Small Fruit and Viticulture* 5, 1–20.

Magee, J.B. and Spiers, J.M. (1995) Influence of mulching systems on yield and quality of southern highbush blueberries, pp. 133–141. In: Gough, R.E. and Korcak, R.F. (eds) *Blueberries: a Century of Research*. Haworth Press, Binghamton, New York.

Mainland, C.M. (1993) Blueberry production strategies. *Acta Horticulturae* 346, 111–116.

Mainland, C.M., Ambrose, J.T. and Garcia, L.E. (1979) Fruit seed development of rabbiteye blueberries in response to pollinator cultivar and gibberellic acid. In: Moore,

J.N. (ed.) *Proceedings of the IV North American Blueberry Research Workers Conference*. University of Arkansas, Fayetteville, Arkansas, pp. 203–211

Mainland, C.M., Kushman, L.J. and Ballinger, W.E. (1975) The effect of mechanical harvesting on yield, quality of fruit and bush damage of highbush blueberry. *Journal of the American Society for Horticultural Science* 100, 129–134.

Martins, N.S., Calado, M.R.A., Pombo, J.A.N. and Mariano, S.J.P.S. (2016) Blueberries field irrigation management and monitoring system using PLC based control and wireless sensor network. In: *2016 IEEE 16th International Conference on Environment and Electrical Engineering (EEEIC)*, Florence, Italy, 7–10 June 2016. IEEE, Piscataway, New Jersey.

Marucci, P.E. (1965) Blueberry pollination. *American Bee Journal* 106, 250–251.

Meader, E.M. and Darrow, G.M. (1947) Highbush pollination experiments. *Proceedings of the American Society for Horticultural Science* 49, 197–204.

Mehra, L.K., MacLean, D.D., Savelle, A.T. and Scherm, H. (2013) Postharvest disease development on southern highbush blueberry fruit in relation to berry flesh type and harvest method. *Plant Disease* 97, 213–221.

Mingeau, M., Perrier, C. and Amèglio, T. (2001) Evidence of drought-sensitive periods from flowering to maturity on highbush blueberry. *Scientia Horticulturae* 89, 23–40.

Moriana, A., Girón, I.F., Martín-Palomo, M.J., Conejero, W., Ortuño, M.F., Torrecillas, A. and Moreno, F. (2010) New approach for olive trees irrigation scheduling using trunk diameter sensors. *Agricultural Water Management* 97, 1822–1828.

Mormile, P., Stahl, N. and Malinconico, M. (2017) The world of plasticulture. In: Malinconico, M. (ed.) *Soil Degradable Bioplastics for a Sustainable Modern Agriculture*. Springer, Berlin, Germany, pp. 1–22.

Morrow, E.B. (1943) Some effects of cross-pollination versus self-pollination in the cultivated blueberry. *Proceedings of the American Society for Horticultural Science* 42, 469–472.

Mulumba, L.N. and Lal, R. (2008) Mulching effects on selected soil physical properties. *Soil and Tillage Research* 98, 106–111.

Muralitharan, M.S., Chandler, S. and Vanstevenick, R.F.M. (1992) Effects of NaCl and Na_2SO_4 on growth and solute composition of highbush blueberry (*Vaccinium corymbosum*). *Australian Journal of Plant Physiology* 19, 155–164.

Nicholas, A.H. (2010) Whitefringed weevils *Naupactus leucoloma* (Boheman), (Coleoptera: Curculionidae) damage sub-surface drip irrigation tape. *Irrigation Science* 28, 353–357.

Nunez, G.H., Olmstead, J.W. and Darnell, R.L. (2015) Rhizosphere acidification is not part of the strategy I iron deficiency response of *Vaccinium arboreum* and the southern highbush blueberry. *HortScience* 50, 1064–1069.

Odneal, M.B. and Kaps, M.L. (1990) Fresh and aged pine bark as soil amendments for establishment of highbush blueberry. *HortScience* 25, 1228–1229.

Ogden, A.B. and van Iersel, M.W. (2009) Southern blueberry production in high tunnels: temperatures, development, yield, and fruit quality during the establishment years. *HortScience* 44, 1850–1856.

Oliveira, P.B., Oliveira, C.M., Lopes da Fonseca, L. and Monteiro, A.A. (1996) Off-season production of primocane-fruiting red raspberry using summer pruning and polyethylene tunnels. *HortScience* 31, 805–807.

Olmstead, J.W., Rodríguez-Armenta, H.P. and Lyrene, P.M. (2013) Using sparkleberry as a genetic source for machine harvest traits for southern highbush blueberry *Hort-Technology* 23, 419–424.

Oyarzún, R., Stöckle, C., and Whiting, M. (2010) Analysis of hydraulic conductance components in field grown, mature sweet cherry trees. *Chilean Journal of Agricultural Research* 70, 58–66.

Ozeki, M. and Tamada, T. (2006) The potentials of forcing culture of southern highbush blueberry in Japan. *Acta Horticulturae* 715, 241–246.

Palma, M.J. and Retamales, J.B. (2017) Cane productivity and fruit quality of highbush blueberry are affected by cane diameter and location within the canopy. *European Journal of Horticultural Science* 82, 159–165.

Pannacci, E., Lattanzi, B. and Tei, F. (2017) Non-chemical weed management strategies in minor crops: a review. *Crop Protection* 96, 44–58.

Paranychianakis, N.V. and Chartzoulakis, K.S. (2005) Irrigation of Mediterranean crops with saline water: from physiology to management practices. *Agriculture, Ecosystems and Environment* 106, 171–187.

Pasa, M.S., Fachinello, J.C., Schmitz, J.D. and Rosa H.F. Jr (2014) Desempenho de cultivares de mirtileiros dos grupos rabbiteye e highbush em função da cobertura de solo. [Performance of rabbiteye and highbush blueberry cultivars as affected by mulching.] *Revista Brasileira de Fruticultura* 36, 161–169.

Patten, K.D., Neuendorff, E.W., Nimr, G.H., Peters, S.C. and Cawton, D.L. (1989) Growth and yield of rabbiteye blueberry as affected by orchard floor management practices and irrigation geometry. *Journal of the American Society for Horticultural Science* 114, 728–732.

Percival, D., Gallant, D. and Harrington, T. (2017) Potential for commercial unmanned aerial vehicle use in wild blueberry production. *Acta Horticulturae* 1180, 233–240.

Pescie, M., Borda, M. Fedyszak, P. and López, C. (2011) Efecto del momento y tipo de poda sobre el rendimiento y calidad del fruto en arándano altos del sur (*Vaccinium corymbosum*) var. O'Neal en la provincia de Buenos Aires. [Effect of the moment and type of pruning on yield and fruit quality in southern highbush blueberry cv. O'Neal in the Buenos Aires Province.] *Revista de Investigación Agropecuaria* 37, 268–274.

Peterson, D.L. and Brown, G.K. (1996) Mechanical harvester for fresh market quality blueberries. *Transactions of the American Society of Agricultural Engineers* 39, 823–827.

Pritts, M.P. and Hancock, J.F. (1992) *Highbush Production Guide*. NRAES-55 Cooperative Extension Bulletin. Cornell University, Ithaca, New York.

Qiaosheng, S., Zuoxin, L., Zhenying, W. and Haijun, L. (2007) Simulation of the soil wetting shape under porous pipe sub-irrigation using dimensional analysis. *Irrigation and Drainage* 56, 389–398.

Reiss, E., Both, A.J., Garrison, S., Kline, W. and Sudal, J. (2004) Season extension for tomato production using high tunnels. *Acta Horticulturae* 659, 153–160.

Renkema, J.M., Lynch, D.H. Cutler, G.C. Mackenzie, K. and Walde, S.J. (2012) Ground and rove beetles (Coleoptera: Carabidae and Staphylinidae) are affected by mulches and weeds in highbush blueberries. *Environmental Entomology* 41, 1097–1106.

Renquist, S. (2008) An evaluation of blueberry cultivars grown in plastic tunnels in Douglas County, Oregon. *International Journal of Fruit Science* 5, 31–38.

Retamales, J.B. and Hanson, E.J. (1989) Fate of ^{15}N-labeled urea applied to mature highbush blueberries. *Journal of the American Society for Horticultural Science* 114, 920–923.

Retamales, J.B., Mena, C., Lobos, G. and Morales, Y. (2015) A regression analysis on factors affecting yield of highbush blueberries. *Scientia Horticulturae* 186, 7–14.

Retamal-Salgado, J., Bastías, R.M., Wilckens, R. and Paulino, L. (2015) Influence of microclimatic conditions under high tunnels on the physiological and productive responses in blueberry 'O'Neal'. *Chilean Journal of Agricultural Research* 75, 291–297

Riedl, H., Johansen, E., Brewer, L. and Barbour, J. (2006) How to reduce bee poisoning from pesticides. Publication no. PNW 591. Oregon State University, Corvallis, Oregon. Available at: https://catalog.extension.oregonstate.edu/sites/catalog/files/project/pdf/pnw591_1.pdf (accessed 6 February 2018).

Rodriguez-Saona, C., Parra, L., Quiroz, A. and Isaacs, R. (2011) Variation in highbush blueberry floral volatile profiles as a function of pollination status, cultivar, time of day and flower part: implications for flower visitation by bees. *Annals of Botany* 107, 1377–1390.

Sadras, V.O. and Trentacoste, E.R. (2011) Phenotypic plasticity of stem water potential correlates with crop load in horticultural trees. *Tree Physiology* 31, 494–499.

Safley, C.D., Cline, W.O. and Mainland, C.M. (2012) Evaluating the profitability of blueberry production. Available at: https://blueberries.ces.ncsu.edu/wp-content/uploads/2012/10/evaluating-the-profitability-of-blueberry-production.pdf?fwd=no (accessed 5 May 2018).

Salamé-Donoso, T.P., Santos, B.M., Chandler, C.K. and Sargent., S.A. (2010) Effect of high tunnels on the growth, yields, and soluble solids of strawberry cultivars in Florida. *International Journal of Fruit Science* 10, 249–263.

Salisbury, F.B. and Ross, C.W. (1991) *Plant Physiology*, 4th edn. Wadsworth Publisher, Belmont, California.

Sampson, B.J. and Cane, J.H. (2000) Pollination efficiencies of three bee (Hymenoptera: Apoidea) species visiting rabbiteye blueberries. *Journal of Economic Entomology* 93, 1726–1731.

Sampson, B.J. and Spiers, J.M. (2002) Evaluating bumble bees as pollinators of 'Misty' southern highbush growing inside plastic tunnels. *Acta Horticulturae* 574, 53–61.

Sampson, B.J., Danka, R.G. and Stringer, S.J. (2004) Nectar robbery by bees *Xylocopa virginica* and *Apis mellifera* contributes to the pollination of rabbiteye blueberries. *Journal of Economic Entomology* 97, 735–740.

Santos, B.M. and Salomé-Donoso, T.P. (2012) Performance of southern highbush blueberry cultivars under high tunnels in Florida. *HortTechnology* 22, 700–704.

Scagel, C.F. and Yang, W.Q. (2005) Cultural variation and mycorrhizal status of blueberry plants in NW Oregon commercial production fields. *International Journal of Fruit Science* 5, 85–111.

Shan, H.Y. (2011) Soil moisture and groundwater recharge. National Chiao Tung University, Hsinchu, Taiwan. Available at: http://www.cv.nctu.edu.tw/chinese/teacher/Ppt-pdf/03Soil%20Moisture%20and%20Groundwater%20Recharge.pdf (accessed 6 February 2018).

Shock, C.C. (2008) Soil water measurement: granular matrix sensors. In: Trimble, S.W. (ed.) *Encyclopaedia of Water Science*, 2nd edn. CRC Press, Boca Raton, Florida, pp. 1058–1062.

Siefker, J.A. and Hancock, J.F. (1987) Pruning effects on productivity and vegetative growth in the highbush blueberry. *HortScience* 22, 210–211.

Smajstrla, A.G. and Harrison, D.S. (2008) *Tensiometers for Soil Moisture Measurement and Irrigation Scheduling.* Publication no. CIR487. University of Florida, IFAS Extension, Gainesville, Florida.

Smith, E., Porter, W., Hawkins, G. and Harris, G. Jr (2016) Blueberry irrigation water quality. Bulletin C1105. University of Georgia Cooperative Extension, Athens, Georgia. Available at: https://secure.caes.uga.edu/extension/publications/files/pdf/C%201105_2.pdf (accessed 26 September 2017).

Soltani, N., Anderson, J.L. and Hamson, A.R. (1995) Growth analysis of watermelon plants grown with mulches and rowcovers. *Journal of the American Society for Horticultural Science* 120, 1001–1009.

Spiers, J.M. (1982) Fertilization, incorporated organic matter, and early growth of rabbiteye blueberries. *Journal of the American Society for Horticultural Science* 107, 1054–1058.

Spiers, J.M. (1986) Root distribution of 'Tifblue' rabbiteye blueberry as influenced by irrigation, incorporated peatmoss, and mulch. *Journal of the American Society for Horticultural Science* 111, 877–880.

Spiers, J.M. (1995) Substrate temperatures influence root and shoot growth of southern highbush and rabbiteye blueberries. *HortScience* 30, 1029–1030.

Steinmetz, Z., Wollmann, C., Schaefer, M. Buchmann, C., David, J., Tröger, J., Muñoz, K., Frör, O. and Schaumann, G.E. (2016) Plastic mulching in agriculture: trading short-term agronomic benefits for long-term soil degradation? *Science of the Total Environment* 550, 690–705.

Storlie, C.A. and Eck, P. (1996) Lysimeter-based crop coefficients for young highbush blueberries. *HortScience* 31, 819–822.

Strik, B.C. (2008) *Growing Blueberries in Your Home Garden.* EC 1304. Oregon State University Extension Service, Corvallis, Oregon. Available at: https://catalog.extension.oregonstate.edu/ec1304/html (accessed 15 September 2017).

Strik, B.C. (2012) Flowering and fruiting on command in berry crops. *Acta Horticulturae* 926, 197–212.

Strik, B.C. (2016) A review of optimal systems for organic production of blueberry and blackberry for fresh and processed markets in the northwestern United States. *Scientia Horticulturae* 208, 92–103.

Strik, B.C. and Buller, G. (2002) Improving yield and machine harvest efficiency of 'Bluecrop' through high density planting and trellising. *Acta Horticulturae* 574, 227–231.

Strik, B.C. and Buller, G. (2005) The impact of early cropping on subsequent growth and yield of highbush blueberry in the establishment years at two planting densities is cultivar dependent. *HortScience* 40, 1998–2001.

Strik, B.C., Buller, G. and Hellman, E. (2003) Pruning severity affects yield, berry weight and hand harvest efficiency of highbush blueberry. *HortScience* 38, 196–199.

Strik, B.C, Vance, A.J. and Finn, C.E. (2017) Northern highbush blueberry cultivars differed in yield and fruit quality in two organic production systems from planting to maturity. *HortScience* 52, 844–851.

Sullivan, D.M., Bryla, D.R. and Costello, R.C. (2014) Chemical characteristics of custom compost for highbush blueberry. In: He, Z. and Zhang, H. (eds) *Applied Manure and*

Nutrient Chemistry for Sustainable Agriculture and Environment. Springer Science+Business Media, Dordrecht, The Netherlands, pp. 293–311.

Swanson, R.H. (1994) Significant historical developments in thermal methods for measuring sap flow in trees. *Agricultural and Forest Meteorology* 72, 113–132.

Takeda, F., Krewer, G., Andrews, E.L., Mullinix, B. Jr and Peterson, D.L. (2008) Assessment of the V45 blueberry harvester on rabbiteye blueberry and southern highbush blueberry pruned to V-shaped canopy. *HortTechnology* 18, 130–138.

Takeda, F., Yang, W.Q., Li, C., Freivalds, A., Sung, K., Xu, R., Hu, B., Williamson, J. and Sargent, S. (2017) Applying new technologies to transform blueberry harvesting. *Agronomy* 7, 33.

Tarara, J.M. (2000) Microclimate modification with plastic mulch. *HortScience* 35, 169–180.

Tertuliano, M., Krewer, G., Smith, J.E., Plattner, K., Clark, J., Jacobs, J., Andrews, E., Stanaland, D., Andersen, P., Liburd, O., Fonsah, E.G. and Scherm, H. (2012) Growing organic rabbiteye blueberries in Georgia, USA: results of two multi-year field studies. *International Journal of Fruit Science* 12, 205–215.

Testi, L., Goldhamer, D.A., Iniesta, F. and Salinas, M. (2008) Crop water stress index is a sensitive water stress indicator in pistachio trees. *Irrigation Science* 26, 395–405.

Topp, G.C., Dow, B., Edwards, M., Gregorich, E.G., Curnoe, W.E. and Cook, F.J. (2000) Oxygen measurements in the root zone facilitated by TDR. *Canadian Journal of Soil Science* 80, 33–41.

Tuell, J.K. and Isaacs, R. (2010) Weather during bloom affects pollination and yield of highbush blueberry. *Journal of Economic Entomology* 103, 557–562.

Tuell, J.K., Ascher, J.A. and Isaacs, R. (2009) Wild bees (Hymenoptera: Apoidea: Anthophila) of the Michigan highbush blueberry agroecosystem. *Annals of the Entomological Society of America* 102, 275–287.

Upadhyaya, M.K. and Blackshaw, R.E. (2007) Non-chemical weed management: synopsis, integration and the future. In: Upadhyaya, M.K. and Blacksaw, R.E. (eds) *Non-chemical Weed Management Principles, Concepts and Technology.* CAB International, Wallingford, UK, pp. 201–209.

van Dalfsen, K.B. and Gaye, M.M. (1999) Yield of hand and mechanical harvesting of highbush blueberries in British Columbia. *Applied Engineering in Agriculture* 15, 393–398.

Vincent, N.J. (1971) The relationship of highbush blueberry (*Vaccinium corymbosum* L.) flower volatile and nectar production to honeybee attractiveness. PhD thesis, Rutgers University, New Brunswick, New Jersey.

Voogt, W., van Dijk, P., Douven, F. and van der Maas, R. (2014) Development of a soilless growing system for blueberries (*Vaccinium corymbosum*): nutrient demand and nutrient solution. *Acta Horticulturae* 1017, 215–221.

Wang, J., Sammis, T.W., Andales, A.A., Simmons, L.J., Gutschick, V.P. and Miller, D.R. (2007) Crop coefficients of open-canopy pecan orchards. *Agricultural Water Management* 88, 253–262.

Whatcom (2005) Compost fundamentals. Washington State University County Extension, Pullman, Washington, DC. Available at: http://www.whatcom.wsu.edu/ag/compost/fundamentals/needs_carbon_nitrogen.htm (accessed 10 December 2010).

Wilk, P., Carruthers, G., Mansfield, C. and Hood, V. (2009) Irrigation and moisture monitoring in blueberries. Primefact 827. NSW Department of Primary Industries,

Orange, New South Wales, Australia. Available at: https://www.dpi.nsw.gov.au/__data/assets/pdf_file/0016/303325/Irrigation-and-moisture-monitoring-in-blueberries.pdf (accessed 6 February 2018).

Williamson, J., Krewer, G., Pavlis, G. and Mainland, C.M. (2006) Blueberry soil management, nutrition and irrigation. In: Childers, N.F. and Lyrene, P.M. (eds) *Blueberries for Growers, Gardeners and Promoters.* Dr Norman F. Childers Publications, Gainesville, Florida, pp. 60–74.

Williamson, J.G. and Darnell, R.L. (1996) Severity and timing of mechanical rejuvenation pruning affects vegetative and reproductive growth of blueberry. *HortScience* 31, 663.

Williamson, J.G., Mejia, L., Ferguson, B., Miller, P. and Haman, D.Z. (2015) Seasonal water use of southern highbush blueberry plants in a subtropical climate. *HortTechnology* 25, 185–191.

Wu, L., Yu, L., Dong, L., Zhu, Y., Li, C., Zhang, Z. and Li, Y. (2006) Comparison of mulching treatments on growth and physiology of highbush blueberry. *Acta Horticulturae* 715, 237–240.

Xie, Z.S. and Wu, X.C. (2009) Studies on substrates for blueberry cultivation. *Acta Horticulturae* 810, 513–520.

Xu, C., Ma, Y. and Chen, H. (2014) Technique of grafting with Wufanshu (*Vaccinium bracteatum* Thunb.) and the effects on blueberry plant growth and development, fruit yield and quality. *Scientia Horticulturae* 176, 290–296.

Yáñez, P., Retamales, J.B., Lobos, G.A. and del Pozo, A. (2009) Light environment within mature rabbiteye blueberry canopies influences flower bud formation. *Acta Horticulturae* 810, 471–474.

Yang, N., Sun, Z.X., Feng, L.S., Zheng, M.Z., Chi, D.C., Meng, W.Z., Hou, Z.Y., Bai, W. and Li, K.Y. (2015) Plastic film mulching for water-efficient agricultural applications and degradable films materials development research. *Materials and Manufacturing Processes* 30, 143–154.

Yang, W.Q., Goulart, B.L., Demchak, K. and Li, Y. (2002) Interactive effects of mycorrhizal inoculation and organic soil amendments on nitrogen acquisition and growth of highbush blueberry. *Journal of the American Society for Horticultural Science* 127, 742–748.

Yeo, J.R., Weiland, J.E., Sullivan, D.M. and Bryla, D.R. (2017) Nonchemical, cultural management strategies to suppress *Phytophthora* root rot in northern highbush blueberry. *HortScience* 52, 725–731.

Yu, P., Li, C., Rains, G. and Hamrita, T. (2011) Development of the berry impact recording device sensing system: hardware design and calibration. *Computers and Electronics in Agriculture* 79, 103–111.

Yu, P., Li, C., Takeda, F., Krewer, G., Rains, G. and Hamrita, T. (2012) Quantitative evaluation of a rotary blueberry mechanical harvester using a miniature instrumented sphere. *Computers and Electronics in Agriculture* 88, 25–31.

7

GROWTH REGULATORS IN BLUEBERRY PRODUCTION

INTRODUCTION

Blueberries are a perennial crop that, when well managed, can produce for 25 years or more in some locations. In order to stay in business, growers need to manage their plantings efficiently. Hand labour can reach up to 80% of the operational costs in a mature blueberry field (Takele *et al.*, 2007). Labour management is key to blueberry crop profitability (Plattner *et al.*, 2008). This involves not only constant monitoring but also timely intervention and detailed assessment of the cost:benefit ratio of management practices. Fruit crop performance can be significantly controlled through genetic and management practices.

Once plants have been established, growers usually find traits that are less than desirable, or environmental factors that frequently reduce crop productivity, diminish quality or somehow affect the profitability of a blueberry field. Among the myriad of management tools available for blueberry culture, plant growth regulators (PGRs) offer opportunities to solve specific problems. They can then become part of the management package and applied whenever the cost:benefit ratio is adequate. The cost of these compounds is generally high so they have to be effective and their impact on plant processes needs to be confirmed (Greene, 2002).

PGRs should probably be called plant bioregulators, as they do not only affect growth. They are 'natural or synthetic compounds applied to plants or plant organs to regulate growth and development' (Petracek *et al.*, 2003). PGRs are generally effective in low concentrations (low dose), have a narrow optimum concentration–response range and must be absorbed by the plant tissue (usually leaves) to induce the desired physiological response. PGRs not only affect the process that they target but also alter the overall physiology of the plant. As a result, there is the need to look for collateral effects or changes, both in the developmental or research phase, and during application in the commercial setting (Bukovac, 2005).

PGRs have been used at all stages of fruit production, including: the nursery, canopy development and growth control, flowering and fruiting, alteration of fruit quality and ripening, stress tolerance and plant–environment interactions. PGRs are usually applied as foliar sprays using water as the carrier solvent. PGR performance depends on the compound applied, but also, importantly, there is a need to provide the proper conditions for the compound to reach the target at a level that can cause the desired changes in plant physiology. Thus, adequate consideration must be given to environmental conditions before, during and after application (Stover and Greene, 2005), as well as plant condition and physiological stage, and application techniques, spray additives and equipment (Bukovac *et al.*, 2002; Bukovac, 2005). Under optimum environmental and plant cultural conditions, the performance of the most active and effectively formulated PGR is determined primarily by the quality of the spray application practice (Bukovac, 2005).

Although our discussion will be centred on the application of PGRs as sprays, other application methods have occasionally been used to deliver PGRs in commercial fruit crop production, such as painting of naphthaleneacetic acid (NAA) for local treatment of pruning cuts, trunk injection of growth retardants (e.g. daminozide, dikegulac and maleic hydrazide) to control shoot growth in fruit crops (Wilkins, 1982), and soil applications of paclobutrazol (Cultar) to inhibit shoot elongation and enhance bud production in blueberries (Ehlenfeldt, 1998).

In this chapter, we will examine the most important considerations regarding the spray application of PGRs, as well as reviewing the factors that affect their performance. The most important current and potential uses of PGRs in blueberry production will be presented, considering not only their desired response but also the most relevant side or collateral effects on plant functioning and fruit quantity/quality. Most emphasis will be on the research that has been done on highbush blueberry (both SHBs and NHBs), but where appropriate, research on related species (lowbush and rabbiteye blueberry, and cranberry) will be presented.

APPLICATION OF PGRS

Efficient and uniform delivery of the desired dose to the intended target is central to maximize the performance of systemic compounds (Bukovac *et al.*, 2002). Spray application of foliarly applied PGRs is dependent on a wide range of interacting variables or events. Among these, the most important are: (i) the effectiveness of the application equipment; (ii) chemical and physical characteristics and formulation of the active ingredient (AI); (iii) atomization of the spray solution/emulsion/suspension; (iv) delivery of the spray uniformly over the intended target; (v) interaction between spray droplets and the plant surface, which leads to retention, droplet drying and residue formation;

(vi) penetration and transport of the AI to the active site; (vii) environmental factors, mainly temperature and relative humidity, before, during and immediately after spray application; and (viii) the plant condition, particularly foliage, with regard to stress, disease or inadequate nutrition. All of these stages and events are interrelated, in the sense that a change in one will usually have a profound effect on another, and may be affected by application variables, plant factors or environmental conditions (Bukovac, 2005). Although this discussion will be focused mainly on the application of PGRs, most of the principles should generally be appropriate for the application of pesticides and nutrients to the canopy of blueberries.

Formulation

The main objective in formulating a systemic ingredient is to structure the compound in a manner that can be readily applied by spraying systems in an aqueous carrier to obtain maximum biological activity. An additional objective in formulating systemic ingredients is to improve plant wetting, coverage and penetration (Bukovac *et al.*, 2002). Most PGRs are commercially available in one formulation. Water is the most commonly used carrier, as it is inexpensive, readily available and an excellent solvent for most PGRs in a wide range of conditions. However, aqueous base solutions have high surface tension and, because of this, are generally ineffective in wetting and spreading compounds across waxy leaves and fruit surfaces such as those of blueberries.

Atomization

Atomization corresponds to the conversion of the spray liquid into a cloud of droplets. Orchard sprayers usually atomize the liquid using nozzles that force the spray liquid through an orifice using high-velocity air or high pressure after passing the spray liquid through a cylinder or a pre-formed plate or disc. Other nozzle systems (e.g. air inclusion, electrostatic, rotary sleeve and spinning disc) have been developed, but they have not been widely adopted by the industry (Bukovac, 2005).

The droplet size population produced by most orchard sprayers ranges from around 100 to 500 μm in diameter. In most droplet spectra, there are a large number of small droplets that contribute little to the total spray volume but significantly to the spray drift. In order to accommodate for variability in plant size and density, training systems, leaf development during the season and different cultivars, growers usually adjust the flow rate to deliver the desired spray volume per tree or per hectare. As flow rate is altered, the droplet size of the spray is changed. Such changes can influence spray penetration into the canopy, as well as retention and coverage.

Spray deposition

Uniform delivery of sprays is difficult to achieve because many factors (e.g. planting systems, plant volumes and densities, seasonal changes in canopy development and surface characteristics) affect the quality and performance of the spray. Spray deposition is a complex process that can be viewed as consisting of the following stages: (i) delivery or transport of the atomized spray; (ii) impaction of spray droplets on the plant surfaces; and (iii) retention by the plant organs.

The most common orchard sprayers are airblast or axial fan sprayers. In most cases, the same sprayer is used for different crops and for applying a range of different chemicals to blueberries throughout the season. In Michigan, blueberry growers commonly use airblast sprayers that propel spray up and through the bushes. Cannon sprayers that project spray across the tops of several blueberry rows are also popular because their field capacity (i.e. ability to treat large areas quickly) is high, and, because fewer passes along the field are needed, the potential for mechanically damaging the developing fruit is reduced (Hanson *et al.*, 2000). However, cannon sprayers may result in irregular deposition and/or off-target drift during windy conditions (VanEe *et al.*, 2000). In a study to evaluate the effect of sprayer type (conventional airblast sprayer versus multifan/nozzle above-row sprayer) on control of fruit rots in mature NHB 'Jersey' blueberries, the nozzle pressures were set at 1.0 and 1.2 kPa for the airblast and above-row sprayers, respectively. The above-row sprayer provided fruit rot control at least equivalent to the airblast sprayer, even though less chemical was applied (Hanson *et al.*, 2000). In most agricultural spray application of chemicals, the largest proportion of the drift is detected at 0–5 m from the source (sprayer) and varies between 8 and 18% of the total volume (Donkersley and Nuyttens, 2011).

Air volume and velocity have a pivotal role in the delivery of the spray cloud. Greater canopy penetration and uniformity of deposition were obtained with airblast sprayers that delivered high-volume and low-velocity air. Unfortunately, both the air and the spray are delivered mainly from a point source and this is the main factor leading to non-uniform spray distribution over the plant (Hanson *et al.*, 2000). In addition, growers usually have limited time to cover their fields and, if high volumes are used, the time employed in covering the whole field is increased, and thus part of the spray cannot be done within the optimum window for maximum effect of the applied compound. When spraying has to be done near harvest, fruit damage and drop by physical contact with the sprayer also need to be taken into consideration.

Research on spray deposition in blueberries is scarce. VanEe *et al.* (2000) divided the canopy of 40-year-old NHB 'Jersey' blueberries into four sections: top, interior, side close to sprayer and side away from sprayer. A black dye and collecting targets were used to measure spray deposition patterns. Their findings confirmed results reported previously for fruit trees (Bukovac, 2005) in that areas of the bushes closer to the spraying equipment received greater coverage than those that were more distant. The effect was more pronounced as the season progressed and more foliage was present on the plant (Fig. 7.1).

A marked decrease in the uniformity of spray coverage occurred during the season near bloom. This change in spray uniformity was correlated with leaf development, as measured in the abovementioned experiment by the proportion of sunlight available in different sections of the canopy throughout the season (Fig. 7.2) (VanEe *et al.*, 2000).

When different pruning severities were imposed (removal of 0, 20 or 40% of the largest canes at the base of the bush), more severe winter pruning

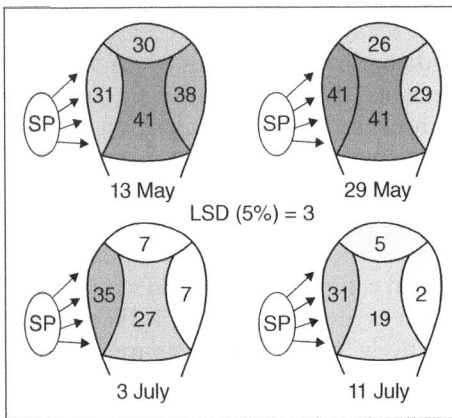

Fig. 7.1. Spray coverage (percentage of surface area of card targets) at different positions in mature NHB 'Jersey' blueberry canopies following application with an airblast sprayer (SP) on four dates between pink bud (13 May) and green fruit (11 July). Data are means across three pruning treatments. The least significant difference (LSD) value refers to comparisons between dates and positions. (Adapted from VanEe *et al.*, 2000.)

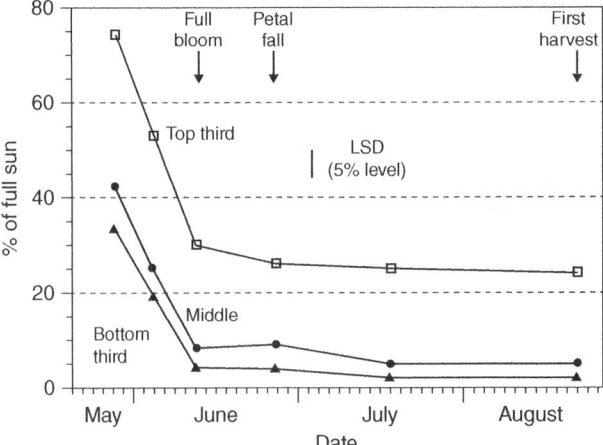

Fig. 7.2. Changes in light levels (percentage of full sun) in the top third, middle and bottom third of mature NHB 'Jersey' blueberry canopies between the first open flowers and the first fruit harvested. Data are means across three pruning treatments. (Adapted from VanEe *et al.*, 2000.)

increased deposition in areas of the canopy that received the least overall coverage. However, pruning tended to have less effect in sections where overall coverage was high (e.g. bottom and near side of the bush) (Fig. 7.3). As growers commonly apply compounds to alternate rows in order to cover a greater area in less time, the impact of this practice was analysed in this study (VanEe *et al.*, 2000). As expected, it was found that the section of the row farthest from the application point of the sprayer had significantly lower amounts of residues. This amounted to about one-fifth of the equivalent section in areas of the field that were sprayed in every row (Fig. 7.4).

PGRs are commonly slightly mobile in plant tissues. The lack of uniformity in spray delivery to fruit crops acquires paramount importance in the case of less mobile compounds such as pesticides and nutrients. This non-uniform distribution raises questions about the merits of sprayer calibration where the focus is placed only on the amount of spray solution delivered through the

Airblast sprayer, LSD (5%) = 3

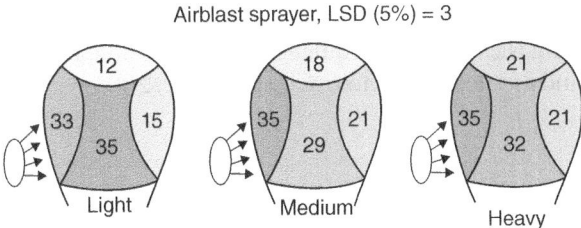

Fig. 7.3. Effect of pruning severity (light, medium or heavy) on spray coverage (percentage of surface area of card targets) at different positions of mature NHB 'Jersey' blueberry canopies following application with an airblast sprayer. Data are means of four treatment dates between pink bud (13 May) and green fruit (11 July). The LSD value refers to comparisons across pruning treatments and position. (Adapted from VanEe *et al.*, 2000.)

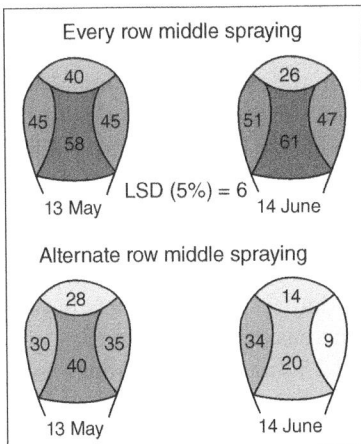

Fig. 7.4. Spray coverage (percentage of surface area of card targets) in different positions of mature NHB 'Jersey' blueberry canopies following application with an airblast sprayer driven down the middle of each row or the middle of alternate rows on two dates. Data are means across three pruning treatments. The LSD value refers to comparisons across spray treatments, date and position. (Adapted from VanEe *et al.*, 2000.)

nozzles under some prescribed sprayer settings and not the uniformity of the spray received by plant organs. This situation may provide a false sense of security, as some sections of the plant may actually be overdosed while others may be undersprayed. Taking a more general view, when the compound has been applied at the proper timing and dose and has not produced the desired effects, non-uniform application might be one of the explanations to be considered. Studies in a range of crops have found that around 60% of tested sprayers had calibration errors greater than 10% (Ayers and Bosley, 2005).

The current practice of recommending a constant dose per unit of ground area ignores the quantity and quality of the target and the biological requirement that PGRs must be absorbed by plant tissues before a response can be induced. Growers recognize that for most PGRs an optimum response is obtained with dilute sprays. However, for a number of practical and economic considerations, dilute spraying is being replaced with lower volumes than required for full plant retention. With high-volume sprayers, retention volume is not critical in determining the performance of the AI, as any additional volume applied will run off. As carrier volume is progressively reduced, spray volume eventually becomes limiting because of inadequate coverage and uniformity, which affects penetration and may lead to phytotoxicity (Bukovac, 2005).

Impaction

The impaction of spray droplets on plant tissues results in either reflection or retention by the surfaces. Droplets that are initially reflected, if not lost, to the surrounding environment may impact again one or more times and eventually be retained. The energy of the movement (kinetic energy) of the impacting droplet will cause it to spread; its surface area will increase, as will the wetting of the plant surface. Part of the kinetic energy is stored in the new liquid–air surface formed. This new surface area tends to return to its lowest state of energy and the droplet retracts. If the kinetic energy exceeds a critical value, the deformed droplet will retract completely, become extended perpendicular to the target surface and will then be reflected from the plant surface. Impaction is an extremely rapid process. The total droplet:surface residence time for a reflecting droplet may be less than 1/1000 of a second. Studies done on different fruit tree crops have found that droplet reflection from leaves is not a significant limitation in spray application to fruit trees (Looney, 1993; Greene, 2002). No data are currently available for blueberries.

The potential surface for impaction of droplets in a fully leafed orchard is larger than in a non-leafed one; hence, in the latter system, droplets have a greater potential to travel through the crop, increasing drift (Donkersley and Nuyttens, 2011). Praat *et al.* (2000) measured 25 times less drift from a fully foliated canopy compared with a dormant one. It is known that downwind ground deposition in orchard cropping systems tends to increase with greater plant height, with decreasing foliage density and with more open space between plants, and varies with different sprayers (Donkersley and Nuyttens, 2011).

Retention

Spray retention is a defining event in foliar application of compounds, as only the portion that is retained will be available for penetration and subsequent biological effect. In the case of PGRs, retention is perhaps the most important event in the PGR spraying process, for it determines the dose available for penetration. The total spray retained from a high-volume application is determined by the total target area wetted and the retention factor (volume per unit area) characteristic of the spray/fruit crop. In low-volume spraying, the volume retained for a given species is linearly related to the volume applied. In both cases, total spray retained per plant increases with the rapid increase in leaf area during the cropping season (see Fig. 7.2). For sprays applied at low volumes, where the spray is retained and remains as discrete droplets on the plant surfaces, reductions in surface tension did not change total retention but significantly increased the affinity of droplets for the surface. Accumulated residues from previous sprays and the presence of some natural product on plant surfaces modify spreading and retention (Bukovac, 2005).

Estimates of the full retention value for a given crop and planting system vary widely. Numerous factors and their interactions make it difficult to estimate the retention accurately. Among these are: (i) composition of the spray solution (e.g. spray additives), which may strongly alter plant surfaces and retention; (ii) spray equipment design and calibration; and (iii) changes in the quantity and quality of the surface area during the season.

Surfactants dramatically alter the form of the deposit. All surfactants lower spray surface tension and improve wetting. Thus, when used, they reduce reflection, increase spreading and, at high-volume spraying, reduce the total dose retained per unit surface area. Another important use of surfactants is to enhance AI penetration. They can do this by altering the solubility of the AI in the spray solution and hence its sorption into the cuticle, and by increasing the droplet–plant surface contact. As surfactants are plasticizers, they soften the crystalline waxes in the cuticle and thus increase the mobility of the agrochemicals across the cuticular membrane (Schönherr *et al.*, 2000). This makes surfactants essential not only for maintenance of long-term physical stability but also for enhancing the biological performance of foliarly applied growth regulator.

Spray formulations typically contain 1–10% of one or more surfactants (Castro *et al.*, 2014). The surfactants used most commonly for PGR application to blueberries are non-ionic and include: X-77, Silwet L-77, Kinetic and Flood (Krewer *et al.*, 2007). Taking into account the complexity of the foliar uptake process, an ideal adjuvant should produce the desired droplet contact area on the leaf surface, keep the AI in a soluble form, and increase the permeability of the cuticle and the plasma membrane. A single adjuvant, whether a surfactant, an ammonium salt or an oil, is unlikely to have all of these properties. Therefore, research on adjuvant mixtures should be further promoted (Wang and Liu, 2007).

Droplet drying and deposit formation

Any non-volatile compound and spray additive present in the spray mixture precipitates out as a residue of varying size, form and consistency as the carrier phase (water) of the droplet evaporates. Rapid drying, particularly of small droplets in low-volume applications, may reduce spreading and penetration. However, it appears that the higher AI concentration in the spray solution compensates by providing a greater driving force for penetration. The AI may be distributed uniformly over the original droplet–plant surface interface, or as masses of various sizes over veins, stomatal cells, damaged areas or specialized structures. Spray additives that are hygroscopic increase the hydration of the deposit, and the AI can diffuse more readily to the AI-depleted deposit–surface interface and replace the AI diffusing into the plant. This can maintain a positive driving force for penetration after the deposit has apparently dried (Bukovac, 2005).

Environmental factors that affect the performance of PGRs

There are various factors that can dramatically alter the degree of plant response to PGRs. Knowledge of these factors is necessary to adjust the spray application so that an adequate amount of the chemical reaches the active sites in the plant cells in order to obtain the desired degree of response. Changes in the environmental conditions in the field result in simultaneous modification of multiple parameters, such as temperature, light and humidity. This situation often makes it difficult to translate results from controlled experiments into protocols that can ensure a given response in the field and provide growers with tools to make adequate decisions in the appropriate time. For ease of discussion, the factors can be separated chronologically as: before application, during application and after application. Most of the research on these matters has been carried out on fruit trees, particularly apples; however, when data are available, information on blueberries or related species will be provided.

Conditions before application
Fruit tree thinners, particularly NAA, have been the most extensively studied PGRs. There are a number of environmental factors previous to the application

of fruit thinners that can alter plant response. Among these are some factors that increase the response, such as: low light intensity, high humidity, frost damage and low temperatures during leaf growth. Conversely, high temperatures and dry conditions prior to thinner application reduce the effectiveness of thinners. Part of the effect is due to the influence of environmental conditions on leaf growth, cuticle development and deposition of epicuticular waxes. In European wild blueberries (*V. myrtillus*, *V. uliginosum* and *V. vitis-idaea*), it was found that cuticle thickness in the adaxial (upper) surface of the leaves was almost double that in the abaxial (lower) side (Semerdjieva *et al.*, 2003). The epicuticular structures of rabbiteye blueberry leaves were dense in new leaves but absent in mature leaves (Freeman *et al.*, 1979). When Konarska (2015) studied the microstructural changes of NHB 'Bluecrop' blueberries, she found that, during fruit development, the thickness of the cuticle increased and its structure changed. She also reported that the surface of NHB 'Bluecrop' fruit was devoid of deep microcracks and contained a small number of stomata that were located primarily near the fruit calyx. These two traits would both limit fruit transpiration and reduce the uptake of sprayed chemicals into the fruit.

Conditions during application
Many label recommendations for PGRs regarding environmental factors (rain, temperature and time of application), as well as the use of surfactants, are based on results from controlled laboratory studies, and are supported by anecdotal field observations. This gives a general trend, but the actual situation in the field may differ from these theoretical models.

The impact of environmental conditions during application has been studied using leaf discs. It was shown that penetration of NAA into pear leaf discs increased with temperature, with a marked increment above 25°C (Bukovac, 2005). The change was attributed to decreases in the cuticular viscosity. No report on these matters has been found for blueberries.

Conditions after application
The effect of environmental conditions after application usually includes a combination of factors. The effect of environment on the plant response to PGRs may result from uptake effects, the influence of PGR conversion to the active form and/or physiological processes through which the PGR action is mediated.

Applications of growth regulators are often made under adverse conditions, with growers debating whether to spray when rainfall is imminent. Field observations suggest that if a spray droplet dries before rain occurs, there is likely to be a substantial amount of PGR activity retained, which in part is due to enhanced uptake as the droplet dries. Wash-off studies typically reveal that PGR activity is reduced if residues are washed from the leaves too soon after application. One such study was done on rabbiteye 'Tifblue' blueberries (NeSmith and Krewer, 1997a). Data on the occurrence of simulated rainfall after 1–72 h of gibberellic acid (GA$_3$) application to 1-year-old plants in the greenhouse showed that there was a rapid GA$_3$ uptake (as indicated by the effect of GA$_3$ on fruit set) 1–4 h after application, with a more gradual uptake over the next 68 h (Fig. 7.5). Thus, if rain occurs after 4 h following the application of GA$_3$ to rabbiteye blueberries, it can be assumed that uptake was mostly completed and there would be no need to reapply the compound.

Most PGRs do not require chemical conversion to be effective because the compound is already in the active form. The most obvious exception is ethephon (a compound assayed for bloom delay in SHB blueberries), which has to decompose and release ethylene to become physiologically active. The plant response to ethephon is highly temperature dependent, as temperature influences both uptake and the rate of ethephon decomposition to release ethylene. It has been observed that the thinning response to ethephon in apples increases linearly with temperature from 8 to 24°C with virtually no thinning at temperatures of 8°C or below (Stover and Greene, 2005).

CURRENT AND POTENTIAL USES OF PGRS IN BLUEBERRY PRODUCTION

In the following sections, a review of the main current and potential uses of PGRs in blueberry production is provided. Where known, both positive and negative collateral effects of their application are provided.

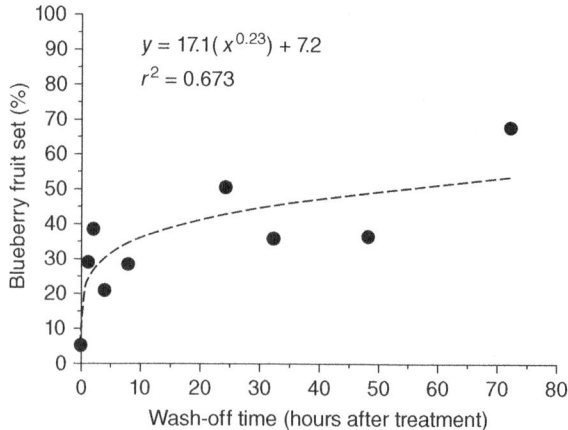

Fig. 7.5. Fruit set of rabbiteye 'Tifblue' blueberries in response to wash-off time following application of GA$_3$. Spray was applied to the point of drip at a concentration of 251 mg GA$_3$/l. (Adapted from NeSmith and Krewer, 1997a.)

Vegetative growth

Mature blueberry plants, especially rabbiteyes, can reach heights in excess of 3 m, making management difficult, particularly application of agrichemicals and mechanical or hand harvesting. Hence, practices that reduce shoot growth but do not adversely affect fruit yield and quality would be helpful for the management of these species.

Paclobutrazol (commercially known as PP333 or Cultar) is a very potent triazole compound and an inhibitor of an early step in the gibberellin (GA) biosynthetic pathway (Rademacher, 2000). Paclobutrazol has a relatively long half-life and limited phloem mobility, and its inhibitory effects can be overcome with exogenous applications of GAs. The compound reduces vegetative growth in several tree fruit and nut species (e.g. apple, pear, grape, walnuts), and is regularly used by avocado growers in Chile (Jorquera *et al.*, 2006). The compound is readily translocated through the xylem, which makes possible to apply it either as a soil drench, a trunk injection or a spray. Among the secondary effects of paclobutrazol application to fruit trees are: increased flower bud initiation, increased fruit colour, reduced fruit acidity in cherry and pear but not in apple, and lower fruit firmness in cherry but no change in apple and greater firmness in pear (Curry and Williams, 1986).

Initial trials were done in the 1980s on rabbiteye blueberries. This species is characterized by its vigorous growth and tall size, which in commercial plantings can commonly exceed 3 m. A single application of paclobutrazol (1, 2, 4 or 8 g AI per plant) was done in 0.5 l water in a circle 15 cm from the bush around the time of bud break (Mainland, 1989). In another trial, soil applications of paclobutrazol (3, 6 or 12 g AI per plant) were carried out at the end of harvest (Spiers, 1988). In the season of application, there were no detectable effects of the treatments. In the following season, both floral and vegetative bud break were delayed from 10 to 15 days by application rates greater than 2 g per bush. At 4 and 8 g per bush, the plant growth pattern was altered, and many buds remained inactive or began growth at unpredictable times during the season. The cane response on individual bushes was markedly irregular. Application rates of 2 g per bush or higher reduced yield, while berry size was reduced at 8 g per bush. In the second season after application, the effect of paclobutrazol was more uniform among canes within a plant, but again all rates that had an effect on vegetative growth caused deleterious effects on the reproductive components. At rates lower than 2 g per plant, there was no effect of paclobutrazol on leaf area or floral or vegetative bud development, and on the rates of photosynthesis or transpiration. Paclobutrazol had no influence on the leaf mineral content of N, P, Ca, Mg, Fe or Cu. Mn was increased by all levels of paclobutrazol, but only the highest level (12 g per plant) leaf levels of K and Zn were increased.

Experiments in NHB blueberries have been focused on the effect of paclobutrazol on hastening fruit production for more rapid evaluation in breeding programmes (Ehlenfeldt, 1998). Soil drenches of 25 ppm paclobutra-zol carried out after harvest inhibited shoot elongation and stimulated earlier and greater flower bud production of 3-year-old NHB 'Bluecrop', 'Bluetta' and 'Jersey' plants. Treatments increased bud numbers by 358–797%, while reducing vegetative bud formation (Table 7.1). This effect caused overcropping and reduced fruit size.

Foliar treatments of paclobutrazol at the end of the growing season on mature NHB 'Bluecrop', 'Blueray' and 'Duke' plants resulted in more moderate effects than a drench application (Table 7.2) (Ehlenfeldt, 1998). Flower bud numbers were increased without affecting vegetative growth. Although bud numbers increased with dose, the number of flowers per bud was not affected by the treatments. 'Bluecrop' and 'Blueray' responded similarly to paclobutra-zol applications, while 'Duke' showed no significant response to any treatment. Part of the lower response of 'Duke' plants might be due to their lower stature in this trial, as the researchers found that a better response was obtained when the compound was applied to plants growing vigorously and in which the inhibition of shoot elongation, to favour carbohydrate accumulation, would trigger flower bud induction.

In the case of soil applications, variable movement of the chemical within the soil profile, due to water movement, soil binding or location of the chemical with respect to the active roots, might explain the differences in response among canes within a bush.

In summary, paclobutrazol may have the potential to be used in commercial blueberry settings for size reduction and increased yields, but insufficient data have been generated to rigorously test this possibility. Experimental results have been variable and much longer-term studies are needed.

Table 7.1. Floral bud production in NHB blueberry cultivars in response to a soil drench of paclobutrazol at 25 ppm. (Adapted from Ehlenfeldt, 1998.)

| Cultivar | Paclobutrazol | Mean no. of floral buds per plant | | Total (% of control) |
		Total	Simple	
'Bluetta'	–	16.7	15.8	–
	+	65.1**	51.3**	390**
'Bluecrop'	–	19.0	18.5	–
	+	68.1**	54.5**	358**
'Jersey'	–	7.6	7.6	–
	+	60.6**	58.1**	797**

Mean values significantly different at **$P \leq 0.01$ (Student's *t*-test).

Table 7.2. Floral buds per plant and number of flowers per bud of 'Bluecrop', 'Blueray' and 'Duke' NHB blueberries treated with foliar applications of paclobutrazol. (Adapted from Ehlenfeldt, 1998.)

Cultivar	Treatment (mg/l)	Floral buds per plant	Flowers per bud
'Bluecrop'	0	11.5[a]	8.9[a]
	5	17.2[ab]	8.8[a]
	10	17.8[ab]	8.7[a]
	50	20.1[b]	9.8[a]
	100	33.0[c]	8.5[a]
	200	31.4[c]	8.9[a]
'Blueray'	0	7.2[a]	8.3[a]
	5	13.1[ab]	7.1[a]
	10	11.8[ab]	7.8[a]
	50	15.8[b]	7.2[a]
	100	30.6[d]	7.0[a]
	200	23.2[c]	8.6[a]
'Duke'	0	3.5[a]	5.8[a]
	10	4.8[a]	5.9[a]
	50	9.5[a]	5.6[a]
	100	7.5[a]	5.4[a]

[a,b,c,d]Means followed by the same letter within columns in a cultivar are not significantly different at $P < 0.01$ (Fisher's protected LSD test).

Enhancement of leaf development

Deciduous fruit crops, including blueberries, when planted in temperate climates, undergo a physiological stage of development every year known as dormancy (Richardson *et al.*, 1974). The release from dormancy and subsequent bud development (flower and vegetative) requires a period of exposure to low temperatures (below 7.2°C), followed by a subsequent rise in spring temperatures (Stringer *et al.*, 2004). In NHB blueberries, vegetative buds usually open about 2 weeks before floral buds; however, some rabbiteye and SHB blueberries have a tendency to open leaf buds after flower buds.

Under climatic conditions in certain blueberry-producing regions (i.e. south-eastern USA, northern Chile, Southern Spain, Peru, Mexico and Portugal), marginal winter chilling occurs. Depending on the cultivar, this can result in delayed canopy development, and may cause early growth cessation and inhibition of shoot, flower and leaf development. Carbohydrate supply at bloom is heavily dependent on the previous season's reserves, but if heavy fruit set occurs before leaf emergence, the carbohydrate supply may be restricted and the development of leaf buds will often be suppressed resulting in low leaf:fruit ratios. Under these conditions, accelerated spring leaf development prior to, or concomitant with, flower opening may increase fruit size and reduce the fruit development period of some poor-leafing cultivars, particularly under marginal chilling (Stringer *et al.*, 2002; Williamson *et al.*, 2002).

Utilization of chill compensation chemicals provides a management option to avoid the effects of insufficient chilling. These chemicals aid in breaking dormancy and promote earlier and greater leaf development (Richardson *et al.*, 1974). Hydrogen cyanamide (HC; Dormex®) has been studied and used extensively in several fruit crops (peaches, apples, grapes and raspberries) for these purposes. When applied at the proper rate and timing to blueberries, it can promote leaf development and avoid deleterious effects on floral buds (Williamson *et al.*, 2001).

Research done on SHB 'Misty' blueberries in the southern USA demonstrated that HC accelerated vegetative bud break if it was applied when 36–40 chilling hours (hours between 0 and 7.2°C) had been accumulated (Williamson *et al.*, 2002). In control plants (no HC), the percentage of vegetative buds that grew remained low throughout the season. HC-treated plants had a large and rapid increase in vegetative bud break (Table 7.3). The response to HC was linear, which implies that vegetative bud break increased proportionally with the increase in HC application rate.

As a secondary effect, HC advanced fruit ripening (Williamson *et al.*, 2002). Significantly more fruit were harvested from HC-treated plants before 1 May. This has important implications in the market price per kg of fruit, as fruit marketed before 1 May can have almost double the price of fruit marketed in late May. In addition, the nearly 0.3 g difference in fruit weight for HC-treated fruit harvested before 1 May also has practical and economical significance. The fruit number was significantly reduced when plants were treated with 1.5% HC (52% of non-treated control), which might have influenced fruit weight. However, HC treatment at 0.75% produced similar fruit size to the

Table 7.3. Effect of HC spray concentration on vegetative bud break of mature SHB 'Misty' blueberries following treatment on 17 December and 6 January (50% flowering on 20 February). (Adapted from Williamson *et al.*, 2002.)

HC concentration (% v/v)	Vegetative buds growing (% of total)		
	29 DBF (22 Jan)	14 DBF (6 Feb)	14 DAF (6 Mar)
0	0	0	2
0.75	33	40	54
1.5	71	74	71
Significance: linear	**	**	**

DAF, days after flowering; DBF, days before flowering.
Linear trend significant at **$P \leq 0.01$.

1.5% HC treatment, but fruit load was similar for 0.75% and the non-treated control. This would imply that most of the effect of HC on fruit size is probably not due to reduced fruit load but to earlier and greater expansion of leaf area (Table 7.4).

Swart (2015) carried out trials in South Africa on 3–4-year-old field-grown SHB 'Bluecrisp', 'Emerald' and 'Star' blueberries for two seasons. Experiments included the application of 1 or 2% HC every 2 weeks in areas where chilling units were between –154 to 197 (year 1) and –65 to 285 (year 2). He concluded that HC was not recommended for use on 'Emerald' and that further studies were needed to identify the best HC application time for 'Star', as the results were inconclusive. For 'Bluecrisp', he found that using either 1 or 2% HC had less effect than the timing of application with respect to promotion of early ripening and increased yield in plants that had experienced insufficient chilling. A 1% HC application could induce similar results as a 2% spray but with a lower risk of flower bud damage. He also discovered that the timing of HC application could be based on the visual appearance of buds and should be carried out before bud scales are open but after at least fair chilling is reached.

Jaldo *et al.* (2009) undertook trials with different doses of HC (0–2.5%) for two seasons (2006: 50 chilling hours; 2007: 520 chilling hours) on 1- and 2-year-old SHBs ('Emerald', 'Jewel', 'Misty', 'O'Neal' and 'Star') in Tucumán, Argentina (27°12′S, 65°33′W). They found that the magnitude of the HC effect on yield and advanced maturity was cultivar dependent and that the lowest HC dose was effective in the season with the lowest number of chilling hours.

The chemical compound N-phenyl-N′-1,2,3-thidiazol-5-ylurea (TDZ; thidiazuron) has been shown to display cytokinin-like activity in apple (Steffens and Stutte, 1989), but initial trials in SHBs have not been promising. TDZ was tested for two seasons at two concentrations (4 or 6% v/v) applied at different phenological stages on 'Bluecrisp' and 'Star' and compared with an untreated

Table 7.4. Effect of HC spray concentration applied on 17 December 1997 on time of harvest and fruit yield (number and weight) of mature 'Misty' SHB blueberries following treatment (full bloom on 20 February). (Adapted from Williamson *et al.*, 2002.)

| | Harvest period | | | | | |
| | 1–30 April | | 1–31 May | | | |
HC rate (% v/v)	Total yield (%)	Mean berry weight (g)	Total yield (%)	Mean berry weight (g)	No. of fruit	Total yield (g)
0	30.8	1.61	69.2	1.31	3582	5050
0.75	52.3	1.93	47.7	1.51	3467	5860
1.5	71.5	1.96	28.5	1.13	1861	3370

control in a commercial field in South Africa (latitude 34°S) (Swart, 2015). The effect on berry ripening, berry size and yield were evaluated. The author concluded that TDZ application could not currently be recommended for these cultivars, as the TDZ concentrations used caused malformation and excessive injury to flowers, especially at later application dates (when reproductive bud scales were open).

In summary, if properly used, HC applied at 1.5% to some poor-leafing SHBs and rabbiteyes can stimulate more rapid leaf development in the spring, resulting in increased fruit size and more concentrated ripening for the first two harvests in blueberry cultivars with poor spring leaf development. In areas where no research on HC has been done, trialling would be advisable to adjust application rates and timing to various cultivars. TDZ should not be used commercially until trialled further on different cultivars using various rates and at a range of phenological stages.

Reproductive growth

Bloom delay
Spring freeze damage is a major problem for blueberries in different regions of the world. Despite efforts to breed for cultivars with a shorter span from bloom to harvest, earliness in the harvest season is usually associated with early blooming. Thus, growers focusing on producing fruit for the early market are more prone to suffer frequently from spring freezes. Ethephon (2-chloroethylphosphonic acid; Ethrel) is a growth regulator that, on contact with plant tissues, releases ethylene, the ripening hormone (Howell *et al.*, 1976). Ethephon was studied for several years and localities to determine if its application in the previous autumn might delay bloom of SHB blueberries in the following season (Krewer *et al.*, 2005; NeSmith, 2005). Ethephon significantly delayed bloom in SHB cultivars (Table 7.5). When measured at bloom (22 February), the selection FL 86-19 and cultivar 'O'Neal' were delayed 1.2 and 1.1 flower bud stages (corresponding to 8–11 days), respectively, by one application of 400 ppm ethephon in the previous autumn (early October). Moreover, ethephon doubled flower density in 'O'Neal' and increased it by over 40% in FL 86-19.

In the SHB 'Sharpblue', bloom delay from ethephon was estimated to be 0–6 days, depending on spray concentration and the stage of flower development in control plants (Table 7.6). At the beginning of bloom (10% anthesis for control), 400 ppm resulted in nearly 6 days' delay in bloom development. By late bloom (90% anthesis for control), the highest concentration (400 ppm ethephon) was nearly 3 days behind the control plants in flower development. The estimated delay in bloom was greater as ethephon concentrations increased, regardless of the level of bloom in the controls. Ethephon-treated plants continued to show delayed blooming with respect to controls in the late

Table 7.5. Effect a single application of ethephon in the autumn (400 ppm in early October) on SHB blueberry (FL 86-19 and 'O'Neal') bloom development and flower bud density in the following growing season (22 February). (Adapted from Krewer *et al.*, 2005.)

Selection/cultivar	Treatment	Flower bud stage[a]	Flower bud density (no. of buds/cm)
FL 86-19	Control	5.3	0.39
	Ethephon	4.1**	0.55**
'O'Neal'	Control	3.5	0.29
	Ethephon	2.6**	0.61**

Value was significantly different from that of the control at **$P \leq 0.01$.
[a]Stage 2, visible swelling, scales separating, flowers still completely enclosed; stage 3, bud scales separated, flower apices visible; stage 4, individual flowers distinguishable, bud scales abscised; stage 5, individual flowers distinctly separated, corollas unexpanded and closed (Spiers, 1978).

Table 7.6. Estimated bloom delay (days) of SHB 'Sharpblue' blueberry plants relative to the control at different stages of bloom development following previous season autumn application of different rates of ethephon. (Adapted from Krewer *et al.*, 2005.)

Open flowers (% of control)	Ethephon concentration (ppm)		
	100	200	400
10	2.5	4.8	5.8
20	2.0	4.3	5.4
30	1.7	3.9	5.0
40	1.5	3.6	4.8
50	1.2	3.3	4.5
60	0.9	3.0	4.2
70	0.7	2.7	3.9
80	0.4	2.4	3.6
90	0.0	1.8	3.1

stages of flower development (more than 60% of flowers open in all treatments). Similar results were obtained with ethephon sprays on rabbiteye 'Climax' blueberries (NeSmith, 2005).

A 2–4-day delay in ripening was observed for treatments that had a 7–10 day delay in flowering (NeSmith, 2005). In some trials, cumulative harvest dates were delayed by 5–9 days for plants treated with the highest ethephon concentration (400 ppm) when compared with control plants (Table 7.7). No harvest delay, relative to controls, was detected for the 100 and 200 ppm concentrations. A negative association between total yield and ethephon

Table 7.7. Effect of autumn ethephon applications to SHB 'Sharpblue' blueberry plants on the cumulative percentage of total yield by harvest date in the year after application. (Adapted from Krewer et al., 2005.)

Ethephon rate (ppm)	Cumulative yield (% of total)				Total yield to 3 June (g)
	26 April	3 May	12 May	19 May	
0	10.6	35.6	65.9	88.0	396
100	7.8	30.5	55.2[a]	86.3	468
200	7.8	30.4	52.2[a]	85.2	305
400	4.6[a]	12.2[a]	31.2[a]	73.3[a]	204[a]
Significance of trend					
Linear	**	*	***	***	***
Quadratic	NS	NS	NS	NS	NS

Trend significant at $*P \leq 0.05$, $**P \leq 0.01$ or $***P \leq 0.001$; NS, not significant.
[a]Value was significantly different from that of the control at $P \leq 0.05$ (Dunnett's test).

concentration was found in one of the two seasons (Table 7.7) (Krewer et al., 2005). The authors attributed yield reduction to poor bloom overlap between treated 'Sharpblue' and untreated 'Misty' pollinizers, as well as poor pollination weather during the period when plants with the treatments were blooming heavily.

Current recommendations for usage of ethephon in South Georgia are to apply two sprays of 400 ppm, with the first application in mid-October (approximately 19 weeks after the end of harvest) and the second in early November (NeSmith, 2005). The studies in Georgia indicate that ethephon can effectively delay bloom (by 7–10 days) in different blueberry cultivars with no apparent phytotoxicity. Depending on temperatures during fruit development, this may translate in a 2–4-day delay of the fruit-ripening date. This smaller impact on harvest date is probably because heat unit accumulation that controls fruit development is slow early in the season. For growers who get a premium price for early fruit, a few days of delay in harvest can have a marked impact on profits (NeSmith, 2005). Individual growers would need to assess the situation in their particular condition (cultivar, weather, market) and balance the potential impact of spring frost versus possible delay in harvest due to ethephon sprays. However, this decision is difficult as the ethephon must be sprayed before the risk of frost is known.

The application of ethephon could have some added benefits. Ethephon application has increased flower bud density both in NHB (Robbins and Doughty, 1984) and rabbiteye (Krewer et al., 2005) blueberries. This increase in flower bud numbers may have the potential to increase yields if the percentage fruit set is not altered. To rigorously test this possibility, it would be important to determine whether other cultivars respond similarly. Ethephon applications may also be useful to synchronize the bloom stages of blueberries for improved pollination and fruit set (Krewer *et al.*, 2005; NeSmith, 2005).

Inhibition of flower bud formation or elimination of flowers in nursery plants

In some conditions, it is desirable to completely eliminate reproductive buds of blueberry plants. It has been demonstrated that young NHB plants often do not establish well if they bear fruit in the first 2–3 years after being planted (Strik and Buller, 2005). In addition, pollen-transmitted viruses (blueberry leaf mottle virus and blueberry shock ilarvirus; Sandoval *et al.*, 1995) can infect these young blueberry plants and permanently decrease their fruiting potential and vigour. Growers can reduce fruit load manually through pruning and fruit thinning. However, this is not usually cost-effective, does not totally prevent viral infection and has less impact on the vegetative/reproductive balance. GA_3 has been shown to effectively inhibit flower bud formation in various fruit species (apple, pear, peach, grapes, sour and sweet cherry; Retamales *et al.*, 2000).

Initial studies showed that GA_3 could reduce flower bud formation in blueberry. When 5–500 ppm GA_3 was sprayed at bloom to increase fruit set on rooted cuttings of NHB 'Coville' blueberries, the return bloom was decreased (by 60–96% of non-sprayed controls) with increasing GA_3 levels (Mainland and Eck, 1969a). However, when similar GA_3 levels were sprayed at bloom on to 5-year-old 'Coville' field plants, there was no effect of GA_3 on return bloom (Mainland and Eck, 1969b). When 2-year-old NHB 'Bluecrop' and 'Elliott' nursery plants grown in containers were treated with 0, 150 or 300 ppm GA_3 at 7, 9, 11 or 13 weeks after full bloom (WAFB), it was found that 300 ppm GA_3 reduced flower bud initiation by as much as 60%. A greater effect was found with the latest application (13 WAFB) and higher GA_3 dose (300 ppm). GA_3 reduced the total number of flowers per bush rather than the number of flowers per bud (Retamales *et al.*, 2000).

Black and Ehlenfeldt (2007) tested different types of GAs (GA_3, GA_4, GA_7 and GA_{4+7}) on flower bud suppression in 1-year-old rooted cuttings of NHB 'Bluecrop' and 'Duke' blueberries. The concentrations ranged from 50 to 600 ppm and were applied during the summer (early July to mid-September in the northern hemisphere). If it is assumed that full bloom would occur during the first week of May, these dates would correspond to 11 WAFB (for early July) and 23 WAFB (for mid- September). When GA_{4+7} was trialled in 'Bluecrop', the greatest degree of flower bud suppression (90%) resulted from applications of 400 ppm repeated weekly from 11 to 21 WAFB. However, these treatments also reduced total vegetative bud number (by 40%) and plant height. There was a dose effect, with the greatest effect from 400 or 600 ppm (average 72 or

73% flower bud suppression for the various timings) versus 42% suppression for 200 ppm GA_{4+7}.

When the effect of timing of GA_{4+7} sprays was evaluated (two sprays per week) the period that corresponded to 13 WAFB was the most effective (71% flower bud suppression) compared with 53–67% in the other timings. This corresponded with the period of greatest effect of GA_3 in experiments reported by Retamales *et al.* (2000). The latest GA_{4+7} applications (early to mid-September in the northern hemisphere, corresponding to 21–22 WAFB) were more effective on 'Duke' than on 'Bluecrop', indicating that flower bud suppression treatments were more effective on the more precocious cultivar, or that a greater proportion of flower buds were being induced at that moment in 'Duke' compared with 'Bluecrop'. When GA_3, GA_4, GA_7 and GA_{4+7} were compared in 'Bluecrop' at similar concentrations (200 ppm) and timing (18–25 August to 1 September), GA_3 appeared to be somewhat more effective than the other GAs; however, compared with the control plants, the overall effect of treatments was low (versus control), which might be due to the low concentration used (Black and Ehlenfeldt, 2007).

Three-year-old NHB 'Aurora', 'Elliott', 'Draper' and 'Liberty' blueberry plants growing in commercial fields in south-west Michigan (latitude 42°54′N) were sprayed two to eight times with 0, 200 or 400 mg/l GA_3 or GA_{4+7}. GA sprays in July and August (around harvest) were more inhibitive to flowering than those in September and October (postharvest) across all cultivars. GA significantly reduced flower bud numbers in three separate studies, but the greatest reduction (49%) required eight applications from July to October. The authors concluded that, under field conditions, GAs would have limited commercial utility for preventing fruiting in NHB blueberries (Lindberg *et al.*, 2014).

In summary, these experiments with young plants indicated that GA_3 and GA_{4+7} have similar effects on suppressing flower bud induction, but in the areas where highbush blueberries are grown, the flower induction process would occur over a long period and, under these conditions, only repeated sprays at 400 ppm for nearly 3 months would ensure a significant suppression of flower bud formation; however, these repeated applications could reduce the total vegetative bud number and plant height.

An alternative option to eliminate flower buds in nursery plants may be the use of HC. Research done mainly with rabbiteye and SHB blueberries has shown that flower buds in the Spiers scale stage 3 (bud scales separated, apices of flowers visible; Spiers, 1978) or beyond will be killed by HC sprays. Trials on the SHB 'Misty' showed that 2% v/v of HC could kill 19–38% of flower buds if applied when 10–30% of flower buds were in stage 3. Bud mortality was related to chilling, with the highest bud mortality for plants with no chilling (Table 7.8). Thus, it appears to be difficult to kill all flower buds in young plants, because, on a given date, many buds are not at the most susceptible stage to be damaged by HC sprays.

Table 7.8. Effect of HC spray concentration (0, 1 or 2% v/v) and pre-treatment chilling levels (0, 150 or 300 h below 7.2°C) on the percentage of flower bud mortality of 2-year-old container-grown SHB 'Misty' blueberries. (Adapted from Williamson *et al.*, 2001.)

HC spray concentration (% v/v)	Pre-treatment chilling (hours <7.2°C)			Significance of trend (linear)
	0	150	300	
0	1.0	1.5	0.1	NS
1	20.0	11.0	2.3	***
2	38.0	26.0	19.0	**
Significance of trend (linear)	***	***	***	

Trend significant at **$P \leq 0.01$ or ***$P \leq 0.001$; NS, not significant.

BALANCING REPRODUCTIVE AND VEGETATIVE GROWTH IN MATURE PLANTS

Inhibition of flower bud formation

In certain growing conditions, avoidance of flower production or removal of flower and/or fruit might be needed to enlarge fruit size or balance fruit load once the plants are established. As fruit are competing with vegetative growth for the available resources (nutrients, carbohydrates), it would be convenient to remove the fruit at the earliest possible date (Looney, 1993). In this context, rather than waiting for the complete expression of reproductive growth at bloom time or thereafter, the idea is to intervene during the induction of reproductive buds, which occurs near the end of summer. As was shown from research done with GA$_3$ in nursery plants, the application of this compound only partially reduces the number of flower buds in blueberry plants (Retamales *et al.*, 2000). Although this is insufficient for nursery plants, it might be adequate for managing crop load in mature field plants. Experiments done on 4-year-old field NHB 'Bluecrop' plants showed that two GA$_3$ applications of 300 ppm at 15 and 18 WAFB markedly reduced the number of fruit (45% less than the non-sprayed control) in the following season and significantly increased fruit size (38% greater than the control) (Table 7.9) (Retamales *et al.*, 2000). Yields would be reduced by 30%, but the greater fruit size might prove advantageous for growers in areas with low frost risk at bloom and who are focused on hand picking for the fresh market, as both harvest efficiency and fruit price should be improved (Strik *et al.*, 2003). Before commercial utilization of this practice, these types of growers should develop trials in different cultivars to establish the best timing and application rates for their specific conditions.

Table 7.9. Effect of the number of applications of GA_3 (300 ppm) on the number of buds per shoot, and the number and weight of fruit at harvest in 4-year-old NHB 'Bluecrop' blueberries growing in Collipulli, Chile (latitude 38°S). Applications were done at 15 and 18 WAFB. The number of buds per shoot was measured on 16 October 1996; the number and weight of fruit at harvest was measured between 17 December 1996 and 16 January 1997. (Adapted from Retamales *et al.*, 2000.)

No. of GA_3 sprays	Buds per shoot	Fruit at harvest	
		Number	Weight per fruit (g)
0	16.6[a]	606[b]	1.3[a]
1	15.7[a]	–	–
2	17.7[a]	332[a]	1.8[b]

[a,b]Mean values within a column with non-identical superscript letters were significantly different at $P \le 0.01$ (LSD test).

Flower and fruit thinning

Larger fruit size is often an advantage for fruit destined for the fresh market, and the economic benefits of treatments that reliably improve average fruit size may be substantial. Part of the reduction of fruit load is usually done at pruning, although both vegetative and reproductive buds are removed with this practice (Strik *et al.*, 2003). The other classical approach to improve average fruit size at harvest has been to reduce the flower and/or fruit number early in the season. In most situations, the use of hand labour implies high costs or delays in implementing the reduction in flower/fruit load, so this is an area where growth regulators have been intensively investigated and used in many fruit crops. As commercially beneficial flower and fruit thinning usually reduces the total crop load (both the number of fruit and the total weight of the harvest), the improvement in fruit size at harvest is accounted for by reduced competition among the fruit for various resources (water, nutrients and carbohydrates). Depending on the intensity and opportunity of the operation, crop reduction might produce larger-sized fruit and a higher return bloom (Looney, 1993).

In the case of blueberries, there have been few trials on thinning agents. Perhaps this is partly due to the recent expansion of the industry where supplies usually were not able to satisfy the demand and the market put less emphasis on fruit quality. However, the situation has been changing rapidly in the last few years as small berries are being directed to the lower-priced processed market and growers are getting higher prices for larger berry sizes (fruit 18–21 mm in diameter is usually classified as large and 22–25 mm in diameter as extra-large).

The first thinning trials were carried out on rabbiteye blueberries, as they commonly produce small-sized fruit. Several compounds that have been used successfully to thin flowers and/or fruitlets in tree fruits (mainly apples) were assayed on 'Tifblue' blueberries in the greenhouse and in the field (Cartagena *et al.*, 1994). In the greenhouse, benzyladenine (BA) at 25 and 75 ppm resulted in 40 and 43% fruit set, GA_3 at 25 and 50 ppm produced 67 and 38% fruit set, NAA at 7.5 and 15 ppm had 55 and 53% fruit set and 1-naphthyl N-methylcarbamate (carbaryl) at 400 and 600 ppm had 40 and 31% fruit set, respectively.

These same compounds were then trialled for two seasons on field-grown 8-year-old 'Tifblue' plants in Mississippi (Cartagena *et al.*, 1994). Applications were carried out at 10 or 20 days after corolla drop, when the berries were on average 5.2 and 8.5 mm in diameter. In these studies, BA at 75 ppm reduced fruit set (39 and 77%, respectively), and a combination of carbaryl (400 ppm) and BA (25 ppm) reduced fruit set (55 and 87%) versus 76 and 97% fruit set for the control in 1991 and 1992, respectively. Overall, combinations of carbaryl and GA_3 reduced fruit set, but the response depended on GA_3 concentration and varied from year to year. GA_3, NAA and carbaryl alone also reduced fruit set, but the results were inconsistent (Table 7.10).

In 1991, greater thinning occurred when sprays were done 10 days after corolla drop (Cartagena *et al.*, 1994). Regarding fruit quality, BA at 25 ppm increased fruit diameter at the first harvest in 1991 (15.6 versus 14.1 mm for

Table 7.10. Fruit set and fruit diameter (first harvest) of mature rabbiteye 'Tifblue' blueberry plants as affected by chemical thinners sprayed 10 days after corolla drop in 1991 and 1992. (Adapted from Cartagena *et al.*, 1994.)

Chemical	Concentration (ppm)	Fruit set (%)		Fruit diameter (mm)	
		1991	1992	1991	1992
Control	0	76.0[d]	96.6[fg]	14.1[de]	16.3[bc]
BA	25	64.9[bcd]	86.6[bc]	15.6[ab]	16.2[cd]
	75	38.7[a]	77.1[ab]	14.4[cde]	15.9[d]
GA_3	25	74.2[d]	78.0[ab]	14.4[cde]	16.7[abcd]
	50	73.1[cd]	95.5[fg]	14.7[bcde]	16.4[abcd]
NAA	7.5	68.7[bcd]	97.2[fg]	14.8[bcde]	16.6[abcd]
	15	69.6[bcd]	90.8[de]	15.1[abcd]	16.4[abcd]
Carbaryl	400	68.8[bcd]	92.4[de]	15.3[abc]	17.1[a]
	600	66.1[bcd]	90.3[de]	15.9[a]	16.7[abc]
Carbaryl + BA	400 + 25	54.7[abc]	87.4[bc]	13.9[e]	16.6[abcd]
	400 + 75	63.9[bcd]	96.3[fg]	14.6[cde]	16.2[cd]
Carbaryl + GA_3	400 + 25	71.2[cd]	92.9[de]	14.5[cde]	16.3[bcd]
	400 + 50	51.7[b]	98.0[fg]	14.2[de]	17.0[ab]

[a,b,c,d,e,f,g]Mean values within a column with non-identical superscript letters were significantly different at $P \leq 0.05$ (*t*-test).

the control), and carbaryl at 400 ppm increased fruit size in 1991 (15.1–15.3 versus 14.1–14.9 mm for the control) as well as in 1992 (16.3 versus 17.1 mm for the control). The increase in fruit diameter was not always related to the degree of thinning, and varied depending on the year and application time (Table 7.10). Yield and return bloom were not influenced by any of the treatments. The results from this study indicated that BA alone or combined with carbaryl could potentially be used for fruit thinning, but the dose and timing need to be refined in further trials in other growing regions and with various plant materials.

Studies carried out in Slovenia on NHB blueberries (Koron and Stopar, 2006) tested the efficacy of ammonium thiosulfate (ATS) and Armothin (AI: alkoxylated fatty alkylamine polymer) at 1.5% v/v, BA at 200 ppm and N-(2-chloro-4-pyridyl)-N'-phenylurea (CPPU) at 10 or 20 ppm applied to mature NHB 'Rancocas' and 'Elliott' plants. ATS and Armothin were applied at full bloom, while BA and CPPU were sprayed 14 days after full bloom (DAFB). In most treatments, there were strong phytotoxic effects with injury to flowers, fruit, shoots and leaves, evidenced by leaf, flower and fruit drop, and reductions in plant growth rate and yield (total fruit number, weight and size). CPPU delayed fruit ripening by 3 weeks (10 ppm) or 4 weeks (20 ppm). Although the fruit number was significantly reduced in several treatments (Table 7.11), the fruit remaining on the plant could not benefit from this effect because the leaves and shoots were damaged and the ability of the plants to generate

Table 7.11. Effect of application of various thinning agents on fruit weight, fruit number and proportion of fruit as related to the control and total yield of NHB 'Rancocas' and 'Elliott' blueberries. (Adapted from Koron and Stopar, 2006.)

Cultivar	Treatment	Fruit weight (g)	Fruit number	Fruit number (% of control)	Total yield per bush (g)
Elliott	Control	0.96[c]	3815	100	3662[c]
	ATS: 1% v/v	0.90[b,c]	2588	67.8	2329[a,b]
	Armothin: 1.5% v/v	0.65[a]	2123	55.6	1380[a,b]
	BA: 200 ppm	0.77[a,b]	1705	44.7	1313[a,b]
	CPPU: 10 ppm	0.99[c]	3564	93.4	3528[c]
	CPPU: 20 ppm	1.02[c]	2943	77.1	3002[b,c]
Rancocas	Control	0.85[b,c,d]	2540	100	2159[c]
	ATS: 1% v/v	0.82[b,c]	1718	67.6	1409[b]
	Armothin: 1.5% v/v	0.55[a]	1622	63.9	892[a,b]
	BA: 200 ppm	0.74[b]	1624	63.9	1202[b]
	CPPU: 10 ppm	0.97[d]	1476	58.1	1432[b]
	CPPU: 20 ppm	0.91[c,d]	457	18.0	416[a]

[a,b,c,d]Mean values within a column and cultivar with non-identical superscript letters were significantly different at $P \leq 0.05$ (Duncan's multiple range test).

carbohydrates was significantly reduced, an effect that lasted for several weeks in some treatments.

A further trial by the same researchers in Slovenia on the NHBs 'Bluecrop' and 'Elliott' utilized lower concentrations of BA (20 or 50 ppm) and CPPU (2 or 5 ppm) applied at late bloom (Koron and Stopar, 2006). No effect on total yield was observed. CPPU at 5 ppm increased fruit diameter and height but not weight in 'Elliott'. At these levels, CPPU did not alter the ripening period in either cultivar.

Preliminary trials on the effect of soybean oil (Golden Natur'l (GN)) applied at bud break (10–11 February) at 0, 6, 9, 12 and 15% v/v were carried out on 3-year-old SHB 'Legacy' plants in Tennessee (Deyton *et al.*, 2005). Soybean oil has been assayed for thinning fruit trees and its application has been found to increase internal CO_2 levels in shoots and fruit, which decreases respiration rates, probably due to feedback inhibition (Myers *et al.*, 1996). Bloom opening was delayed by 2–6 days with sprays at a concentration above 9% v/v GN, with higher concentrations causing greater bloom delay. Concentrations of 0, 6 and 9% GN produced 0, 30 and over 70% flower bud mortality, respectively, at 36 days after treatment. Plants treated with 12 and 15% v/v GN had an estimated 24 and 13% of the normal crop load, respectively (compared with untreated control plants). A similar trial of GN used young plants of various SHB cultivars in North Carolina (Deyton *et al.*, 2005). Treatments of 0, 6, 9 and 12% v/v GN were sprayed on 5 March. These concentrations of GN did not affect flower bud mortality, crop load or berry size across several SHB cultivars. From these preliminary trials, it appears that concentrations between 6 and 9% may be adequate for flower bud delay and thinning. High bud mortality will, of course, reduce the yield, but determination of an oil dose–response (bud mortality) relationship may provide a means of chemically thinning blueberries in the future.

In conclusion, further trials need to be developed in a range of cultivars with different compounds and concentrations before recommendations can be established for the use of thinning agents in blueberries. The need to increase fruit size for the fresh market should activate research in this area in the near future.

Fruit set improvement

Fruit set is the first step in fruit development; it is established during and soon after fertilization. Seed-bearing plants have a unique double fertilization event with two pollen nuclei fertilizing the embryo and the endosperm (McAtee *et al.*, 2013). Adequate commercial yields in blueberry require at least 60% fruit set (Eck, 1988). Rabbiteye blueberries often exhibit low fruit set in low-chill regions (NeSmith, 2005). Less than 40% fruit set has been measured in 'Brightwell', 'Climax' and 'Tifblue' in Georgia (Davies and Buchanan, 1979; NeSmith and

Adair, 2004) as well as in 'Bluegem' in Florida (Davies and Buchanan, 1979). GA_3 has been studied extensively to increase fruit set in rabbiteye blueberries (Cano-Medrano and Darnell, 1998; Merino *et al.*, 2002; NeSmith, 2002 and 2005; NeSmith and Krewer, 1992, 1997b; NeSmith *et al.*, 1995, 1999).

The current recommendation for Georgia is to make two GA_3 applications of 150–200 ppm, the first at floral stage 6 of the Spiers (1978) scale of flower development (corollas completely expanded and open, when 40–50% of flowers would be open and about 10% of petals should have fallen), followed by a second application 14 days later (Krewer *et al.*, 2007). If application is made too early (flower stage 5), bee activity and pollination are reduced and the flowers do not open normally. This scheme should increase fruit set in most rabbiteye blueberry cultivars from 0–9% up to a range of 30–75% (NeSmith, 2005).

GA_3 has also been used to set rabbiteye blueberry fruit following a freeze. Freeze-damaged flowers may never open properly or be receptive to bee pollination, so the application of GA_3 can save a harvest. The impact of a freeze on fruit set will vary with stage of bloom, cultivar, wind and temperature conditions, and the duration of low temperatures (NeSmith *et al.*, 1999). A bloom temperature below $-3.3°C$ is likely to kill flowers at Spiers floral stages 5 and 6. Blossom temperatures in the range of 0 to $-3.3°C$ will cause partial damage to flowers at floral stages 5 (individual flowers distinctly separated, corollas unexpanded and closed) and 6 (corollas completely expanded). At this stage of development, it has been shown that fruit set can be significantly increased (2 versus 38%) if GA_3 is applied at 250 ppm immediately after the freeze and a second application is made 10–18 days later (Krewer *et al.*, 2007). Under these conditions, fruit size in GA_3-treated plants was 66% ('Brightwell') and 75% ('Tifblue') of open-pollinated controls. The reduction of fruit size caused by GA_3 sprays was partially due to increased fruit set (i.e. increased competition for carbohydrates) and reduced seed numbers within GA_3-treated fruit (NeSmith *et al.*, 1995). Some of the problems encountered with the use of GA_3 for these purposes are: (i) there is some variability in response among seasons and cultivars; (ii) the application of GA_3 can result in high numbers of pigmy fruit (less than 1 g), a problem that is especially marked when GA_3 is applied to rabbiteye 'Climax' and some SHB cultivars; and (iii) in some cases, stressed plants set too much fruit, which causes them to have poor vegetative growth and low return bloom (Krewer *et al.*, 2007).

Research on both lowbush and highbush blueberries has shown that GA_3 (50–500 ppm) can allow the development of seedless (parthenocarpic) berries (Barker and Collins, 1965; Mainland and Eck, 1969a), but no field studies have been published regarding the application of GA_3 to improve fruit set in highbush blueberry plants after a freeze. Fruit set levels similar to or slightly higher than those of hand-pollinated treatments (62%) were obtained with 500 ppm GA_3 or a combination of an auxin (50 ppm NAA) and 50 ppm GA_3 to greenhouse-grown 'Coville' NHB. Parthenocarpic fruit was on average around 60% of the size of pollinated (seeded) berries (Mainland and Eck, 1969b).

Although some SHB blueberries have shown low fruit set (19% in 'Millennia' and 22% in 'Bluecrisp'), other cultivars ('Bladen', 'O'Neal', 'Palmetto', 'Reveille', and 'Sharpblue') are reported to have fruit set over 50% (Lang and Danka, 1991; Williamson and NeSmith, 2007). Fruit set in NHB cultivars was found to be 56 and 66% for 'Bluecrop' and 'Patriot', respectively (MacKenzie, 1997). Thus, GA_3 applications to increase fruit set in highbush fields would only be needed when conditions such as low bee activity or spring freeze damage to flowers generate conditions for low fruit set; otherwise, the heavier fruit load in the season when GA_3 is applied would tend to produce small fruit and may reduce flowering in the following year.

Fruit size enlargement

As research in the 1990s showed that GA_3 increased fruit set in rabbiteye blueberries but had a deleterious impact on fruit size, there was a need for another compound that could enlarge fruit size with a neutral or small effect on fruit set. Previous trials in other fruit crops (kiwifruit, apples, table grapes, olives and persimmon) had demonstrated that the synthetic cytokinin CPPU could markedly enhance fruit size when applied near the time of bloom (Antognozzi *et al.*, 1993a,b; Greene, 1989; Reynolds *et al.*, 1992; Sugiyama and Yamaki, 1995). This compound has now been tested extensively on rabbiteye blueberries both in the greenhouse and in the field in different seasons, and with different cultivars and concentration/timing combinations (Merino *et al.*, 2002; NeSmith and Adair, 2004; NeSmith, 2005, 2008; Serri and Hepp, 2006; Williamson and NeSmith, 2007). In some trials, large increases in fruit set (up to three times) and size (up to 35%) were found. The optimum window of application of CPPU for rabbiteye blueberries was defined as 7–21 days after 50% bloom (stage 6), with the highest success being from an application made around 14 ± 3 days after 50% bloom (NeSmith, 2008). In rabbiteye cultivars ('Bluebelle', 'Brightwell', 'Climax', 'Powderblue', 'Premier' and 'Tifblue'), this application caused an average increase of 5–25% in berry size and a nearly 20% increase in fruit set (NeSmith and Adair, 2004; NeSmith, 2005, 2008). The effect on fruit set was more pronounced in poor fruit set situations, such as when there was little overlap in bloom date among cultivars and low bee activity (NeSmith, 2008).

In SHB blueberries, different combinations of CPPU (5, 10 or 15 ppm) applied 7, 10, 14 and/or 20 days after stage 5 (or after 50% bloom) have been tried in a number of cultivars: 'Bladen', 'Bluecrisp', 'Georgia Gem', 'Legacy', 'Magnolia', 'Millennia', 'O'Neal', 'Palmetto', 'Reveille', 'Santa Fe', 'Sharpblue' and 'Star' (Williamson and NeSmith, 2007). CPPU field applications increased fruit set in Georgia (15–100%) but not in Florida, although in Georgia the fruit set was reduced in 'Bladen' (30%) and no effect was found in some CPPU treatments applied to 'O'Neal' and 'Reveille'. Individual berry weight was generally increased (10–40%), although no effect of CPPU sprays on fruit size was obtained in Georgia for 'Palmetto' and 'Georgia Gem', or for some CPPU treatments in 'Millennia' and 'O'Neal' (Table 7.12; Williamson and NeSmith, 2007).

When fruit number was calculated based on yield per plant and berry weight, it was found that SHB cultivars reacted differently to CPPU application. In 'Star', fruit number was little affected, with the lowest fruit numbers reaching around 80% that of control plants for 10 ppm treatments, while in 'Sharpblue' and 'Santa Fe', the CPPU treatments reduced fruit count by 25–56% and 54–78%, respectively. The highest fruit drop was obtained with two applications (10 and 20 DAFB), with the most responsive dose being 5 ppm for 'Santa Fe' and 10 ppm for 'Sharpblue' (Williamson and NeSmith, 2007).

The delay in fruit ripening that was reported previously in rabbiteye blueberries (NeSmith and Adair, 2004) was also observed in SHBs. As SHBs are aimed at early markets, this delay in harvest could reduce profitability. The proportions of ripe berries before 10 May in the control treatment (no CPPU) in 'Reveille', 'Bladen', 'Georgia Gem' and 'Palmetto' were 1–3, 10–15, 35 and 72%, respectively, while in the plants treated with 10 ppm CPPU applied 10–14 days after 50% bloom, these proportions were less than 1, 1–5, 27 and 53%, respectively. The greatest delay in fruit harvest appeared to be associated with those treatments that increased mean berry weight the most. Field observations indicated that the delay was most noticeable when young emerging leaves were burned by spray applications (Williamson and NeSmith, 2007).

The inclusion of a surfactant (Silwet L-77, 0.5% v/v) did not seem to consistently affect the level of fruit set in response to CPPU applications in SHBs. In

Table 7.12. Effect of CPPU applied to mature plants on fruit yield and mean berry fresh weight of three SHB blueberry cultivars grown in Florida. Control plants were sprayed with water and surfactant (Triton B-1956 at 0.05% v/v) at 14 DAFB. (Adapted from Williamson and NeSmith, 2007.)

CPPU treatment	Yield (g per plant)			Berry weight (g per berry)		
	'Sharpblue'	'Star'	'Santa Fe'	'Sharpblue'	'Star'	'Santa Fe'
Control	2750[a]	3656[d]	2066[a]	1.21[b]	1.04[b]	1.18[c]
5 ppm, 14 DAFB	2479[ab]	5259[a]	2013[a]	1.46[a]	1.57[a]	1.48[ab]
5 ppm, 14 DAFB + 5 ppm, 20 DAFB	1873[ab]	4911[ab]	1282[a]	1.34[ab]	1.53[a]	1.36[b]
10 ppm, 14 DAFB	1895[ab]	4089[cd]	1848[a]	1.47[a]	1.46[a]	1.47[ab]
10 ppm, 14 DAF + 10 ppm, 20 DAFB	1520[b]	4443[bc]	2007[a]	1.51[a]	1.59[a]	1.50[a]

[a,b,c,d]Mean values within a column with non-identical superscript letters were significantly different at $P \leq 0.05$ (Duncan's multiple range test).

general, the addition of a surfactant had a greater negative than positive impact on fruit set, size and phytotoxicity (Williamson and NeSmith, 2007).

Serri and Hepp (2006) conducted trials in Chile on NHB 'Elliott' and 'Lateblue' blueberries. They applied 10 ppm CPPU at 10–15 days after 50% bloom and obtained a 20 and 50% increase in fruit weight in 'Elliott' and 'Lateblue', respectively. Within each cultivar, fruit numbers were similar in the CPPU and control treatments (no CPPU applied), suggesting that there were similar increases in yield for both cultivars in response to CPPU. The CPPU-treated plants had a 3- and 15-day delay in reaching 50% fruit harvested for 'Elliott' and 'Lateblue', respectively, perhaps due to higher fruit numbers. The authors did not describe any phytotoxicity derived from CPPU sprays.

Considering that in Chile several localities and cultivars have an extended bloom season, there was a need to study the effect of repeated CPPU sprays. Trials were therefore carried out over two seasons in south-central Chile (latitude 36°S) on the NHB 'Duke' in which the number of applications (one, two or three), dose (0, 5 or 10 ppm CPPU) and time of application (3, 10 and/or 17 DAFB) were trialled (Retamales *et al.*, 2014). The results showed that CPPU treatments did not affect individual fruit weight (1.14 g for the control versus 1.11–1.45 g for CPPU-treated plants) or fruit weight loss postharvest after 20 days at 4°C (Table 7.13). Only the application of 10 ppm CPPU both at 3 and 17 DAFB significantly increased fruit yield per plant with respect to control (32.5% greater). Fruit diameter was both positively and negatively affected by CPPU treatments compared with the control, with the highest positive impact over the control at an application of 10 ppm at 10 DAFB (7.6% greater than control) and 10 ppm CPPU applied 3 and 17 DAFB (4.5% greater than control). There were also positive and negative impacts on soluble solids, with the highest positive effect over the control with the use of 10 ppm CPPU applied at 17 DAFB (7.3% greater than control) and the application of 10 ppm CPPU sprayed 3, 10 and 17 DAFB (4.5% greater than control). The beneficial effects of CPPU extended to the postharvest period, with greater wax deposition and, probably as a consequence, a significant and consistent reduction in both fruit rotting and weight loss. Overall, 10 ppm applied at both 3 and 17 DAFB to 'Duke' appeared to be the most promising, as it increased both fruit yield and diameter, and had soluble solids and postharvest behaviour similar to that of the control plants (Retamales *et al.*, 2014).

In general, in the SHB trials done in the USA (Williamson and NeSmith, 2007) and the experiments on 'Duke' in Chile (Retamales *et al.*, 2014), the inclusion of two application dates was beneficial, probably because this allowed a greater proportion of flowers to be at the appropriate stage for maximum CPPU effect. However, it appears (at least for 'Duke' under the growing conditions in central Chile) that three applications of 10 ppm CPPU was excessive, as this reduced yield by 27.0% with respect to the control, due to both reduced numbers of fruit (17.0%) and smaller fruit weight (27.3%) with respect to the control plants. The fact that CPPU treatments did not have deleterious effects

Table 7.13. Effect of the application of CPPU, at different dosages, times and number of applications, on mature NHB 'Duke' plants. (Adapted from Retamales *et al.*, 2014.)

Treatment		Yield per plant		Fruit size		Soluble solids	
Date (DAFB)	Dose (ppm)	Weight (g)	% of control	Diameter (mm)	% of control	°Brix	% of control
Control	–	5745[bcd]	–	13.2[fg]	–	11.0[cd]	–
3	5	6172[abc]	7.4	13.2[fg]	0.0	10.8[cde]	–1.8
10	5	6873[abc]	19.6	13.1[fg]	–0.8	10.3[ef]	–6.4
17	5	7081[ab]	23.3	13.3[ef]	0.8	10.9[cd]	–0.9
3 and 10	5	6978[ab]	21.5	13.5[cde]	2.3	10.7[de]	–2.7
3 and 17	5	5948[abc]	3.5	13.6[bcd]	3.0	11.2[bc]	1.8
10 and 17	5	5161[cd]	–10.2	13.4[cdef]	1.5	10.7[cde]	–2.7
3, 10 and 17	5	6449[abc]	12.3	13.0[gh]	–1.5	10.7[de]	–2.7
3	10	5722[bcd]	–0.4	12.9[h]	–2.3	10.8[cde]	–1.8
10	10	6120[abc]	6.5	14.2[a]	7.6	10.8[cd]	–1.8
17	10	5342[bcd]	–7.0	13.3[def]	0.8	11.8[a]	7.3
3 and 10	10	5851[bcd]	1.8	13.6[bc]	3.0	11.1[bcd]	0.9
3 and 17	10	7614[a]	32.5	13.8[b]	4.5	11.2[bc]	1.8
10 and 17	10	6118[abc]	6.5	13.2[efg]	0.0	9.8[f]	–10.9
3, 10 and 17	10	4177[d]	–27.3	12.8[h]	–3.0	11.5[ab]	4.5
Significance		**		***		***	

Trend significant at **$P \le 0.01$ or ***$P \le 0.001$.
[a,b,c,d,e,f,g,h]Mean values within a column with non-identical superscript letters were significantly different at $P \le 0.05$ (Duncan's multiple range test).

on postharvest weight loss (6.7% loss for the control and 5–6.6% for CPPU-treated fruit) is interesting, as there was, in some cases, a greater fruit expansion (diameter), which could have reduced epidermal wax thickness and might have affected this barrier to water loss from the fruit.

Some detrimental effects observed when using CPPU in blueberry are: (i) some inconsistent responses to the PGR in different seasons, zones and cultivars; (ii) foliage and blossom/fruit burn under some circumstances; (iii) development of shorter internodes in some cases; and (iv) delayed fruit maturity, generally in the range of 3–7 days. In the case of SHBs, reduced spray volumes seem to avoid or minimize phytotoxicity (NeSmith, 2005; Williamson and NeSmith, 2007).

In summary, CPPU is advancing towards commercial use in blueberry management in different countries. Research has shown that for most cultivars and producing regions, both fruit size and fruit set benefit the most when

CPPU is applied at 5–10 ppm between 7 and 21 days after bloom. However, any use should be on a trial basis until more research is done in different conditions and with various plant materials.

Harvest regulation: advanced maturity and enhanced abscission

A concentrated harvest period is a beneficial trait for machine harvesting and is a criterion in breeding blueberry cultivars. The harvest period for individual NHB cultivars ranges from 3 to 6 weeks in temperate climates, but in the case of SHBs growing in subtropical climates this period can extend beyond 3 months. This prolonged harvest period for any given cultivar presents a problem for the mechanical harvesting of the fruit. Plant structures (including immature fruit) are damaged with each pass of the harvester and green fruit are removed. Even when fruit are hand harvested, greater efficiency in labour could be obtained if more fruit were available to harvest at a given time. Thus, it would be beneficial to determine management practices that concentrate fruit ripening in highbush blueberries (Eck, 1970).

Application of ethephon to several fruit crops (e.g. sour cherries, apples) has enhanced ripening. Fruit drop in rabbiteye 'Tifblue' blueberries was increased with 100 ppm ethephon applied when the first berries were maturing. Four days after treatment, the control plants had 2% berry drop compared with 26% in ethephon-treated plants. By 8 days after treatment, the control had 4% fruit drop, while ethephon-treated plants reached 33% fruit drop (Ban *et al.*, 2007).

Ethephon and methyl jasmonate (MeJa) were investigated in 4–6-year-old commercial fields in Georgia and Florida to determine their potential for increasing fruit detachment during harvest of rabbiteye ('Climax', 'Powderblue' and selection T-451) and SHB ('Farthing', 'O'Neal' and 'Star') blueberries (Malladi *et al.*, 2011). Ethephon induced the abscission of mature and immature berries. MeJa applications (at a concentration of at least 20 mM) generally induced rapid and extensive fruit abscission, often within 1 day of treatment. Fruit drop induced by MeJa was attenuated by the coapplication of aminoethoxyvinylglycine, an ethylene biosynthesis inhibitor, suggesting that MeJa induced fruit detachment was partly due to its effects on ethylene biosynthesis. MeJa applications caused leaf yellowing and necrosis of leaf tips and margins, especially at high rates of application (at least 20 mM). Both ethephon and MeJa applications detached fruit with their pedicels. The authors concluded that, although ethephon and MeJa have the potential to be used as harvest aids in blueberry, the rates of application require further optimization to minimize potential phytotoxicity. Additionally, effective de-stemming of the berries may be essential if these compounds are to be used as harvest aids.

In another study of the postharvest effects of MeJa, rabbiteye 'Climax' blueberries were fumigated with 0, 1, 10 or 100 μmol/l MeJa for 10 h and stored at 4°C for 14 and 28 days (Yang *et al.*, 2015). The results showed that, after 28 days of storage, the 10 μmol/l MeJa dose significantly inhibited fruit decay, better maintained firmness and increased fruit soluble solids. It also increased the activities of superoxide dismutase and peroxidase enzymes at 14 days but not at 28 days of storage. In addition, it induced the generation of O_2 and malondialdehyde. The authors concluded that the 10 μmol/l MeJa treatment could be useful in prolonging the storage potential of blueberry.

Experiments done in NHB 'Weymouth' blueberries showed that all concentrations of ethephon induced a greater proportion of the total fruit produced to be ready to harvest at first picking, compared with the non-sprayed control (Table 7.14) (Eck, 1970). Within 72 h of application, a marked influence on ripening was noticeable to the eye. In 'Blueray', ethephon treatments did not differentiate from the control in the first picking and there was a clear dose–response in this cultivar. Applications of 1920 ppm ethephon resulted in a greater percentage of fruit harvested at first picking than at 480 ppm. In both cultivars, the ripening effect seemed to disappear after the second picking. For the first two pickings, fruit size was significantly reduced by the ethephon applications, where fruit with the two highest doses (1920 and 3840 ppm) had 20–21% less weight than the control fruit in 'Weymouth'. The two highest ethephon concentrations were also detrimental for fruit size in 'Blueray'; however, fruit weight was only affected in the second harvest and the reduction amounted to 17% with respect to the untreated control. The internal fruit

Table 7.14. Effect of concentration of ethephon applied to NHB 'Weymouth' and 'Blueray' blueberries at 2 weeks before anticipated harvest on the proportion (%) of fruit by weight harvested at different pickings. (Adapted from Eck, 1970.)

Ethephon (ppm)	Percentage of total yield by weight					
	Harvests of 'Weymouth'			Harvests of 'Blueray'		
	First	Second	Third + fourth	First	Second	Third
0	18[a]	45[a]	37	47[ab]	39[b]	14
240	30[b]	30[b]	40	47[ab]	32[abc]	21
480	32[b]	20[c]	48	41[a]	35[ab]	24
960	34[bc]	23[bc]	43	64[ab]	26[bc]	10
1920	40[bc]	27[bc]	33	70[b]	20[d]	10
3840	46[c]	18[c]	36	69[b]	18[d]	13
Significance	**	**	NS	**	**	NS

Trend significant at **$P \leq 0.01$; NS, non-significant at $P \leq 0.05$.
[a,b,c,d]Mean values within a column with non-identical superscript letters were significantly different at $P \leq 0.05$ (Duncan's multiple range test).

quality of both NHB cultivars was affected by ethephon sprays. While the pH was increased and titratable acidity was decreased significantly by all ethephon concentrations, only the highest dose of ethephon reduced soluble solids following application (Eck, 1970). Research carried out on rabbiteye 'Tifblue' blueberries has shown that fruit firmness was reduced from 220 to 50 g/mm^2 after 4 days of treatment with 100 ppm ethephon (Ban *et al.*, 2007).

In another trial, aimed at establishing the joint influence of ethephon concentration and time of application, fruit of greenhouse-grown NHB 'Morrow' were dipped directly into ethephon (2000, 4000 or 8000 ppm) at different stages of growth (8, 22, 28, 36, 43 or 48 DAFB) (Warren *et al.*, 1973). The earliest ripening was obtained with 8000 ppm applied late in stage II of fruit growth (28 DAFB). The harvest period was shortest (less than a week) when treatments were done in stage III (starting at 36 DAFB). As in the study by Eck (1970), berry weight decreased with higher ethephon rates, but little difference was noted with respect to the stage of development at the time of treatment. An increase in ethephon concentration resulted in greater acidity, a reduction in soluble solids and a decrease in the soluble solids:acid ratio at each time of application until stage III of development. The researchers concluded that the optimum time of application was between the end of stage II and the beginning of stage III (Warren *et al.*, 1973).

Experiments done in the 1960s and 1970s demonstrated that ethephon at 1.12 kg/ha applied foliarly 2 weeks before harvest would, on most occasions, double the anthocyanin levels in cranberries (Eck, 1972). The changes in anthocyanin levels were fully expressed 8 days after treatment (Fig. 7.6). These trials also showed that the final yield was not affected by ethephon applications. Research on rabbiteye 'Tifblue' blueberries showed that 100 ppm ethephon applied at the onset of first berry coloration increased anthocyanin levels

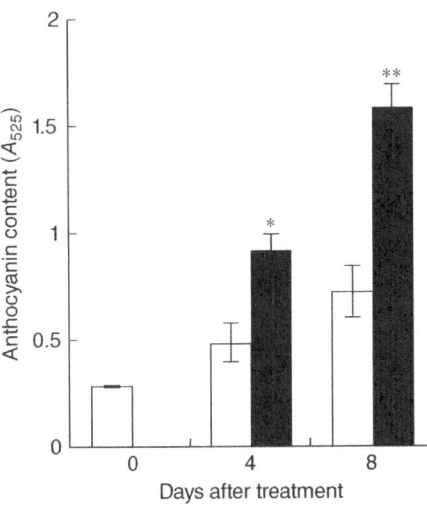

Fig. 7.6. Effect of ethephon (filled bars) on anthocyanin content (as determined by absorbance at 525 nm (A_{525})) of rabbiteye 'Tifblue' blueberries compared with a non-sprayed control (open bars). Values are means with standard error represented by the vertical bars. The mean value was significantly different from that of the control at *$P<0.05$ or **$P<0.01$. (Adapted from Ban *et al.*, 2007.)

by similar magnitudes to those reported in the cranberry experiments (Ban *et al.*, 2007).

It has been suggested that ethephon could be useful when sprayed after large fruit has been hand harvested as a means to promote more uniform ripening of the fruit left on the plant (Ban *et al.*, 2007). Trials on NHBs by Robbins and Doughty (1984) showed that 2000 ppm ethephon applied right after the first harvest increased the number of flower buds and number of flowers for the next season in 'Bluecrop' but not in 'Stanley'.

Studies on the use of ethephon in blueberries were initiated nearly 30 years ago. Despite this long history of research, this management tool has not been implemented commercially. The increasing emphasis on blueberries as a source of anthocyanins and the need for greater efficiency in blueberry production might provide a new opportunity for this compound. The results presented in this section demonstrate that ethephon is capable of advancing maturity in highbush blueberries, but its influence appears to be cultivar dependent and of limited duration. The internal quality of the berry was also altered. The changes that ethephon causes in acidity, pH and firmness, and a possible reduction in size, would be disadvantageous; however, the increase in anthocyanin is a beneficial effect that might warrant further study. As the cultivars that have been studied had dissimilar behaviour, information will be needed on the response to ethephon in a greater number of cultivars, including some recently released commercial cultivars.

CONCLUSIONS

The application of PGRs to blueberries is an issue that has received less attention than for other fruit crops. The data that are available suggest that there is uneven distribution of sprayed compounds within the canopy. This might lead to ineffectiveness of the PGRs applied or phytotoxicity due to overdoses in certain areas of the plant. Various environmental factors (temperature, light quality and intensity, rain and relative humidity) influence the effectiveness of PGRs.

The characteristics and effects of various PGRs applied for different purposes have been reviewed, and the main and collateral effects, both positive and negative, presented. It can be concluded that growth regulators may be a valuable tool in blueberry management and that their use should acquire greater importance as blueberry planting expands into areas less ideal for this crop and as the markets increase the quality standards for the fruit. Such is the case for the effect of CPPU on fruit size and for ethephon on anthocyanins. The various growth regulators that have been tested to overcome deficiencies in the management of blueberries would fall into three classes: (i) those that are used commercially on a regular basis; (ii) those that show promise but need further trials in different growing regions and with a greater diversity of plant materials; and (iii) those that have been investigated but have not produced adequate results. The application of GA_3 for improving fruit set or to set fruit after a frost can be included in the first group. Along with this, the use of HC for leaf development is a tool that has been used with adequate results in different regions, and the use of ethephon to enhance leaf development can, in most conditions, be safely adopted by growers. In the second group, the use of GAs, particularly GA_3, for inhibition of flowering will require further trials. A similar situation exists with the application of HC for flower thinning. Finally, the use of CPPU for fruit enlargement, along with some collateral effects on fruit abscission, needs additional research. These applications focus on crop load regulation and fruit size, a subject that should demand greater attention from the blueberry industry in the coming years. The third group includes those applications that have been tested in other species or for other purposes, but which, when trialled in highbush blueberries, have not produced adequate results. In this group can be included fruit thinners (NAA, BA, ATS, Armothin, soybean oils and others) and paclobutrazol for reducing plant size, as well as ethephon for maturity enhancement.

REFERENCES

Antognozzi, E., Famiani, F., Palliotti, A. and Tombesi, A. (1993a) Effects of CPPU (cyto-kinin) on kiwifruit productivity. *Acta Horticulturae* 329, 150–152.

Antognozzi, E., Proietti, P. and Boco, M. (1993b) Effect of CPPU (cytokinin) on table olive cultivars. *Acta Horticulturae* 329, 153–155.

Ayers, P.D. and Bosley, B. (2005) Sprayer calibration fundamentals. Fact sheet no. 5.003. Colorado State University Extension, Fort Collins, Colorado. Available at: http://www.ext.colostate.edu/pubs/farmmgt/05003.html (accessed 7 February 2018).

Ban, T., Kugishima, M., Ogata, T., Shiozaki, S., Horiuchi, S. and Ueda, H. (2007) Effect of ethephon (2-chloroethylphosphonic acid) on the fruit ripening characters of rabbiteye blueberry. *Scientia Horticulturae* 112, 278–281.

Barker, W.G. and Collins, W.B. (1965) Parthenocarpic fruit set in lowbush blueberry. *Proceedings of the American Society for Horticultural Science* 87, 229–233.

Black, B.L. and Ehlenfeldt, M.K. (2007) Foliar applications of GA_{4+7} reduce flowering in highbush blueberry. *HortScience* 42, 555–558.

Bukovac, M.J. (2005) Maximizing performance of plant growth regulators by improv-ing spray application. *HortTechnology* 15, 222–231.

Bukovac, M.J., Cooper, J.A., Whitmoyer, R.E. and Brazee, R.D. (2002) Spray application plays a determining role in performance of systemic compounds applied to foliage of fruit plants. *Acta Horticulturae* 594, 65–75.

Cano-Medrano, R. and Darnell, R.L. (1998) Effect of GA_3 and pollination on fruit set and development in rabbiteye blueberry. *HortScience* 33, 632–635.

Cartagena, J.R., Matta, F.B. and Spiers, J.M. (1994) Chemical fruit thinning of *Vaccinium ashei* Reade. *Journal of the American Society for Horticultural Science* 119, 1133–1136.

Castro, M.J.L., Ojeda, C. and Fernández-Cirelli, A. (2014) Advances in surfactants for agrochemicals. *Environmental Chemical Letters* 12, 85–95.

Curry, E.A. and Williams, M.W. (1986) Effect of paclobutrazol on fruit quality: apple, pear and cherry. *Acta Horticulturae* 179, 743–754.

Davies, F.S. and Buchanan, D.W. (1979) Fruit set and bee activity in four rabbiteye blueberry cultivars. *Proceedings of the Florida State Horticultural Society* 92, 246–247.

Deyton, D., Sams, C.E., Ballington, J.R. and Cummins, J. (2005) Bloom delay and fruit thinning of blueberry with soybean oil. *HortScience* 40, 1057.

Donkersley, P. and Nuyttens, D. (2011) A meta analysis of spray drift sampling. *Crop Protection* 30, 931–936.

Eck, P. (1970) Influence of Ethrel upon highbush blueberry fruit ripening. *HortScience* 5, 23–25.

Eck, P. (1972) Cranberry yield and anthocyanin content as influenced by ethephon, SADH and malathion. *Journal of the American Society for Horticultural Science* 97, 213–214.

Eck, P. (1988) *Blueberry Science*. Rutgers University Press, New Brunswick, New Jersey.

Ehlenfeldt, M.K. (1998) Enhanced bud production in highbush blueberry (*Vaccinium corymbosum* L.) in response to paclobutrazol. *HortScience* 33, 75–77.

Freeman, B., Albrigo, L.G. and Biggs, R.H. (1979) Cuticular waxes of developing leaves of blueberry, *Vaccinium ashei* Reade cv. Bluegem. *Journal of the American Society for Horticultural Science* 104, 398–403.

Greene, D.W. (1989) CPPU influences 'McIntosh' apple crop load and fruit characteristics. *HortScience* 24, 94–96.

Greene, D.W. (2002) Development of new plant growth regulators from a university perspective. *HortTechnology* 12, 71–74.

Hanson, E., Hancock, J., Ramsdell, D.C., Schilder, A., VanEe, G. and Ledebuhr, R. (2000) Sprayer type and pruning affect the incidence of blueberry fruit rots. *HortScience* 35, 235–238.

Howell, G.S., Stergios, B.G., Stackhouse, S.S. and Bittenbender, H.C. (1976) Ethephon as a mechanical aid for highbush blueberries. *Journal of the American Society for Horticultural Science* 101, 111–115.

Jaldo, H.E., Berettoni, A.R., Ale, J.G. and Forns, A.C. (2009) Effect of hydrogen cyanamide (HC) on fruit ripening and yield of southern highbush blueberries in northwestern of Argentina. *Acta Horticulturae* 810, 869–876.

Jorquera, C., Cautín, R. y Olaeta, J.A. (2006) *Uso del Paclobutrazol y Prohexadione de Calcio en Palto* (Persea americana *Mill.*) cv. Hass. [*Use of Paclobutrazol and Prohexadione Ca in Avocado* (Persea americana *Mill.*) cv. Hass.] Taller de Licenciatura, Pontificia Universidad Católica de Valparaíso, Valparaíso, Chile.

Konarska, A. (2015) Morphological, anatomical, and ultrastructural changes in *Vaccinium corymbosum* fruits during ontogeny. *Botany* 93, 589–602.

Koron, D. and Stopar, M. (2006) Effect of thinners on yield, fruit size and ripening time of highbush blueberry. *Acta Horticulturae* 715, 273–278.

Krewer, G., NeSmith, D.S., Williamson, J.G., Maus, B. and Mullinix, B. (2005) Ethephon for bloom delay of rabbiteye and southern highbush blueberries. *Small Fruit Review* 4, 43–57.

Krewer, G., Cline, B. and NeSmith, D.S. (2007) Southeast regional blueberry horticulture and growth regulator guide. Southern Region Small Fruit Consortium, North Carolina State University, Raleigh, North Carolina. Available at: http://www.

smallfruits.org/assets/documents/ipm-guides/3_21_07SR_BlueberryHortGuide. pdf (accessed 6 February 2018).

Lang, G.A. and Danka, R.G. (1991) Honey-bee-mediated cross- versus self-pollination of 'Sharpblue' blueberry increases fruit size and hastens ripening. *Journal of the American Society for Horticultural Science* 116, 770–773.

Lindberg, W., Hanson, E. and Lobos, G.A. (2014) Partial inhibition of flowering in young highbush blueberries with gibberellins. *Ciencia e Investigación Agraria* 41, 349–356.

Looney, N.E. (1993) Improving fruit size, appearance and other aspects of fruit crop 'quality' with plant bioregulating chemicals. *Acta Horticulturae* 329, 120–127.

MacKenzie, K.E. (1997) Pollination requirements of three highbush blueberry (*Vaccinium corymbosum* L.) cultivars. *Journal of the American Society for Horticultural Science* 122, 891–896.

Mainland, C.M. (1989) Managing the growth and fruiting of rabbiteye (*Vaccinium ashei* Reade) blueberries with pruning and growth regulators. *Acta Horticulturae* 241, 195–200.

Mainland, C.M. and Eck, P. (1969a) Fruit and vegetative responses of the highbush blueberry to gibberellic acid under greenhouse conditions. *Journal of the American Society for Horticultural Science* 94, 19–20.

Mainland, C.M. and Eck, P. (1969b) Fruiting response of the highbush blueberry to gibberellic acid under field conditions. *Journal of the American Society for Horticultural Science* 94, 21–23.

Malladi, A., Vashisth, T. and Johnson, L.K. (2011) Ethephon and methyl jasmonate affect fruit detachment in rabbiteye and southern highbush blueberry. *HortScience* 47, 1745–1749.

McAtee, P., Karim, S., Schaffer, R. and David, K. (2013) A dynamic interplay between phytohormones is required for fruit development, maturation, and ripening. *Frontiers in Plant Science* 4, 1–7.

Merino, R., Serri, H. and Holzapfel, J. (2002) Effect of GA_3 and CPPU on fruit of 'Tifblue' rabbiteye blueberry (*Vaccinium ashei* R.). *Acta Horticulturae* 574, 239–243.

Myers, R.E., Deyton, D.E. and Sams, C.E. (1996) Applying soybean oil to dormant peach trees alters internal atmosphere, reduces respiration, delays bloom, and thins flower buds. *Journal of the American Society for Horticultural Science* 121, 96–100.

NeSmith, D.S. (2002) Response of rabbiteye blueberry (*Vaccinium ashei* Reade) to the growth regulators CPPU and gibberellic acid. *HortScience* 37, 666–668.

NeSmith, D.S. (2005) Use of plant growth regulators in blueberry production in the southeastern U.S.: a review. *International Journal of Fruit Science* 5, 41–52.

NeSmith, D.S. (2008) Effect of timing of CPPU applications on rabbiteye blueberries. *HortScience* 43, 1446–1448.

NeSmith, D.S. and Adair, H.M. (2004) Rabbiteye blueberry field trials with growth regulator CPPU. *Small Fruit Review* 3, 183–191.

NeSmith, D.S. and Krewer, G. (1992) Flower bud stage and chill hours influence the activity of GA_3 applied to rabbiteye blueberry. *HortScience* 27, 316–318.

NeSmith, D.S. and Krewer, G. (1997a) Response of rabbiteye blueberry (*Vaccinium ashei* Reade) to gibberellic acid rate. *Acta Horticulturae* 446, 337–342.

NeSmith, D.S. and Krewer, G. (1997b) Effect of bee pollination and GA_3 on fruit size and maturity of three rabbiteye blueberry cultivars with similar fruit densities. *HortScience* 34, 1106–1107.

NeSmith, D.S., Krewer, G., Rieger, M. and Mullinix, B. (1995) Gibberellic acid-induced fruit set of rabbiteye blueberry following freeze and physical injury. *HortScience* 30, 1241–1243.

NeSmith, D.S., Krewer, G. and Lindstrom, O.M. (1999) Fruit set of rabbiteye blueberry (*Vaccinium ashei* Reade) after subfreezing temperatures. *Journal of the American Society for Horticultural Science* 124, 337–340.

Petracek, P.D., Silverman, F.P. and Greene, D.W. (2003) A history of commercial plant growth regulators in apple production. *HortScience* 38, 937–942.

Plattner, K., Fonsah, E.G., Escalante, C., Kliewer, G., Andersen, P.C., Liburd, O. and Tertuliano, M. (2008) A plan for economic evaluation of organic blueberry production in Georgia. In: *Southern Agricultural Economics Association Annual Meeting*, Dallas, Texas, 2–6 February 2008. Available at: http://ageconsearch.umn.edu/bitstream/6805/2/sp08pl01.pdf (accessed 25 September 2010).

Praat, J.P., Maber, J. and Manktelow, D.W.L. (2000) The effect of canopy development and sprayer position on spray drift from a pipfruit orchard. *New Zealand Plant Protection* 53, 241–247.

Rademacher, W. (2000) Growth retardants: effects of gibberellin biosynthesis and other metabolic pathways. *Annual Review of Plant Physiology and Plant Molecular Biology* 51, 501–531.

Retamales, J.B., Hanson, E.J. and Bukovac, M.J. (2000) GA_3 as a flowering inhibitor in blueberries. *Acta Horticulturae* 527, 147–152.

Retamales, J.B., Lobos, G.A., Romero, S., Godoy, R. and Moggia, C. (2014) Repeated applications of CPPU on highbush blueberry cv. Duke increase yield and enhance fruit quality at harvest and during postharvest. *Chilean Journal of Agricultural Research* 74, 157–161.

Reynolds, A.G., Wardle, D.A., Zurowski, C. and Looney, N.E. (1992) Phenylureas CPPU and thidiazuron affect yield components, fruit composition, and storage potential of four seedless grape selections. *Journal of the American Society for Horticultural Science* 117, 85–89.

Richardson, E.A., Seely, S.D. and Walker, D.R. (1974) A model for the estimation of completion of rest for 'Redhaven' and 'Elberta' peach trees. *HortScience* 9, 331–332.

Robbins, J.A. and Doughty, C.C. (1984) Flower bud formation, flowering, and fruiting in highbush blueberry in response to chlormequat, ethephon, and daminozide. *HortScience* 19, 100–102.

Sandoval, C.R., Ramsdell, D.C. and Hancock, J.F. (1995) Infection of wild and cultivated *Vaccinium* spp. with blueberry leaf mottle nepovirus. *Annals of Applied Biology* 126, 457–464.

Schönherr, J., Baur, P. and Uhlig, B.A. (2000) Rates of cuticular penetration of 1-naphthylacetic acid (NAA) as affected by adjuvants, temperature, humidity and water quality. *Plant Growth Regulation* 31, 61–74.

Semerdjieva, S.I., Phoenix, G.K., Hares, D., Gwynn-Jones, D., Callaghan, T.V. and Sheffield, E. (2003) Surface morphology, leaf and cuticle thickness of four dwarf shrubs from the sub-Artic heath following long-term exposure to enhanced levels of UV-B. *Physiologia Plantarum* 117, 289–294.

Serri, H. and Hepp, R. (2006) Effect of the growth regulators CPPU on fruit quality and fruit ripening of highbush blueberries. *Acta Horticulturae* 715, 279–282.

Spiers, J.M. (1978) Effect of stage of bud development on cold injury in rabbiteye blueberry. *Journal of the American Society for Horticultural Science* 103, 452–455.

Spiers, J.M. (1988) Response of 'Tifblue' rabbiteye blueberry to soil-applied paclobutrazol. *HortScience* 23, 837–839.

Steffens, G.L. and Stutte, G.W. (1989) Thidiazuron substitution for chilling requirement in apple cultivars. *Journal of Plant Growth Regulation* 8, 801–808.

Stover, E.W. and Greene, D.W. (2005) Environmental effects on the performance of foliar applied plant growth regulators: a review focusing on tree fruits. *HortTechnology* 15, 214–221.

Strik, B. and Buller, G. (2005) The impact of early cropping on subsequent growth and yield of highbush blueberry in the establishment years at two planting densities is cultivar dependant. *HortScience* 40, 1998–2001.

Strik, B., Buller, G. and Hellman, E. (2003) Pruning severity affects yield, berry weight, and hand harvest efficiency of highbush blueberry. *HortScience* 38, 196–199.

Stringer, S.J., Spiers, J.M., Braswell, J. and Marshall, D.A. (2002) Effects of hydrogen cyanamide application rates and timing on fruit and foliage of 'Climax' rabbiteye blueberry. *Acta Horticulturae* 574, 245–251.

Stringer, S.J., Marshall, D.A., Sampson, B.J. and Spiers, J.M. (2004) The effects of chill hour accumulation on hydrogen cyanamide efficacy in rabbiteye and southern highbush blueberry cultivars. *Small Fruit Review* 3/4, 339–347.

Sugiyama, N. and Yamaki, Y.T. (1995) Effects of CPPU on fruit set and fruit growth in Japanese persimmon. *Scientia Horticulturae* 80, 337–343.

Swart, P. (2015) Harvest scheduling of southern highbush blueberries (*Vaccinium corymbosum* L. interspecific hybrids) in a climate with moderate winter chilling. MSc thesis, University of Stellenbosch, Stellenbosch, South Africa.

Takele, E., Faber, B., Gaskell, M., Nigatu, G. and Sharabeeen, I. (2007) Sample costs to establish and produce blueberries in San Luis Obispo, Santa Barbara, and Ventura counties, conventional production, 2007. University of California Cooperative Extension, Napa, California. Available at: https://coststudyfiles.ucdavis.edu/uploads/cs_public/44/77/44772592-bf29-4bed-ba46-047d07f09096/blueberry_sc2007.pdf (accessed 8 February 2018).)

VanEe, G., Ledebuhr, R., Hanson, E., Hancock, J. and Ramsdell, D.C. (2000) Canopy development and spray deposition in highbush blueberry. *HortTechnology* 10, 353–359.

Wang, C.J. and Liu, Z.Q. (2007) Foliar uptake of pesticides – present status and future challenge. *Pesticide Biochemistry and Physiology* 87, 1–8.

Warren, J.M., Ballinger, W.E. and Mainland, C.M. (1973) Effects of ethephon upon fruit development and ripening of highbush blueberry in the greenhouse. *HortScience* 8, 504–507.

Wilkins, R.M. (1982) The potential for the controlled delivery of plant growth regulators. In: J.S. McLaren (ed.) *Chemical Manipulation of Crop Growth and Development*. Butterworths Scientific, London, pp. 111–126.

Williamson, J.G. and NeSmith, D.S. (2007) Effects of CPPU applications on southern highbush blueberries. *HortScience* 42, 1612–1615.

Williamson, J.G., Maust, B.E. and NeSmith, D.S. (2001) Timing and concentration of hydrogen cyanamide affect blueberry bud development and flower mortality. *HortScience* 36, 922–924.

Williamson, J.G., Krewer, G., Maust, B.E. and Miller, E.P. (2002) Hydrogen cyanamide accelerates vegetative budbreak and shortens fruit development period of blueberry. *HortScience* 37, 539–542.

Yang, H.Y., Wu, W.L. Li, W.L., Yao, B., Wei, Y.L. and Lu, L.F. (2015) Effect of methyl jasmonate treatments on physiological changes of blueberry fruit. *Journal of Food Safety and Quality* 6, 4483–4488.

BLUEBERRY DISEASES AND PESTS, THEIR MANAGEMENT AND CULTIVAR RESISTANCE

INTRODUCTION

Blueberries are routinely subject to a wide array of diseases. Probably the most widespread problems in blueberry are caused by mummy berry (*Monilinia vaccinii-corymbosi* (Reade)), blueberry stunt phytoplasma, blueberry shoestring virus, blueberry shock virus, tomato ringspot virus, blueberry scorch virus, blueberry necrotic ring blotch virus, stem blight (*Botryosphaeria* spp.), stem canker (*Botryosphaeria corticis* Demaree and Wilcox), *Phytophthora* root rot (*Phytophthora cinnamomi* Rands), *Phomopsis* canker and twig blight (*Phomopsis vaccinii* Shear), bacterial leaf scorch (*Xylella fastidiosa*), *Botrytis* (*Botrytis cinerea* Pers.: Fr.), *Alternaria* fruit rot (*Alternaria* spp.) and anthracnose fruit rot (*Colletotrichum gloeosporioides* (Penz.) Penz. and Sacc.).

Most of the above diseases are widespread, although mummy berry and viral diseases are most prevalent in areas that grow NHBs, and stem blight, bacterial leaf scorch, cane canker and *Phytophthora* root rot are most common in rainy, hot climates where SHBs are grown. Leaf spots and fungal-induced defoliation are also a problem in the south-eastern USA. Rabbiteye blueberries have somewhat different disease susceptibilities from highbush, but can be affected by *Botrytis* blossom and twig blight, stem blight, blueberry stunt and mummy berry, and several defoliating fungal diseases. For many of these diseases, resistant cultivars are available (Table 8.1). The symptoms of the major blueberry diseases are summarized in Appendix 8.1.

A number of insects and arthropods do significant damage to highbush blueberries including the blueberry maggot (*Rhagoletis mendax* Curran), blueberry gall midge (*Dasineura oxycoccana* Johnson), blueberry bud mite (*Acalitus vaccinii* Keifer), flower thrips (*Frankliniella* spp.), Japanese beetle (*Popillia japonica* Newman), sharp-nosed leaf hopper (stunt vector) (*Scaphytopius magdalensis* Prov.), blueberry aphid (the vector of blueberry shoestring and blueberry scorch viruses) (*Illinoia pepperi* Mac. G.), spotted-wing *Drosophila* (*Drosophila suzukii* (Matsumura)), cranberry fruit worm (*Acrobasis vaccinii* Riley), cherry

Table 8.1. Reported resistance in highbush and rabbiteye cultivars to various diseases (see text for more details and references).

Disease or pest	Most resistant cultivars	Level of resistance
Alternaria fruit rot	NHB: 'Aurora', 'Brigitta', 'Draper', 'Elliott'	High
Anthracnose fruit rot	NHB: 'Aurora', 'Bluejay', 'Brigitta', 'Draper', 'Legacy', 'Toro'	Moderate to high
	SHB: 'Blueridge', 'Sharpblue'	High
	Rabbiteye: 'Bluebell', 'Centurion', 'Homebell', 'Powderblue', 'Southland'	Moderate
Bacterial leaf scorch	SHB: 'Emerald', 'Millennia', 'V5', 'Windsor'	Moderate to high
Blueberry scorch: eastern strain	NHB: 'Jersey'	High
Blueberry scorch: western strain	NHB: see Table 8.2	High
Blueberry shock	NHB: 'Bluecrop', 'Legacy', 'Toro'	High
	SHB: 'Bladen', 'Harding'	High
Blueberry shoestring	NHB: 'Bluecrop', 'Bluejay', 'Northland'	Moderate
Blueberry stunt	Rabbiteye: 'Premier', 'Tifblue '	High resistance to vector
Fusicoccum canker	NHB: 'Ama', 'Bluetta', 'Goldtraube', 'Hardyblue', 'Heerma', 'Patriot', 'Spartan'	High
Leaf spots	SHB: 'Bladen', 'Reveille'	High
Mummy berry shoot blight	NHB: 'Jersey', 'Duke', 'Bluejay', 'Elliott', 'Lateblue', 'Spartan'	Moderate to high
	SHB: 'Reveille'	High
	Rabbiteye: 'Coastal', 'Delite', 'Centurion', 'Walker'	Moderate
Mummy berry fruit rot	NHB: 'Bluejay', 'Brigitta', 'Rubel', 'Reka'	High
	SHB: 'Reveille'	High
Necrotic ringspot	NHB: 'Jersey'	High
Phomopsis twig blight and canker	NHB: 'Bluecrop', 'Elliott', 'Bluetta', 'Rubel',	Moderate to high
	SHB: 'Bluechip', 'Cape Fear', 'Reveille'	Moderate
Phytophthora root rot	SHB: 'Emerald', 'Primadonna', 'Santa Fe', 'Springhigh'	High
Powdery mildew	NHB: 'Berkeley', 'Bluecrop', 'Coville', 'Earliblue', 'Rancocas'	Moderate to high

continued

Table 8.1. *continued.*

Disease or pest	Most resistant cultivars	Level of resistance
Stem blight	NHB: 'Weymouth'	Moderate
	SHB: 'Cape Fear', 'Murphy', 'O'Neal', 'Springhigh', 'Santa Fe'	High
	Rabbiteye: 'Powderblue', 'Premier'	Moderate
Stem or cane canker	SHB: 'Croatan', 'Emerald', 'Jewel', 'Millennia', 'O'Neal', 'Primadonna', 'Reveille', 'Santa Fe', 'Sapphire', 'Sebring', 'Springhigh', 'Springwide', 'Windsor'	High, depending on race of fungus
Red ringspot	NHB: 'Bluecrop', 'Jersey', 'Rubel'	Moderate to high
	Rabbiteye: 'Woodard'	High

fruit worm (*Grapholita packardi* Zell) and the plum curculio (*Conotrachelus nenuphar* Herbst). Flower thrips, blueberry bud mite and the blueberry gall midge are particular problems in the south-eastern USA. Rabbiteye blueberries generally suffer from fewer major pests than highbush types; however, significant damage is caused by cranberry fruit worm, blueberry gall midge and blueberry bud mites in rabbiteye blueberries. Appendix 8.2 summarizes the characteristic symptoms of pest damage in blueberries.

A number of weed species compete with blueberries for water, light and nutrients. Annual weeds tend to be most troublesome in new plantings. Some common annual weeds include pigweeds (*Amaranthus* spp.), common ragweed (*Ambrosia artemisiifolia*), common lambsquarters (*Chenopodium album*) and various annual grasses (*Digitaria sanguinalis*, *Setaria* spp.). Perennial weeds become more troublesome as plantings age. Examples include Canada thistle (*Cirsium arvense*), field bindweed (*Convolvulus arvensis*), field horsetail (*Equisetum arvense*) and quackgrass (*Agropyron repens*), as well as woody species such as blackberry (*Rubus* spp.), poison ivy (*Toxicodendron radicans*) and Virginia creeper vine (*Parthenocissus quinquefolia*).

COMMON FUNGAL DISEASES OF HIGHBUSH AND RABBITEYE BLUEBERRY

Primary data sources on symptoms and spread are Pritts and Hancock (1992), Cline and Schilder (2006) and Polashock *et al.* (2016). Pictures of these diseases are available online (http://www.canr.msu.edu/blueberries/pest_management/diseases).

Fungi primarily affecting fruit

Alternaria fruit rot

Alternaria fruit rot (*Alternaria* spp.) causes a fruit rot that begins at the blossom end of the berries as sunken lesions covered by blackish, dark green sporulations. It is the most common postharvest rot of blueberries. The fungus overwinters on old, dried-up berries and peduncles, and its spores are dispersed by wind-blown rain and splashing. Fungicides help control this disease, but the most effective methods of control are cultural, including removal of infected wood, harvesting dry fruit before they are overripe and rapid cooling after harvest. Most *Alternaria* rot occurs at the stem scar after harvest. Hancock *et al.* (2008) found the most resistant NHB cultivars to be 'Aurora', 'Brigitta', 'Draper' and 'Elliott'.

Anthracnose fruit rot or ripe rot

Anthracnose fruit rot or ripe rot (*Colletotrichum acutatum*) is a widespread problem wherever blueberries are grown. As diseased fruit ripen, their blossom end softens and puckers with orange spore masses. The spores are readily transferred by rainfall and where fruit touch. The fungus overwinters on twigs, canes and bud scales, and sporulates in the spring during bloom. Fungicide sprays are commonly used to control this disease, beginning at bloom and continued at 7–14-day intervals throughout the green-fruit stage. Pruning can help reduce contamination by removing diseased twigs and opening the canopy. Disease levels are also reduced by the timely harvesting of ripe berries and cooling fruit immediately after harvest. Postharvest rots can be severe when the fruit is not handled properly after harvest.

Blueberry cultivars vary in their level of resistance to anthracnose fruit rot. Polashock *et al.* (2005) found the NHBs 'Brigitta', 'Elliott' and 'Legacy' to be the most resistant, while Hancock *et al.* (2008) found 'Bluejay', 'Brigitta', 'Draper', 'Elliott', and 'Toro' to be the most resistant. Wise *et al.* (2010) listed 'Draper', 'Elliott' and 'Little Giant' as resistant, and 'Aurora' as moderately resistant. Ehlenfeldt *et al.* (2005) found 'Blue Ridge', 'Elliott', 'Legacy', 'Little Giant' and 'Sharpblue' to have the highest levels of resistance to fruit and foliar infection. Anco and Ellis (2016) suggested that 'Brigitta', 'Elliott', 'Hannah's Choice', 'Legacy', 'Little Giant' and 'Reveille' have some resistance, while 'Bluecrop', 'Bluetta', 'Blueray', 'Chanticleer', 'Coville', 'Jersey', and 'Spartan' are highly susceptible.

Among rabbiteye cultivars, Smith *et al.* (1996) reported that 'Bluebell', 'Centurion', 'Homebell', 'Southland' and 'Woodard' were the most resistant. Daykin and Milholland (1984) found 'Morrow' and 'Powderblue' to be most resistant in a screen of eight highbush and rabbiteye cultivars.

Botrytis fruit rot

Botrytis fruit rot (*Botrytis cinerea* Pers.: Fr.) is a widespread disease in highbush and rabbiteye blueberries. Infected berries are covered by a fluffy grey mould, which develops pre- and postharvest. The fungus also causes a blossom cluster blight where flowers turn brown and are covered by a grey mould (see section on *Botrytis* blossom and twig blight, below, for more details on disease development and spread). The disease is favoured by freeze injury to flowers and cool wet weather during bloom. Effective fungicides are available. Little information is available on the resistance of NHB and SHB blueberries to *Botrytis* fruit rot; however, Smith *et al.* (1996) found 'Menditoo' and 'Premier' to be among the most susceptible rabbiteye cultivars, while 'Southland', 'Tifblue' and 'Woodard' were the most resistant.

Mummy berry

Mummy berry (*Monilinia vaccinii-corymbosi* (Reade)) is probably the most common disease of highbush and rabbiteye blueberries. Infected floral inflorescences are purple-brown; whitish-grey spore masses are found on the midrib of infected leaves and at the base of infected blossoms. Berries shrivel and turn a pinkish colour as they ripen. These 'mummy berries' fall to the ground, and mushroom-like apothecia (brown and cup-shaped) germinate the following spring and produce ascospores. The spores are discharged during wet weather and are dispersed by wind to new leaf and flower shoots. These shoots become blighted and produce conidia, which are spread to open flowers by wind and insects. Fungicides are effective against this disease, with sprays being applied from bud break through bloom on a 7–14-day cycle. Overwintering mummies can also be eliminated by hand raking or by burying them with a 2.5 cm layer of mulch in the autumn or winter. Dedej *et al.* (2004) found that honeybee hives equipped with dispensers containing the biocontrol agent *Bacillus subtilis* significantly reduced the amount of mummy berry disease.

Cultivars vary widely in their susceptibility to mummy berry, and there is no strong association between resistance to the shoot blight and fruit infection stages. Stretch *et al.* (1995) found among NHB cultivars that 'Bluejay', 'Duke', 'Elliott' and 'Jersey' were the most resistant to the shoot blight stage. Ehlenfeldt and Stretch (2000) found resistance to shoot blight to be weaker in rabbiteyes than in highbush cultivars, but they deemed the rabbiteyes 'Centurion', 'Coastal', 'Delite' and 'Walker' to be the least susceptible. Stretch and Ehlenfeldt (2000) found the most resistant NHB cultivars to the fruit infection stage to be 'Bluejay', 'Brigitta' and 'Reka'. Cline and Schilder (2006) suggested that 'Croatan' and 'Jersey' are highly susceptible to both phases of the disease, while 'Reveille' is quite resistant. They described 'Rubel' as being very susceptible to shoot blight but relatively resistant to fruit infection. Wise *et al.* (2010) listed the NHBs 'Bluejay', 'Duke', 'Elliott', 'Lateblue', and half-highs 'Northblue' and 'Northsky' as resistant, and NHBs 'Jersey' and 'Spartan' as moderately resistant.

Fungi primarily affecting canes

Botrytis blossom and twig blight

Botrytis blossom and twig blight (*Botrytis cinerea*) is widespread in highbush and rabbiteye blueberries, with rabbiteyes generally being more severely affected (particularly 'Tifblue'). It begins as a blossom cluster blight where flowers turn brown and are covered by a grey mould. About 1–2 weeks after blossom infection, leaves can develop irregular necrotic areas. Berries can also become infected. Infected berries are covered by a fluffy grey mould. The fungus overwinters on infected plant material and is spread by airborne spores. Cool, wet conditions favour the spread of this disease.

Protective fungicide sprays can be used to reduce *Botrytis* infection when long periods of cool, moist weather are predicted; however, such control is often unsatisfactory (Smith *et al.*, 2012). Fungicides have to be applied at the proper stage of flower development for effective control. 'Tifblue' flowers are most susceptible when they are in full bloom (Smith, 1998). Two strategies that help reduce disease incidence are to prune off infected twigs in the winter and to make sure that the bush canopy is open. The use of bumble bees as vectors of the fungicidal biological control agents Prestop (*Gliocladium catenulatum*) and Mycostop (*Streptomyces griseoviridis*) has also been shown to be useful against blueberry blossom blight (Smith *et al.*, 2012).

A limited amount of information is available on cultivar susceptibility to *Botrytis* blossom blight. Among rabbiteyes, Smith (1998) determined that 'Gulfcoast' and 'Tifblue' were generally more susceptible than 'Climax' and 'Premier'. In another study, Smith (1999) determined that 'Magnolia' and 'Tifblue' were more susceptible than 'Climax', 'Jubilee' and 'Premier', In both of these studies, susceptibility was positively associated with floral development. Of the three rabbiteye cultivars most often grown in North Carolina ('Powderblue', 'Premier' and 'Tifblue'), 'Powderblue' is notably more susceptible to *Botrytis* outbreaks during bloom, and is routinely sprayed with fungicides.

Phomopsis canker and twig blight

Of *Phomopsis* canker and twig blight (*Phomopsis vaccinii* Shear), *Phomopsis* canker is most common in the cooler production regions such as Michigan, while twig blight is most prevalent in hotter, moist production regions such as North Carolina. *Phomopsis* canker appears first on young stems as 2.5–5 cm long reddish-brown areas that develop into elongated, flattened cankers that are covered by small, pimple-like pycnidia. Infected stems typically wilt, and their leaves turn brown during the heat of summer. Winter injury and harvester damage offer the most common entry points for infection. In *Phomopsis* twig blight, flower buds are infected as they open, leaving a trail of brown, dead blossoms down the fruiting shoot. The symptoms are most visible at the green-fruit stage. This fungus also causes a fruit rot distinguished by very soft berries that split and leak juice, but most cultivars are resistant to this phase of the disease (unless fruit are allowed to hang on the bush for too long).

P. vaccinii overwinters on dead twigs and canes, and splashing rain spreads the conidiospores. Fungicides can be used to control twig blight; they are usually sprayed from bud break to bloom, on a 7–14-day schedule. Disease spread is minimized by removal of diseased canes during the winter.

There is wide variation among highbush and rabbiteye cultivars in resistance to *Phomopsis* canker and twig blight. Baker *et al.* (1995) found 'Bluetta' and 'Elliott' to be the least susceptible of nine NHB cultivars to *Phomopsis* canker. In a screen of 50 blueberry cultivars, Polashock and Kramer (2006) found by far the strongest resistance to twig blight in half-high ('Chippewa' and 'Northsky') and lowbush ('Blomidon', 'Chignecto' and 'Cumberland') cultivars. Among the rabbiteye and highbush cultivars screened, 'Rubel' was the least susceptible and 'Duke', 'Emerald', 'Hannah's Choice', 'Legacy' and 'Powderblue' were the most susceptible. Cline and Schilder (2006) suggest that the highbush cultivars 'Harrison', 'Jersey' and 'Murphy' are highly susceptible to twig blight, while 'Croatan' and 'O'Neal' are moderately susceptible, and 'Bluechip', 'Cape Fear' and 'Reveille' are relatively resistant. Wise *et al.* (2010) listed 'Bluetta' and 'Elliott' as being resistant to *Phomopsis* twig blight and canker, and 'Bluecrop' and 'Rubel' as moderately resistant.

Fusicoccum (godronia) canker

Fusicoccum (godronia) canker (*Fusicoccum putrefaciens*) is a serious disease in the cooler production areas. Symptoms generally appear on the lower third of 1–2-year-old stems, first as small, reddish areas (like a bull's-eye), which then develop into elliptical, brownish-purple lesions 2.5–15 cm long. Infected canes wilt and die back. The fungus overwinters in diseased wood, and new infections occur throughout the growing season whenever it rains. Wounds are not necessary for infection. To control disease spread, infected canes should be removed and destroyed, and monthly applications of fungicides made throughout the growing season.

In a survey of 31 NHB cultivars in Norway, Strømeng and Stensvand (2001) found 'Goldtraube' and 'Hardyblue' to be the most resistant, 'Ama', 'Bluetta', 'Heerma', 'Patriot' and 'Spartan' to have low to moderate susceptibility, 'Berkeley', 'Bluecrop', 'Duke' and 'Ivanhoe' to have moderate to high susceptibility, and 'Blueray', 'Collins', 'Earliblue' and 'Jersey' to be the most susceptible. Wise *et al.* (2010) rated NHB 'Rancocas' as resistant in Michigan and 'Coville' and 'Rubel' as moderately resistant.

Stem blight

Stem blight (*Botryosphaeria* spp.) is most common in rainy, hot climates where SHBs and rabbiteyes are grown. It is the most important disease limiting establishment of blueberries in the south-eastern USA (Cline and Schilder, 2006). *Botryosphaeria dothidea*, *B. obtusa* and *B. ribis* are the species most commonly associated with stem blight. The disease enters young canes through wounds and results in the death of canes and ultimately the whole bush; young plants are most susceptible. It begins as a rapid wilt-down of leaves on individual branches and spreads downwards until the whole cane is dead. An infected stem cut longitudinally displays a light-brown discoloration under the bark. Symptoms usually appear soon after harvest and get worse as the season progresses.

The fungus overwinters in dead and dying canes of a wide range of woody host plants. Spores are carried by wind and rain from infected wood throughout most of the year, except in mid-winter. Removal of diseased canes is the most effective method of reducing spread of the disease in established plantings. It is also important to avoid droughty, sandy soils and heavy muck soils. Fungicides are of only limited value in controlling this disease.

There appears to be a wide range in resistance among blueberries to stem blight. In a survey of 50 blueberry cultivars, Polashock and Kramer (2006) found using an attached stem assay that half-high ('Chippewa', 'Northblue' and 'Northsky') and lowbush ('Putte') cultivars had much higher resistance to stem blight than highbush cultivars. 'Weymouth' was the most resistant highbush, while 'Bluecrop', 'Blueray', 'Duke' and 'Ozarkblue' were among the most susceptible. The rabbiteye 'Powderblue' was also highly susceptible. In detached stem assays conducted by Smith (2004, 2009) of mostly southern rabbiteye and highbush cultivars, 'Brightwell', 'Bluecrisp', 'Emerald', 'Misty' 'Ozarkblue', 'Sapphire' and 'Star' were among the most resistant, while 'Alapaha', 'Austin', 'Legacy', 'O'Neal', 'Reveille' and 'Tifblue' were highly susceptible.

Cline and Schilder (2006) indicated that some of the most susceptible SHB cultivars are 'Bluechip' and 'Bounty', while 'Cape Fear' 'Murphy' and 'O'Neal' are very resistant. They also suggested that the SHBs 'Bladen', 'Croatan', 'Harrison' and 'Reveille' and the rabbiteyes 'Powderblue' and 'Premier' are susceptible, but losses of plants to disease are low enough (10–20%) that fields can be established successfully. 'Legacy' is reported to be susceptible but can be grown successfully where plant health is maintained by proper site selection, fertility, pruning and irrigation (Bill Cline, North Carolina State University, Raleigh, North Carolina, personal communication). In various patent descriptions, P.M. Lyrene indicated that his SHB releases 'Santa Fe' and 'Springhigh' have high resistance to stem blight, while 'Emerald' and 'Primadonna' are medium to high.

Stem or cane canker

Stem or cane canker (*Botryosphaeria corticis* Demaree and Wilcox) is a significant problem in rainy, hot climates where cultivated blueberries are grown. The disease attacks only young, vigorously growing shoots. Spores are released when it is wet from April to September. Disease symptoms first appear as raised, red bumps and develop over 4–6 months into cankers that are swollen and have deep cracks running through them. The most susceptible cultivars ('O'Neal', 'Weymouth' and 'Wolcott') can have a series of these cankers running the whole length of the stem.

Fungicides are not particularly effective in controlling this disease, making clean propagation, sanitation, maintenance of high plant vigour and the use of resistant cultivars critical. It is also highly beneficial to plant in areas isolated from infected plantations. Diseased canes should be religiously removed and extreme care taken to propagate only from healthy sources. Even visibly clean stems can be infected if taken from fields where the disease is present.

There are multiple races of stem canker. 'Croatan' is resistant to most of the races found in North Carolina, while 'O'Neal' and 'Reveille' are resistant to some of them (Ballington *et al.*, 1993). In his patent applications, P.M. Lyrene indicated that the SHBs 'Emerald', 'Millennia', 'Primadonna', 'Santa Fe', 'Sapphire', 'Sebring', 'Springhigh', 'Springwide' and 'Windsor' are highly resistant to the races found in Florida.

Fungi affecting leaves

Leaf spots

Several leaf spot diseases infect blueberries in the south-eastern USA, with the most serious being anthracnose leaf spot (*Gloeosporium minus*) and *Septoria* leaf spot (*Septoria albopunctata*).

Anthracnose leaf spot is an extremely common foliar disease of blueberries in the south-eastern USA and often results in defoliation and reduced yield. Symptoms appear first as small reddish flecks of colour and develop into large brown lesions 1–2.5 cm in diameter, with a bull's-eye pattern. Stems with infected leaves eventually turn brown and then grey and then die. In some cultivars such as 'Jersey', stem dieback of up to 50 cm can occur (Cline and Schilder, 2006).

Septoria leaf spot is a widespread problem in highbush and rabbiteye blueberries across all of the south-eastern USA. Infected leaves have numerous small, purple spots from 3–6 mm in diameter, with white to tan centres. Stem lesions can also be found, primarily on young plants and at the base of mature bushes. Plants can be completely defoliated by *Septoria* leaf spot in wet years.

Some SHB cultivars such as 'Bladen' and 'Reveille' are resistant to these two diseases, although fungicide applications are required with most cultivars. Others, such as 'Star', are quite susceptible. At least one fungicide application is generally recommended before harvest and then every 2 weeks until the end of summer. The common practice of summer topping of bushes also helps control these diseases (see Chapter 6, this volume) by removing older infected leaves and stimulating the growth of new, vigorous shoots.

Leaf rust

Leaf rust (*Naohidemyces vaccinii*, formerly *Pucciniastrum vaccinii*) is common in the south-eastern USA and is observed occasionally in the eastern USA, Argentina, New Zealand, Spain and Australia (Cline and Schilder, 2006). It is the primary defoliating fungus in Florida. The disease appears as reddish-brown spots on leaves, which become yellow and drop prematurely. The undersides of the leaves have yellow and orange clusters of spores. The disease is most severe in areas where its alternative host, the hemlock tree, lives; however, the disease can successfully overwinter on evergreen blueberry leaves in warm climates where there are no hemlocks. Levels of disease are generally not sufficient to warrant fungicide sprays, but in areas of heavy disease pressure, fungicides should be used to help retain leaves after harvest (Williamson and Miller, 2002). This may be particularly critical in evergreen production systems. The NHB cultivars 'Bluecrop', 'Collins', 'Earliblue' and 'Weymouth' are resistant (https://blogs. cornell.edu/newfruit/files/2017/01/BB-leaf-rust-fast-fact-11wug92.pdf).

Powdery mildew

Powdery mildew (*Microsphaera vaccinii*) is widespread and attacks all cultivars of highbush and rabbiteye blueberries, but its overall economic impact is usually minimal. As the season progresses, small (3–6 mm), irregular reddish-brown spots covered with a faint white mould appear on leaves that become somewhat distorted. In the late summer and autumn, round black fruiting bodies (0.8–1.6 mm) develop in the web-like structures. These 'cleistothecia' are where the causal fungus overwinters. Fungicides can be used to control this disease, although the economic benefit of spraying for powdery mildew is suspect. The fungicides applied for control of fruit rots and other leaf diseases are also effective against powdery mildew. Wise *et al.* (2010) rated 'Berkeley', 'Earliblue' and 'Rancocas' as resistant and 'Bluecrop' and 'Coville' as moderately resistant.

Fungal diseases primarily affecting roots

Armillaria root rot

Armillaria root rot (*Armillaria mellea*) is a common disease of woody plants across the world. Diseased bushes are low in vigour and generally decline gradually, although they die very fast after their trunks are girdled by the disease. The most obvious symptoms are mushrooms that appear along the base of crowns and white, fan-shaped fungal growth under the bark at soil level. The disease spreads down rows as healthy roots come into contact with diseased ones. *Armillaria* root rot is most important where new blueberry fields are planted in old, diseased woodlots with stumps and roots, and in plantings mulched with infected wood chips. Fungicides are poor at controlling this disease: the best defence is to plant on sites without old tree stumps and roots, or to wait until they decay naturally.

Phytophthora root rot

Phytophthora root rot (*Phytophthora cinnamomi* Rands) is most common in rainy, hot climates where SHBs are grown. Rabbiteyes are generally more tolerant than highbush cultivars, but are still affected; rabbiteye 'Tifblue' is one of the most tolerant. The disease is often associated with heavy soils on sites with poor drainage. Diseased bushes characteristically have reduced vigour with wilting leaves, premature yellowing and reddening of leaves, and eventually shoot dieback and defoliation. The disease is spread by zoospores that swim to the root tips and invade them. The best control is to avoid soils with very poor drainage and to use raised beds. *Phytophthora* root rot has been a particular problem in SHBs grown in pine bark beds that are saturated with water.

In one study, Smith (2006) found 'Bluecrisp', 'Jewel', 'Jubilee', 'Misty', 'O'Neal', 'Southmoon' and 'Star' to be more vigorous in infested soil than other SHB cultivars. In another study by Smith (2012), the SHBs 'Gulfcoast', 'Southmoon' and 'Springhigh' were the most vigorous in diseased soil, although no cultivar thrived. P.M. Lyrene indicated in his patent applications that 'Emerald', 'Primadonna', 'Santa Fe' and 'Springhigh' have high resistance to *Phytophthora* root rot.

COMMON BACTERIAL DISEASES OF HIGHBUSH AND RABBITEYE BLUEBERRIES

Bacterial leaf scorch

Bacterial leaf scorch (*Xylella fastidiosa*) was identified recently in Georgia and is rapidly becoming a major problem in the south-eastern USA (Brannen *et al.*, 2007; Harmon and Hopkins, 2009; Oliver *et al.*, 2015). It is caused by the same organism that causes Pierce's disease of grapes. The first symptom is a burn at the tips of leaves that resembles drought damage or fertilizer burn. The symptoms are caused by a blockage of the xylem by the bacteria and induced plant products. The scorching can start on individual stems but eventually becomes uniformly distributed throughout the bush. Leaves eventually abscise, and young stems take on a yellow colour. Stem dieback does not occur until the later stages of the disease. The plant eventually dies when its leaves drop. It is likely that insect vectors transmit the bacterium, possibly leafhoppers and spittle bugs.

To control this disease, care should be taken to propagate only from *Xylella*-free plants. Once diseased plants are identified in the field, they should be removed and destroyed. Insecticides are available that can be used against leaf hoppers, and soil-applied neonicotinoid products may also help in the spring. However, *X. fastidiosa* diseases are difficult to control because the host range of the bacterium and leafhopper vectors includes hundreds of plant species, and the leafhopper vectors are highly mobile and lay hundreds of eggs (Andersen *et al.*, 2016). Among SHB cultivars, 'V1' (FL 86-19) is probably the most susceptible, although 'Star' and numerous other cultivars readily get infected with the disease (Brannen *et al.*, 2007; Chang *et al.*, 2009). 'Emerald', 'Millennia', 'V5' and 'Windsor' have at least field resistance to this bacterium. Rabbiteye blueberries vary in their resistance to bacterial leaf scorch but in general are more tolerant than SHBs. Rabbiteye 'Powderblue' is known to be susceptible, whereas 'Premier' is resistant.

Crown gall

Crown gall (*Agrobacterium tumefaciens*; synonym *Rhizobium radiobacter*) enters the roots through wounds, and forms tumours or galls on roots and sometimes the crown. Infected plants can be stunted. It is generally associated with diseased nursery stock. Infected plants should be removed and destroyed by burning, and grasses should be grown in diseased fields for 2–3 years before replanting.

COMMON VIRAL DISEASES OF HIGHBUSH AND RABBITEYE BLUEBERRY

Primary sources on symptoms and spread of common blueberry viral diseases are Pritts and Hancock (1992), Cline and Schilder (2006), Schilder and Miles (2008), Martin *et al.* (2012) and Polashock *et al.* (2016). Pictures of these diseases are available online (http://www.blueberries.msu.edu/diseases.html).

Blueberry leaf mottle

Blueberry leaf mottle virus is important locally in Michigan and is most prevalent in NHBs 'Jersey' and 'Rubel'. In 'Rubel', the tops of bushes are killed back by the disease and there is little renewal growth. Leaves become deformed and mottled. 'Jersey' bushes also have stem dieback, but it is less severe and little leaf deformity occurs, although the leaves are smaller and paler green. This disease can be confused with necrotic leafspot, as both diseases cause shoot dieback, but leaves of mottle-diseased bushes do not have necrotic spots. The disease is spread between bushes by infected pollen, and it can take up to 3–4 years for symptoms to appear.

Mosaic disease

Mosaic disease is found to a limited extent in most NHB production areas. The causal agent has not been fully elucidated but an ophiovirus (viral group characterized by an elongated and highly filamentous and flexible nucleocapsid with helical symmetry) has been detected in all mosaic disease infected samples (Thekke-Veetil *et al.*, 2014, 2015). The symptoms are bright yellow and green mottling of leaves with some red streaking. Symptoms can disappear and reappear on stems across years.

Blueberry shoestring

Blueberry shoestring virus is most prevalent in areas that grow NHBs. Diseased bushes have strap-like leaves that are often misshapen into twisted and crescent shapes. New stems also have narrow, elongated reddish streaks that are 0.6–2.5 cm long (or sometimes longer). Fruit on infected canes have a reddish-purple colour, and flowers can have a pink tinge or reddish streak. The disease is vectored by the blueberry aphid, *I. pepperi*. The disease has a 2–4-year latent period, and symptoms are often distributed haphazardly within the bush. Hancock *et al.* (1986) found that all NHBs were susceptible to rub inoculation with the virus, although 'Bluejay' had the lowest infection rate. Acquaah *et al.* (1995) found after rub inoculation that NHBs 'Blueray' (46.3%) and 'Elliott' (50%) had the highest infection rates, followed by rabbiteye 'Climax' (36.3%) and SHB 'O'Neal' (12.5%). The lowest rates of infection were found in SHBs 'Georgia Gem' (2.5%) and 'Misty' (2.5%), rabbiteye 'Brightwell' (0.0%) and NHB 'Bluecrop' (2.5%). Wise *et al.* (2010) described 'Bluecrop', 'Bluejay' and 'Northland' as field resistant.

Blueberry scorch

Blueberry scorch virus is most prevalent on the coasts of North America, but reports have also come from Michigan, Italy and the Netherlands. In New Jersey, it is referred to as Sheep Pen Hill disease and is caused by a strain unique from that of the Pacific Northwest. In the most susceptible cultivars, the primary symptom is a sudden necrosis of both flowers and leaves during bloom (although a few leaves survive in the infected zone). As leaves yellow in the autumn, they can display reddish areas around their veins. The most tolerant cultivars either show mild chlorosis and some distortion of new shoots or no symptoms at all (Table 8.2). Only one or two shoots per bush show symptoms initially before the virus spreads throughout the plant. Significant yield declines and death occurs over 3–5 years.

Scorch is vectored by aphids and spreads in a circular pattern from the original point of infection; it does not take more than a few years before all plants in a field are infected. Infected bushes should be removed and burned and insecticides applied to control the aphid vector. Cultivars show a wide range of responses to the North Coast strain of blueberry scorch virus

Table 8.2. Reaction of different highbush blueberry cultivars to the western and eastern scorch virus strains. (Adapted from Bristow *et al.*, 2000.)

Level of reaction	West coast strain	East coast strain
Flower and leaf necrosis, general decline	'Berkeley', 'Bluejay', 'Collins', 'Darrow', 'Earliblue', 'Elliott', 'Jersey', 'Northland', 'Pemberton', 'Rubel', 'Spartan'	All cultivars except 'Jersey'
Marginal chlorosis and/ or pale green leaves	'Bluecrop', 'Legacy', 'Olympia'	
Symptomless	'Bluechip', 'Bluegold', 'Bluetta', 'Coville', 'Duke', 'Hardiblue', 'Lateblue', 'Nelson', 'Northblue', 'Northcountry', 'Northsky', 'Nui', 'O'Neal', 'Puru', 'Reka', 'Sierra', 'Sunrise', 'Toro'	'Jersey'

(Table 8.2), whereas NHB 'Jersey' is the only cultivar that appears unaffected by the East Coast strain (Martin, 2006).

Blueberry shock

Blueberry shock virus has been reported throughout the Pacific Northwest, but also in Nova Scotia (Canada) and California, Pennsylvania, New York and Michigan. The primary symptom, similar to blueberry scorch, is a sudden necrosis of flowers and leaves during bloom. Unlike scorch, a second flush of leaf growth occurs that appears normal, although there are no fruit. Symptoms disappear in subsequent years and crops become normal.

The virus is spread by pollen. The current recommendation in the Pacific Northwest is to let the virus run its course through a field, recognizing that there will be a 1–2-year crop loss, which is better than replanting and the subsequent 4–6 years to return to full production (Martin *et al.*, 2012).

There appears to be variation among cultivars in susceptibility and rate of spread (Finn *et al.*, 2016). In Oregon, 'Baby Blues', 'Bladen', 'Bluecrop', 'Darrow', 'Harding', 'Lateblue', 'Legacy', 'Razz', 'Toro' and all rabbiteye cultivars have tested negative for blueberry shock virus for 10 years or more while growing among many known virus-positive plants. The NHBs 'Berkeley', 'Bluegold', 'Brigitta', 'Nui' and 'Spartan' tested positive in the year following their first bloom, while 'Bluecrop' and 'Toro' showed variable responses. The virus spread rapidly in 'Berkeley', 'Bluegold', 'Bluetta', 'Duke', 'Earliblue' and 'Liberty', but slowly in 'Bluecrop' and 'Blueray'.

Blueberry necrotic ring blotch

Blueberry necrotic ring blotch virus appeared very recently in the southeastern USA and has become widespread. Infected plants develop irregular red or brown ring spots on the upper and lower surfaces of leaves, with or without green centres, depending on the cultivar. Eventually the rings join together and cover the entire leaf surface, possibly leading to defoliation. The vector for this virus is probably an eriophyid mite (Burkle *et al.*, 2012). There is likely cultivar variation in resistance or tolerance, as some SHBs ('O'Neal', 'Star' and FL 86-19) show much more extensive symptoms than others. The virus does not persist in plants after natural defoliation in the autumn, and is not transmitted through vegetative propagation (Robinson *et al.*, 2016).

Blueberry red ringspot

Blueberry red ringspot virus is found mostly in the eastern USA but has been reported across the range of highbush blueberries. The most apparent symptoms are reddish-brown ringspots with green centres (3–6 mm in diameter) on stems and red to purple spots on the upper surfaces of leaves. The fruit can also have red blotches; infected bushes of 'Ozarkblue' produce deformed fruit. The vector has not been fully elucidated, but mealybug has been implicated in Michigan (Schilder and Miles, 2008). Infected cultivars have variable yield reductions from an apparent full crop to 25% crop loss. Diseased bushes should be removed and destroyed.

The NHBs 'Bluetta', 'Blueray', 'Coville', 'Darrow', 'Earliblue' and 'Rubel' are known to be susceptible, while 'Bluecrop' and 'Jersey' are resistant (Pritts and Hancock, 1992). The rabbiteye 'Woodard' is also likely to be resistant (Ehlenfeldt *et al.*, 1993).

Necrotic ringspot disease

Necrotic ringspot disease is caused by two viruses: tobacco ringspot virus and tomato ringspot virus (Martin *et al.*, 2012). The symptoms of these two diseases are so similar that the only way to differentiate them is through diagnostic tests. Necrotic ringspot disease is an important one in the north-western USA and has been found in Arkansas, Michigan, New York, Canada and Chile. It causes brown necrotic spots on older leaves (1.6–3.1 mm in diameter); these spots can fall out, leaving 'shot holes'. Diseased bushes become stunted, and gradually decline in yields over several years. Both diseases are spread by the dagger nematode in a circular fashion at a rate of about 1 m per annum. The virus has a wide array of hosts, including apples, grapes and raspberries. Weeds such as dandelion, chickweed and narrow-leaved plantain can act as hosts. Control consists of removing diseased bushes and fumigating before planting new fields.

Cultivars susceptible to both viruses include NHBs 'Collins', 'Elliott', 'Jersey', 'Pemberton' and 'Rubel'. 'Bluecrop' has some resistance to tomato ringspot virus but is highly susceptible to tobacco ringspot virus. Necrotic ringspot symptoms have not been reported in rabbiteye blueberries.

COMMON PHYTOPLASMA DISEASES OF HIGHBUSH AND RABBITEYE BLUEBERRIES

Blueberry stunt phytoplasma

Blueberry stunt phytoplasma is a widespread problem in NHBs and rabbiteyes. Leaves on diseased plants are smaller and cup downwards with yellowish margins but have veins that remain green. The disease causes a severe stunting of bushes and the fruit ripen slowly, with reduced size. The sharp-nosed leaf hopper (*Scaphytopius magdalensis*) vectors the disease. The leaf hopper overwinters in roots and stems. For control, all infected plants should be removed and destroyed, and leaf hopper activity should be monitored so that insecticides can be applied when they become active. The rabbiteye cultivars 'Premier' and 'Tifblue' are resistant to the vector (Ballington *et al.*, 1993).

OTHER LOCALLY IMPORTANT DISEASES OF HIGHBUSH AND RABBITEYE BLUEBERRY

Alternaria leaf spot

Alternaria leaf spot (*Alternaria tenuissima*) is somewhat common in the southeastern USA and Argentina. The leaf spots are circular to irregular-shaped, brownish-grey lesions with red borders; they are most common on the lower leaves of bushes. The same fungus causes twig blight in Argentina (Wright *et al.*, 2004) and one of the most important worldwide fruit rots. Fungicides are available to control this disease.

Bacterial canker

Bacterial canker (*Pseudomonas syringae*) is found primarily in the Pacific Northwest. Diseased canes have reddish-brown (to black) cankers that can extend the full length of a cane and girdle stems. The cane dies above the canker. Diseased stems should be pruned out and excessive autumn growth prevented by proper fertilization. Copper fungicides can be used in the spring and autumn to help control this disease.

Blueberry fruit drop-associated virus

Blueberry fruit drop-associated virus is a novel DNA virus isolated from NHB 'Bluecrop' plants (Diaz-Lara and Martin, 2016). It was first detected in British Columbia (Canada) in the late 1990s and in a single field in Washington State in 2012. Infected bushes abort almost all of their fruit about 3 weeks prior to harvest, when the berries are about 3–5 mm in diameter. This virus has been proposed as a new member of the family *Caulimoviridae*.

Cylindrocladium rot

Cylindrocladium rot (*Calonectria illicicola*) is a pathogen sometimes important in propagation beds in the south-eastern USA. It is often called 'peanut blight' because the same pathogen affects peanuts. The fungus kills softwood and hardwood cuttings in a circular fashion. There are orange, spherical fruiting bodies on the dead stems. The use of clean rooting medium is the most effective preventative measure.

Neofusicoccum stem canker

Neofusicoccum stem canker (*Neofusicoccum* spp.) is a disease of growing importance in all blueberry production regions in Chile. Diseased plants have reddish-brown necrotic lesions that extend the full length of the canes, often on only one side of the bush. Twigs above the lesions die back, and the whole cane eventually succumbs. Pruning wounds are thought to be an important entry point for the disease. Fungicides can be used to protect wounds against the pathogen (Latorre *et al.*, 2013). A number of highbush and rabbiteye cultivars are known to be susceptible, including 'Bluecrop', 'Brightwell', 'Brigitta', 'Duke', 'Elliott', 'Misty' and 'O'Neal' (Espinoza *et al.*, 2009). 'Elliott' may be the most susceptible.

Phomopsis soft rot

Phomopsis soft rot (*P. vaccinii*) is of minor importance, except on the NHB 'Harrison', which is no longer planted. Infected berries split open and leak juices, primarily during postharvest handling. This is the same fungus that causes blueberry twig blight and cane canker, and the fungicides used to control these diseases have some efficacy against the fruit rot stage.

Pestalotiopsis cane canker

Pestalotiopsis cane canker (*Pestalotiopsis* spp. and *Truncatella angustata*) is a common problem in Chile (Espinoza *et al.*, 2008). Diseased plants have light-brown necrotic lesions on the canes ringed with a reddish line. The twigs die back and eventually the whole plant collapses. Pruning wounds are a key entry point for the disease, and fungicides may help control these pathogens, although recommendations have not been developed. After twig inoculations, 'Brightwell' and 'O'Neal' appeared to be the most susceptible cultivars, 'Bluecrop' and 'Misty' the least susceptible, and 'Brigitta', 'Duke' and 'Elliott' were intermediate.

Red leaf

Red leaf (*Exobasidium vaccinii*) is found mostly on lowbush blueberries but occasionally appears on NHBs grown in the mid-western and eastern USA. The primary symptoms are terminal leaves that pucker and turn red in mid-summer; the undersides of these leaves are covered by cream-coloured layers of spores. Eventually, the affected areas dry up. Infected bushes do not recover and should be destroyed. Fungicides can be used to protect healthy bushes if a high level of disease is present.

Exobasidium fruit spot or green spot

Exobasidium fruit spot or green spot (*E. vaccinii*) is recognized as a serious disease of blueberry fruit in the south-eastern USA. Affected berries develop green unripe spots that do not ripen, producing a highly visible green or pink spot on an otherwise ripe berry. In severe cases, the crop has traditionally been rendered unharvestable and abandoned in the field. In recent years, effective control strategies have been developed using *Exobasidium*-active fungicides that are applied during the late-dormant period and bloom, followed by two to three applications post-bloom in conditions of high disease pressure (Brannen *et al.*, 2016).

Witches broom

Witches broom (*Pucciniastrum goeppertianum*) is a minor disease of highbush blueberries and is usually found in northern temperate regions near fir trees. It is much more common in lowbush blueberries. Diseased plants have broom-like masses of spongy shoots with short internodes and small leaves. The most effective control measures are to plant at least 500 m from fir trees and to kill diseased plants with herbicide.

INSECT PESTS OF HIGHBUSH AND RABBITEYE BLUEBERRIES

Primary sources for this section are Pritts and Hancock (1992), Liburd and Arevalo (2006) and the Blueberries webpage of Michigan State University Extension (http://www.canr.msu.edu/blueberries/pest_management/insects), where pictures of blueberry insect pests can be found.

Insect pests of flowers and fruit

Flower thrips

Flower thrips (*Frankliniella bispinosa* (Morgan), *Frankliniella occidentalis* (Pergrande) and *Frankliniella tritici* (Fitch)) are particular problems in the south-eastern USA on SHB and rabbiteye blueberries. These species are most active during bloom, feeding on ovaries, pollen and corollas of the flowers and resulting in reduced pollination and seed set. Their damage on fruit appears as small, round, necrotic areas. Chilli thrips (*Scirtothrips dorsalis* Hood) have recently been recorded attacking SHB and rabbiteye blueberries in Florida. They feed on young green tissues, leaves and fruits. Thrips belonging to the genera *Frankliniella* and *Scirtothrips* have a wide range of hosts and travel in wind currents. They can be monitored by tapping flowers over white boards, placing blooms in ziplock plastic bags or using white sticky traps (Liburd *et al.*, 2009; Sial, 2016). Biological control agents have been used to control thrips in other crops, but their effectiveness has not been demonstrated in blueberry (Arevalo *et al.*, 2009). A number of biologically based pesticides are available for their control.

Blueberry maggot

Blueberry maggot (*R. mendax* Curran) is the most serious pest of blueberries in the eastern half of the USA. Eggs are generally laid under the skin of ripening or ripe berries, although some females oviposit into 'full green' fruit. Only one egg is laid per fruit, and the maggot hatches 2–7 days later. The maggots are initially colourless and become whitish. The fruit with maggots are very soft and often have a small hole where the eggs were inserted. After feeding in the fruit for about 20 days, the larvae fall to the ground to pupate.

The blueberry maggot overwinters in the soil under bushes for 1–2 years, depending on how much chilling the pupae receive over the winter. The adult is a little smaller than a housefly and has a black and dark-grey body; there are distinctive black bands on its wings. It also has whitish markings on its thorax and thin bands of white on its abdomen.

The blueberry maggot is commonly monitored using yellow sticky traps baited with ammonium acetate. When flies are detected, insecticides are generally applied every 7–10 days throughout the season to prevent egg laying. Integrated pest management programmes have also been developed where pesticide application ceases after two applications if no additional flies are detected (Burrack and Littlejohn, 2012).

Liburd *et al.* (1998) evaluated 18 highbush cultivars for infestation by the blueberry maggot fly over three seasons. They found significantly lower numbers of maggots in berries of the early-ripening cultivars 'Bluetta' and 'Earliblue'. Of the later-ripening ones, 'Herbert' and 'Northland' had the lowest number of maggots in their berries.

Cherry fruit worm

Cherry fruit worm (*Grapholita packardi* Zeller) is a major pest in the mid-Atlantic and mid-western USA. Eggs are laid on developing berries and leaves at about petal drop. The larvae enter the berries at the calyx cup and feed within them.

They may move between berries but do not web them together. The larvae are initially white with black heads, and become pink with brown heads. After larval development, the larvae leave the berries and hibernate in burrows on weed stems or pruned blueberry stubs. They pupate in the early spring and emerge during bloom. The adult is a dark-grey moth with a wingspan of about 9.5 mm; the wings have chocolate-brown bands. Traps can be used to track emergence and the abundance of adults. Insecticide sprays are commonly applied in areas where infestations are heavy.

Cranberry fruit worm

Cranberry fruit worm (*Acrobasis vaccinii* Riley) is a widespread problem in the eastern half of North America. Eggs are deposited in the calyx cup of berries and, on hatching, the larvae bore into the fruit, usually near the stem. The larvae are green with a dark head and feed on multiple berries, webbing the berries together and leaving their frass behind. The presence of the frass and webbing can be used to separate cranberry fruit worm from cherry fruit worm damage. The insect overwinters in a cocoon made of silk and soil particles; the larvae pupate in the spring. The adults are small with dark greyish-brown wings with two distinctive white patches on each forewing. Traps (one per hectare) are used to monitor for this pest. Pesticides are applied when the moths begin flight, and a degree-day model has recently been developed for Michigan for timing these sprays.

In a 6-year field trial of ten highbush cultivars in which infestation by cranberry fruit worm was assessed, 'Duke' had the highest percentage of fruit clusters with larvae in three of the six years, perhaps because it was the earliest cultivar evaluated (Van Timmeren and Isaacs, 2009). 'Legacy', 'Rubel' and 'Toro' had the lowest levels of infestation in two of the six years.

Cranberry weevil or blueberry blossom weevil

Cranberry weevil or blueberry blossom weevil (*Anthonomus musculus* Say) is most serious in New Jersey and Massachusetts but is only a sporadic problem. The weevils sometimes feed on developing buds, but they are most active when the inflorescences begin to open. They sometimes clip the flower pedicel, which dangles and eventually drops off the bush. Their damage also appears as tiny holes drilled into flower buds and corollas; infected flowers do not open and turn purple before falling to the ground. Leaf buds can also be attacked, completely destroying the buds or leaving small round holes in the earliest-developing leaves.

The grub is small, legless, white and C-shaped with a brown head. The adult weevil is small (1.5–2.5 mm long) and brown with whitish markings. Weevils are monitored by the beating tray method or counting the number of individual punctures in flower clusters. Control methods are required if there are more than five adults per bush or more than one puncture is found per five flower clusters (Liburd and Arevalo, 2006).

Plum curculio

Plum curculio (*Conotrachelus nenuphar* Herbst) is an important pest of fruit crops that occasionally causes economic damage in blueberry, primarily in the mid-Atlantic and southern USA. The oviposition wound on the fruit is a diagnostic crescent-shaped scar that remains visible throughout the season. The larva is a white grub, 6 mm long, with no legs and a brown head. It feeds on a single fruit, which may fall to the ground. The larvae leave the fruit after feeding and pupate in the ground. Early cultivars are most likely to be harvested while larvae are still in the fruit; however, fruit infested with plum curculio is usually so badly damaged that the berries drop prematurely rather than being harvested along with sound fruit.

Adults emerge in the mid-summer to autumn, and overwinter under debris. The adult is rarely seen but can be identified as a small weevil, 6 mm long, with a long snout. The surface of the insect is predominantly brown and wrinkled, with grey, white and black specks. The adults 'play dead' when disturbed. Traps are available to monitor plum curculio and help to time insecticide sprays.

Japanese beetle

Japanese beetle (*Popillia japonica* Newman) is a major pest of blueberries in the eastern USA where it is an introduced pest. It is a shiny deep-green beetle, with dark-brown wing covers and an abdomen with white tufts along its sides. Damage appears as skeletonized leaves and scarred fruits. Leaf feeding is generally not a significant concern unless population numbers are extremely high; however, fruit feeding can significantly reduce quality and may serve as an entry point for disease. These beetles also hang tightly on to berries and can ultimately contaminate packaged fruit.

The larvae prefer to feed on the roots of grasses and as a result are more common in sodden fields. The grubs are C-shaped and cream coloured with brown heads and have three pairs of legs. Adults emerge in the summer as berries are beginning to ripen and are active for 6–12 weeks. Adults can be monitored beginning in mid-May with traps baited with pheromone, although the placement of traps in crop fields can actually attract more beetles to a field.

The grubs are the most susceptible stage for control using insecticides. Removal of grassy areas in and around fields during July and August can significantly reduce populations (Szendrei *et al.*, 2005). To help suppress populations biological control agents are also available such as the nematode *Heterorhabditis bacteriophora* and the bacteria *Bacillus thuringiensis* and *Bacillus popillae* (milky spore).

There may be some differences among cultivars in their susceptibility to Japanese beetle damage. When Van Timmeren and Isaacs (2009) evaluated the susceptibility of ten highbush cultivars to Japanese beetle feeding in laboratory and field trials, they found 'Brigitta' to be the most susceptible to fruit feeding, while 'Elliott' and 'Legacy' were the least susceptible to foliage feeding.

Insect pests of buds

Blueberry gall midge or cranberry tip worm

Blueberry gall midge or cranberry tip worm (*Dasineura oxycoccana* Johnson) is a widespread pest of highbush and rabbiteye blueberries, and can become particularly problematic in the south-eastern USA (Lyrene and Payne, 1995). The adult is a tiny fly, with long legs, transparent wings and globular antennae. The larvae go through several colour changes from transparent to orange. Females lay eggs in floral and vegetative buds. Flower buds dry up and fall apart soon after infestation. Developing vegetative shoots are killed, resulting in a tip burn that can be confused with frost damage. Mature larvae fall to the ground to pupate. There can be several generations produced each year. This pest can be monitored by examining shoots for the percentage of buds infested or by placing shoot tips into ziplock bags and monitoring for larval emergence (Sarzynski and Liburd, 2003; Yang, 2005). For control, insecticides can be sprayed during early bud development.

The rabbiteyes 'Brightwell' and 'Powderblue' are highly resistant to flower bud damage; 'Aliceblue', 'Beckyblue', 'Bonita', 'Climax', 'Tifblue' and 'Woodard' are moderately susceptible, and 'Premier' and 'Windy' are highly susceptible (Lyrene and Payne, 1995). Most SHB cultivars are highly resistant to flower bud damage, although they do suffer considerable vegetative damage (Lyrene, 2007). 'Climax' is one of the most susceptible cultivars to vegetative bud damage; infested plants can be almost leafless in the spring.

Blueberry bud mite

Blueberry bud mites (*Acalitus vaccinii* Keifer) have been considered a particular problem in the south-eastern USA, but are becoming increasingly important in all blueberry production regions, particularly Michigan. The blueberry bud mite is too tiny to be seen by the naked eye, but is whitish, elongated and conical with eight legs near its head. Heavily infested buds are reddish in colour and have rough bumps on their outer scales. As the buds open, the flowers desiccate and become distorted with distinctive red blisters (Weibelzahl and Liburd, 2015). The resulting flowers often do not set fruit, and the fruit that do develop have rough skins. Bud mites do not cause the vegetative tip burn associated with gall midge damage.

Monitoring can be done in the late summer and autumn as buds are being set or in early spring to identify infestations. Fields can be sampled by taking ten randomly selected shoots and examining the top five fruiting buds on each shoot for a total of 50 buds per field. While miticides are available for control of this pest, management can be challenging because of its small size and the difficulty in getting miticide residues into the tiny cracks and crevices of the buds it inhabits.

Timely pruning of old canes helps control this disease, along with horticultural oils and miticides. In southern states, the practice of hedging (mowing) bushes after harvest significantly reduces bud mite damage the following year. Some cultivars, particularly 'Rubel', are sensitive to the mite's feeding; others show few symptoms. Among four highbush cultivars in the field, Isaacs and Gajek (2003) found 'Burlington' and 'Rubel' to be the most highly infested, while 'Bluecrop' and 'Jersey' were the least susceptible. Recent observations indicate that 'Liberty' is also highly susceptible, while 'Draper' is resistant (Rufus Issacs and Pat Edger, Michigan State University, East Lansing, Michigan, personal communication).

Insect pests of foliage

Blueberry aphid

Blueberry aphid (*I. pepperi* MacG) is bright green and is usually found on new succulent leaves and stem tips. It is of greatest importance in Michigan due to its role as a vector of blueberry shoestring virus. The largest individuals can be 4 mm in length. The adults give birth to live young without mating, and several generations of live-bearing females are produced each year, leading to very high densities by mid-season. Aphids overwinter as eggs on bushes. The feeding activity of the aphids produces honeydew, which supports the growth of a black sooty mould. The primary economic damage of the blueberry aphid is as a vector for blueberry shoestring virus. Aphids can be monitored by searching the succulent lower shoots on bushes weekly after bloom. Natural enemies usually keep aphid populations suppressed, but if fields are infected with virus or are composed of susceptible cultivars, both broad-spectrum and selective insecticides are available for their control. Several species of parasitic wasps (*Praon* and *Aphidius* spp.) and predatory insects attack aphids and their eggs (Isaacs *et al.*, 2008), so insecticides should be used that have lower toxicity to beneficial insects.

A wide range of densities of blueberry aphids was found on 18 NHB cultivars, but no immunity was identified (Hancock *et al.*, 1982). 'Bluejay', 'Bluehaven', Bluetta' and 'Northland' supported the lowest numbers, while 'Coville', 'Darrow', 'Lateblue', 'Jersey' and 'Spartan' carried the highest numbers.

Leaf rollers

Three species of leaf roller are common in the USA – red-banded (*Argyrotaenia velutinana* Walker), fruit-tree (*Archips argyrospila* Walker) and oblique-banded (*Choristoneura rosaceana* Harris) leaf roller – while the orange tortrix leaf roller (*Argyrotaenia citrana*) is most common in the Pacific Northwest. The fruit-tree leaf roller adult is metallic brown with dark-brown spots on its wings. The oblique-banded leaf roller is tan with chocolate-coloured bands on its wings. The red-banded leaf roller has a complex pattern of colours on its wings including patches of brown, orange, tan and silver. The orange tortrix moth has wings that are pale yellowish-brown to grey in colour with darker mottling.

Leaf rollers construct a shelter by rolling leaves with silk, and then pupate within their shelters. They sometimes tie flowers and green fruit together with silk. The larvae feed on flowers and the surface of berries, although their major importance is as a contaminant of harvested blueberries. They are easily dislodged from their shelters. Natural predators normally keep leaf roller numbers in check, although chemical insecticides are an option if numbers are too high. Pheromone traps are available to determine adult emergence, and growing-degree models have been developed to predict egg hatch, larval development and optimal timing for control (Schilder *et al.*, 2015).

Sharp-nosed leaf hoppers

Sharp-nosed leaf hoppers (*Scaphytopius magdalensis* Prov., *Scaphytopius acutus* and *Scaphytopius frontalis*) are widespread pests that do not cause direct injury to blueberry bushes but vector the phytoplasma that causes blueberry stunt disease. The sharp-nosed leaf hopper overwinters in blueberry leaves on the ground as an egg. Eggs hatch in the spring, and the insect goes through five sedentary nymphal instars before becoming an adult in mid-summer. The dark brownish-black adults can travel great distances. Insecticides are available for leaf hopper control, and their activity periods can be tracked using yellow sticky boards. Resistance to the sharp-nosed leaf hopper has been found in *Vaccinium virgatum* and *V. elliottii*, but not in wild or cultivated *V. corymbosum* (Meyer and Ballington, 1990). The rabbiteyes 'Premier' and 'Tifblue' are resistant to the vector (Ballington *et al.*, 1993).

Spotted-wing *Drosophila*

Spotted-wing *Drosophila* (*Drosophila suzukii*) is a new, major pest of blueberries that attacks many other berry crops, stone fruit, grapes and some pome fruit crops. Its spotted wings are characteristic of the species. It is native to Asia and arrived in California in 2008, rapidly spreading across all fruit production regions of the USA, Canada, Mexico and Europe (EPPO, 2017). Spotted-wing *Drosophila* oviposit with a serrated ovipositor into intact fruit prior to harvest, and within a few days the fruit flesh starts to break down, leading to collapse of the fruit. If not controlled, harvested fruit can carry the larvae inside to distribution centres where it will be rejected. To effectively manage this pest, fields must be monitored with traps regularly and if spotted-wing *Drosophila* are found, effective insecticides must be applied when the fruit are ripening or ripe (van Timmeren and Isaacs, 2013; Isaacs *et al.*, 2015).

White-marked tussock moth

White-marked tussock moth (*Orgyia leucostigma* (J.E. Smith)) is most common as a pest in the north-eastern USA and Canada near woodlots. Mature larvae are large (30 mm long) with distinctive coloration and hairs. They have a bright-red head with a yellowish body, a pair of upright pencil tufts of black hairs on the prothorax, and four white to yellowish brush-like tufts of hairs on the back towards the head. The hairs can irritate the skin of the harvesting crew. Female moths lay large, hairy masses of eggs on blueberry branches. Frequent pruning and good weed management reduce the numbers of these moths, but if populations reach damaging levels, monitoring and control guidelines are available (Isaacs and van Timmeren, 2009).

Other locally important insect pests of highbush and rabbiteye blueberry

Blueberry leaf beetle
Blueberry leaf beetle (*Colaspis pseudofavosa* Riley) is most commonly found in the southern USA in poorly managed fields. The adults are shiny black and 4 mm long. The adults feed on leaves and skeletonize them, but they do most of their damage to younger leaves. High infestation during a cropping season can interfere with next year's yield. After several seasons of high infestations, bushes can be killed. Insecticides can help control this insect, but well-maintained fields rarely have significant infestations.

Blueberry spanworm
Blueberry spanworm (*Itame argillacearia* Packard) is a minor pest, most commonly found in the northern USA on lowbush blueberries. Adult blueberry spanworms have grey-brown wings; females have dark spots on the wings, while males are mostly uniform in colour. First-instar larvae are tan or grey with black spots, and mature larvae are yellow-orange with a line of black spots. In the early season, the larvae feed on flower buds and blossoms, and then move to leaves. They feed at night and hide in leaf litter during the day. Lowbush blueberries can be monitored for this insect by sweeping the foliage with nets; insecticides are sprayed when population numbers are high.

Citrus thrips
Citrus thrips (*Scirtothrips citri* (Moulton)) have become a problem in California where blueberries are planted next to citrus. Citrus thrips feed on the new flush growth of blueberry plants, which causes stunting and probably affects yield and fruit quality. They are found in the blueberry canopy from May to early October. Regular insecticide sprays are being developed to control this insect (http://ipm.ucanr.edu/PMG/r57300111.html).

Leaf-footed bugs
Leaf-footed bugs (*Leptoglossus* spp.) are common in the southern USA, generally where little pesticide is being sprayed. They are usually controlled by natural enemies. These bugs are brown, about 2 cm long, and their hind legs are shaped like a leaf. They damage fruit by poking holes into them. If population numbers become problematic, insecticides can be used to control them.

Scale

Scale (several species) is generally found in older fields on old wood, and can reduce bush vigour. They feed on the phloem and produce honeydew that supports sooty moulds. The scales are small, waxy dots, 2–3 mm wide, on stems covering a yellow insect. Population sizes are generally held in check by good pruning practices and several natural enemies.

White grubs

White grubs (*Cyclocephala longula*) have recently become a major problem in SHB blueberries in California (Haviland and Hernandez, 2011). They feed on plant roots, stunt plants and sometimes kill young, newly planted bushes. The grubs pupate in May and fly at dark from mid-June to mid-July; egg hatch occurs in mid-July. The nematode *H. bacteriophora* and the insecticide imidacloprid are effective in controlling the grub.

European grapevine moth

The European grapevine moth (*Lobesia botrana* (Denis & Schiffermüller)) is a pest that is affecting blueberries in fields located in Chile's central zone (latitude 33–36°S) (González, 2015). It was originally introduced in 2008 to grapes in central Chile from Argentina. To control this pest and avoid its spread to other fruit-growing areas has required an intensive programme that includes systematic trapping, pesticide applications, use of parasitoids (*Coccigomymus fuscipes* (Brétes); Hymenoptera, Ichneumonidae) and sexual confusion. The moth reduces both yields and fruit quality in all areas of the world where it is present. The larvae feed directly on the fruit, which increases the susceptibility of the fruit to fungi, particularly to *Botrytis* and *Aspergillus* spp. *L. botrana* is a polyphagous insect with a wide range of alternative hosts, which include plums, prunes, sweet cherries, almonds and kiwi (González, 2015; CFIA, 2016). This allows survival of the insect during periods when blueberry fruits are not available for feeding. Blueberry growers in Chile are required to spray insecticides regularly to control this pest. In order to allow the import of Chilean blueberries, USA phytosanitary authorities demand fumigation of the fruit with methyl bromide.

NEMATODE PESTS OF HIGHBUSH AND RABBITEYE BLUEBERRIES

The most common nematodes found on blueberry plants are root-lesion (*Pratylenchus* spp.), dagger (*Xiphinema* ssp.) and stubby-root (*Paratrichodorus* ssp.) nematodes. The dagger nematode vectors the disease necrotic ringspot. All nematodes are unsegmented roundworms that are almost invisible without magnification. They range in size from 1 mm (stubby-root nematodes) to 2.5 mm (dagger nematodes). Specialized laboratory procedures are necessary for their isolation and identification. To test for nematodes, soil samples should be taken in June–July in the northern hemisphere (Pritts and Hancock, 1992).

The impact of nematodes in established plantings is not generally thought to be great; however, high nematode populations probably slow the growth of new plantings. It is not known what nematode levels cause economic damage, but pre-plant fumigation is recommended in the late summer or early autumn the year before planting.

COMMON WEEDS OF HIGHBUSH AND RABBITEYE BLUEBERRY FIELDS

Pictures of the most serious weed problems of blueberries are available online (http://www.canr.msu.edu/blueberries/pest_management/weeds).

There are numerous weeds regularly found in blueberry fields (Table 8.3). Weeds can be divided into three groups: (i) annuals, which live less than 1 year; (ii) biennials, which live up to 2 years; and (iii) perennials, which live more than 2 years. Within these broad categories are grasses (monocots), broadleaf plants (dicots), sedges and horsetails. There are both woody perennials and herbaceous plants.

There are two types of annuals: (i) cool season (or winter) annuals that germinate in late summer or autumn, are dormant during the winter, flower in spring or early summer and then die (e.g. chickweed); and (ii) warm season (or summer) annuals that germinate during spring or summer, flower and then die at the end of that growing season (e.g. crabgrass and foxtail). Biennials remain in a vegetative stage during the first season and after winter, they bolt, flower, set seed and then die (e.g. wild carrot and bull thistle). Perennial weeds live for many seasons and flower more than once. Their perennial structures (rhizomes, stolons, crowns, entire plants, nutlets and/or roots) survive from year to year (e.g. nutsedge, quackgrass, dandelion and dewberry). Sedges are perennial, and while they look superficially like grasses, they belong to the family *Cyperaceae* rather than the family *Poaceae*, and have stems with three vertical rows. Horsetails are not flowering plants. They belong to the ancient family *Equisetum* and reproduce asexually via spores.

Management of weeds in blueberries is important for a number of reasons: (i) they compete with plants for water, nutrients and light; (ii) some serve as alternative hosts for insects and diseases; (iii) weeds growing close to blueberries reduce air flow, which can favour fungal growth and harbouring of insects; (v) they provide habitat for vertebrate pests such as voles; (vi) they can compete for pollinators during bloom; and (vii) some produce fruit that can contaminate harvested blueberries.

Probably the most important weed management strategy is to eliminate all perennial weeds before planting. Once perennial weeds become established in a blueberry field, they become extremely difficult to remove. The key to perennial

Table 8.3. Most common weeds found in highbush and rabbiteye blueberry fields.

Type	Common name	Species name
Annual broadleaf	Annual sow thistle	*Sonchus oleraceus*
	Barnyard grass	*Echinochloa crus-galli*
	Black nightshade	*Solanum ptycanthum*
	Chickweed	*Stellaria media*
	Cocklebur	*Xanthium strumarium*
	Fall panicum	*Panicum dichotomiflorum*
	Galinsoga	*Galinsoga quadriradiata*
	Giant foxtail	*Setaria faberi*
	Jimsonweed	*Datura stramonium*
	Lambsquarters	*Chenopodium album*
	Pigweed	*Amaranthus* spp.
	Purslane	*Portulaca oleracea*
	Ragweed	*Ambrosia artemisiifolia*
	Shepherd's purse	*Capsella bursa-pastoris*
	Smooth hawksbeard	*Crepis capillaris*
	Velvetleaf	*Abutilon theophrasti*
Biennial broadleaf	Bull thistle	*Cirsium vulgare*
Perennial broadleaf	Bindweed, field	*Convolvulus arvensis*
	Bindweed, hedge	*Calystegia sepium*
	Bittersweet, oriental	*Celastrus orbiculatus*
	Canada thistle	*Cirsium arvense*
	Cinquefoil	*Potentilla* spp.
	Common catsear	*Hypochaeris radicata*
	Crabgrass	*Digitaria ischaemum*
	Creeping buttercup	*Ranunculus repens*
	Dandelion	*Taraxacum officinale*
	Dock	*Rumex crispus*
	Goldenrod	*Solidago canadensis*
	Hemp dogbane	*Apocynum cannabinum*
	Horse nettle	*Solanum carolinense*
	Marestail	*Conyza canadensis*
	Milkweed, common	*Asclepias syriaca*
	Nightshade, bitter	*Solanum dulcamara*
	Pokeweed	*Phytolacca americana*
	Red sorrel	*Rumex acetosella*
	Smartweed	*Polygonum pensylvanicum*
	Smilax (greenbrier)	*Smilax* spp.
	Stinging nettle	*Urtica dioica*

continued

Table 8.3. *continued.*

Type	Common name	Species name
	Trumpet creeper	*Campsis radicans*
	Virginia creeper	*Parthenocissus quinquefolia*
	White clover	*Trifolium repens*
	White heath aster	*Aster pilosus*
Perennial grass	Johnson grass	*Sorghum halepense*
	Quackgrass	*Agropyron repens*
	Reed canary grass	*Phalaris arundinacea*
Woody perennial	Blackberry	*Rubus* spp.
	Dewberry	*Rubus* spp.
	Elderberry	*Sambucus canadensis*
	Grapevine	*Vitis* spp.
	Himalayan blackberry	*Rubus armeniacus*
	Poison ivy	*Toxicodendron radicans*
	Sassafras	*Sassafras albidum*
	Sumac	*Rhus* spp.
Other	Horsetail	*Equisetum arvense*
	Yellow nutsedge	*Cyperus esculentus*

weed control is to eliminate them as much as possible during the year prior to planting, using a combination of cultivation and herbicide application (before they go to seed). Growing rye or other cover crops for 1 or 2 years prior to planting blueberries will also reduce the number of weeds.

Weeds can be controlled by cultivation, mulching and herbicide sprays (Majek, 2006; Wise *et al.*, 2010). Shallow cultivation can be used to control weeds in the row middles, although perennial grass sods such as tall or hard fescue are often planted instead to suppress weeds. Fabric weed barriers are often used to control weeds in the first few years after planting. Mulches are commonly employed within the rows to smother emerging weeds and prevent germination. Heavy mulches such as sawdust or wood chips are applied annually at depths of 7.5–10 cm (see Chapter 6, this volume, for more information on mulches). Straw mulches are not used, as they decompose quickly and provide favourable environments for field mice and voles. Perennial ryegrass is sometimes planted within rows for weed suppression.

The herbicides employed, and their time of application, depend on the weed species present. Some pre-emergence herbicides are used in the spring before the germination of the weed seeds; other post-emergence ones are used in the middle and latter part of the season. Pre-emergence herbicides are applied to the soil surface, and rainfall or irrigation is necessary to move the herbicide into the soil. Post-emergence herbicides kill weeds through the leaves and must be applied carefully to avoid contact with the blueberry. They work best when the weeds are actively growing, and most need a rain-free period of at least 1–8 h after application to do their work. Some post-emergence herbicides kill only the tissues that come in contact with them (e.g. paraquat), while others (systemic herbicides) are translocated in the plant (e.g. glyphosate).

It is critical that established fields are scouted regularly during the season to remove perennial weeds before they become established. It is also important to avoid using single herbicides repeatedly, as this can lead to an increase in resistant weeds. Specific herbicide recommendations can be found on local extension websites (e.g. University of Florida: http://edis.ifas.ufl.edu/wg016; Washington State University: http://whatcom.wsu.edu/ipm/manual/blue/weed-management.html; Michigan State University: http://www.canr.msu.edu/blueberries/:pest_management/weeds).

CONCLUSIONS

The most widespread disease problems in highbush blueberries are mummy berry, blueberry stunt phytoplasma, blueberry shoestring virus, blueberry shock virus, tomato ringspot virus, blueberry scorch virus, stem blight, cane canker, *Phytophthora* root rot, *Phomopsis* canker, *Botrytis*, *Alternaria* fruit rot and anthracnose fruit rot. Most of these diseases are widespread, although mummy berry and the virus diseases are most prevalent in areas that grow NHBs, and stem blight, cane canker and *Phytophthora* root rot are most common in rainy, hot climates where SHBs are grown. Resistant or tolerant genotypes have been described for most of these diseases, but the genetics of resistance has only been determined for *Phytophthora* root rot, *Phomopsis* canker, cane canker and stem blight. Chemical control strategies exist for most of the fungal diseases, but not for the viruses. Several insects do significant damage to highbush blueberries including blueberry maggot, blueberry gall midge, blueberry bud mite, flower thrips, Japanese beetle, sharp-nosed leafhopper (stunt vector), blueberry aphid (shoestring and blueberry scorch virus vector), cranberry fruit worm, cherry fruit worm, plum curculio and spotted-wing *Drosophila*. Flower thrip, blueberry bud mite and gall midge infestations are a particular problem in the south-eastern USA. Little variation in resistance has been reported to most of these pests in *Vaccinium* spp., except for sharp-nosed leafhopper, blueberry aphid, bud mite and gall midge. Chemical control strategies have been developed for most of these pests, although effective control of the bud mite and spotted-wing *Drosophila* is problematic. Many weed species compete with blueberries, including pigweeds, common ragweed, common lambsquarters, annual and perennial grasses, Canada thistle, field bindweed, horsetail, blackberry, poison ivy and Virginia creeper vine. Most of these can be controlled by timely cultivation, mulching and herbicide sprays.

REFERENCES

Acquaah, T., Ramsdell, D.C. and Hancock, J.F. (1995) Resistance to blueberry shoe-string virus in southern highbush and rabbiteye cultivars. *HortScience* 30, 1459–1460.

Anco, D.J. and Ellis, M.A. (2016) Fruit rots of blueberry: *Alternaria*, anthracnose, and *Botrytis*. Ohioline, Ohio State University Extension, Columbus. Available at: http://ohioline.osu.edu/factsheet/HYG-3213-11 (accessed 3 November 2017).

Andersen, P., Mizell, R. and Brodbeck, B. (2016) Blueberry leaf scorch: what can be learned from other *Xylella fastidiosa*-mediated diseases. In: *Proceedings of the XI International Blueberry Symposium*. International Society for Horticultural Science. University of Florida, Gainesville, Florida.

Arevalo, H.A., Fraulo, A.B. and Liburd, O.E. (2009) Management of flower thrips in blueberries in Florida. *Florida Entomologist* 92, 14–17.

Baker, J.B., Hancock, J.F. and Ramsdell, D.C. (1995) Screening highbush blueberry cultivars for resistance to *Phomopsis* canker. *HortScience* 30, 586–588.

Ballington, J.R., Rooks, S.D., Milholland, R.D., Cline, W.O. and Meyers, J.R. (1993) Breeding blueberries for pest resistance in North Carolina. *Acta Horticulturae* 346, 87–94.

Brannen, P., Scherm, H. and Allen, R. (2016) Management of *Exobasidium* leaf and fruit spot disease of blueberry in the southeastern United States. In: *Proceedings of the XI International Blueberry Symposium*. International Society for Horticultural Science. University of Florida, Gainesville, Florida.

Brannen, P.M., Krewer, G., Boland, B., Horton, D. and Chang, C.J. (2007) Bacterial leaf scorch of blueberry. Available at: http://www.smallfruits.org/assets/documents/crops/blueberries/pest-BlueberryXylella.pdf (accessed 22 February 2018).

Bristow, P.R., Martin, R.R. and Windom. G.E. (2000) Transmission, field spread, cultivar response, and impact on yield in highbush blueberry infected with *Blueberry scorch virus*. *Phytopathology* 90, 474–479.

Burkle, C., Olmstead, J.W. and Harmon, P.F. (2012) A potential vector of *Blueberry necrotic ring blotch virus* and symptoms on various host genotypes. *Phytopathology* 102, S4.17.

Burrack, H.J. and Littlejohn, K. (2012) Rethinking blueberry maggot (*Rhagoletis mendax*) distribution and abundance in North Carolina: when area wide management is unintentional. *International Journal of Fruit Science* 12, 106–113.

Canadian Food Inspection Agency (CFIA) (2016) Phytosanitary import requirements to prevent the introduction of *Lobesia botrana*, the European grapevine moth. CFIA, Ottawa, Ontario, Canada. Available at: http://inspection.gc.ca/plants/plant-pests-invasive-species/directives/horticulture/d-13-03/eng/1448986060402/1448986061775 (accessed 3 November 2017).

Chang, C.J., Donaldson, R., Brannen, P., Krewer, G. and Boland, R. (2009) Bacterial leaf scorch, a new blueberry disease caused by *Xylella fastidiosa*. *HortScience* 44, 413–417.

Cline, W.O. and Schilder, A. (2006) Identification and control of blueberry diseases. In: Childers, N.F. and Lyrene, P.M. (eds) *Blueberries for Growers, Gardeners and Promoters*. Dr Norman F. Childers Publications, Gainesville, Florida, pp. 115–138.

Daykin, M.E. and Milholland, R.D. (1984) Infection of blueberry fruit by *Colletotrichum gloeosporioides*. *Plant Disease* 68, 948–950.

Dedej, S., Delaplane, K.S. and Scherm, H. (2004) Effectiveness of honey bees in delivering the biocontrol agent *Bacillus subtilis* to blueberry flowers to suppress mummy berry disease. *Biological Control* 31, 422–427.

Diaz-Lara, A. and Martin, R.R. (2016) Blueberry fruit drop-associated virus: a new member of the family *Caulimoviridae* isolated from blueberry exhibiting fruit-drop symptoms. *Plant Disease* 100, 2211–2214.

Ehlenfeldt, M.K. and Stretch, A.W. (2000) Mummy berry blight resistance in rabbiteye blueberry cultivars. *HortScience* 35, 1326–1328.

Ehlenfeldt, M.K., Stretch, A.W. and Draper, A.D. (1993) Sources of genetic resistance to red ringspot virus in a breeding population of blueberry. *HortScience* 28, 207–208.

Ehlenfeldt, M.K., Polashock, J.J., Stretch, A.W. and Kramer, M. (2005) Leaf disk infection by *Colletotrichum acutatum* and its relation to fruit rot in diverse blueberry germplasm. *HortScience* 41, 270–271.

European and Mediterranean Plant Protection Organization (EPPO) (2017) Distribution of *Drosophila suzukii*. EPPO, Paris, France. Available at: https://gd.eppo.int/taxon/DROSSU/distribution (accessed 3 November 2017).

Espinoza, J.G., Briceño, E.X., Keith, L.M. and Latorre, B.A. (2008) Canker and twig dieback of blueberry caused by *Pestalotiopsis* spp. and a *Truncatella* sp. in Chile. *Plant Disease* 92, 1407–1414.

Espinoza, J.G., Briceño, E.X., Chávez, E.R., Úrbez-Torres, J.R. and Latorre, B.A. (2009) *Neofusicoccum* spp. associated with stem canker and dieback of blueberry in Chile. *Plant Disease* 93, 1187–1194.

Finn, C.E., Mackey, T., Postman, J. and Martin, R. (2016) Identifying blueberry germplasm that is slow to get *Blueberry shock virus* in the Pacific Northwest United States. In: *Proceedings of the XI International Blueberry Symposium.* International Society for Horticultural Science. University of Florida, Gainesville, Florida.

González, R.H. (2015) *Lobesia botrana (D.&S.) y Otras Polillas Plagas de la Vid en Chile (Lepidoptera: Tortricidae).* [*Lobesia botrana* and other moth pest of grapes in Chile (Lepidoptera: Tortricidae).] Serie Ciencias Agronómicas no. 22. Facultad de Ciencias Agronómicas, Universidad de Chile, Santiago, Chile.

Hancock, J.F., Schulte, N.L., Siefker, J.H., Pritts, M.P. and Roueche, J.M. (1982) Screening highbush blueberry cultivars for resistance to the aphid *Illinoia pepperi*. *HortScience* 17, 362–363.

Hancock, J.F., Morimoto, K.M., Schulte, N.L., Martin, J.M. and Ramsdell, D.C. (1986) Search for resistance to blueberry shoestring virus in highbush blueberry cultivars. *Fruit Varieties Journal* 40, 56–58.

Hancock, J., Callow, P., Serçe, S., Hanson, E. and Beaudry, R. (2008) Effect of cultivar, controlled atmosphere storage, and fruit ripeness on the long-term storage of highbush blueberries. *HortTechnology* 18, 199–205.

Harmon, P.F. and Hopkins, D.L. (2009) First report of bacterial leaf scorch caused by *Xylella fastidiosa* on southern highbush blueberry in Florida. *Plant Disease* 93, 1220.

Haviland, D.R. and Hernandez, N.M. (2011) Development of management programs for white grubs in California blueberries. *International Journal of Fruit Science* 12, 114–123.

Isaacs, R. and Gajek, D. (2003) Abundance of blueberry bud mite (*Acalitus vaccinii*) in Michigan blueberries, and variation in infestation among common highbush blueberry varieties. *IOBC/WPRS Bulletin* 26, 127–132.

Isaacs, R. and van Timmeren, S. (2009) Monitoring and temperature-based prediction of the whitemarked tussock moth (Lepidoptera: Lymantriidae) in blueberry. *Journal of Economic Entomology* 102, 637–645.

Isaacs, R., Schilder, A., Miles, T. and Longstroth, M. (2008) Michigan blueberry facts: blueberry aphid and blueberry shoestring virus (E3050). Michigan State University, East Lansing, Michigan. Available at: http://msue.anr.msu.edu/resources/michigan_blueberry_facts_blueberry_aphid_and_blueberry_shoestring_virus_e30 (accessed 22 February 2018).

Isaacs, R., Wise, J., Garcia-Salazar, C. and Longstroth, M. (2015) SWD management recommendations for Michigan blueberry. Michigan State University Extension, East Lansing, Michigan. http://www.ipm.msu.edu/uploads/files/SWD/SWDManagementMIBlueberriesJune2015.pdf (accessed 22 February 2018).

Latorre, B.A., Torres, R., Silva, T. and Elfar, K. (2013) Evaluation of the use of wound-protectant fungicides and biological control agents against stem canker (*Neofusicoccum parvum*) of blueberry. *Ciencia e Investigación Agraria* 40, 547–557.

Liburd, O.E. and Arevalo, H.A. (2006) Insects and mites in blueberries. In: Childers, N.F. and Lyrene, P.M. (eds) *Blueberries for Growers, Gardeners and Promoters*. Dr Norman F. Childers Publications, Gainesville, Florida, pp. 99–110.

Liburd, O.E., Alm, S.R and Casagrande, R.A. (1998) Susceptibility of highbush blueberry cultivars to larval infestation by *Rhagoletis mendax* (Diptera: Tephritidae). *Environmental Entomology* 27, 817–821.

Liburd, O.E., Sarzynski, E.M., Arévalo, H.A. and MacKenzie, K. (2009) Monitoring and emergence of flower thrips species in rabbiteye and southern highbush blueberries. *Acta Horticulturae* 810, 251–258.

Lyrene, P.M. (2007) Breeding southern highbush blueberries. *Fruit Breeding Reviews* 30, 353–406.

Lyrene, P.M. and Payne, J.A. (1995) Blueberry gall midge: a new pest of rabbiteye blueberries. *Journal of Small Fruit Production* 3, 111–124.

Majek, B.A. (2006) Weeds in blueberries. In: Childers, N.F. and Lyrene, P.M. (eds) *Blueberries for Growers, Gardeners and Promoters*. Dr Norman F. Childers Publications, Gainesville, Florida, pp. 86–98.

Martin, R.R. (2006) Blueberry scorch virus. Descriptions of Plant Viruses, Association of Applied Biologists, Warwick, UK. Available at: http://www.dpvweb.net/dpv/showdpv.php?dpvno=415 (accessed 6 September 2010)

Martin, R.R., Polashock, J.J. and Tzanetakis, I.E. (2012) New and emerging viruses of blueberry and cranberry. *Viruses* 4, 2831–2852.

Meyer, J.R. and Ballington, J.R. (1990) Resistance of *Vaccinium* spp. to the leafhopper *Scaphytopius magdalensis* (Homoptera: Cicadellidae). *Annals of the Entomological Society of America* 83, 515–520.

Oliver, J.E., Cobine, P.A., and de la Fuente, L. (2015) *Xylella fastidiosa* isolates from both subsp. *multiplex* and *fastidiosa* cause disease on southern highbush blueberry (*Vaccinium* sp.) under greenhouse conditions. *Phytopathology* 105, 855–862.

Polashock, J.J. and Kramer, M. (2006) Resistance of blueberry cultivars to *Botryosphaeria* stem blight and *Phomopsis* twig blight. *HortScience* 41, 1457–1461.

Polashock, J.J., Ehlenfeldt, M.K., Stretch, A.W. and Kramer, M. (2005) Anthracnose fruit rot resistance in blueberry cultivars. *Plant Disease* 89, 33–38.

Polashock, J.J., Caruso, F.L., Averill, A.L. and Schilder, A.C. (2016) *Compendium of Blueberry, Cranberry and Lingonberry Diseases and Pests*, 2nd edn. APS Press, St Paul, Minnesota.

Pritts, M.P. and Hancock, J.F. (1992) *Highbush Blueberry Production Guide*. Publication no. NRAES-55. Northeast Regional Agricultural Engineering Service, Cooperative Extension, Ithaca, New York.

Robinson, T.S., Scherm, H, Brannen, P.N., Allen, R. and Deom, C.M. (2016) Blueberry necrotic ring blotch virus in southern highbush blueberry: Insights into *in planta* and in-field movement. *Plant Disease* 100, 1575–1579.

Sarzynski, E.M. and Liburd, O.E. (2003) Techniques for monitoring cranberry tipworm (Diptera: Cecidomyiidae) in rabbiteye and southern highbush blueberries. *Journal of Economic Entomology* 96, 1821–1827.

Schilder, A., Isaacs, R., Hanson, E., and Cline, B. (2015) *A Pocket Guide to IPM Scouting in Highbush Blueberries (E2928)*. Michigan State University Extension, East Lansing, Michigan.

Schilder, A.M. and Miles, T.D. (2008) Michigan blueberry facts: virus and viruslike diseases of blueberries (E3048). Michigan State University Extension, East Lansing, Michigan. Available at: http://msue.anr.msu.edu/resources/michigan_blueberry_facts_virus_and_viruslike_diseases_of_blueberries_e3048 (accessed 22 February 2018).

Sial, A. (2016) Monitoring and managing of thrips in blueberries. UGA Blueberry Blog, University of Georgia, Athens, Georgia. Available at: http://blog.caes.uga.edu/blueberry/2016/03/thrips/ (accessed 24 November 2016).

Smith, B.J. (1998) Botrytis blossom blight of southern blueberries: cultivar susceptibility and effect of chemical treatments. *Plant Disease* 82, 924–927.

Smith, B.J. (1999) Susceptibility of southern blueberry cultivars to *Botrytis* blossom blight. *Fruit Varieties Journal* 53, 48–52.

Smith, B.J. (2004) Susceptibility of southern highbush blueberry cultivars to *Botryosphaeria* stem blight. *Small Fruits Review* 3, 193–201.

Smith, B.J. (2006) Phytophthora root rot and botryosphaeria stem blight: important diseases of southern highbush blueberries in the southern United States. *Acta Horticulturae* 715, 473–479.

Smith, B.J. (2009) Botryosphaeria stem blight of southern blueberries: cultivar susceptibility and effect of chemical treatments. *Acta Horticulturae* 810, 385–394.

Smith, B.J. (2012) Survival of southern highbush blueberry cultivars in phytophthora root rot infested fields in south Mississippi. *International Journal of Fruit Science* 12, 146–155.

Smith, B.J., Magee, J.B. and Gupton, C.L. (1996) Susceptibility of rabbiteye blueberry cultivars to postharvest diseases. *Plant Disease* 80, 215–218.

Smith, B.J., Sampson, B.J. and Walter, M. (2012) Efficacy of bumble bee disseminated biological control agents for control of botrytis blossom blight of rabbiteye blueberry. *International Journal of Fruit Science* 12, 156–168.

Stretch, A.W. and Ehlenfeldt, M.K. (2000) Resistance to the fruit infection phase of mummy berry disease in highhush blueberry cultivars. *HortScience* 35, 1271–1273.

Stretch, A.W., Ehlenfeldt, M.K. and Brewster, V. (1995) Mummy berry blight resistance in highbush blueberry cultivars. *HortScience* 30, 589–591.

Strømeng, G.M. and Stensvand, A. (2001) Susceptibility of highbush blueberry (*Vaccinium corymbosum* L.) cultivars to godronia canker (*Godronia cassandrae* f.sp. *vaccinii*) in Norway. *Gartenbauwissenschaft* 66, 78–84.

Szendrei, Z., Mallampalli, N. and Isaacs, R. (2005) Effect of tillage on abundance of Japanese beetle, *Popillia japonica* Newman (Col., Scarabaeidae), larvae and adults in highbush blueberry fields. *Journal of Applied Entomology* 129, 258–264.

Thekke-Veetil, T., Ho, T., Keller, K.E., Martin, R.R., and Tzanetakis, I.E. (2014) A new ophiovirus is associated with blueberry mosaic disease. *Virus Research* 189, 92–96.

Thekke-Veetil, T., Polashock, J.J., Marn, M.V., Plesko, I.M., Schilder, A.C., Keller, K.E., Martin, R.R. and Tzanetakis, I.E. (2015) Population structure of blueberry mosaic associated virus: evidence of reassortment in geographically distinct isolates. *Virus Research* 201, 79–84.

van Timmeren, S. and Isaacs, R. (2009) Susceptibility of highbush blueberry cultivars to cranberry fruitworm and Japanese beetle. *International Journal of Plant Science* 9, 23–34.

van Timmeren, S., and Isaacs, R. (2013) Control of spotted wing drosophila, *Drosophila suzukii*, by specific insecticides and by conventional and organic crop protection programs. *Crop Protection* 54, 126–133.

Weibelzahl, E. and Liburd, O.E. (2015) Blueberry bud mite, *Acalitus vaccinii* (Keifer) on southern highbush blueberry in Florida. Publication no. ENY-858. University of Florida IFAS Extension, Gainesville, Florida. Available at http://edis.ifas.ufl.edu/in844 (accessed 7 November 2017).

Williamson, J.G. and Miller, E.P. (2002) Early and mid-fall defoliation reduces flower bud number and yield of southern highbush blueberry. *HortTechnology* 12, 214–216.

Wise, J.C., Gut, L.J., Isaacs, R., Schilder, A.M.C., Sundin, G.M., Zandstra, B., Beaudry, R. and Lang, G. (eds) (2010) *Michigan Fruit Management Guide (E0154)*. Michigan State University Extension, East Lansing, Michigan.

Wright, E.R., Rivera, M.C., Esperón, J., Cheheid, A. and Rodríguez Codazzi, A. (2004) Alternaria leaf spot, twig blight, and fruit rot of highbush blueberry in Argentina. *Plant Disease* 88, 1383–1385.

Yang, W.Q. (2005) *Blueberry Gall Midge: A Possible New Pest in the Northwest*. Publication no. EM8889. Oregon State University, Corvallis, Oregon.

APPENDIX 8.1. SYMPTOMS OF MAJOR BLUEBERRY DISEASES

Fruit rots

- Fruit have sunken lesions at the blossom end that are covered by dark, greenish-black masses of spores; widespread problem – *Alternaria* fruit rot.
- Fruit have sunken lesions at the blossom end that are covered with orange spore masses; widespread problem – anthracnose fruit rot.
- Fruit shrivel and turn a pinkish colour as they ripen; widespread problem – mummy berry.
- Fruit are covered by a fluffy grey mould; widespread problem – *Botrytis* fruit rot.

Spring shoot blights

- Infected floral inflorescences are purple-brown; whitish-grey spore masses are found on the midrib of infected leaves and at the base of infected blossoms; widespread problem – mummy berry.
- Flowers turn brown and are covered by a grey mould; stems initially are not affected, later becoming dark brown or black, noticeably darker than *Phomopsis* twig blight; widespread problem – *Botrytis* blossom and twig blight.
- Trail of brown, dead blossoms and leaves along the fruiting shoot, all leaves are dead in infected zone, wood around bud is brown and necrotic; most common in hotter, moister production regions – *Phomopsis* twig blight.
- Sudden, complete necrosis of both flowers and leaves during bloom; some leaves are still alive in the infected zone; no fungal masses; most prevalent on the coasts of North America, but reports have also come from Michigan, Italy and the Netherlands – blueberry scorch.
- Sudden, complete necrosis of flowers and leaves during bloom; some leaves are alive in infected zone; no fungal masses; second flush of leaf growth occurs that appears normal; generally restricted to the Pacific Northwest but also reports from Michigan – blueberry shock.

Cane diebacks/reduced bush vigour

- Dead canes have elongated, flattened cankers that are covered by small, pimple-like pycnidia; most common in cooler production regions – *Phomopsis* canker.

- Lower third of 1–2-year-old stems have small, reddish areas (like a bull's-eye) that develop into elliptical, brownish-purple lesions 2.5–15 cm long; most common in the coolest production regions – *Fusicoccum* canker.
- No apparent canker, but infected, dying stems show a pecan-brown discoloration under the bark; widespread problem in highbush and rabbiteye blueberries in rainy, hot climates – stem blight.
- Raised, red bumps appear and develop over 4–6 months into cankers that are swollen, with deep cracks running through them; most common in rainy, hot climates – stem canker.
- Bushes have reduced vigour with wilting leaves that prematurely yellow and redden; most common in hot, moist climates where SHBs are grown – *Phytophthora* root rot
- Bushes are low in vigour and have mushrooms at the base of crowns, as well as fan-shaped fungal growth under the bark at soil level; locally important worldwide in replanted woodlots – *Armillaria* root rot.
- Extensive dieback of stems and little or no crop; leaves appear malformed with small circular, necrotic spots ranging from 1.5 to 4.5 mm in diameter; spots are also found on young stems, and flower clusters sometimes appear deformed; most important in the north-western USA and found in Michigan, New York, Canada and Chile – tomato ringspot.
- Extensive dieback of stems and little or no crop; leaves have brown necrotic spots on older leaves 1.5–3 mm wide that fall out, leaving 'shot holes'; locally important in the northern USA, Canada and Chile – necrotic ringspot.
- Extensive dieback of stems and little or no crop; leaves are reduced in size and cup downward with yellowish margins but have veins that remain green; widespread problem in NHBs and rabbiteyes – blueberry stunt.
- Leaves abscise and young stems take on a yellow colour; earlier symptoms are a leaf burn that first appears on individual stems but eventually affects all canes; stem dieback does not occur until the later stages of disease – bacterial leaf scorch.

Leaf discolorations and distortions

- Irregular reddish-brown spots on leaves that are covered with a faint white mould; infected leaves are somewhat distorted; common worldwide problem – powdery mildew.
- Reddish-brown spots on leaves that become yellow and drop prematurely; the undersides of leaves have yellow and orange spores associated with the lesions; whole plant may be defoliated; common in the south-eastern USA and found in the eastern USA, Argentina, Spain and Australia – leaf rust.
- Small, reddish flecks of colour appear initially and develop into large, brown lesions 12.5–25 mm in diameter, with a bull's-eye pattern; whole

plant may be defoliated; widespread problem in the south-eastern USA – *Gleosporium* or anthracnose leaf spot.

- Numerous small, purple spots, 1.5–4.5 mm in diameter, on leaves that have white to tan centres; whole plant may be defoliated; widespread problem in the south-eastern USA – *Septoria* leaf spot.
- Leaves are strap-like and often misshapen into twisted and crescent shapes; new stems have narrow reddish streaks; most important in Michigan but found elsewhere in NHBs – blueberry shoestring.
- Leaves appear malformed with small, circular, necrotic spots ranging from 1.6 to 4.7 mm in diameter; spots are also found on young stems, and flower clusters sometimes appear deformed; leads to extensive shoot dieback; most important in the north-western USA but found elsewhere in highbush blueberries – tomato ringspot.
- Leaves have brown necrotic spots on older leaves 1–3 mm wide that fall out, leaving 'shot holes'; leaves may become rosetted, and stem dieback may occur; leads to extensive shoot dieback; occasionally found in the northern USA, Canada and Chile – necrotic ringspot.
- Reddish-brown ring spots 3–6 mm in diameter on stems and red to purple spots on the upper surfaces of leaves; most common in eastern USA – red ringspot.
- Irregular red or brown ring spots on the upper and lower surfaces of leaves, with or without green centres; rings may join together and cover the entire leaf surface leading to defoliation; widespread in south-eastern USA – blueberry necrotic ring blotch.
- Leaves are mottled and often malformed; leaves can be strap-like or rosetted; most important in Michigan – blueberry leaf mottle.
- Bright yellow and green mottling of leaves with some red streaking; found to a limited extent in most NHB production areas – mosaic.
- Leaves are reduced in size and cup downward with yellowish margins but have veins that remain green; leads to extensive shoot dieback; widespread problem in NHBs and rabbiteyes – blueberry stunt.
- Leaves have a scorching at their tips, resembling drought or fertilizer burn; the scorching starts on individual stems but eventually becomes uniformly distributed throughout the bush; leaves eventually abscise and young stems take on a yellow colour – bacterial leaf scorch.

APPENDIX 8.2. CHARACTERISTIC SYMPTOMS OF BLUEBERRY PEST DAMAGE

Flowers

- Brown feeding damage on corollas and ovaries by tiny insects; reduced pollination and fruit set; most important in the eastern USA – flower thrips.
- Flower pedicels are clipped and dangling; tiny holes are drilled into flower buds and corollas; adult weevil is small, 1.5–2.5 mm long, with a pronounced snout and brown with whitish markings; sporadic problem mostly in New Jersey and Massachusetts – cranberry weevil.

Fruit

- Larvae begin colourless and turn to whitish; no webbing attached to fruit; adult fly has distinctive black bands on its wings; the most serious pest of blueberries in the eastern half of the USA – blueberry maggot.
- Larvae enter berries at the calyx cup; they are initially white with black heads and become pink with brown heads; no webbing attached to fruit; adult is a dark-grey moth with chocolate-brown markings; major pest in the mid-Atlantic and mid-western USA – cherry fruit worm.
- Fruit have a diagnostic crescent-shaped scar where the eggs were laid; no webbing attached to fruit; larva is a white grub, 6 mm long, with no legs and a brown head; adult is a small weevil with a long snout; primarily a problem in the southern USA – plum curculio.
- Larvae bore into the fruit near the stem; they are green with a dark head and feed on multiple berries; berries are webbed together and covered with frass; adults have dark, greyish-brown wings and two white markings on each forewing; widespread problem in the eastern half of North America – cranberry fruit worm.
- Fruit collapse due to the feeding of small white maggots; primarily on the central coast of California and in the Pacific Northwest (but now reported in Florida, North Carolina and the Great Lakes region) – spotted-wing *Drosophila*.

Leaves and buds

- Leaves are skeletonized and covered with clusters of shiny, deep-green bee-tles with dark-brown wing covers and abdomens with white tufts along their sides; major pest of blueberries in the eastern USA – Japanese beetle.
- Flower buds dry up and disintegrate; terminal vegetative buds are killed and blacken, or produce very short shoots with only a few highly distorted leaves; SHBs in the south-eastern USA may be almost leafless in the spring; a widespread pest of highbush and rabbiteye blueberries, particularly in the south-eastern USA – blueberry gall midge.
- Heavily infested flower buds are reddish in colour and have rough bumps on their outer scales; they desiccate and produce distorted flowers with distinctive red blisters; vegetative buds are not attacked; more of a problem in the south-eastern USA on rabbiteyes and SHBs but widely distributed – blueberry bud mite.
- Young leaves and succulent new stems are covered with small, bright-green aphids; feeding activity produces honeydew, which supports the growth of a black sooty mould; vector of blueberry shoestring virus; of greatest importance in Michigan – blueberry aphid.
- Leaves are rolled together with silk; flowers and green fruit can also be tied together with silk; common problem in eastern and north-western USA – leaf rollers.
- Leaves show feeding damage; very distinctive larvae are present that have a bright-red head, a yellowish body, and tufts of hairs that project out from the head and line both sides of the body; the hairs can irritate the skin; female moths lay large, hairy masses of eggs on blueberry branches; most common as a pest in the north-eastern USA and Canada near woodlots – white-marked tussock moth.

9

Pre- and Postharvest Management of Fruit Quality

INTRODUCTION

About two-thirds of highbush blueberry world production is marketed as fresh fruit, and consumption of blueberries has doubled in the last 12 years. This expansion was triggered, at least in part, by the discovery in the 1990s that blueberries are one of the fruit species with highest antioxidant content (Prior *et al.*, 1998). Antioxidants have been found to generate diverse positive impacts on human health, and the highest benefits are obtained by the consumption of fresh fruit. Consequently, the concept of quality in blueberries has changed to include not only size, sugar (soluble solids (SS)), acids (titratable acidity (TA)) and firmness but also the concentration of antioxidants.

The growth in demand for fresh blueberry fruit has required not only greater production of fruit during the summer but also year-round supply from the southern hemisphere. To meet this demand, the postharvest life of the fresh fruit has had to be expanded from days to weeks. Numerous management practices have a marked effect on the quality and postharvest life of the fruit, from site selection to the choosing of cultivars and the types of cultural practices. In this chapter, we define the attributes of fruit quality, followed by an analysis of the factors that affect quality before harvest. Finally, we describe the main factors that influence the postharvest life of the fruit and the approaches that can be used to maintain fruit quality, with a focus on the fresh market.

FRUIT MATURATION AND QUALITY

Maturity at harvest is the most important factor that determines fruit quality and postharvest life. The quality of blueberry fruit cannot be improved after it is picked; therefore, it is important to harvest the fruit when its development is optimum for handling and consumption. Generally, fruit develop the highest sugar and most intense flavours if allowed to ripen fully on the plant (Beaudry, 1992). Blueberries generally ripen rapidly, usually going from 50% pink to fully blue in 2–3 days, and then require only several more days to develop full flavour and sweetness (Forney, 2009). As a rule, immature fruit are more subject to dehydration and bruising during handling and storage, and present inferior quality when they become overripe. However, overripe fruit are likely to become soft and mealy soon after harvest, with an insipid flavour, and they are more prone to decay.

The maturity of fruit on the bush at any one time can vary greatly and probably influences postharvest life. Bravo (2017) measured the effect of fruit maturity and fruit positioning within the canopy on fruit quality after 30 and 45 days refrigerated storage. Fruit of NHB 'Brigitta' and 'Duke' plants from rows planted in north–south orientation were harvested from the east and west sides of bushes when they first became blue, or were left on the plant for 6 additional days. The percentage of blue fruit and the total number of fruit were higher on the east side of the plant and they were softer. The delay in harvest significantly increased the amount of soft and very soft fruit, which greatly affected their storage life.

Depending on cultivar, there can be a trade-off between when the fruit are most flavourful and when they have the longest postharvest life. Less mature blueberries of most cultivars have a greater storage potential than more mature fruit; however, fully mature berries of many cultivars have a sweeter flavour. Lobos *et al.* (2014) measured the effect of delaying harvest on fruit quality and storage life of the late NHBs 'Aurora', 'Elliott' and 'Liberty' as their bushes progressed from 30% to 60% blue fruit. In all three cultivars, as fruit ripened there was a steady decline in TA while SS remained stable; this generated a higher SS:TA ratio, which indicated that the fruit were becoming sweeter. This was supported by taste panel perceptions of greater sweetness associated with fruit that was harvested later. There was also a significant overall reduction in fruit firmness and storage life as the percentage of blue fruit at harvest increased; however, a significant interaction for storage life between fruit maturity and cultivar was observed, indicating that the fruit of some cultivars were less influenced by overall crop ripeness than others. The storage life of 'Elliott' was significantly influenced by fruit ripeness, while 'Aurora' and 'Liberty' were not. 'Liberty' produced the highest percentage of sound fruit of the three cultivars, and its fruit stored the longest, suggesting that 'Aurora' and 'Liberty' fruit can be left longer on the bush before harvesting than 'Elliott' without significantly damaging storage life. This would allow 'Aurora' and 'Liberty' to develop a sweeter flavour before shipping. The physicochemical and sensory evaluations were highly correlated, suggesting that TA, SS and firmness could be used as predictors of consumer preferences.

MEASURING FRUIT QUALITY

Quality is defined as the degree of excellence or superiority in a combination of attributes, properties or characteristics that give each commodity value in terms of human food (Kader, 1999). The relative importance of each quality component depends on the commodity and its intended use (e.g. fresh or processed) and varies among producers, retailers and consumers. To the producers, it is important that a certain commodity has high yield, is easy and cheap to harvest, and can withstand long-distance shipping to markets in good condition. Appearance, quality, firmness, uniformity, susceptibility to decay and shelf-life are important to wholesale and retail marketers.

In general, the consumer's initial purchase is based on fruit appearance (including freshness) and firmness (texture). However, subsequent purchases are dependent on the consumer's satisfaction, given mainly by flavour and quality, which are related to the fruit's SS (mainly sugars), TA (organic acids), SS:TA ratio, flesh firmness and nutritional quality (Kader, 1999). Beaudry (1992) suggested an overall set of quality standards for blueberries (Table 9.1).

Fruit quality is commonly established by instrumental assessment of each characteristic of the fruit (size, colour, SS, pH and acidity), as well as sensory evaluation including taste panels. However, there are some drawbacks in measuring individual quality parameters. The methods often require sample preparation, are destructive, time-consuming and expensive, and generally focus on only a few aspects of fruit quality. Particular care must also be taken to analyse samples under the same conditions. For example, when fruit temperature (4.4–38°C) was studied for its effect on firmness readings independent of maturity stage, the firmness of berries decreased when they were warmed and increased when they were cooled (16% difference between the extreme temperatures). There is also inherent variability among plants in a field and within each bush, and sample collection may not appropriately estimate the population of fruit within a field.

Table 9.1. Recommended quality standards for blueberry fruit. (Adapted from Beaudry, 1992.)

Attribute	Level or range
pH	2.25–4.25[a]
Acidity (primarily citric acid)	0.3–1.3% w/w
SS	>10% w/w
SS:TA ratio	10–33
Firmness	>70 g for a 1 mm deformation
Size	>10 mm
Colour	Blue, <0.5% w/w anthocyanin

[a]Perkins-Veazie *et al.* (1995) established a pH level of less than 3.5.

Systems have been developed to measure simultaneously a number of fruit maturity variables. Guidetti *et al.* (2009) used visible–near-infrared spectroscopy (Vis-NIR; 450–980 nm) to evaluate correlations between SS, firmness, ascorbic acid and total content of anthocyanins and polyphenols from fresh and homogenized samples of NHB 'Brigitta' and 'Duke' blueberries. Correlation coefficients between Vis-NIR predicted values and instrumental levels ranged from 0.80 and 0.92 for predictive models that were established using fresh samples. The highest correlations between predicted and instrumental values were for flavonoid and anthocyanin contents ($r > 0.90$), while for fresh samples the highest correlations were for anthocyanin content in 'Duke' and SS and firmness in 'Brigitta'.

Studies using Vis-NIR for evaluating the impact of artificially induced changes in source–sink relationships (through leaf girdling and leaf plucking) on the maturation of intact fruit of SHB blueberries showed that Vis-NIR, in conjunction with statistical methods, was able to detect changes in pigmentation, SS, water and reflective index (Mowat and Poole, 1998). After studying the use of NIR (750–2500 nm) and mid-infrared (MIR; 4000–400 cm^{-1}) spectroscopy in NHBs 'Brigitta' and 'Duke', Sinelli *et al.* (2008) concluded that ascorbic acid concentrations could not be adequately assessed through NIR or MIR. However, both NIR and MIR were adequate for estimating SS concentrations. In addition, NIR models adequately predicted total concentrations of phenols, flavonoids and anthocyanins.

Alterations in volatiles assessed by gas chromatography–mass spectrometry have been used to detect and discriminate among diseases of several crops. An electronic nose (E-nose) is an array of electronic gas sensors tuned to a cross-section of volatiles that is capable of indirectly measuring volatiles emanating from the fruit. This technique was applied successfully to detect the presence of and differentiate among three postharvest diseases (*Botrytis cinerea*, *Colletotrichum gloeosporioides* and *Alternaria* spp.) of 'Brightwell' rabbiteye blueberries with 90% overall correct classification (Li *et al.*, 2010).

Consumer preference studies

In general, the perception of taste by consumers includes sugars and acids, while in the concept of flavour, aroma is also included in the above parameters. In a study of consumer preferences by Gilbert *et al.* (2014), the sensory parameters that were by far most important to consumers were sweetness and intensity of flavour. They preferred a fruity berry that 'is so sweet . . . no sugar is needed'. Mild- and tart-flavoured berries were much less favoured. Consumers also strongly desired a 'crispy berry that pops in your mouth' but that was not so firm it did not give. They showed a strong dislike for mealy, pasty berries with lots of seeds and those with a tough chewy skin. They also disliked mushy berries and meaty ones with little juice. Large berries were favoured over small ones, with a strong positive rating for 'the biggest berries they ever saw' and 'larger than marbles'. Consumer's feelings about berry colour were mixed: dark-blue-coloured berries received the highest rating, but bright blue ones were also favourably rated. Gilbert *et al.* (2015) had consumer panellists rate their overall liking, texture, sweetness, sourness and flavour intensity of 19 SHB cultivars and selections over 3 years. Overall liking of fruit was significantly correlated with sweetness, texture and overall flavour, while sourness had a significantly negative relationship with overall liking.

Saftner *et al.* (2008) utilized different groups of visual, textural and flavour-related quality parameters in a sensory evaluation of ten highbush and two rabbiteye blueberries. Visual quality comprised appearance, blue colour and fruit size. Textural quality included bursting energy, skin toughness, texture during chewing and juiciness, while flavour-related quality was broken down into sweetness, tartness, sweet/tart balance, blueberry flavour and overall eating quality. Flavour-related characteristics better predicted consumer preferences for overall quality than the other parameters in this study, although textural and visual characteristics also contributed. In general, cultivars with lower compression firmness had low sensory scores for bursting energy and texture during chewing. For the 12 cultivars used in the study, overall eating quality was most highly correlated with flavour acceptability ($r = 0.87$) and blueberry-like flavour intensity ($r = 0.85$). Within the range of fruit size and colour evaluated, the results indicated that fruit size is a better indicator of sensory visual quality than the acceptability and intensity of fruit colour.

In a study of the juice of six cultivars, rabbiteye blueberries differed significantly from the flavour of SHB blueberries (Bett-Garber *et al.*, 2015). SHB berries had a significantly more intense blueberry flavour and a sweeter taste. Blueberry flavour correlated with oxalic acid, citric acid and antioxidant capacity. A sweet taste was positively correlated with glucose, total sugars, oxalic acid, citric acid, anthocyanidins, and Brix value, and negatively correlated with sucrose, quinic acid, and total acids. It negatively correlated with quinic acid and total acids. Sour taste correlated positively with total acids and TA, and correlated negatively with pH and Brix:TA ratio.

CULTIVAR COMPARISONS OF INDIVIDUAL FRUIT QUALITY PARAMETERS

Fruit size

There is considerable variability in the size of rabbiteye and highbush cultivars (Saftner *et al.*, 2008; Gilbert *et al.*, 2014; Gündüz *et al.*, 2015). Large fruit are easier and cheaper to harvest (Strik *et al.*, 2003) and have greater consumer appeal than small fruit. Saftner *et al.* (2008) found that individual fruit weight values were highly correlated with scores for panel acceptability of size ($r = 0.67$), appearance ($r = 0.62$) and overall eating quality ($r = 0.47$). These results indicate that larger berries are preferred for fresh consumption.

Weight loss

The maximum weight loss that can occur during storage before blueberries become unsaleable has been estimated to be 5–8% (Sanford *et al.*, 1991). Machine-harvested rabbiteye blueberries stored at 3°C had a weight loss of 0.2% per week, which was about double the rate of change in weight loss of hand-harvested fruit. When Perkins-Veazie *et al.* (1995) evaluated eight SHB clones (plus the NHB 'Bluecrop' and rabbiteye 'Climax'), the weight loss after 21 days of storage at 5°C ranged from 3.6% (in clone A109) to 6.6 or 6.7% (in 'Climax' and 'Gulfcoast'). Miller *et al.* (1993) showed a marked effect of packing material on weight loss of SHB 'Sharpblue'. After 3 weeks at 1°C plus 2 days at 16°C, fruit packed manually in fibre-pulp cups had significantly higher weight loss than those packed automatically in polystyrene cups (7.4 versus 1.0%). In controlled-atmosphere storage, Alsmairat *et al.* (2011) found that moisture loss was only 0.6–2.3% over an 8-week storage period and was dependent on cultivar and storage atmosphere.

Fruit maturity and subsequent cuticle development plays an important role in maintaining the water status of fruit. Moggia *et al.* (2016) measured the effect of maturity at harvest (75% blue, 100% blue and overripe fruit) on cuticular triterpene content, firmness and weight loss in NHB 'Brigitta' and 'Duke' fruit that were bagged in macro-perforated low-density polyethylene bags. 'Duke' fruit softened faster and were more prone to dehydration than 'Brigitta' samples, and overripe fruit were less firm after storage. Weight loss and softening rates were highly correlated with ursolic acid content at harvest.

Aroma

Blueberries do not have as strong a characteristic aroma as strawberries and apples. However, over 100 volatile compounds have been identified in highbush blueberries, including low-molecular-weight esters, alcohols, aldehydes, and acylic and cyclic terpenes (Hirvi and Honkanen, 1983). As fruit ripen, the concentration of aroma volatiles increases rapidly, closely following pigment formation.

Significant variation in the number and quantity of volatile compounds has been reported in cultivars of highbush and rabbiteye blueberry. These compounds vary greatly across cultivar, developmental state and harvest date (Du *et al.*, 2011; Gilbert *et al.*, 2013, 2015). The most widespread aromatic compounds are: *trans*-2-hexenol, *trans*-2-hexenal, linalool, α-terpineol, geraniol, limonene, *cis*-3-hexen-1-ol, nerol, 1-penten-3-ol, hexanal and 1,8-cineole (Parliament and Kolor, 1975; Hirvi and Honkanen, 1983; Horvat and Senter, 1985; Baloga *et al.*, 1995). Among these volatile compounds, trained flavour panellists have determined that a blueberry aroma is produced by a minimum mixture of *trans*-2-hexenal, *trans*-2-hexenol and linalool (Parliament and Kolor, 1975) and *cis*-3- hexen-1-ol and geraniol (Horvat and Senter, 1985).

Saftner *et al.* (2008) found that total aromatic volatile concentrations were not correlated with sensory scores for flavour, overall eating quality or any other sensory characteristic. However, in another study associating likability among SHB cultivars with specific aromatic compounds, Gilbert *et al.* (2015) reported that β-caryophyllene oxide, 2-heptanone and neral were significantly associated with sweetness. The volatiles 2-undecanone and 3-methyl-1-butanol were significantly associated with increases in blueberry flavour intensity. Sourness was positively associated with linalool and 1,8-cineole, two of the compounds that have been associated with typical blueberry flavour. E-2-Hexenal was negatively associated with liking and sweetness.

Scar

The size and wetness of the scar is an important attribute in blueberries, as it is considered to be the principal point of entry of microorganisms and is associated with about 90% of fruit decay (Cappellini and Ceponis, 1977). Blueberries do not present an abscission zone, and final fruit separation from the pedicel is brought about by mechanical rupture of the vascular system and the epidermis. As a consequence, when fruit is removed from the plant at harvest, there is a wound in both the vascular tissue and the epidermis (Gough and Litke, 1980).

The size of the fruit scar can also have a significant impact on fruit water loss during storage. In a comparison of three germplasm lines of NHB blueberries, the fruit scar was shown to account for an average of 25% of the moisture lost at 20°C and 45% at 0°C (Moggia *et al.*, 2017a). While the stem scar covered only 0.19–0.74% of the fruit surface area, on an area basis its rate of transpiration was hundreds of times greater than that of the cuticle. Fruit categorized as having a large stem scar generally had a greater rate of water loss and were less firm than fruit categorized as having a medium- or small-sized stem scar, although there was considerable variability. One line exhibited a 75% lower rate of water loss from its stem scar than would be predicted based on its scar diameter. This implies that beyond the physical effect of scar size, a genetically controlled component may influence fruit water loss in a different manner.

The scar size itself is to a large extent a genetically controlled trait. The stem scar diameter varied by as much as 50% among eight clones of SHB blueberries (ranging from 1.46 mm in A109 to 2.20 mm in MS108)

(Perkins-Veazie *et al.*, 1995). From field observations, there is the perception that, within a cultivar, scar size is positively correlated with fruit size (i.e. larger fruit usually have larger scars). However, Ballington *et al.* (1984) reported no association between scar diameter and fruit size among various cultivars of *Vaccinium* spp. The data of Perkins-Veazie *et al.* (1995) also indicated little relationship between fruit size and scar diameter. Moggia *et al.* (2017a) found that fruit with a large scar were larger (greater in weight, length and diameter) than those with a medium or small scar for two of the families studied but not the third.

Colour

Highbush blueberry colour is a highly complex attribute affected by chemical and physical parameters. The light blue colour of fresh blueberries is determined by the amount of waxy 'bloom' (quantity and structure) and the anthocyanin content of the skin. Most of the anthocyanin pigments are formed during the 6 days following the development of red colour (Woodruff *et al.*, 1960). Blue colour development can take place off the plant, but further sugar accumulation is very limited.

The wax reflects and refracts light, causing the light-coloured appearance on the skin surface (Albrigo *et al.*, 1980). Sapers *et al.* (1984) studied the wax ultrastructure of ripe NHB berries under the electron microscope and distinguished two types of cultivar: one, represented by 'Blueray' and 'Burlington', that had upright and flat wax platelets over a layer of continuous wax or annealed patches of wax, and another, typified by 'Elliott', that had few if any platelets and instead had an extreme degree of patchiness.

The whitish material or 'bloom' on the surface of the fruit is a rather loose, thin and fragile wax deposit. The fragile nature of the wax makes it sensitive to even gentle rubbing, brushing and bouncing of the fruit. This relationship has made preservation of the waxy bloom during handling an important goal.

Under laboratory conditions, a colorimeter is commonly used to obtain objective colour measurements of blueberry fruit. The instrument provides three variables: L (with lower values provided by darker fruit), a (where negative and positive values indicate predominance of green and red, respectively) and b (expressing the blue component or blueness). In a study on ten highbush and two rabbiteye blueberry cultivars, Saftner *et al.* (2008) found that variations in green (negative a) and red (positive a) chromas were not as large as that of the blue chroma (Table 9.2).

Research on 11 highbush cultivars showed that, in fresh blueberries, there was a close relationship between the L value and visual assessments of waxy bloom. When numerical scores were assigned to bloom, a regression analysis of this relationship yielded a correlation coefficient of $r = 0.75$ (values can range from -1 to 1). Samples that scored high for bloom also tended to have

Table 9.2. Surface colour measurements (*L*, *a* and *b*; see text for details), fruit weight in fresh fruit and aromatic volatile concentration of blueberry extracts from ten NHB and two rabbiteye blueberry cultivars listed by harvesting season. Total volatile concentration is reported in detector area response units of picoamps (pA). (Adapted from Saftner *et al.*, 2008.)

| Cultivar | Surface colour | | | Fruit weight (g) | Aromatic volatile concentration (pA) |
	L	*a*	*b*		
NHB					
'Chanticleer'	24.52[c,d,e]	−0.09[a,b,c]	−4.90[b,c]	2.09[c,d,e]	1397[g]
'Duke'	27.21[a,b,c]	−0.13[a,b,c]	−5.36[b,c]	2.38[b,c]	2739[a,b,c]
'Hannah's Choice'	26.28[a,b,c,d]	−0.03[a,b]	−5.11[b,c]	2.85[a]	1768[f,g]
'Weymouth'	21.94[d,e]	−0.04[a,b]	−3.94[a,b]	1.56[f]	3375[a]
'Berkeley'	30.84[a,b]	−0.70[b,c]	−5.10[b,c]	2.48[a,b,c]	3035[a,b]
'Bluecrop'	27.33[a,b,c]	0.09[a]	−5.48[b,c]	2.41[b,c]	2243[c,d,e,f]
'Bluegold'	30.12[a,b]	−0.66[b,c]	−5.70[c]	2.57[a,b]	2555[b,c,d]
'Coville'	27.51[a,b,c]	−0.01[a,b]	−5.30[b,c]	2.24[b,c,d]	2450[b,c,d,e]
'Elliott'	31.73[a]	−0.78[c]	−4.79[b,c]	1.94[d,e]	1994[d,e,f,g]
'Lateblue'	26.16[a,b,c,d]	−0.32[a,b,c]	−3.88[a,b]	2.45[b,c]	1820[e,f,g]
Rabbiteye					
'Coastal'	20.29[e]	0.11[a]	−2.39[a]	1.84[e,f]	2432[b,c,d,e]
'Montgomery'	27.73[a,b,c]	−0.37[a,b,c]	−2.30[a]	1.66[f]	1764[f,g]

[a,b,c,d,e,f,g]Mean values within a column with non-identical superscript letters were significantly different at *P*≤0.05 (Tukey's honestly significant difference test).

slightly higher negative values of *b* (an indication of greater blueness), with a correlation coefficient of $r = -0.62$ (Sapers *et al.*, 1984). The surface *L* values were also negatively correlated with sensory scores for intensity of blue colour ($r = -0.62$) and significantly correlated with sensory scores for acceptability of appearance ($r = 0.46$). Chromaticity *b* values were correlated with sensory scores for intensity of blue colour ($r = 0.48$) and negatively correlated with sensory scores for acceptability of appearance ($r = -0.41$). These results would indicate that consumer preferences are for brighter, less intensively blue-coloured fruit.

Firmness

Fruit firmness is an important characteristic in blueberries, as it relates to consumer appeal and postharvest decay of the fruit (NeSmith *et al.*, 2002). Fruit with higher firmness can better withstand harvesting (especially mechanical) and subsequent shipping. Fruit of firmer cultivars can be left on the bush longer (hanging potential), allowing more flexibility in the timing of harvests (Ehlenfeldt, 2005). According to Beaudry (1992), the eating texture of the blueberry is affected by a number of factors including skin thickness, pulp firmness and the presence of stone cells.

Most of the mechanical methods used to measure firmness determine the force needed to puncture, penetrate or deform the fruit (Chiabrando *et al.*, 2009). In blueberry, the softening that occurs at ripening is linked to the enzymatic digestion of cell-wall components such as pectin, cellulose and hemicellulose. Blueberry seems to be unusual in the sense that fruit softening occurs without substantial modifications in the size of pectin molecules. In contrast, hemicellulose levels decrease in blueberry as ripening progresses (Vicente *et al.*, 2007). Total water-soluble pectin, which constitutes much of the middle lamella, decreases linearly as the fruit passes from green to blue. This degradation of the middle lamella and cell wall is directly responsible for the loss of firmness.

In blueberries, species ancestry has not been consistently related to firmness; however, cultivars with higher firmness often possess a greater percentage of *V. darrowii*. Conversely, cultivars with softer fruit often have a higher percentage of lowbush (*V. angustifolium*) ancestry (Ehlenfeldt and Martin, 2002). Saftner *et al.* (2008) reported that the NHBs 'Chanticleer', 'Lateblue' and 'Weymouth' were softer than nine other NHB cultivars and also had low sensory scores for bursting energy and texture during chewing. The largest difference (16–24%) was between 'Hannah's Choice' (the firmest cultivar) and 'Chanticleer', 'Coastal', 'Lateblue' and 'Weymouth', which were the softest. In general, compression firmness values were most tightly correlated to sensory scores for juiciness ($r = 0.48$), followed by bursting energy (crispness, $r = 0.44$) and texture during chewing ($r = 0.33$), and were not significantly associated with sensory scores for intensity of skin toughness. The rather weak correlation between compression firmness and sensory scores for texture during chewing might be due to the abundance and/or size of stone cells and seeds, which were not considered in this study.

Blueberry firmness is more greatly affected by changes in maturity than by differences among cultivars (Beaudry, 1992), and slight differences in maturity can have profound influences on the relationship between instrumental firmness and sensory textural scores. Ballinger *et al.* (1973) measured firmness (resistance to compression) of NHB 'Berkeley' fruit at different maturity stages. They showed that small, green, unripe fruit were extremely firm (57.8 g/0.1 mm of compression) and that firmness dropped sharply up to the red colour stage; however, firmness varied little as berries passed from the red-purple (9.6–9.8 g/0.1 mm) to the completely blue stage (7.6–8.8 g/0.1 mm). Similar trends were observed by Vicente *et al.* (2007) in NHB 'Duke' fruit. Research by Ballinger *et al.* (1973) showed that fruit from later harvests had lower firmness readings. When the NHBs 'Croatan', 'Morrow', 'Murphy' and 'Wolcott' were harvested 11 days apart, the firmness was, on average, from 7.8 to 6.4 g/0.1 mm lower in later than in earlier harvests.

In a 3-year trial in California to evaluate the instrumental and sensory quality of SHB cultivars, Bremer *et al.* (2008) reported (Table 9.3) that 'Jewel' and 'O'Neal' had the lowest average firmness (13.6 g/0.1 mm), while 'Misty' and 'Reveille' had the highest (18.1 g/0.1 mm), and 'Emerald' and 'Star' were intermediate (17.0 g/0.1 mm). As with other quality attributes (SS, TA, SS:TA ratio, firmness and antioxidant capacity), there was a significant interaction between cultivars and season, indicating that for a given characteristic the variability for a certain cultivar was dependent on the environmental conditions of the specific growing season. Differences in climate, soils, cultural practices and other environmental factors may cause firmness of some cultivars to vary from region to region and from season to season. Ballinger *et al.* (1973) showed that firmness varied by 14.6% in successive seasons, and that five out of six NHB cultivars showed the same trend. Donahue *et al.* (2000) reported that smaller fruit in lowbush blueberry always had higher firmness readings.

Ehlenfeldt (2005) found the following when the firmness and holding ability of 19 highbush blueberry cultivars were studied during postharvest: (i) the seven softest cultivars were released before 1953, which he partially attributed to the incorporation of firmer-fruited species material into *V. corymbosum*; (ii) at weekly intervals, most cultivars showed single-digit decreases in firmness from the first harvest, except for 'Legacy', which had a 15% increase; (iii) when harvest was delayed to wait for a greater proportion of blue fruit, firmness in all cultivars decreased (ranging from –1 to –15%) at week 2, and decreased even further at week 3 (by as much as –23% for 'Chanticleer'), with 'Legacy' again departing from the rest as it showed almost no decrease; and (iv) holding ability cannot be predicted based solely on initial firmness, as 'Legacy' was not the

Table 9.3. Ranges in quality attributes of six SHB blueberry cultivars grown in the San Joaquin Valley (California) for three seasons, from 2005 to 2007. (Adapted from Bremer *et al.*, 2008.)

Cultivar	SS (% w/w)	TA (% w/w)	SS:TA ratio	Firmness (g/0.1 mm)[a]	TEAC (µmol TE/g FW)
'Emerald'	11.6–12.3	0.60–0.90	13.2–20.0	16.2–18.2	13.2–19.1
'Jewel'	10.9–12.3	0.67–1.00	11.4–18.1	12.1–14.8	10.3–11.7
'Misty'	11.1–13.7	0.57–0.83	16.6–20.6	15.5–21.2	17.4–21.9
'O'Neal'	10.8–11.8	0.27–0.77	14.5–40.5	12.8–14.8	11.7–13.6
'Reveille'	13.3–15.8	0.70–0.80	18.1–22.9	15.9–21.6	13.8–20.7
'Star'	11.1–12.9	0.67–0.77	16.4–17.4	17.0–19.7	12.1–12.7

TEAC, Trolox equivalent antioxidant capacity (oxygen radical absorbance capacity (ORAC) assay); TE, Trolox equivalents; FW, fresh weight.
[a]Firmness was converted from initial readings of lbs per 4 mm of compression.

firmest fruit initially but showed a better ability to hold and possibly increase its firmness.

Fruit firmness can be reduced dramatically by how the fruit is handled. When NHB 'Wolcott' blueberry fruit were dropped on hard boards, they softened (bruised) in proportion to the distance of the fall. Multiple drops of small distances (10.2 cm) softened the blueberries as much as large distances (40.8 cm) if the total distance of the increments was the same. Regardless of cultivar, size, ripeness or initial firmness, the firmness of blueberries after a standard fall can be predicted if their initial firmness is known (Ballinger *et al.*, 1973). Similarly, when lowbush blueberries were dropped on to a moving smooth conveyor belt from heights of 40, 80, 120 or 160 cm, increasing height resulted in greater loss of fruit firmness, as measured by instrumental and sensory methods. Studies on rabbiteye 'Brightwell' blueberries established that the greatest loss in firmness was caused by machine harvesting (20–30%), followed by a 10–15% loss in firmness due to grading and sorting.

Miller *et al.* (1993) found that fruit firmness declined during the postharvest storage of SHB 'Sharpblue' fruit. As blueberries from later harvests had a more rapid decrease in fruit quality during storage than the fruit from earlier harvests, Miller *et al.* (1993) suggested that fruit from this cultivar should be picked as soon as possible after reaching marketable maturity. Sanford *et al.* (1991) concluded that in lowbush the storage temperature had a greater influence on fruit firmness than bruising at harvest. Berries held at 0°C were the firmest, while fruit held at 5°C showed a disproportional decrease in firmness, and each additional rise in storage temperature (up to 20°C) resulted in incremental decreases in fruit firmness. Paniagua *et al.*, (2013) worked with rabbiteye 'Centurion' blueberries and found that up to the point where berries had lost 1.3% of their weight there was no association between weight loss and firmness. Beyond this point weight loss had a high and consistent influence on berry softening. For every percentual point of weight loss there was a 0.08 N loss in berry firmness.

Perkins-Veazie *et al.* (1995) showed that fruit firmness at harvest is not a good indicator of firmness after storage. For example, compared with other clones, SHB 'O'Neal' fruit had a high epidermal and stem scar firmness before storage but was intermediate to low in firmness after storage. Fruit from SHB selection G616 were of similar firmness to the NHB cultivar 'Bluecrop' and rabbiteye cultivar 'Climax' before storage, but were the softest fruit of all the blueberries after storage (Table 9.4). For all blueberries tested, epidermal firmness decreased after storage (average 25.4% for SHB cultivars). Except for rabbiteye 'Climax' and SHB A109, all fruit had reduced firmness at the stem scar following storage.

Giongo *et al.* (2013) used a novel texture analyser, TA.XT*plus* (Stable MicroSystems, Godalming, UK) to assess the changes in mechanical profile of 49 different highbush and half-high cultivars and selections during postharvest storage. They used a storage index based on six mechanical parameters

Table 9.4. Blueberry fruit epidermal and stem scar firmness before and after 21 days of storage at 5°C plus 1 day at 20°C. Epidermal/stem scar firmness was measured with a gram gauge penetrometer adapted with a 0.3 mm wire. (Adapted from Perkins-Veazie *et al.*, 1995.)

	Epidermal firmness (g/0.1 mm)			Stem scar firmness (g/0.1 mm)			Average firmness (g/0.1 mm)		
Cultivar/ selection	Before storage	After storage	Firmness change (%)	Before storage	After storage	Firmness change (%)	Before storage	After storage	Firmness change (%)
A109	5.9[b]	4.3[b]	−27[a,b]	4.1[f]	4.0[b,c]	2[c]	5.0[d]	4.2[b]	17[b]
'Cape Fear'	6.1[b]	4.9[a]	−20[b]	6.0[a,b]	5.1[a]	15[b,c]	6.1[b]	5.0[a]	17[b]
'Cooper'	6.1[b]	4.1[b,c]	−33[a,b]	5.5[c]	4.5[b]	18[a,b]	5.8[b,c]	4.3[b]	26[a,b]
'Gulfcoast'	6.8[a]	4.8[a]	−29[a,b]	6.4[a]	5.2[a]	19[a,b]	6.6[a]	5.0[a]	24[a,b]
G616	5.1[c]	3.7[c]	−27[a,b]	4.7[d,e]	3.6[c]	23[a,b]	4.9[d]	3.7[c]	26[a,b]
'MS108'	5.9[b]	4.9[a]	−17[b]	5.9[b,c]	5.2[a]	12[b,c]	5.9[b,c]	5.1[a]	14[b]
'O'Neal'	6.7[a]	4.4[b]	−34[a]	6.4[a]	4.4[b]	31[a]	6.6[a]	4.4[b]	33[a]
'Sierra'	5.8[b]	4.9[a]	−16[b]	5.5[c]	4.9[a]	11[b,c]	5.7[c]	4.9[a]	13[b]
'Bluecrop'	5.1[c]	4.3[b]	−16[b]	4.8[d]	4.3[b]	10[b,c]	5.0[d]	4.3[b]	13[b]
'Climax'	5.8[b]	4.4[b]	−24[b]	4.3[e,f]	4.4[b]	−2[c]	5.1[d]	4.4[b]	13[b]

[a,b,c,d,e,f]Mean values within a column with non-identical superscript letters were significantly different at $P \leq 0.05$ (least significant difference test).

to determine that 'Aurora', 'Draper', NZ6 and 'Ozarkblue' had the highest storage potential, while 'Chippewa', 'Duke', 'HardyBlue', and 'Jersey' had the lowest.

Blaker and Olmstead (2015) compared the bioyield force of fruit from seven SHB blueberry genotypes having standard ('Springhigh', 'Star', Windsor') or crisp (FL 06-561, FL 06-562, FL 98-325 and 'Sweetcrisp') textures and compared these values with their dry weight and levels of alcohol-insoluble residue, uronic acid and neutral sugars on separated flesh and skin tissue. They found that the bioyield force of the standard-texture genotypes was significantly less than all four crisp genotypes; however, these differences were not associated with dry weight, alcohol-insoluble residue, uronic acid or neutral sugars. These results indicate that there is a measurable phenotypic difference between crisp- and standard-texture blueberries, but quantitative differences among total cell-wall material, pectins and neutral sugars are not responsible for these differences in perception.

Moggia *et al.* (2016) studied the effect of maturity at harvest (75% blue, 100% blue or overripe) on firmness and weight loss of NHB 'Brigitta' and 'Duke' fruit after cold storage and associated these parameters with cuticular triterpene content. 'Duke' fruit softened faster and became more dehydrated than 'Brigitta' samples, and maturity level in both cultivars was negatively associated with firmness after storage. Levels of ursolic acid were significantly correlated with rates of weight loss and softening.

In a study of the rabbiteye 'Brilliant', Chen *et al.* (2015) found that fruit firmness declined during storage concomitantly with an increase in water-soluble pectins and a decrease in sodium carbonate-soluble pectins, cellulose and hemicellulose. Blueberries stored at low temperature (5°C) were firmer than those held at 10°C, which was probably due to a lower water-soluble pectin content and a higher content of sodium carbonate-soluble pectins, cellulose and hemicellulose. At the lower temperature, there were lower activities of the cell-wall-degrading enzymes polygalacturonase, cellulase, galactosidase and mannosidase.

Soluble solids

As berries approach maturity and pass from the red stage to the blue stage, the total sugars increase, mainly due to an increase in reducing sugars. Woodruff *et al.* (1960) established that the largest increase occurs in the first 6 days after red coloration of the berries. SS levels in rabbiteye and highbush blueberries tend to vary from year to year, among different locations and across cultivars (Saftner *et al.*, 2008; Gilbert *et al.*, 2014; Gündüz *et al.*, 2015). To get an adequate SS level for the consumer is quite critical in blueberries to reach the desired SS level at harvest, as the fruit does not accumulate sugars after harvest.

Beaudry (1992) suggested that perception of sweetness may be affected by other factors, such as TA. Highbush blueberry cultivars such as 'Bluegold' and 'Chanticleer' with high SS (13.0–13.2% w/w) were not perceived as particularly sweet in sensory evaluation, while the opposite occurred with cultivars having low SS, such as 'Coville', 'Duke' and 'Lateblue' (Saftner *et al.*, 2008). This may indicate that a difference in SS by itself does not reflect the perception of fruit sweetness by the consumers.

Kader *et al.* (2003) showed in strawberries that anthocyanins and phenolic compounds, which are even more prominent in blueberries, strongly refract light and contribute up to 32% to SS readings obtained from a refractometer. They found that removal of anthocyanins and phenolic compounds before measuring SS with a refractometer increases the reliability of SS as an indicator of sweetness.

After evaluating SHB cultivars in California for 3 years, Bremer *et al.* (2008) reported that SS values tended to be more stable than TA or firmness (Table 9.3). When comparing cultivars, they found that 'Reveille' had the highest average for SS (14.4% w/w), 'O'Neal' had the lowest (11.4% w/w) and the other cultivars were intermediate (11.7–12.3% w/w). Hancock *et al.* (2008) found the average content of SS among highbush cultivars to range from 9.5% w/w ('Bluecrop') to 12.7% w/w ('Brigitta'). Gilbert *et al.* (2015) found that perceived sweetness was positively correlated with total sugar and fructose concentration. When Kushman and Ballinger (1963) studied different harvest schedules (3-, 6-, 9- and 12-day intervals) in NHB 'Wolcott' blueberries they found that harvest interval had little influence on SS levels, although an increase in sugars (mainly reducing) was obtained with longer intervals. In their study on the effect of storage on the quality of SHB fruit, Perkins-Veazie *et al.* (1995) found that levels of SS did not differ greatly among clones and between fresh fruit and that stored for 21 days at 5°C plus 1 day at 20°C, except for fruit of the rabbiteye 'Climax' (included as a standard), which increased significantly after storage. Although weight loss can concentrate sugars, the magnitude of weight loss of 'Climax' fruit was less than that of the SHB 'Gulf-coast', which showed no change in SS.

Titratable acidity

Organic acids are important for flavour in rabbiteye and highbush blueberries, and TA varies greatly across year, location and cultivar (Saftner *et al.*, 2008; Gilbert *et al.*, 2014; Gündüz *et al.*, 2015). The composition of organic acids is a distinguishing characteristic among *Vaccinium* spp. In highbush blueberries, the predominant organic acid is usually citric (average 75%; range 38–90% w/w), while the proportions of malic, quinic and succinic acids are 3, 5 and 17%, respectively. The most important organic acids in rabbiteye fruit are succinic and malic acids (50 and 34%, respectively), while citric acid accounts for only 10% (Ehlenfeldt *et al.*, 1994).

As the combination of citric and malic acids gives a sour taste and succinic acid provides a bitter taste, the composition of organic acids affects sensory quality (Bremer *et al.*, 2008). Acid profile differences may also have a bearing on other important factors such as fruit colour development, decay susceptibility, and insect and bird predation (Ehlenfeldt *et al.*, 1994). Greater fruit acidity enhances the colour strength of anthocyanins, as observed by Sapers *et al.* (1984) in samples of NHB 'Coville' and 'Elliott' fruit.

Both environmental and developmental factors affect acidity levels in blueberry fruit. Research on NHB blueberries established that acidity falls sharply in the first 6 days after fruit reach red coloration (Woodruff *et al.*, 1960). Kushman and Ballinger (1963) studied different harvest schedules (3-, 6-, 9- and 12-day intervals) in the NHB 'Wolcott' and found that total TA tended to decrease as the season progressed and as the harvest interval was lengthened. In the case of SHBs, a 3-year evaluation in California determined that TA varied by up to 50% among seasons for 'Emerald', 'Jewel' and 'O'Neal' (Bremer *et al.*, 2008). 'O'Neal' had a significantly lower average (0.55% citric acid) than the rest which averaged 0.70–0.80% citric acid (Table 9.3). Bremer *et al.* (2008) concluded that blueberries with very low TA (0.3% w/w), despite high SS concentrations between 10 and 12% w/w, were not acceptable to consumers.

Statistical analysis of samples collected at different maturity stages in 11 highbush cultivars showed that 86% of the total variability in TA among cultivars could be explained by genetic differences (Sapers *et al.*, 1984). Perkins-Veazie *et al.* (1995) found that TA values of SHB cultivars ranged from 0.54 to 1.13%, with an average of 0.84%. Hancock *et al.* (2008) found TA in NHB cultivars to range from 0.90% ('Jersey') to 2.10% ('Bluegold').

When the association between instrumental and sensory quality was studied by Saftner *et al.* (2008) in highbush and rabbiteye blueberries, it was found that TA was inversely correlated with pH ($r = -0.76$), but, as also reported by Rosenfeld *et al.* (1999), it was not related to scores of tartness or to any other flavour-related sensory evaluation. The authors concluded that the apparent lack of correlation between TA and flavour-related evaluations may suggest that there is an optimal acid concentration needed in blueberry fruit for enhanced flavour. However, part of the explanation might be related to the type of acid present in each cultivar.

During storage, the average acidity dropped slightly in four out of eight SHBs (Perkins-Veazie *et al.*, 1995) and in NHBs (Perkins-Veazie *et al.*, 1995; Chiabrando *et al.*, 2009). Similar decreases in acidity during storage were found by Smittle and Miller (1988) in 'Woodard' rabbiteye blueberry during 21 days of storage at 5°C; however, Miller and Smittle (1987) found little change in acidity of 'Climax' and 'Woodard' rabbiteye blueberries during 21 days of storage at 3°C. Acids are one of the energy reserves of the fruit, being used in respiration and converted to simpler molecules such as CO_2 and water. Acids decrease as a result of respiration, but water loss in the fruit might increase the concentration of acids (Echeverría *et al.*, 2009). As a consequence, TA would change during storage, depending on the rates of respiration and water loss.

SS:TA ratio

Low SS:TA ratios have been associated with good keeping quality (Ballinger and Kushman, 1970). An SS:TA ratio of 6.5 or lower was recommended by Galletta (1975) as desirable in highbush blueberry cultivars for resistance to postharvest decay organisms, and this ratio has not been challenged since for recently released cultivars. Based on the relationship between SS:TA ratios and relative keeping quality of blueberries, Galletta *et al.* (1971) established three classes: (i) cultivars with SS:TA values lower than 18, which possess good keeping quality; (ii) cultivars with SS:TA values between 18 and 32, which have medium keeping quality; and (iii) cultivars with SS:TA values higher than 32, in which the keeping quality would be low.

The SS:TA ratio varies greatly across year, location and cultivars of rabbiteye and highbush blueberries (Saftner *et al.*, 2008; Gilbert *et al.*, 2014; Gündüz *et al.*, 2015). Evaluation of six SHB cultivars in California showed that there was high variability among seasons in SS:TA ratio (near 50%) for 'Emerald', 'Jewel' and 'O'Neal' (Bremer *et al.*, 2008). Perkins-Veazie *et al.* (1995) studied SHB clones in Arkansas; their results indicated that at harvest most of the clones, excluding 'Cape Fear' (SS:TA = 18.7) had SS:TA ratios lower than 18, which is recommended for the longest storage life. Following 21 days of storage at 5°C plus 1 day at 20°C, only 'O'Neal' (SS:TA = 22.9) and the rabbiteye 'Climax' (SS:TA = 26.2) had an SS:TA ratio higher than 18 (Perkins-Veazie *et al.*, 1995). Hancock *et al.* (2008) found SS:TA ratios in NHB blueberries to range from 5.9 ('Bluegold') to 12.8 ('Jersey').

The SS:TA ratio increases from unripe green (approx. 3) to the fully blue stage (approx. 20) and then remains at that level (Castrejón *et al.*, 2008). This is because TA declines from the unripe green (about 3% w/w) to the fully blue stage (about 0.5% w/w), while SS increase from the unripe green (approx. 9% w/w) to the 100% ripe stage (approx. 15% w/w), and both sugars and acidity change little later on.

SS:TA ratios differ among blueberry types; the NHB cultivars 'Croatan', 'Bluecrop', 'Morrow' and 'Weymouth' had an average value of 4.7, while the rabbiteye blueberries 'Callaway', 'Garden Blue', 'Homebell' and 'Tifblue' reached a value of 10.5 (Ballington *et al.*, 1984). The values for SS:TA ratio in various SHBs across three seasons varied from 11.4 to 40.5 (Table 9.3).

There seems to be strong environmental and management effects on SS:TA ratios, as Saftner *et al.* (2008) reported values of 24.9 and 20.1 for the NHBs 'Bluecrop' and 'Weymouth', respectively (Table 9.5), while the SS:TA ratios published by Ballington *et al.* (1984) were the surprisingly low 5.0 for 'Bluecrop' (SS = 7.93; TA = 1.59) and 3.7 for 'Weymouth' (SS = 7.88; TA = 2.10). Part of the difference might be due to the fact that in the latter study, all of the fruit was collected in one pass including all stages of ripeness. Research comparing successive harvests of different NHB cultivars showed that 'Coville' and 'Elliott' were consistently high in acidity: the SS:TA ratio and anthocyanin levels remained constant for successive harvests. However, first harvests of 'Berkeley', 'Bluetta', 'Collins' and 'Earliblue' were higher in acidity than second harvests, although the SS:TA ratios were still within the ripe range (Sapers *et al.*, 1984).

The organic acid concentration influences the perception of sweetness. Each reduction of 0.1% (as a proportion of total fruit weight) is equivalent to an increase of 1% in perceived sweetness. During the ripening of highbush blueberries, citric acid declines from about 1.2 to 0.6% of total fruit weight, corresponding to a perceived sweetness increase of 6% (Beaudry, 1992).

pH

Saftner *et al.* (2008) found that the pH values of blueberry extracts correlated with scores for intensity of flavour ($r = 0.56$) and acceptability of flavour ($r = 0.51$), as well as overall eating quality ($r = 0.48$). The pH values for NHB blueberries ranged from 2.5 for 'Elliott' and 'Lateblue' to 3.4 for 'Chanticleer' (Table 9.5). Chiabrando *et al.* (2009) determined that the pH of NHB 'Bluecrop' and 'Coville' fruit increased from 2.8 at harvest to 3.3 after 35 days of cold storage. Good storage quality is associated with pH values lower than 3.5 (Perkins-Veazie *et al.*, 1995).

When evaluating the relationship between pH and acidity for a range of blueberry progenies, Galletta *et al.* (1971) established that, for a change of 1 pH unit, the acidity reflected a fourfold change in concentration.

There does not appear to be a consistent pattern of change in pH values during storage. Perkins-Veazie *et al.* (1995) reported that fruit pH values for SHB fruit at harvest were in the range of 3.01 (for selection MS108) to 3.43 (for 'Cape Fear') with an average of 3.24. When these fruits were again measured after 21 days of storage at 5°C plus 1 day at 20°C, it was found that the average pH had increased slightly, and the range varied from 3.12 (for MS108) to 3.47 (for selection G616). The pH of NHB 'Bluecrop' fruit changed from 3.39 at harvest to 3.51 after storage (Perkins-Veazie *et al.*, 1995). pH was reported to decrease after cold storage at 5 and 3°C for rabbiteye blueberries 'Climax' and 'Woodard' (Smittle and Miller, 1988) and for the SHB 'O'Neal' in

Table 9.5. Compression firmness of whole fruit and SS, TA and pH values from ten NHB and two rabbiteye blueberry cultivars listed by harvesting season. (Adapted from Saftner *et al.*, 2008.)

Cultivar	Compression firmness (N)	SS (% w/w)	TA (% w/w)	SS:TA ratio	pH
NHB					
'Chanticleer'	1.56[b,c,d]	13.0[a]	0.40[b,c]	32.3[a,b]	3.4[a]
'Duke'	1.67[a,b,c]	10.9[b,c,d]	0.43[b,c]	25.5[b,c,d]	3.0[a,b,c]
'Hannah's Choice'	1.86[a]	12.3[a,b]	0.45[b,c]	27.3[a,b,c]	3.3[a,b]
'Weymouth'	1.51[c,d]	11.2[b,c,d]	0.56[b,c]	20.1[c,d]	2.8[b,c]
'Berkeley'	1.54[c,d]	11.5[b,c,d]	0.44[b,c]	26.6[b,c,d]	3.1[a,b]
'Bluecrop'	1.64[a,b,c]	11.5[b,c,d]	0.46[b,c]	24.9[b,c,d]	3.1[a,b]
'Bluegold'	1.71[a,b,c]	13.2[a]	0.64[b]	20.9[c,d]	3.1[a,b,c]
'Coville'	1.66[a,b,c]	10.8[c,d]	0.58[b,c]	18.7[d]	3.0[a,b,c]
'Elliott'	1.64[a,b,c]	11.3[b,c,d]	1.27[a]	9.0[e]	2.5[c]
'Lateblue'	1.40[d]	10.6[d]	1.22[a]	8.9[e]	2.5[c]
Rabbiteye					
'Coastal'	1.37[d]	12.2[a,b,c]	0.35[c]	35.6[a]	3.0[a,b,c]
'Montgomery'	1.76[a,b]	11.3[b,c,d]	0.58[b,c]	19.5[c,d]	2.8[b,c]

[a,b,c,d,e]Mean values within a column with non-identical superscript letters were significantly different at $P \leq 0.05$ (Tukey's honestly significant difference test).

fruit stored in a modified atmosphere (Echeverría *et al.*, 2009). The latter attributed this drop in pH to CO_2 diffusion into the fruit tissues.

Vitamin C

Vitamin C levels vary widely among blueberry cultivars (Gündüz *et al.*, 2015). In a comparison of 24 NHB cultivars in Michigan, vitamin C values ranged from 16.3 to 34.3 mg per 100 g fresh weight with 'Draper' and 'Lateblue' having the lowest values and 'Bluetta', and 'Rubel' the highest. In a comparison of seven NHB cultivars in Michigan and Oregon, vitamin C values ranged from 16.2 to 22.6 mg per 100 g fresh weight, with 'Draper' and 'Reka' having the lowest values and 'Elliott' and 'Legacy' the highest. In Georgia (USA), in a comparison of seven rabbiteye and 11 SHB cultivars, the SHB 'Primadonna' had the highest level of vitamin C (31.0 mg per 100 g), while the SHB 'O'Neal' and rabbiteye 'Climax' had the lowest (16.3 and 16.4 mg per 100 g, respectively).

Antioxidant capacity

There is strong evidence that the antioxidants present in fruits and vegetables protect lipids, proteins and nucleic acids against oxidative damage initiated by free radicals. It has been established that free radicals play a major role in cancer, heart, vascular and neurodegenerative diseases (Howard *et al.*, 2003).

Among 41 fruits and vegetables tested for their antioxidant capacity using an assay for oxygen radical absorbance capacity (ORAC), blueberries had the highest value. Although various kinds of antioxidants have been identified in fruit, anthocyanins and other phenolic compounds have received the greatest attention (You *et al.*, 2011). Blueberry fruits contain an array of phenolics, including anthocyanins, quercetin, kaempferol, myricetin, chlorogenic acid and procyanidins, which contribute to antioxidant capacity. Up to 60% of the total phenolic content in highbush blueberries is accounted for by anthocyanins (Kalt *et al.*, 2003). Anthocyanins are responsible for the bright orange, red and blue colours in fruit and are dependent on environmental pH values (You *et al.*, 2011). Indeed, anthocyanins change their colour with pH: they appear red in acidic, violet in neutral and blue in basic aqueous solution (Yoshida *et al.*, 2009).

Blueberry fruit have the highest concentration of antioxidants and phenolics in the skin, more than double those of the seeds (Table 9.6). For a given weight, the total amount of skin or surface area increases as the berry size decreases. Various authors have found a highly inverse relationship between fruit size and antioxidant activity (Connor *et al.*, 2002b; Moyer *et al.*, 2002; Howard *et al.*, 2003).

There is considerable variation among blueberry cultivars in antioxidant capacity. Ehlenfeldt and Prior (2001) determined the ORAC, phenolic and

Table 9.6. ORAC, phenolic and anthocyanin levels in different parts of SHB 'Reveille' and 'Bladen' blueberries. (Adapted from Mainland and Tucker, 2002.)

Cultivar	Berry part	ORAC (μmol TE/g FW)	Phenolics (mg/g)	Anthocyanins (mg/g)
'Reveille'	Whole berry	16	2.9	1.0
	Seeds	28	5.6	1.4
	Skin	66	9.8	9.1
'Bladen'	Whole berry	34	5.2	2.3
	Seeds	59	14.2	1.6
	Skin	166	27.4	12.7
Overall average	Whole berry	25	4.1	1.7
	Seeds	44	9.9	1.5
	Skin	116	18.6	10.9

TE, Trolox equivalents; FW, fresh weight.

anthocyanin concentrations in fruit of 87 highbush blueberry cultivars and found that in SHBs values ranged from 4.6 ('Avonblue') to 22.3 ('Sharpblue') μmol Trolox equivalents (TE)/g fresh weight (FW). In NHBs, the range was 5.5 ('Berkeley') to 30.5 ('Elliott') and 31.1 ('Rubel'). In a comparison of 20 NHB, seven rabbiteye and 11 SHB cultivars, Gündüz *et al.* (2015) found the highest antioxidant capacity in NHB 'Rubel', 'Elliott' and 'Lateblue'. Heritability estimates in blueberry progenies were 0.43, 0.46 and 0.56 for antioxidant capacity, total phenolics and total anthocyanins, respectively (Connor *et al.*, 2002a).

As well as genotype, the antioxidant capacity can be affected by location, growing season, cultural management, maturity, and postharvest handling and storage. Howard *et al.* (2003) found significant main effects for growing season and genotype × growing season for ORAC, total antioxidants and fruit weight. Similarly, Connor *et al.* (2002b) evaluated nine NHB cultivars in three locations (Michigan, Minnesota and Oregon) for two seasons and found a significant genotype × environment interaction for antioxidant activity. Although differences in overall mean antioxidant activity among locations occurred, there was no significant change in rank among locations. In contrast, in a 3-year evaluation of SHBs in California, Bremer *et al.* (2008) found that antioxidant capacity varied significantly among cultivars but not among seasons (Table 9.3). 'Misty' had the highest average ORAC value (19.7 μmol TE/g FW) and 'Jewel' the lowest (11.0 μmol TE/g FW).

Delaying fruit harvest can have a marked positive influence on levels of anthocyanins in blueberry fruit. Fruit maturity had a significant effect on antioxidant activity, total phenolic content and anthocyanin content, and bush ripeness × fruit maturity interactions were significant (Connor *et al.*, 2002c). Similar results were reported by Prior *et al.* (1998) who found that when the fruit of the rabbiteye cultivars 'Brightwell' and 'Tifblue' were left on the bush for an extended time (49 days) after first becoming blue, the antioxidant activity was much higher (124% for 'Brightwell' and 64% for 'Tifblue') than when the berries had first become blue (Table 9.7). In another study done by Mainland and Tucker (2002), the levels of antioxidants in highbush blueberries remained constant or decreased from the ripe to the overripe stage, while harvesting overripe berries in rabbiteye caused a 10% increase in ORAC values (Table 9.8).

Levels of antioxidant capacity are not always tightly associated with phenolic or anthocyanin content. Studies on the NHBs 'Bluegold', 'Brigitta' and 'Nelson' showed that anthocyanin content was substantially higher in fruit of more advanced stages of ripeness (fully blue versus 5–50 or 50–95% blue). In contrast, the phenolic content and ORAC values were lower in riper fruit (Kalt *et al.*, 2003). In contrast, Castrejón *et al.* (2008) found in the NHBs 'Berkeley', 'Bluecrop', 'Reka' and 'Puru' that ORAC values and phenolic contents were higher during early maturation (100% whitish green) and stabilized from 60% blue on. In another experiment, berries of the NHB 'Elliott' were harvested from plants at two levels of bush ripeness (30–50 and 60–80% of ripe berries on plants) and separated into three maturity classes on the basis of percentage fruit colour. The authors found that the level of bush ripeness had no significant effect on antioxidant activity, total phenolic content or anthocyanin content; however, fruit maturity as well as bush ripeness × fruit maturity interactions had a significant effect on these three traits (Connor *et al.*, 2002c).

Kalt *et al.* (1999) found that there was a slight increase in both anthocyanins and ORAC values at 20°C but not at other temperatures (0, 10 or 30°C). Anthocyanins continued to be synthesized during storage at 20°C, although the rate of pigment formation declined after about 4 days. Less anthocyanin pigment was formed in the least ripe fruit. After 8 days of storage at 20°C, the anthocyanin concentration of fruit harvested at 5–50 or 50–95% blue exceeded that of ripe fruit (Kalt *et al.*, 2003). Berries of the NHBs 'Bluecrop', 'Bluegold', 'Brigitta', 'Elliott', 'Legacy', 'Liberty', 'Jersey' 'Little Giant' and 'Nelson' were stored from 3 to 7 weeks at 5°C and none of the cultivars showed

Table 9.7. ORAC, phenolic and anthocyanin levels in rabbiteye blueberries 'Tifblue' and 'Brightwell' for two stages of maturity (just ripe = just blue; overripe = beginning to soften). (Adapted from Prior *et al.*, 1998.)

Cultivar	Maturity stage	ORAC (μmol TE/g FW)	Phenolics (mg per 100 g FW)	Anthocyanins (mg per 100 g FW)
'Tifblue'	Just ripe	23.0	3.6	0.9
	Overripe	37.8	4.1	1.5
'Brightwell'	Just ripe	15.3	2.7	0.6
	Overripe	34.3	4.6	1.6

Table 9.8. ORAC, phenolic and anthocyanin levels in highbush ('Croatan', 'Reveille' and 'Bladen') and rabbiteye ('Tifblue' and 'Powderblue') blueberries of three stages of maturity (pre-ripe = slight red on scar; ripe = fully ripe; overripe = beginning to soften). (Adapted from Mainland and Tucker, 2002.)

Cultivar	Maturity stage	ORAC (μmol TE/mg FW)	Phenolics (mg/g)	Anthocyanins (mg/g)
'Croatan'	Pre-ripe	24	3.6	1.2
	Ripe	26	3.6	1.6
	Overripe	23	3.8	1.4
'Reveille'	Pre-ripe	12	1.9	0.3
	Ripe	16	2.6	0.7
	Overripe	15	2.5	0.6
'Bladen'	Pre-ripe	24	3.3	0.7
	Ripe	43	5.2	1.9
	Overripe	34	4.1	2.0
Average highbush	Pre-ripe	20	2.9	0.7
	Ripe	28	3.8	1.4
	Overripe	24	3.5	1.3
'Tifblue'	Pre-ripe	8	1.2	0.2
	Ripe	16	2.5	0.7
	Overripe	21	3.2	0.9
'Powderblue'	Pre-ripe	13	2.1	0.3
	Ripe	21	3.4	1.0
	Overripe	26	4.6	1.4
Average rabbiteye	Pre-ripe	11	1.7	0.3
	Ripe	23	3.0	0.9
	Overripe	25	3.9	1.2

a significant change in antioxidant activity during storage (Connor *et al.*, 2002c). However, berries of the NHB 'Elliott' with 50–75% fruit coloration, harvested from bushes with 60–80% mature fruit, showed a significant increase in antioxidant activity, total phenolic content and anthocyanin content during the first 3 weeks of storage (Connor *et al.*, 2002c). In contrast, Remberg *et al.* (2003) reported that antioxidant capacity (ferric reducing antioxidant power (FRAP) values) of the NHBs 'Aron', 'Bluecrop', 'Hardyblue', 'Patriot' and 'Putte' decreased considerably (by 24–34%) during 4 weeks of cold storage (1 or 8°C) in a controlled atmosphere (10% CO_2 and 10% O_2; note that CO_2 and O_2 concentrations in this chapter are expressed in percentages on a v/v basis and are assumed to be equivalent to kilopascals (kPa)). It appears that increases in anthocyanins can be obtained only at high temperatures

(20°C), which are not compatible with optimization of other quality parameters during postharvest that are important for consumers.

The impact of cultivation type on antioxidants is controversial. While Wang S.Y. *et al.* (2008b) found that NHB 'Bluecrop' fruit grown organically yielded significantly higher total phenolics, total anthocyanins and antioxidant activity (ORAC) and that the cultural method changed the concentrations of the antioxidants present in the fruit, You *et al.* (2011) reported that although there were significant differences among various rabbiteye cultivars ('Climax', 'Powderblue', 'Tifblue' and 'Woodard') in total phenolics, total anthocyanins and ORAC values, the levels of these compounds did not differ significantly between organic and conventional cultivation. Similarly, Sablani *et al.* (2010) compared organic and conventional cultivation of highbush 'Duke' and 'Reka' blueberries, and found that total antioxidant content, phenolic content and total antioxidant activity of berries were not altered by the agricultural production system. Phytochemicals from the NHBs 'Bluecrop', 'Bluejay', 'Brigitta', 'Darrow' and 'Patriot' grown in conventional and organic cultivation in the Black Sea Region (Turkey) were compared in a trial published by Çelik *et al.* (2013). They reported that, while blueberries grown in conventional cultivation exhibited higher tartaric acid, citric acid, ascorbic acid, glucose and TE antioxidant capacity, fruit grown organically showed higher total phenolics, total monomeric anthocyanins, FRAP and malic acid.

Fruit splitting

Splitting in highbush blueberries is observed most frequently when plants receive a large amount of rainfall just before harvest. Additionally, drought-stressed rabbiteye blueberries are more likely to sustain rain-related splitting (Lyrene and Crocker, 1991; Austin, 1994). Water absorbed through the epidermis of the skin, as well as from the roots, contributes to splitting. Splits in blueberries are usually oblong wounds in the fruit skin that may range from a small, shallow crack in the skin alone to, more commonly, deep wounds that penetrate into the fruit pulp. Deeper wounds suggest that splitting occurs not only at the epidermis but also from deep within the fruit (Marshall *et al.*, 2008).

Splitting has been researched extensively in sweet cherries, tomatoes and grapes. The factors contributing to splitting in cherries include cultivar differences, water temperature, length of the wetting period, SS, fruit firmness and turgor, and elasticity of the skin (Khadivi-Khub, 2015). Absorption of external water through the fruit skin has directly or indirectly been demonstrated to cause cracking in cherries.

SHB and rabbiteye blueberry cultivars are susceptible to rain-induced splitting, but the severity differs among cultivars (Marshall *et al.*, 2002). Marshall *et al.* (2006) carried out a survey in which rabbiteye blueberry growers in Louisiana and Mississippi rated cultivars on the observed severity of fruit splitting using a scale of 1 (no splitting) to 5 (severe splitting). Among the three most widely planted cultivars, 'Premier' exhibited the least splitting (1.2 rating) followed by 'Climax' (2.2) and 'Tifblue' (3.1). They also found that fruit splitting reduced marketable fruit and thus profit by 14–30%. In general, firmness measured as either deformation or elasticity correlated with splitting tendencies in rabbiteye and SHB blueberries (Marshall *et al.*, 2008).

Fruit decay

Fresh fruits are prone to fungal contamination in the field, during harvest, transport and retail, and in the consumer's hands. Fruits contain high levels of sugars and other nutrients that support microbial growth, and their low pH makes them particularly susceptible to fungal spoilage because most bacteria that would compete with fungi prefer a near-neutral pH (Almenar *et al.*, 2007). Some fungi cause spoilage in the field, while others proliferate and cause most of their damage after harvest. Fungal spoilage of fruits depends on the cultivar, as well as methods of harvesting, handling, transport and postharvest storage (Ballinger *et al.*, 1978). Woodruff and Dewey (1959) described the deterioration events for highbush blueberries and concluded that most fruit breakdown during storage was physiological, with fungal infection and growth occurring adventitiously on the debilitated tissues.

Blueberries are more resistant to fungal spoilage than the other berry crops. In a study on the level of contamination in fresh samples of berries and citrus, Tournas and Katsoudas (2005) found that the contamination level (percentage of contaminated berries per sample) differed among the various berry types. The highest mean contamination level (82%) was observed in raspberries, closely followed by blackberries and strawberries, while the lowest percentage (38%) occurred in blueberries. Although blueberries are known to be susceptible to decay, this lower contamination level in blueberries compared with other berries may be related to their smooth, hard skin, which makes them less susceptible.

The most common fungi isolated from blueberries are *B. cinerea* Pers. ex Fr. and *Alternaria*, followed by *Fusarium* spp., *Penicillium* spp., yeasts, *Cladosporium* spp., *Trichoderma* spp. and *Aureobasidium* spp. (Tournas and Katsoudas, 2005). For more information on the organisms causing fruit spoilage in blueberries, see Chapter 8 (this volume).

Differences in the fruit quality of highbush and rabbiteye cultivars

In a comparison of three rabbiteye and two NHB cultivars, the rabbiteyes averaged significantly higher skin toughness determined by puncture tests;

however, sensory panellists did not perceive differences ($P > 0.05$) (Silva *et al.*, 2005). The rabbiteye cultivars had significantly lower levels of pectic acid but significantly higher levels of protopectin. The rabbiteyes also had significantly higher levels of neutral detergent fibre, lignin, hemicellulose and cellulose.

When Gündüz *et al.* (2015) compared SHB with rabbiteye cultivars grown in Georgia (USA), they found the rabbiteye blueberries as a group had significantly higher SS, pH and phenolic content than the SHBs but lower TA and fruit weight. However, many cultivars of both types fell into the range of the other one. In other comparisons by these researchers of NHB, SHB and rabbiteye blueberries in Oregon, no significant differences were observed for any of the traits except for fruit weight, although only two rabbiteye cultivars were evaluated.

In other work comparing blueberry types for their phytochemical composition in Georgia (USA), Sellappan *et al.* (2002) found that rabbiteye cultivars had higher average levels of TA, total phenolic content and antioxidant capacity than SHBs. A number of rabbiteye cultivars had higher levels for these properties than any of the SHBs, although some of the SHB selections had values higher than the rabbiteye average values. In a Brazilian study, Rodrigues *et al.* (2011) found that the NHB 'Bluecrop' had lower antioxidant, total phenolic acid and TA levels than all of the rabbiteye cultivars. When Ballington *et al.* (1984) evaluated four NHB and four rabbiteye cultivars for SS, TA and SS:TA ratio, they found that cultivar average values of both types fell within the range of each other. Perkins-Veazie *et al.* (1995) also evaluated eight SHBs, one NHB ('Bluecrop') and one rabbiteye ('Climax') for SS, TA, SS:TA ratio, pH and total anthocyanins. The values in 'Bluecrop' and 'Climax' were within the range of values found among the SHB cultivars.

The considerable overlap found among cultivars of all three blueberry types suggests that genetic barriers do not exist among the various types of blueberries to breeding cultivars with comparable sugar, acid and phytochemical properties. This is not surprising, as SHBs were developed from NHBs, and rabbiteye is a close relative.

FACTORS THAT INFLUENCE STORAGE LIFE OF BLUEBERRIES

Cultivar

Prange and DeEll (1997) stated that virtually all postharvest quality factors are under genetic control. Therefore, from a quality and postharvest standpoint, cultivar selection is the most important management decision in blueberry production. The variable storage life among and within cultivars is a result of inherent factors determining fruit quality, as well as their interaction with the growing conditions and storage environments (Forney, 2009). Fruit of different cultivars vary in size, colour, texture and flavour, as well as in storage potential (Connor *et al.*, 2002b).

In one comparison of the long-term storability (0–5°C, 2% O_2 and 8% CO_2) of nine NHB cultivars, Hancock *et al.* (2008) found that 'Bluegold', 'Brigitta' and 'Legacy' were the best in storage, reaching 4–7 weeks. In another comparison of 17 cultivars, they found 'Brigitta' to store the longest (8 weeks) followed by 'Aurora' and 'Draper'.

Climate

The factors that control photosynthesis (i.e. light, temperature, CO_2 and rainfall) are the major environmental controllers of berry fruit quality. Once the site has been selected, there is little control over these external variables, except where protected blueberry cultivation is realized. However, the plant microclimate can be controlled by various practices: planting density, training, pruning, irrigation, application of growth regulators and fertilization (Prange and DeEll, 1997). These variables can affect air movement and solar penetration within the plant. As explained in Chapter 4 (this volume) on photosynthesis, the goal is to capture a large proportion of sunlight during the season and partition an important amount of carbohydrates towards reproductive growth.

Sudden or excess exposure to sun can cause sunburn. Sunburn either may produce evident damage to the fruit that will render them unsuitable for marketing or will alter their physiology and diminish their storage potential. Light intensity above photosynthetic saturation levels can increase fruit temperature and may result in fruit damage and loss of firmness (Sams, 1999).

High temperatures during ripening can have various undesirable effects. Warm berries are softer than cool berries (Sams, 1999) and more readily become dark during handling (Lyrene, 2006). The force needed to detach a ripe berry from the plant is lower when the temperatures are cool and the berry is fully turgid. Hot weather during harvesting makes the berries of some highbush blueberry cultivars taste bland, but the same high temperatures may make rabbiteye berries sweeter (Lyrene, 2006).

Rain during harvesting can adversely affect fruit quality of highbush blueberries because it delays the harvest, washes off fungicides, moistens stem scars, and splits and softens berries, all of which can also have an impact on the incidence of fungal diseases. The problem is exacerbated if high temperatures occur concurrently with rain (Pritts and Hancock, 1992). Frequent rains during harvesting may dilute flavours and reduce berry sweetness (Lyrene, 2006).

Nutrition

Although all essential elements are needed for adequate yield and high fruit quality, N, K and Ca are the nutrients most often linked with fruit quality in blueberries.

Ballinger and Kushman (1969) found that the application of N to high-bush blueberry increased the fruit:leaf ratio and decreased fruit size and acidity. High levels of N can indirectly influence fruit quality by increasing shoot growth, which will impair pesticide distribution within the canopy, increase the risk of disease on fruit, delay maturation and delay fruit drying after rainfall (Hart *et al.*, 2006). Excess N has been associated with softer fruit (Sams, 1999).

Although P deficiencies are rare in blueberries (Hart *et al.*, 2006), low levels of P in highbush blueberries have been associated with fewer leaves, a high fruit:leaf ratio and small fruit size (Ballinger and Kushman, 1969). Low P has been reported to result in a loss of firmness, particularly in fruit that are low in Ca (Sams, 1999). Townsend (1973) observed that in one of three years, fruit size in highbush blueberries decreased with the application of P fertilizers.

Adequate K nutrition has been associated with increased yields, fruit size, SS and ascorbic acid concentrations, improved fruit colour, increased shelf-life and better shipping quality of many horticultural crops (Lester *et al.*, 2010). However, like N, K fertilization can result in a decrease in firmness or crispness, as measured by a decrease in resistance to compression (Sams, 1999). In fruit crops in general, pH regulation has been found to be associated with organic acid levels and fruit K content. As the most abundant and mobile cation, K is generally associated with high fruit acidity (Prange and DeEll, 1997). Ballinger and Kushman (1969) found that TA increased in highbush blueberries with higher K levels.

Ca is the element that has received most attention with regard to its beneficial impact on fruit quality and postharvest life of the fruit (Sams, 1999). However, the experimental effects of Ca sprays on fruit quality have been inconsistent. Ballinger and Kushman (1969) found that soil applications of Ca actually increased the fruit:leaf count and decreased fruit size, while Hanson (1995) found no effects of preharvest Ca sprays on fruit quality. Stückrath *et al.* (2008) reported that 30 ml/l of a fertilizer containing 120 g Ca^{2+}/l applied 12 times in the season at 4–19-day intervals to NHB 'Elliott' blueberries significantly improved fruit Ca levels and texture (associated to the presence of low-methoxyl pectins) and influenced fruit colour measurements (L, b and chroma). Hanson and Berkheimer (2004) applied calcitic limestone (1100 kg/ha) or calcium sulfate (550 kg/ha) for five seasons to mature NHB 'Jersey' plants. The treatments increased soil pH and Ca levels but had inconsistent effects on Ca levels in leaves and fruit. Ca applications did not alter berry yield, size or firmness, or fruit rot incidence. Angeletti *et al.* (2010) applied calcium sulfate (600 kg/ha) for one season to 'O'Neal' and 'Bluecrop' highbush blueberries and found reduced fruit softening, which was attributed to a 10% increase in Ca content in cell walls. Ca treatments also lowered the fruit respiration rate and weight but did not affect colour, anthocyanins, acidity or sugar levels compared with control plants (no Ca).

It has been shown that the Ca supply to the fruit depends not only on the provision of the element by the soil but, perhaps more importantly, also on the capacity of the plant to absorb the element (root growth), as well as the competition for Ca between reproductive and vegetative tissues (fruits versus shoots). This model fits with the observation that the removal of excess vegetation (i.e. summer pruning) can shift more water, and with it more Ca, to the fruit. This greater influx of Ca to the fruit would improve fruit quality (see Chapter 5, this volume, for more details on Ca nutrition).

Plant water status

A deficiency or excess of water can influence the postharvest quality of berry crops. Management of water often poses a dilemma between yield and postharvest quality (Prange and DeEll, 1997). As 80–90% w/w of the blueberry fruit is water, fruit growth is highly dependent on water availability (Sargent *et al.*, 2006). Lobos *et al.* (2016) studied the effect of regulated deficit irrigation on yield, physiological parameters and fruit quality of mature NHB 'Brigitta' fruit in Colbún (Chile; latitude 35°41′S) during two seasons, and in South Haven, Michigan (latitude 42°21′N) for one season. Irrigation treatments replaced 50, 75 or 100% (control) of actual evapotranspiration (ET_a). Severe water deficit (50% ET_a) decreased fruit quality (berry size, TA, SS and weight) and increased oxidative stress during both seasons in Colbún. In contrast, mild water stress (75% ET_a) resulted in similar fruit yields and quality (firmness, fruit size, TA, SS and berry weight) as the 100% ET_a treatment, but with higher water productivity and intermediate antioxidant capacity. The grower has to establish an adequate level of water that will allow normal fruit growth, without reaching levels of excess water with the consequent reduced oxygenation of the root system. A mild water shortage can reduce crop yield and fruit size but may benefit some quality attributes such as concentration of antioxidants, which are highest in the skin of the fruit (Mainland and Tucker, 2002). Monitoring and maintenance of adequate water levels are discussed in Chapter 6 (this volume).

Canopy management

Ballinger and Kushman (1969) concluded that the fruit:leaf count (F:L) ratio influences highbush fruit quality to a greater degree than mineral nutrition. A high F:L ratio results in later ripening, lower SS and smaller berries. During the harvest season, as berries are harvested, the SS increase when the F:L ratio drops to a level of 1:1 to 2:1. These authors suggested that this was due to fewer fruit competing for carbohydrates from the sources (leaves). As described in Chapter 4 (this volume), light availability affects not only flower bud induction but also fruit quality. In rabbiteye blueberries, it has been found that fruits picked from shaded parts of the canopy have similar weight but are less blue than those exposed to the sun (Patten *et al.*, 1987).

Harvest and handling methods

Blueberries are very susceptible to mechanical damage and bruise easily. This results at least in a loss of firmness that leads to reduced fruit quality and shelf-life (Xu *et al.*, 2015). Bruises appear as internal browning in the flesh of damaged fruit due to tissue breakage and oxidation of phenolic compounds (Studman, 1997; Opara and Pathare, 2014). The berries may be bruised at numerous points in the commercial mechanical harvesting and handling operation. The fruit are first removed from the plant by vigorous shaking, which can result in contact with neighbouring stems and other berries. The detached berries then fall as far as 2.5 m to the catching plates at the bottom of the harvester. On their way, they impact on plant structures and/or components of the machine. The berries then roll off the catching plates on to a conveyor, which usually moves the fruit through a forced-air system to separate the fruit from other plant tissues. The berries are then dropped at different heights on to other berries in a fruit lug. This lug may be transferred a few times before arriving at the packing shed (Dale *et al.*, 1994).

The proportion of marketable fruit is generally much lower for mechanically harvested than for hand-harvested fruit. Research with over-the-row harvesters by Mainland *et al.* (1975) in North Carolina determined that machine harvesters operating in mature highbush blueberries decreased the yield of marketable ripe fruit by 19–44%. Compared with commercially hand-harvested fruit, machine-harvested fruit was 10–30% softer, and when held for 7 days at 21°C, the fruit developed 11–41% more decay. In the case of rabbiteye blueberries, Austin and Williamson (1977) reported that hand-harvested fruit were 29–37% firmer, and after 7–11 days at 15.5°C, machine-harvested lots had more than twice the amount of soft and unmarketable fruit. In rabbiteye blueberries, respiration rates at ambient temperature were 31.1% higher for machine-harvested berries than for hand-harvested fruit (Nunez-Barrios *et al.*, 2005). Additional research in rabbiteye blueberries showed that the magnitude of the effect of machine harvesters was dependent on cultivar (Miller and Smittle, 1987). After machine harvesting, the berries of 'Climax', which at harvest were firmer, less acidic and had a lower SS:TA ratio than 'Woodard', developed less decay and had a longer inherent shelf-life than those of 'Woodard'. These findings are in contrast to the work of Galletta *et al.*

(1971), in which a higher SS:TA ratio was significantly associated with decreasing shelf-life.

Shaker-bar frequencies and harvest time during the day have a marked influence on harvested fruit quality. Howell *et al.* (1976) reported that, by decreasing vibration frequency, the amount of bruising on highbush blueberries was reduced. A decrease in shaker-bar frequency during high-turgor harvest times (night and early morning) has been found to foster improved fruit quality of blueberries. When rabbiteye 'Tifblue' blueberries were machine harvested at different times of the day (06:00, 09:00, 12:00 or 15:00), it was found that the number of mature berries remaining on the plants decreased with later harvest times during the day (Patten *et al.*, 1988). Harvesting in the early morning when there was dew on the fruit did not have detrimental effects on fruit storability relative to fruit that was harvested dry. However, the authors did not evaluate the possible negative impact of wet surfaces on fruit waxy 'bloom'. The effect of harvest time during the day on packout and fruit quality after storage (14 or 28 days at 5°C) was inconsistent between and within years.

A harvester (V45) from the USDA that required bushes to be divided into a V-shape during the shaking operation was studied for its effects on fruit quality. Trials on the NHBs 'Bluecrop' and 'Elliott' harvested with the V45 showed that internal quality and firmness were better than those using the commercial rotary harvester and as good as the hand-harvested fruit (Brown *et al.*, 1996; Peterson *et al.*, 1997). Based on these promising results, the V45 was evaluated in 6-year-old rabbiteye 'Brightwell' and 'Powderblue' blueberries, as well as in 3-year-old SHB FL 86-19 and 'Star' bushes. Plants had to be pruned to remove 30–50% of the canopy and open the middle, resulting in V-shaped plants. The V45 caused little cane damage. In rabbiteye blueberries, internal fruit damage and skin splitting were less in V45-harvested fruit than in fruit harvested by a sway harvester, and were nearly the same as those of hand-harvested fruit. However, in the SHB FL 86-19, the V45 detached a lower proportion of blue fruit and excessive amounts of immature and stemmed fruit. The percentage bloom coverage in the fruit harvested with the V45 was intermediate between hand-harvested and sway-machine-harvested fruit for both rabbiteye 'Brightwell' and SHB FL 86-19 (Takeda *et al.*, 2008).

The fruit are dropped at multiple points in packing lines, which can cause bruising. Xu *et al.* (2015) measured the impacts along packing lines with a Blueberry Impact-Recording Device (BIRD; Yu *et al.*, 2011) and found that the highest impacts occurred at the transfer points. The impacts were most severe at the final handling step, when the sensor was dropped into the hopper above the clamshell filler or collecting trays. These impacts were significantly associated with levels of bruising in laboratory analysis. The authors concluded that the potential for bruise damage was high wherever blueberries were dropped 30 cm or more on to inclined stainless steel surfaces. Padding at these locations significantly reduced fruit bruising.

To compare the bruise susceptibility of three firm-textured blueberries ('Farthing,' 'Sweetcrisp' and FL 05-528) and a soft-textured one ('Scintilla'), Yu *et al.* (2014) dropped fruit and the BIRD sensor from different heights on to hard plastic and padding material. The soft-textured fruit of 'Scintilla' was more susceptible to bruising when dropped 120 cm on to the hard surface (76% bruise incidence) than fruit of the more firm-textured blueberries (31–68% bruise incidence). The BIRD sensor measurements were significantly correlated with fruit bruising incidence on the hard plastic surfaces but not on the padded ones.

Moggia *et al.* (2017b) compared levels of internal browning in NHB 'Brigitta' and 'Duke' fruit of variable firmness after mechanical impact. Hand-picked fruit were segregated into soft (less than 1.60 N), medium (1.61–1.80 N) and firm (1.81–2.00 N) categories, dropped 32 cm on to a hard plastic surface and then held in refrigerated storage for several weeks. The fruit of 'Duke' softened the fastest during storage, and there was a significant correlation between firmness and internal browning, although internal browning was not significantly different between dropped and undropped fruit. The fruit of 'Brigitta' showed a significant relationship between firmness and internal browning, and there were marked differences in internal browning between dropped and undropped fruit.

POSTHARVEST CONDITIONS

Maximizing quality and extending the market life of fresh blueberries adds value to the fruit by enabling access to new markets (Forney, 2009). The maximum quality and storage life of the fruit have already been determined when the fruit are harvested. Success in achieving the maximum storage life of blueberry fruit is dependent on slowing down the degradative processes following ripening (senescence) and limiting the progress of decay. Two principles can be used to guide postharvest decision making: (i) the fruit is alive and responsive to the environment; and (ii) the fruit's quality potential never increases after the fruit has been picked (Sargent *et al.*, 2006).

Fruit quality loss during postharvest handling is primarily the result of decay, physiological breakdown, physical abuse and dehydration. The fruit must be of high initial quality to maximize postharvest life. Following harvest, blueberry fruit must be cooled and held near 0°C and at a relative humidity of about 95% for maximum storage life (Forney, 2009). Controlled or modified atmospheres are techniques that can be used in conjunction with low temperatures to enhance storage life and reduce decay of blueberries. With the expansion of the blueberry industry and the increased demand for high-quality fresh fruit throughout the year, the use of postharvest technologies to optimize marketing of high-quality fresh fruit is of utmost importance.

Cooling after harvest

Harvested blueberries should be cooled as soon as possible. Sargent *et al.* (2006) recommended that if blueberries are going to be shipped, they should be cooled to 1°C within 4 h of harvest. Beaudry (1992) reported that reducing temperatures rapidly from field levels to 0°C increased the shelf-life of highbush blueberries by as much as eight- to tenfold relative to non-cooled fruit (Table 9.9) and causes an eightfold reduction in respiration rate. Research on rabbiteye blueberries has shown that immediate refrigeration at 1°C for hand-harvested fruit will give a gain of 35% in firmness compared with fruit left at 22°C for 8 days. Surprisingly, refrigeration had a marginal effect for machine-harvested fruit with a 9% gain in firmness compared with ambient temperature (Nunez-Barrios *et al.*, 2005). After 8 days, machine-harvested and hand-harvested fruit stored at 22°C had equivalent firmness.

A reduction in respiratory rate signals a slowdown in the overall metabolism of the fruit, which influences softening, tissue breakdown and pigment synthesis. At a temperature of 26.7°C, blueberry respiration can produce as much as 6100 kcal heat/t/day. Unless this heat is removed by cooling, it can elevate fruit temperature by as much as 14.4°C (Boyette *et al.*, 1993). The respiration rate of blueberries at 26.7°C is nearly 20 times the rate at 4.5°C. In

Table 9.9. Effect of O_2 concentration and temperature on the shelf-life of NHB 'Bluecrop', 'Jersey' and 'Elliott' blueberry fruit stored in modified-atmosphere packages, as judged by visual rating only. CO_2 levels, although not reported, would be approximately one-quarter of the gradient in O_2 between the package interior and air. (Adapted from Beaudry, 1992.)

Cultivar	Temperature (°C)	Average postharvest life (days)[a]			
		Air	Intermediate	Optimal	Anaerobic
'Bluecrop'	0	49	68	88	106
	5	27	35	44	56
	25	7	10	13	13
'Jersey'	0	28	31	77	80
	5	16	32	38	53
	25	8	8	12	12
'Elliott'	0	48	57	62	75
	5	32	35	35	40
	25	6	7	8	9

[a]Target O_2 concentrations are: air, 21%; intermediate, 15%; optimal, the lowest O_2 tolerated without causing fermentation (approx. 2–4%); and anaerobic, levels that induce fermentation (less than 2%).

other words, blueberries held at 4.5°C have nearly 20 times the shelf-life of those held at 26.7°C.

Pallets of fruit need to be cooled as soon as possible, preferably through forced-air cooling. In still air, the average cooling rate of pallets of blueberries is slow because heat is transferred from the interior only by conduction, and all of the materials surrounding the fruit reduce the cooling rate (Boyette *et al.*, 1993). With room cooling alone, it requires more than 36 h to cool blueberries in the centre of a pallet to 4.5°C. Forced-air cooling can rapidly drop fruit temperature and reduce the metabolic activity of the fruit, delay the softening and thus reduce decay susceptibility. Depending on the circumstances, the rate of cooling has been found to be four to ten times (Vicente *et al.*, 2005) or 16–20 times (Boyette *et al.*, 1993) faster with forced-air cooling.

Forced-air cooling is accomplished by exposing packages (field lugs or trading packages) to higher air pressure on one side than on the other. This pressure differential forces the cool (1°C) and moist (90–95% relative humidity) air through the packages, which removes heat more effectively from the berries. To obtain proper air movement, it is necessary to adequately stack the packages in order to minimize any spaces that might force the air to pass around rather than through the containers, reducing cooling efficiencies. Recommendations call for 5–8% of the lateral surface and 3–5% of the total surface in the bottom to remain void in order to ensure adequate air movement through the packages (Vicente *et al.*, 2005). Cooling with chilled water (hydrocooling) is very effective for blueberries that will be processed. However, it damages the bloom of the fruit, making it impractical for the fresh market (Sargent *et al.*, 2006).

Cooling during storage

After the berries have been cooled, it is recommended that they are held at 0–1°C and 85–95% relative humidity (Vicente *et al.*, 2005). Under these conditions, blueberries can maintain an acceptable condition for 2–3 weeks or more depending on cultivar (Vicente *et al.*, 2005; Schotsmans *et al.*, 2007). The critical temperature for freezing blueberries is −1.3°C. Overall, berries with a higher SS content are less likely to freeze.

The effect of storage temperature differs considerably among cultivars. For instance, Bounous *et al.* (1997) found that the mass loss of NHBs stored at 1°C for 3 weeks was 2.5% for 'Dixi', 21% for 'Darrow' and 25% for 'Coville'. However, firmness at harvest was 28% greater for rabbiteye 'Climax' than for 'Woodard' blueberries, and the difference increased to 38% after 2 weeks of storage at 3°C (Miller and Smittle, 1987). Cultivars can vary in their response to cooling. NeSmith *et al.* (2005) found that the rate of firmness loss was similar among rabbiteye blueberry cultivars at 1 and 12°C, except for 'Premier', which lost firmness more rapidly at the higher temperature. At 22°C,

'Brightwell' had the lowest rate of firmness drop. The greatest difference in rate of firmness loss among cultivars was found at 32°C, where 'Powderblue' had a sixfold increase over 'Brightwell'. The extent of mass loss in response to temperature increases followed a similar trend to firmness loss, but there were fewer differences among cultivars (NeSmith *et al.*, 2005).

Storing pallets of fruit at the optimal temperature of 0°C can sometimes result in condensation forming inside the overwrapped package, which is unacceptable for most receivers (Beaudry, 1992). This condensation can be reduced by maintaining the fruit temperature at or above the dew point. Condensation can influence fungal decay, although its impact is generally thought to be minimal. However, some containers allow greater moisture loss than others. Almenar *et al.* (2008) compared decay levels of NHB 'Elliott' fruit packed in the standard commercial clamshells made of polyethylene terephthalate (PET) or of the experimental biodegradable polylactide (PLA), and found that fungal development after 18 days at 10°C was lower in PET than in PLA containers (5 versus 11%). This effect could be attributed to the greater moisture loss in PET containers (Beaudry, 1992). Tasting panels showed that based on flavour, texture, external appearance and overall quality, consumers could distinguish between blueberries from different packages, and they preferred those packaged in PLA containers (Almenar *et al.*, 2010).

Nunes *et al.* (2004) studied the effects of various temperatures (0, 5, 10, 15 or 20°C) on the quality of NHB 'Patriot' fruit and concluded that: (i) a single quality factor cannot be used to express loss of quality of blueberries over the normal physiological range of temperatures; and (ii) prediction of blueberry shelf-life calculated from data from the literature on respiration rates at various temperatures is not precise unless the type of cultivar and the quality of the fruit at harvest, as well as environmental factors involved, are well known and the limiting quality factor is closely related to the overall metabolic rate.

Controlled and modified atmospheres

Modified atmosphere (MA) and controlled atmosphere (CA) are used as supplements to temperature management for extending the postharvest life of berries (Vicente *et al.*, 2005). This technology involves altering the normal concentrations of O_2 and CO_2. The difference between a MA and CA is that the control of O_2 and CO_2 levels is active under a CA system, while the control is passive under a MA (Sargent *et al.*, 2006). O_2 and CO_2 are biologically active molecules important in metabolic processes in plants. In the case of many climacteric fruits, O_2 and CO_2 alter ripening not only through their influence on respiration per se but largely through their inhibitory effects on the action of ethylene, the ripening hormone (Beaudry, 1999). Conditions of low O_2 and high CO_2 slow the decline in quality by inhibiting ripening and, at sufficiently high CO_2 concentrations (i.e. anaerobic conditions), by suppressing the activity of aerobic organisms that cause decay (Beaudry, 1992) (Table 9.9). For blueberries, the impact of CO_2 on decay suppression tends to be of greater importance than the effects of reduced O_2.

The primary use of MAs/CAs for blueberries is in long-distance marine transport. A wide array of systems is available, from a simple one that keeps the fresh-air exchange gate closed if the CO_2 concentration is below a certain set limit (AFAM+; Thermo King, Minnesota) to more sophisticated systems that have an initial gas flush followed by active ventilation controls (TransFresh Tectrol CA; Transfresh Corporation, California) or membrane separationn systems with supplemental CO_2 injection (EverFresh; Carrier Corp., Florida). Another system consists of a large bag that is wrapped around a pallet and sealed with tape (TransFresh Tectrol MA; Transfresh Corporation, California); the bag is then pierced with a nozzle through which CO_2 can be injected to attain an atmosphere of 5–10% O_2 and 10–15% CO_2. The nozzle is then removed and the perforation is sealed (Bounous *et al.*, 1997; Sargent *et al.*, 2006).

One problem that faces the industry is the deleterious effects that occur once the CA containers are opened and the fruit is subject to dramatic changes in temperature and gas composition. MA packaging (MAP) has the potential to alter O_2 and CO_2 regimes throughout the marketing chain. A package should maintain the atmospheric composition over the range of temperatures commonly encountered between harvest and consumption. Poor temperature control, however, can cause package O_2 levels to drop low enough to induce anaerobic respiration and generate off-flavours. MAP is designed to generate a physiologically adequate O_2 partial pressure inside the package by matching total respiratory O_2 uptake of the packaged product to the total permeation through the film (Beaudry *et al.*, 1992).

The most appropriate concentrations of CO_2 to store blueberries range from 10 to 12% (Sargent *et al.*, 2006), as decay organisms are controlled and the physiological breakdown is slowed down at this concentration (Beaudry, 1992). Sargent *et al.* (2006) reported that lowering the O_2 concentration would have little benefit in extending storage life, and an O_2 concentration that is too low (less than 2% O_2) may inhibit flavour development or cause the development of off-flavours. The atmosphere surrounding the fruit during storage can have marked effects on fruit quality and condition depending on *Vaccinium* spp., cultivar, harvest date, handling immediately after harvest, storage conditions and packaging (Schotsmans *et al.*, 2007).

Alsmairat *et al.* (2011) tested the impact of storage atmospheres on nine highbush cultivars in which the CO_2 and O_2 percentages totalled 21%. Fruit firmness, skin reddening and decay declined, and the proportion of fruit with internal discoloration tended to increase as CO_2 concentrations increased. Cultivar effects were far more pronounced than atmospheric effects. 'Brigitta', 'Duke', 'Legacy', 'Liberty' and 'Toro' appeared well suited to extended CA storage, 'Elliott' stored moderately well and 'Jersey', 'Nelson' and 'Ozarkblue' stored poorly.

   Work on five NHB cultivars found lower decay levels after 4 weeks under CA storage (10% O_2, 10% CO_2) when fruit were stored at 1°C, but the opposite trend was observed in two of the cultivars ('Hardyblue' and 'Patriot') when the fruit were at 8°C (Remberg *et al.*, 2003). Rabbiteye blueberries stored for 28 days at 1.5°C had 5% decay in both regular storage (RS) and CA storage (2.5% O_2, 15% CO_2) for 'Maru', but in 'Centurion', CA storage significantly decreased the incidence of decay from 8 to 1.5% (Schotsmans *et al.*, 2007).

   Schotsmans *et al.* (2007) found no effect of storage conditions (RS versus CA) on weight loss and shrivelling of rabbiteye 'Centurion' and 'Maru' blueberries after 6 weeks at 1.5°C. Beaudry *et al.* (1998) reported that the mass loss of NHB 'Berkeley' and 'Bluecrop' fruit was reduced by CA storage (2% O_2, 8% CO_2 for 21 days) compared with RS. After 6 weeks of storage at 1°C, weight loss was 10% lower in NHB 'Darrow' and 25% lower in NHB 'Coville' stored under MA (19% CO_2) compared with RS (Bounous *et al.*, 1997).

   Forney *et al.* (2003) found that firmness increased in NHB 'Burlington' fruit when stored at 0% CO_2, and that softening was only slight at 10% CO_2 but increased at higher CO_2 levels (15, 20 or 25%) after 3 and 6 weeks at 0°C. Softening occurred concomitant with flesh discoloration. Schotsmans *et al.* (2007) found greater softening of rabbiteye 'Centurion' and 'Maru' blueberries in CA storage (2.5% O_2, 15% CO_2) than RS at 1.5°C. Rodríguez and Zoffoli (2016) found that CO_2 levels higher than 8% or O_2 lower than 2% induced fruit softening in NHB 'Brigitta' after 30–45 days at 0°C. They found that induction of fruit softening by CO_2 was cultivar dependent, with 'Duke', 'Legacy' and 'O'Neal' being extremely sensitive, as significant softening was triggered at 6% CO_2.

   Smittle and Miller (1988) reported that total sugar accumulation for the rabbiteye cultivars 'Climax' and 'Woodard' decreased in storage, with a higher decrease in RS compared with CA storage. Similar trends were found by Schotsmans *et al.* (2007) for 'Maru' and 'Centurion' rabbiteye blueberries.

   Remberg *et al.* (2003) found that changes in acidity were not consistent among highbush blueberry cultivars after 4 weeks of storage in RS or CA (10% O_2, 10% CO_2). The TA of rabbiteye 'Centurion' blueberries changed little during RS, whereas a significant increase occurred during CA storage of this cultivar. In contrast, Smittle and Miller (1988) reported that the pH and acidity of rabbiteye blueberries were not affected by storage duration or atmosphere composition.

   Paniagua *et al.* (2014) compared three cooling delays (0, 12 or 24 h at 10°C), three atmosphere concentrations (air, 10% CO_2 + 2.5% O_2 and 10% CO_2 + 20% O_2) and two storage temperatures (0 and 4°C) for their impact on the final quality of highbush 'Brigitta' and rabbiteye 'Maru' fruit. Delays in the time of cooling had a small effect on final weight but a large effect on firmness and rot incidence. Atmospheres with 10% CO_2 significantly reduced decay, most dramatically at the lowest O_2 concentration, although low O_2 also tended to soften the fruit. The authors suggested that to achieve optimal postharvest storage for blueberries, minimizing temperature variability in the supply chain is important, as well as finding the optimal combination of high CO_2 and low O_2 for each cultivar.

Alternative methods for reducing postharvest spoilage

Currently the most common approaches taken to reduce postharvest spoilage of blueberries are based on controlling the rate of fruit ripening and pathogen growth using low temperatures, preventative fungicide applications and MA storage (Vicente *et al.*, 2005). Fruit rot diseases are best controlled using several integrated strategies. Success is reached when all the tools available are used to produce and maintain a quality berry. These include preharvest, harvest and postharvest procedures (du Jardin, 2015). Among the preharvest control methods are cultivar selection, fungicide applications, and pruning to open the canopy to allow better air circulation and fungicide coverage. Frequent harvesting to remove all ripe fruit will drastically reduce fruit rots (Cline, 1997), and prevention of infection through proper sanitation is effective to control decay.

Among the postharvest measures to reduce microbial spoilage of fruits are careful culling, storage at low temperatures or under CA/MA, and application of various chemicals and physical treatments (Miller *et al.*, 1994). The most common postharvest method to reduce or slow down decay is cooling. NHB 'Bluecrop' and 'Bluetta' blueberries pre-cooled at 2°C had 60–80% less decay than berries that were not pre-cooled when held for 24 h at 21°C following a 3-day simulated transit period at 10°C (Hudson and Tiedjen, 1981). Although the growth of many fungi is slower at 1°C, postharvest decays such as anthracnose (*Gloeosporium* spp.), grey mould rot (*B. cinerea* Pers. ex Fr.) and *Alternaria* spp. result in spoilage (Miller *et al.*, 1994).

Several disease control strategies are needed simultaneously to avoid severe losses after harvest. Among these strategies, a fungicide application at the end of the maturation period before harvest in conjunction with sulfur dioxide (SO_2) treatments applied immediately after harvest have been reported as effective in controlling spoilage, either to packed fruit alone or in combination with a CO_2-enriched atmosphere (Cantín *et al.*, 2012; Rivera *et al.*, 2013). Rodríguez and Zoffoli (2016) reported trials in which SHB 'Legacy' fruit were fumigated for 30 min at 20°C with SO_2 at a concentration–time product of 100–150 μl/l/h, followed by packaging in hermetically sealed bags with two perforations of 3 mm^2 (MAP 2) or in a perforated (0.3% ventilation area) low-density polyethylene bag and storage for 45 days at 0°C. Decay, weight loss, and the percentage of dehydrated and soft fruit were reduced effectively when the SO_2-fumigated fruit were stored in either MAP 2 or the perforated bag. Decay caused by *B. cinerea* was particularly well controlled by SO_2. The authors concluded that the SO_2-fumigated fruit, put into a perforated bag was a better option than packaging the fruit into MAP bags, where the CO_2 steady-state concentrations that controlled decay were extremely close to the level that injured the fruit.

In recent years, emphasis has been focused on developing alternatives to fungicide sprays. Due to health risks, there has been increasing concern regarding the use of synthetic fungicides in fruit production and their presence in the environment (Sharpe *et al.*, 2009; Wang *et al.*, 2010). These alternatives can be classified as chemical or physical treatments. Among the chemical treatments, the trend is towards the use of natural products. In blueberries, there are reports on the effects of isothiocyanates, ozone, high-O_2 atmospheres, essential oils and hexanal. Allyl isothiocyanate is a natural volatile compound that is present in plants belonging to the *Brassicaceae* family and is responsible for the pungent taste of mustard and horseradish. Application of allyl isothiocyanate to NHB 'Duke' blueberries retarded blueberry decay by nearly 90% during storage at 10°C, but the treatment decreased total phenolics and total anthocyanins and reduced antioxidant activities (Wang *et al.*, 2010). The sensory quality of the fruit was not evaluated. Ozone has been reported to have strong antimicrobial effects against fungi and other pathogens. It rapidly inactivates microorganisms by reacting with intracellular enzymes, nucleic acids, and components of cell envelopes and spore coats. In 2001, the US Food and Drug Administration approved ozone for use on food. In blueberry, ozone treatments (450 or 6000 ppb for 48 h at 20°C) reduced the growth of fungal spores of *B. cinerea* without deleterious effects on fruit quality; however, ozone had little effect on growth of *Sclerotinia sclerotiorum* and the overall incidence of decay was not reduced (Sharpe *et al.*, 2009). The authors attributed this result to the high susceptibility of blueberries to fungal infection, and to the presence of latent infections that occurred at bloom and were not affected by the low penetration of gaseous ozone treatment.

NHB 'Duke' fruit placed for 9–35 days at 5°C in high-O_2 atmospheres (40, 60, 80 or 100% O_2) showed a decreased incidence of decay with increasing O_2 concentrations over 40% (Zheng *et al.*, 2003). TA, SS and surface colour were little affected. O_2 levels between 60 and 100% promoted increases in total phenolics and total anthocyanins.

Essential oils are aromatic oily extracts obtained from plant tissues. Various components of essential oils have been identified to be effective in inhibiting microbial growth. Increasing evidence has shown that some essential oils also possess antioxidant properties (Ruberto and Baratta, 2000). After 4 weeks at 10°C all seven essential oils tested inhibited NHB 'Duke' fruit decay development (16–72%) compared with control. The effect was attributed to the antimicrobial capability of these compounds, which would act through disruption of cellular membrane functions and interference with active sites of enzymes and cellular metabolism. Sugar and organic acid components were improved by oil applications. Although there was usually an increment in antioxidant activity, the reduction in decay was not correlated with a promotion of antioxidant activity. This suggests that potential antimicrobial activity against pathogens causing spoilage is largely dependent on the potency of a particular compound in inhibiting the microbes and less on its effects on antioxidant promotion (Wang C.Y. *et al.*, 2008a).

Application of edible coatings has shown great potential to maintain fruit quality by reducing moisture and gas transfer, decreasing microbial growth, and retarding fruit ripening and senescence. Of particular interest in food-preservative agents is the application of chitosan owing to its non-toxic, biodegradable, antibacterial and film-forming properties (Qiu *et al.*, 2014). Chitosan is a deacetylated form of the biopolymer chitin, produced naturally and industrially. Due to its multifunctional properties, chitosan has been shown to be an effective postharvest fungicide and preservative for various fruits. Jiang *et al.* (2016) applied L-chitosan and H-chitosan to rabbiteye blueberries at rates of between 0.2 and 2 mg/l and found that spore germination and mycelial growth of *B. cinerea* were significantly inhibited by chitosan in a concentration-dependent mode. Application of a chitosan coating exhibited a positive effect on the changes in weight loss, firmness, total phenolics and anthocyanins as storage time increased.

Hexanal (hexanaldehyde) is an alkyl aldehyde used in the flavour industry to produce fruity flavours. It is a natural plant volatile with antifungal properties that have been reported to reduce postharvest diseases. When NHB 'Brigitta', 'Burlington' and 'Duke' blueberry fruit were treated with hexanal vapour at 0.9 μl/l of air for 24 h, there was a 50–70% reduction in decay in treated fruit compared with the control. Marketable fruit in all three cultivars was 20–40% greater following hexanal treatments after 12 weeks of storage compared with controls (Song *et al.*, 2010). The volatile nature of hexanal complicates its commercial use, so Almenar *et al.* (2007) developed a method to encapsulate hexanal into cyclodextrins (naturally occurring molecules produced from starch) in order to control *in vitro* postharvest pathogens of berry fruits. They found that 1.1, 1.3 and 2.3 μl/l of air was necessary to prevent growth of *Colletotrichum accutatum*, *B. cinerea* and *Alternaria alternata*, respectively.

The physical treatments trialled to control decay in blueberries include gamma and UV radiation and hot-water dips. When various doses (0–3 kGy) of gamma radiation were tried on rabbiteye 'Climax' blueberries, it was found that irradiation generated softer berries and greater decay as the dose was increased. The irradiated berries had a lower consumer preference and reduced fresh market quality (Miller *et al.*, 1994). Alternatively, UV radiation (UV-C at 0–4 kJ/m^2) was tested as a means to extend shelf-life of NHB 'Bluecrop' and 'Collins' blueberries; while weight loss and firmness were found not to be affected by radiation treatment, the decay incidence of ripe rot was decreased by 10% with 1–4 kJ/m^2, while antioxidants (total anthocyanins, total phenolics and FRAP) were usually higher in treated fruit of both cultivars, with higher radiation levels needed to obtain significant effects on 'Bluecrop'

(Perkins-Veazie *et al.*, 2008). Hot-water dips are effective physical treatments for fungal pathogen control in various fruit species as most fungal spores and latent infections are either on the surface or in the first few layers under the epidermis of the fruit. Fan *et al.* (2008) found that *B. cinerea* and *Colletotrichum* spp. were the main spoilage microorganisms in NHB 'Burlington' blueberries. These fungi were effectively controlled by hot-water treatments (60°C for 15 or 30 s). Although the heat treatments diminished weight loss and the proportion of shrivelled and split berries, they also reduced the bloom of fruit, most likely by melting the surface wax. This limits the likelihood of this technique being used in decay control of fresh fruit.

In summary, although there are some emerging technologies that could complement the benefits of low-temperature storage, fungicides and MAs, there remain many aspects that should be understood before these are adopted extensively. The feasibility and limitations of these options must be evaluated on a commercial scale (Vicente *et al.*, 2005).

CONCLUSIONS

Blueberry production and marketing have increased markedly in the last decade. Maturity at harvest is the most important factor that determines postharvest life and final fruit quality. The highest quality fruit (higher sugar, better postharvest, maximum health benefits and more intense flavours) are those allowed to ripen fully on the plant. Fruit quality can be established by external appearance, sensory attributes, nutritional content and microbiological condition.

Sensory studies have determined that the overall eating quality of blueberries is most tightly correlated with flavour acceptability and blueberry-like flavour intensity. Blueberries do not possess a strong characteristic aroma, but over 100 volatile compounds have been identified in their fruit. The size of the stem scar explains about 90% of fruit decay. Fruit colour depends on the components on and within the fruit skin and is determined primarily by the extent of the waxy bloom (quantity and structure).

Fruit firmness also relates to consumer appeal and to postharvest decay of the fruit. Although SS levels vary among seasons and locations, their values were more stable than TA or firmness. SS:TA ratio values can predict postharvest life of the fruit. The keeping quality will be good for clones with a SS:TA value less than 18 and low for fruit with SS:TA greater than 32. The pH values of blueberry extracts correlate with scores of intensity of flavour and acceptability of flavour, as well as overall eating quality.

Blueberry fruit contain various phenolics, mainly anthocyanins, which contribute to antioxidant capacity. Anthocyanins account for up to 60% of the total phenolic content in highbush blueberries. Besides genotype, the antioxidant capacity can be affected by location, growing season, maturity, and cultural and postharvest handling. There are contrasting results on the impact of cultivation type on antioxidants.

The variable storage life among and within cultivars is a result of inherent factors determining fruit quality, as well as their interaction with growing conditions and storage environments. Light, temperature, nutrients (especially K and Ca), CO_2 and water are the major environmental controllers of berry fruit quality. Fruit quality loss during postharvest handling is mainly the result of decay, physiological breakdown, physical abuse and dehydration. Several methods and practices have been developed to extend the postharvest life of the fruit.

Harvested blueberries should be cooled as soon as possible with forced air to reduce the respiration rate and to slow the ripening process and the decline in quality. Berries stored at 0–1°C and 85–95% relative humidity can maintain an acceptable condition for several weeks. CA or MA, as well as MAP, used in conjunction with low temperatures enhance storage life and reduce decay of blueberries. The most appropriate concentrations to store blueberries are in the range 10–12% CO_2 and 2–3% O_2. Lowering O_2 below 2% may cause the development of off-flavours.

REFERENCES

Albrigo, L.G., Lyrene, P.M. and Freeman, B. (1980) Waxes and other surface characteristics of fruit and leaves of native *Vaccinium elliotti* Chapm. *Journal of the American Society for Horticultural Science* 105, 230–235.

Almenar, E., Auras, R., Rubino, M. and Harte, B. (2007) A new technique to prevent the main post harvest diseases in berries during storage: inclusion complexes β-cyclodextrin–hexanal. *International Journal of Food Microbiology* 118, 164–172.

Almenar, E., Samsudin, H., Auras, R., Harte, B. and Rubino, M. (2008) Postharvest shelf life extension of blueberries using a biodegradable package. *Food Chemistry* 110, 120–127.

Almenar, E., Samsudin, H., Auras, R. and Harte, J. (2010) Consumer acceptance of fresh blueberries in bio-based packages. *Journal of the Science of Food and Agriculture* 90, 1121–1128.

Alsmairat, N., Contreras, C., Hancock, J.F., Callow, P. and Beaudry, R. (2011) Use of combinations of commercially relevant O_2 and CO_2 partial pressures to evaluate the sensitivity of nine highbush blueberry cultivars to controlled atmospheres. *HortScience* 46, 74–79.

Angeletti, P., Castagnasso, H., Micelli, E., Terminiello, L., Concellón, A., Chaves, A. and Vicente, A.R. (2010) Effect of preharvest calcium applications on postharvest quality, softening and cell wall degradation of two blueberry (*Vaccinium corymbosum*) varieties. *Postharvest Biology and Technology* 58, 98–103.

Austin, M.E. (1994) *Rabbiteye Blueberries: Development, Production and Marketing.* Agscience, Auburndale, Florida.

Austin, M.E. and Williamson, R.E. (1977) Comparison of harvest methods of rabbiteye blueberries. *Journal of the American Society for Horticultural Science* 102, 454–456.

Ballinger, W.E. and Kushman, L.J. (1969) Relationship of nutrition and fruit quality of Wolcott blueberries grown in sand culture. *Journal of the American Society for Horticultural Science* 94, 329–335.

Ballinger, W.E. and Kushman, L.J. (1970) Relationship of stage of ripeness to composition and keeping quality of highbush blueberries. *Journal of the American Society for Horticultural Science* 95, 239–242.

Ballinger, W.E., Kushman, L.J. and Hamann, D.D. (1973) Factors affecting the firmness of highbush blueberries. *Journal of the American Society for Horticultural Science* 98, 583–587.

Ballinger, W.E., Maness, E.P. and McClure, W.F. (1978) Relationship of stage of ripeness and holding temperature to decay development of blueberries. *Journal of the American Society for Horticultural Science* 103, 130–134.

Ballington, J.R., Ballinger, W.E., Swallow, W.H., Galletta, G.J. and Kushman, L.J. (1984) Fruit quality characterization of 11 *Vaccinium* species. *Journal of the American Society for Horticultural Science* 109, 684–689.

Baloga, D.W., Vorsa, N. and Lawter, L. (1995) Dynamic headspace gas chromatography-mass spectrometry analysis of volatile flavor compounds from wild diploid blueberry species. *ACS Symposium Series* 596, 235–247.

Beaudry, R.M. (1992) Blueberry quality characteristics and how can they be optimized. *Annual Report of the Michigan State Horticultural Society* 122, 140–145.

Beaudry, R.M. (1999) Effect of O_2 and CO_2 partial pressure on selected phenomena affecting fruit and vegetable quality. *Postharvest Biology and Technology* 15, 293–303.

Beaudry, R.M., Cameron, A.C., Shirazi, A. and Dostal-Lange, D.L. (1992) Modified-atmosphere packaging of blueberry fruit: effect of temperature on package O_2 and CO_2. *Journal of the American Society for Horticultural Science* 117, 436–441.

Beaudry, R.M., Moggia, C.E., Retamales, J.B. and Hancock, J.F. (1998) Quality of 'Ivanhoe' and 'Bluecrop' blueberry fruit transported by air and sea from Chile to North America. *HortScience* 33, 313–317.

Bett-Garber, K.L., Lea, J.M., Watson, M.A., Grimm, C.C., Lloyd, S.W., Beaulieu, J.C., Stein-Chisholm, R.E., Andrzejewski, B.P. and Marshall, D.A. (2015) Flavor of fresh blueberry juice and the comparison to amount of sugars, acids, anthocyanidins, and physicochemical measurements. *Journal of Food Science* 80, S881–S827.

Blaker, K.M. and Olmstead, J.W. (2015) Cell wall composition of the skin and flesh tissue of crisp and standard texture southern highbush blueberry genotypes. *Journal of Berry Research* 5, 9–15.

Bounous, G., Giacalone, G., Guarinone, A. and Peano, C. (1997) Modified atmosphere storage of highbush blueberries. *Acta Horticulturae* 446, 197–203.

Boyette, M.D., Estes, E.A., Mainland, C.M. and Cline, W.O. (1993) Postharvest cooling and handling of blueberries. Publication no. AG-413-7. North Carolina Cooperative Extension Service, Raleigh, North Carolina. Available at: https://content.ces.ncsu.edu/postharvest-cooling-and-handling-of-blueberries (accessed 27 February 2018).

Bravo, C. (2017) Influencia de la posición de la fruta dentro de la planta y de su estado de madurez a cosecha, sobre la postcosecha de arándanos (*Vaccinium corymbosum* L.). [Influence of fruit position within the plant and its degree of maturity on the postharvest of blueberries.] MgSc thesis, Universidad de Talca, Talca, Chile.

Bremer, V., Crisosto, G., Molinar, R., Jimenez, M., Dollahite, S. and Crisosto, C.H. (2008) San Joaquin Valley blueberries evaluated for quality attributes. *California Agriculture* 62, 91–96.

Brown, G.K., Schulte, N.L., Timm, E.J., Beaudry, R.M., Peterson, D.L., Hancock, J.F. and Takeda, F. (1996) Estimates of mechanization effects on fresh blueberry quality. *Applied Engineering in Agriculture* 12, 21–26.

Cantín, C.M., Minas, I.S., Goulas, V., Jimenez, M., Manganaris, G.A., Michailides, T.J. and Crisosto, C.H. (2012) Sulfur dioxide fumigation alone or in combination with CO_2-enriched atmosphere extends the market life of highbush blueberry fruit. *Postharvest Biology and Technology* 67, 84–91.

Cappellini, R.A. and Ceponis, M.J. (1977) Vulnerability of stem-end scars of blueberry fruits to postharvest decays. *Phytopathology* 67, 118–119.

Castrejón, A.D.R., Eichholz, I., Rohn, S., Kroh, L.W. and Huykens-Keil, S. (2008) Phenolic profile and antioxidant activity of highbush blueberry (*Vaccinium corymbosum*) during fruit maturation and ripening. *Food Chemistry* 109, 564–572.

Çelik, H., Özgen, M. and Saraçoğlu, O. (2013) Organik ve standart olarak yetiştirilen bazidotless yüksek boylu maviyemiş (*Vaccinium corymbosum* l.) çeşitlerinin fitokimyasal içerikleri ile antioksidan kapasitelerinin Karşılaştırılması. [Comparison of phytochemicals and antioxidant capacities of some standard and organically grown highbush blueberries (*Vaccinium corymbosum* L.).] *Tarim Bilimleri Dergisi* 18, 167–176.

Chen, H., Cao, S., Fang, X., Mu, H., Yang, H., Wang, X., Xu, Q. and Gao, H. (2015) Changes in fruit firmness, cell wall composition and cell wall degrading enzymes in postharvest blueberries during storage. *Scientia Horticulturae* 188, 44–48.

Chiabrando, V., Giacalone, G. and Rolle, L. (2009) Mechanical behaviour and quality traits of highbush blueberry during postharvest storage. *Journal of the Science of Food and Agriculture* 89, 989–992.

Cline, W.O. (1997) Fruit rots diseases of blueberry. North Carolina State University Plant Pathology Extension, Raleigh, North Carolina. Available at: http://www.ces.ncsu.edu/depts/pp/notes/Fruit/blueberryinfo/berryrots.htm (accessed 1 February 2011).

Connor, A.M., Luby, J.J. and Tong, C.B.S. (2002a) Variation and heritability estimates for antioxidant activity, total phenolic content, and anthocyanin content in blueberry progenies. *Journal of the American Society for Horticultural Science* 127, 82–88.

Connor, A.M., Luby, J.J., Tong, C.B.S., Finn, C.E. and Hancock, J.F. (2002b) Genotypic and environmental variation in antioxidant activity, total phenolic content, and anthocyanin content among blueberry cultivars. *Journal of the American Society for Horticultural Science* 127, 89–97.

Connor, A.M., Luby, J.J., Hancock, J.F., Berkheimer, S. and Hanson, E.J. (2002c) Changes in fruit antioxidant activity among blueberry cultivars during cold-temperature storage. *Journal of Agricultural and Food Chemistry* 50, 893–898.

Dale, A., Hanson, E.J., Yarborough, D.E., McNicol, R.J., Stang, E.J., Brennan, R., Morris, J.R. and Hergert, G.B. (1994) Mechanical harvesting of berry crops. *HortReviews* 16, 255–382.

Donahue, D.W., Benoit, P.W., Lagasse, B.J. and Buss, W.R. (2000) Consumer and instrumental evaluation of Maine wild blueberries for the fresh pack market. *Postharvest Biology and Technology* 19, 221–228.

du Jardin, P. (2015) Plant biostimulants: definition, concept, main categories and regulation. *Scientia Horticulturae* 196, 3–14.

Du, X.F., Plotto, A., Song, M., Olmstead, J. and Rouseff, R. (2011) Volatile composition of four Southern highbush blueberry cultivars and effect of growing location and harvest date. *Journal of Agricultural Food Chemistry* 59, 8347–8357.

Echeverría, G., Cañumir, J. and Serri, H. (2009) Postharvest behavior of highbush blueberry fruits cv. O'Neal cultivated with different organic fertilization treatments. *Chilean Journal of Agricultural Research* 69, 391–399.

Ehlenfeldt, M.K. (2005) Fruit firmness and holding ability in highbush blueberry – implications for mechanical harvesting. *International Journal of Fruit Science* 5, 83–91.

Ehlenfeldt, M.K. and Martin, R.B. Jr (2002) A survey of fruit firmness in highbush blueberry and species-introgressed blueberry cultivars. *HortScience* 37, 386–389.

Ehlenfeldt, M.K. and Prior, R.L. (2001) Oxygen radical absorbance capacity (ORAC) and phenolic and anthocyanin concentrations in fruit and leaf tissues of highbush blueberry. *Journal of Agricultural and Food Chemistry* 49, 2222–2227.

Ehlenfeldt, M.K., Meredith, F.I. and Ballington, J.R. (1994) Unique organic acid profile of rabbiteye vs. highbush blueberries. *HortScience* 29, 321–323

Fan, L., Forney, C.F., Song, J., Doucette, C., Jordan, M.A., McRae, K.B. and Walker, B. (2008) Effect of hot water treatments on quality of highbush blueberries. *Journal of Food Science* 73, 292–297.

Forney, C.F. (2009) Postharvest issues in blueberry and cranberry and methods to improve market life. *Acta Horticulturae* 810, 785–798.

Forney, C.F., Jordan, M.A. and Nicholas, K.U.K.G. (2003) Effects of CO_2 on physical, chemical, and quality changes in 'Burlington' blueberries. *Acta Horticulturae* 600, 587–593.

Galletta, G.J. (1975) Blueberries and cranberries. In: Janick, J. and Moore, J.N. (eds) *Advances in Fruit Breeding.* Purdue University Press, West Lafayette, Indiana, pp. 154–196.

Galletta, G.J., Ballinger, W.E., Monroe, R.J. and Kushman, L.J. (1971) Relationships between fruit acidity and soluble solids levels of highbush blueberry clones and fruit keeping quality. *Journal of the American Society for Horticultural Science* 86, 758–762.

Gilbert, J.L., Schwieterman, M.L., Colquhoun, T.A., Clark, D.G. and Olmstead, J.W. (2013) Potential for increasing Southern highbush blueberry flavor acceptance by breeding for major volatile components. *HortScience* 48, 835–843.

Gilbert, J.L., Olmstead, J.W., Colquhoun, T.A., Levin, L.A., Clark, D.G. and Moskowitz, H.R. (2014) Consumer-assisted selection of blueberry fruit quality traits. *HortScience* 49, 864–873.

Gilbert, J.L., Guthart, M.J., Gezan, S.A., de Carvalho, M.P., Schwieterman, M.L., Colquhoun, T.A., Bartoshuk, L.M., Sims, C.A., Clark, D.G. and Olmstead, J.W. (2015) Identifying breeding priorities for blueberry flavor using biochemical, sensory, and genotype by environment analyses. *PLOS One* 10, e0138494.

Giongo, L., Poncetta, P., Loretti, P. and Costa, F. (2013) Texture profiling of blueberries (*Vaccinium* spp.) during fruit development, ripening and storage. *Postharvest Biology and Technology* 76, 34–39.

Gough, R.E. and Litke, W. (1980) An anatomical and morphological study of abscission in highbush blueberry fruit. *Journal of the American Society for Horticultural Science* 105, 335–341.

Guidetti, R., Berghi, R., Bodria, L., Spinardi, A., Mignani, I. and Folini, L. (2009) Prediction of blueberry (*Vaccinium corymbosum*) ripeness by a portable Vis-NIR device. *Acta Horticulturae* 810, 877–885.

Gündüz, K., Serçe, S. and Hancock, J.F. (2015) Variation among highbush and rabbiteye cultivars of blueberry for fruit quality and phytochemical characteristics. *Journal of Food Composition and Analysis* 38, 69–79.

Hancock, J., Callow, P., Serce, S., Hanson, E. and Beaudry, R.M. (2008) Effect of cultivar, controlled atmosphere storage, and fruit ripeness on the long-term storage of highbush blueberries. *HortTechnology* 18, 199–205.

Hanson, E.J. (1995) Preharvest calcium sprays do not improve highbush blueberry (*Vaccinium corymbosum* L.) quality. *HortScience* 30, 977–978.

Hanson, E.J. and Berkheimer, S.F. (2004) Effect of soil calcium applications on blueberry yield and quality. *Small Fruits Review* 3, 133–139.

Hart, J., Strik, B., White, L. and Yang, W. (2006) *Nutrient Management for Blueberries in Oregon.* Publication No. EM 8918. Oregon State University Extension Service, Corvallis, Oregon.

Hirvi, T. and Honkanen, E. (1983) The aroma of blueberries. *Journal of the Science of Food and Agriculture* 34, 992–998.

Horvat, R.J. and Senter, S.D. (1985) Comparison of volatile constituents from rabbiteye blueberries (*Vaccinium ashei*) during ripening. *Journal of Food Science* 50, 429–431.

Howard, L.R., Clark, J.R. and Brownmiller, C. (2003) Antioxidant capacity and phenolic content in blueberries as affected by genotype and growing season. *Journal of the Science of Food and Agriculture* 83, 1238–1247.

Howell, G.S., Stergois, B.G., Stackhouse, S.S., Bittenbender, H.C. and Burton, C.L. (1976) Ethephon as a mechanical aid for highbush blueberries (*Vaccinium australe* Small). *Journal of the American Society for Horticultural Science* 101, 111–115.

Hudson, D.E. and Tiedjen, W.H. (1981) Effects of cooling rate on shelflife and decay of highbush blueberries. *HortScience* 16, 656–657.

Jiang, H., Sun, Z., Jia, R., Wang, X. and Huang, J. (2016) Effect of chitosan as an antifungal and preservative agent on postharvest blueberry. *Journal of Food Quality* 39, 516–523.

Kader, A. (1999) Fruit maturity, ripening, and quality relationships. *Acta Horticulturae* 485, 203–208.

Kader, A., Hess-Pierce, B. and Almenar, E. (2003) Relative contributions of fruit constituents to soluble solids content measured by a refractometer. *HortScience* 38, 383 (abstract).

Kalt, W., Forney, C.F., Martin, A. and Prior, R.L. (1999) Antioxidant capacity, vitamin C, phenolics, and anthocyanins after fresh storage of small fruits. *Journal of Agricultural and Food Chemistry* 47, 4638–4644.

Kalt, W., Lawand, C., Ryan, D.A.J., McDonald, J.E., Donner, J. and Forney, C.F. (2003) Oxygen radical absorbing capacity, anthocyanin and phenolic content of highbush blueberries (*Vaccinium corymbosum* L.) during ripening and storage. *Journal of the American Society for Horticultural Science* 128, 917–923.

Khadivi-Khub, A. (2015) Physiological and genetic factors influencing fruit cracking. *Acta Physiologiae Plantarum* 37, 1718.

Kushman, L.J. and Ballinger, W.E. (1963) Influence of season and harvest interval upon quality of 'Wolcott' blueberries grown in eastern North Carolina. *Proceedings of the American Society for Horticultural Science* 83, 395–405.

Lester, G.E., Jifon, J.L. and Makus, D.J. (2010) Impact of potassium nutrition on postharvest fruit quality: melon (*Cucumis melo* L) case study. *Plant and Soil* 335, 117–131.

Li, C., Krewer, G.W., Ji, P., Scherm, H. and Kays, S.J. (2010) Gas sensor array for blueberry fruit disease detection and classification. *Postharvest Biology and Technology* 55, 144–149.

Lobos, G.A., Callow, P.W. and Hancock, J.F. (2014) The effect of delaying harvest date on fruit quality and storage of late highbush blueberry cultivars (*Vaccinium corymbosum* L.). *Postharvest Biology and Technology* 87, 133–139.

Lobos, T.E., Retamales, J.B., Ortega-Farías, S., Hanson, E.J., López-Olivari, R. and Mora, M.L. (2016) Pre-harvest regulated deficit irrigation management effects on postharvest quality and condition of *V. corymbosum* fruits cv. Brigitta. *Scientia Horticulturae* 207, 152–159.

Lyrene, P.M. (2006) Weather, climate and blueberry production. In: Childers, N.F. and Lyrene, P.M. (eds) *Blueberries for Growers, Gardeners and Promoters*. Dr Norman F. Childers Publications, Gainesville, Florida, pp. 14–20.

Lyrene, P.M. and Crocker, T.E. (1991) *Commercial Blueberry Production in Florida*. University of Florida IFAS, Gainesville, Florida.

Mainland, C.M. and Tucker, J.F. (2002) Blueberry health information – some new, mostly review. *Acta Horticulturae* 574, 39–43.

Mainland, C.M., Kushman, L.J. and Ballinger, W.E. (1975) The effect of mechanical harvesting on yield, quality of fruit and bush damage of highbush blueberry. *Journal of the American Society for Horticultural Science* 100, 129–134.

Marshall, D.A., Spiers, J.M. and Braswell, J.H. (2006) Splitting severity among rabbiteye (*Vaccinium ashei* Reade) blueberry cultivars in Mississippi and Louisiana. *International Journal of Fruit Science* 6, 77–81.

Marshall, D.A., Spiers, J.M. and Stringer, S.J. (2008) Blueberry splitting tendencies as predicted by fruit firmness. *HortScience* 43, 567–570.

Marshall, D.A., Spiers, M.J. and Curry, K.J. (2002) Incidence of splitting in 'Premier' and 'Tifblue' rabbiteye blueberries. *Acta Horticulturae* 574, 295–303.

Miller, W.R. and Smittle, D.A. (1987) Storage quality of hand- and machine-harvested rabbiteye blueberries. *Journal of the American Society for Horticultural Science* 112, 487–490.

Miller, W.R., McDonald, R.E. and Crocker, T.E. (1993) Quality of two Florida blueberry cultivars after packaging and storage. *HortScience* 28, 144–147.

Miller, W.R., Mitcham, E.J., McDonald, R.E. and King, J.R. (1994) Postharvest storage quality of gamma-irradiated 'Climax' rabbiteye blueberries. *HortScience* 29, 98–101.

Moggia, C., Graell, J., Lara, I., Schmeda-Hirschmann, G., Thomas-Valdés, S. and Lobos, G.A. (2016). Fruit characteristics and cuticle triterpenes as related to postharvest quality of highbush blueberries. *Scientia Horticulturae* 211, 449–457.

Moggia, C., Beaudry, R.M., Retamales, J.B. and Lobos, G.A. (2017a). Variation in the impact of stem scar and cuticle on water loss in highbush blueberry fruit argue for the use of water permeance as a selection criterion in breeding. *Postharvest Biology and Technology*, 132, 88–96.

Moggia, C., Graell, J., Lara, I., González, G. and Lobos, G.A. (2017b) Firmness at harvest impacts postharvest fruit softening and internal browning development in mechanically damaged and non-damaged highbush blueberries (*Vaccinium corymbosum* L.). *Frontiers in Plant Science* 8, 535.

Mowat, A.D. and Poole, P.R. (1998) Non-destructive discrimination of post-harvest fruit properties using visible-near infrared spectroscopy. *Acta Horticulturae* 464, 496–498.

Moyer, R.A., Hummer, K.E., Finn, C.E., Frei, B. and Wrolstad, R.E. (2002) Anthocyanins, phenolics, and antioxidant capacity in diverse small fruits: *Vaccinium*, *Rubus* and *Ribes*. *Journal of Agricultural and Food Chemistry* 50, 519–525.

NeSmith, D.S., Prussia, S., Tetteh, M. and Krewer, G. (2002) Firmness losses of rabbiteye blueberries (*Vaccinium ashei* Reade) during harvesting and handling. *Acta Horticulturae* 574, 287–293.

NeSmith, D.S., Nunez-Barrios, A., Prussia, S.E. and Aggarwal, D. (2005) Postharvest berry quality of six rabbiteye blueberry cultivars in response to temperature. *Journal of the American Pomological Society* 59, 13–17.

Nunes, M.C.N., Emond, J.P. and Brecht, J.K. (2004) Quality curves for highbush blueberry as a function of the storage temperature. *Small Fruits Review* 3, 423–438.

Nunez-Barrios, A., NeSmith, D.S., Chinnan, M. and Prussia, S.E. (2005) Dynamics of rabbiteye blueberry fruit quality in response to harvest method and postharvest handling temperature. *Small Fruits Review* 4, 73–81.

Opara, U.L. and Pathare, B. (2014) Bruise damage measurement and analysis of fresh horticultural produce – a review. *Postharvest Biology and Technology* 91, 9–24.

Paniagua, A.C., East, A.R., Hindmarsh, J.P. and Heyes, J.A. (2013) Moisture loss is the major cause of firmness change during postharvest storage of blueberry. *Postharvest Biology and Technology* 79, 13–19.

Paniagua, A.C., East, A.R. and Heyes, J.A. (2014) Interaction of temperature control deficiencies and atmosphere conditions during blueberry storage on quality outcomes. *Postharvest Biology and Technology* 95, 50–59.

Parliament, T.H. and Kolor, M.G. (1975) Identification of the major volatile components of blueberry. *Journal of Food Science* 40, 762–763.

Patten, K.D., Neuendorff, E.W. and Nimr, G. (1987) Sunlight and leaf area effects on the fruit development of rabbiteye blueberries. *HortScience* 22, 1095 (abstract).

Patten, K.D., Neuendorff, E.W. and Nimr, G. (1988) Quality of 'Tifblue' rabbiteye blueberries and efficiency of machine harvesting at different times of the day. *Journal of the American Society for Horticultural Science* 113, 953–956.

Perkins-Veazie, P., Clark, J.R., Collins, J.K. and Magee, J. (1995) Southern highbush blueberry clones differ in postharvest fruit quality. *Fruit Varieties Journal* 49, 46–52.

Perkins-Veazie, P., Collins, J.K. and Howard, L. (2008) Blueberry fruit response to postharvest application of ultraviolet radiation. *Postharvest Biology and Technology* 47, 280–285.

Peterson, D.L., Wolford, S.D., Timm, E.J. and Takeda, F. (1997) Fresh market quality blueberry harvester. *American Society of Agricultural Engineers* 40, 535–540.

Prange, R.K. and DeEll, J.R. (1997) Preharvest factors affecting postharvest quality of berry crops. *HortScience* 32, 824–830.

Prior, R.L., Cao, G., Martin, A., Sofic, E., McEwen, J., O'Brien, C., Lischner, N., Ehlenfeldt, M., Kalt, W., Krewer, G. and Mainland, C.M. (1998) Antioxidant capacity as influenced by total phenolic and anthocyanin content, maturity, and variety of *Vaccinium* species. *Journal of Agricultural and Food Chemistry* 46, 2686–2693.

Pritts, M.P. and Hancock, J.F. (1992) *Highbush Blueberry Production Guide*. Publication no. NRAES-55. Northeast Regional Agricultural Engineering Service, Cooperative Extension, Ithaca, New York.

Qiu, M., Wu, C., Ren, G.R., Liang, X., Wang, X.Y. and Huang, J.Y. (2014) Effect of chitosan and its derivatives as antifungal and preservative agents on postharvest green asparagus. *Food Chemistry* 155, 105–111.

Remberg, S.F., Haffner, K. and Blomhoff, R. (2003) Total antioxidant capacity and other quality criteria in blueberries cvs. 'Bluecrop', 'Hardyblue', 'Patriot', 'Putte' and 'Aron' after storage in cold store and controlled atmosphere. *Acta Horticulturae* 600, 595–598.

Rivera, S.A., Zoffoli, J.P. and Latorre, B.A. (2013) Determination of optimal sulfur dioxide time and concentration product for postharvest control of gray mold of blueberry fruit. *Postharvest Biology and Technology* 83, 40–46.

Rodrigues, E., Poerner, N., Rockenbach, I.I., Gonzaga, L.V., Mendes, C.R. and Fett, R. (2011) Phenolic compounds and antioxidant activity of blueberry cultivars grown in Brazil. *Ciencia e Tecnologia de Alimentos* 31, 911–917.

Rodríguez, J. and Zoffoli, J.P. (2016) Effect of sulfur dioxide and modified atmosphere packaging on blueberry postharvest quality. *Postharvest Biology and Technology* 117, 230–238.

Rosenfeld, H.J., Meberg, K.R., Haffner, K. and Sundell, H.A. (1999) MAP of highbush blueberries: sensory quality in relation to storage temperature, film type and initial high oxygen atmosphere. *Postharvest Biology and Technology* 16, 27–36.

Ruberto, G. and Baratta, M.T. (2000) Antioxidant activity of selected essential oil components in two lipid model systems. *Food Chemistry* 69, 167–174.

Sablani, S.S., Andrews, P.K., Davies, N.M., Walters, T., Saez, H., Syamaladevi, O.M. and Mohekar, O.R. (2010) Effect of thermal treatments on phytochemicals in conventionally and organically grown berries. *Journal of the Science of Food and Agriculture* 90, 769–778.

Saftner, R., Polashock, J., Ehlenfeldt, M.K. and Vinyard, B. (2008) Instrumental and sensory quality characteristics of blueberry fruit from twelve cultivars. *Postharvest Biology and Technology* 49, 19–26.

Sams, C.E. (1999) Preharvest factors affecting postharvest texture. *Postharvest Biology and Technology* 15, 249–254.

Sanford, K.A., Lister, P.D., McRae, K.B., Jackson, E.D., Lawrence, R.A., Stark, R. and Prange, R.K. (1991) Lowbush blueberry quality changes in response to mechanical damage and storage temperature. *Journal of the American Society for Horticultural Science* 116, 47–51.

Sapers, G.M., Burgher, A.M., Phillips, J.G., Jones, S.B. and Stone, E.G. (1984) Color and composition of highbush blueberry cultivars. *Journal of the American Society for Horticultural Science* 109, 105–111.

Sargent, S.A., Brecht, J.K. and Forney, C.F. (2006) Blueberry harvest and postharvest operations: quality maintenance and food safety. In: Childers, N.F. and Lyrene, P.M. (eds) *Blueberries for Growers, Gardeners and Promoters*. Dr Norman F. Childers Publications, Gainesville, Florida, pp. 139–151.

Schotsmans, W., Molan, A. and MacKay, B. (2007) Controlled atmosphere storage of rabbiteye blueberries enhances postharvest quality aspects. *Postharvest Biology and Technology* 44, 277–285.

Sellappan, S., Akoh, C.C. and Krewer, G. (2002) Phenolic compounds and antioxidant capacity of Georgia-grown blueberries and blackberries. *Journal of Agricultural and Food Chemistry* 50, 2432–2438.

Sharpe, D., Fan, L., McRae, K., Walker, B., MacKay, R. and Doucette, C. (2009) Effects of ozone treatment on *Botrytis cinerea* and *Sclerotinia sclerotiorum* in relation to horticultural product quality. *Journal of Food Science* 74, 250–257.

Silva, J.L., Marroquin, E., Matta, F.B., Garner, J.O. and Stojanovic, J. (2005) Physicochemical, carbohydrate and sensory characteristics of highbush and rabbiteye blueberries. *Journal of the Science of Food and Agriculture* 85, 1815–1821.

Sinelli, N., Spinardi, A., Di Egidio, V., Mignani, I. and Casiraghi, E. (2008) Evaluation of quality and nutraceutical content of blueberries (*Vaccinium corymbosum* L.) by near and mid-infrared spectroscopy. *Postharvest Biology and Technology* 50, 31–36.

Smittle, D. and Miller, W. (1988) Rabbiteye blueberry storage life and fruit quality in controlled atmospheres and air storage. *Journal of the American Society for Horticultural Science* 113, 723–728.

Song, J., Fan, L., Forney, C., Campbell-Palmer, L. and Fillmore, S. (2010) Effect of hexanal vapour to control postharvest decay and extend shelf-life of highbush blueberry fruit during controlled atmosphere storage. *Canadian Journal of Plant Science* 90, 359–366.

Strik, B., Buller, G. and Hellman, E. (2003) Pruning severity affects yield, berry weight and hand harvest efficiency of highbush blueberry. *HortScience* 38, 196–199.

Stückrath, R., Quevedo, R., de la Fuente, L., Hernández, A. and Sepúlveda, V. (2008) Effect of calcium foliar application on the characteristics of blueberry fruit during storage. *Journal of Plant Nutrition* 31, 849–866.

Studman, C. (1997) Factors affecting the bruise susceptibility of fruit. In: Jeronimidis, G. and Vincent, J.F.V. (eds) *Proceedings of the 2nd International Conference of Plant Biomechanics*, University of Reading, Reading UK, pp. 273–281.

Takeda, F., Krewer, G., Andrews, E.L., Mullinix, B. and Peterson, D.L. (2008) Assessment of the V45 blueberry harvester on rabbiteye blueberry and southern highbush blueberry pruned to V-shaped canopy. *HortTechnology* 18, 130–138.

Tournas, V.H. and Katsoudas, E. (2005) Mould and yeast flora in fresh berries, grapes and citrus fruits. *International Journal of Food Microbiology* 105, 11–17.

Townsend, L.R. (1973) Effect of N, P, K and Mg on the growth and productivity of the highbush blueberry. *Canadian Journal of Plant Science* 53, 161–168.

Vicente, A.R., Civello, P.M., Martínez, G.A., Powell, A.L.T., Labavitch, J.M. and Chaves, A.R. (2005) Control of postharvest spoilage in soft fruit. *Stewart Postharvest Review* 4, 1–9.

Vicente, A.R., Ortugno, C., Rosli, H., Powell, A.L.T., Greve, L.C. and Labavitch, J.M. (2007) Temporal sequence of cell wall disassembly events in developing fruits. 2. Analysis of blueberry (*Vaccinium* species). *Journal of Agricultural and Food Chemistry* 55, 4125–4130.

Wang, C.Y., Wang, S.Y. and Chen, C.-T. (2008a) Increasing antioxidant activity and reducing decay of blueberries by essential oils. *Journal of Agricultural and Food Chemistry* 56, 3587–3592.

Wang, S.Y., Chen, C.-T., Sciarappa, W., Wang, C.Y. and Camp, M.J. (2008b) Fruit quality, antioxidant capacity, and flavonoid content of organically and conventionally grown blueberries. *Journal of Agriculture and Food Chemistry* 56, 5788–5794.

Wang, S.Y., Chen, C.-T. and Yin, J.-J. (2010) Effect of allyl isothiocyanate on antioxidants and fruit decay in blueberries. *Food Chemistry* 120, 199–204.

Woodruff, R.E. and Dewey, D.H. (1959) A possible harvest index for 'Jersey' blueberries based on the sugar and acid contents of the fruit. *Quarterly Bulletin of Michigan State University, Agricultural Experimental Station* 42, 340–349.

Woodruff, R.E., Dewey, D.H. and Sell, H.M. (1960) Chemical changes of 'Jersey' and 'Rubel' blueberry fruit associated with ripening and deterioration. *Proceedings of the American Society for Horticultural Science* 75, 387–401.

Xu, R., Takeda, F., Krewer, G. and Li, C. (2015) Measure of mechanical impacts in commercial blueberry packing lines and potential damage to blueberry fruit. *Postharvest Biology and Technology* 110, 103–130.

Yoshida, K., Mori, M. and Kondo, T. (2009) Blue flower color development by anthocyanins: from chemical structure to cell physiology. *Natural Products Report* 26, 884–915.

You, Q., Wang, B., Chen, F., Huang, Z., Wang, X. and Luo, P.G. (2011) Comparison of anthocyanins and phenolics in organically and conventionally grown blueberries in selected cultivars. *Food Chemistry* 125, 201–208.

Yu, P., Li, C., Rains, G. and Hamrita, T. (2011) Development of the berry impact recording device sensing system: hardware design and calibration. *Computers and Electronics in Agriculture* 79, 103–111.

Yu, P., Li, C., Takeda, F. and Krewer, G. (2014) Visual bruise assessment and analysis of mechanical impact measurement in southern highbush blueberries. *Applied Engineering in Agriculture* 30, 29–37.

Zheng, Y., Wang, C.Y., Wang, S.Y. and Zheng, W. (2003) Effect of high-oxygen atmospheres on blueberry phenolics, anthocyanins, and antioxidant capacity. *Journal of Agricultural and Food Chemistry* 51, 7162–7169.

INDEX

Note: Page numbers in **bold** type refer to figures
Page numbers in *italic* type refer to tables

CABI – who we are and what we do

This book is published by **CABI**, an international not-for-profit organisation that improves people's lives worldwide by providing information and applying scientific expertise to solve problems in agriculture and the environment.

CABI is also a global publisher producing key scientific publications, including world renowned databases, as well as compendia, books, ebooks and full text electronic resources. We publish content in a wide range of subject areas including: agriculture and crop science / animal and veterinary sciences / ecology and conservation / environmental science / horticulture and plant sciences / human health, food science and nutrition / international development / leisure and tourism.

The profits from CABI's publishing activities enable us to work with farming communities around the world, supporting them as they battle with poor soil, invasive species and pests and diseases, to improve their livelihoods and help provide food for an ever growing population.

CABI is an international intergovernmental organisation, and we gratefully acknowledge the core financial support from our member countries (and lead agencies) including:

Discover more

To read more about CABI's work, please visit: **www.cabi.org**

Browse our books at: **www.cabi.org/bookshop**,
or explore our online products at: **www.cabi.org/publishing-products**

Interested in writing for CABI? Find our author guidelines here:
www.cabi.org/publishing-products/information-for-authors/

Printed and bound by CPI Group (UK) Ltd, Croydon, CR0 4YY

11/01/2026

14804832-0003